ESOTERIC COSMOLOGY AND MODERN PHYSICS

BODO BALSYS

UNIVERSAL DHARMA
PUBLICATIONS
SYDNEY, AUSTRALIA

ISBN 978-0-9923568-9-7

© 2020 Balsys, Bodo

3rd Edition, 2024

All rights reserved, including those of translation into other languages. No part of this book may be reproduced, stored in a retrieval system, or transmitted in any form, or by any means, electronic, mechanical, photocopying, recording or otherwise, without the written permission of the publisher.

The Mother of the World, watercolour by the author. Photo by Juric.P.

Dedication

Thanks to my students, past, present and future, and in particular to those that have helped in the production of this book.

Oṁ

Obeisance to the Gurus!
To the Buddhas of the three times.
To the Council of Bodhisattvas, *mahāsattvas*.
To them I pledge allegiance.

Oṁ Hūṁ! Hūṁ! Hūṁ!

Contents

Preface xi

1. Divine Causation, Preliminary Precepts 1
   - The primacy of Mind 1
   - The great Symbol 5
   - Darkness, *ākāśa* and *svabhāva* 12
   - *Skandhas* and *saṃskāras* 18
   - The *ālayavijñāna*, *śūnyatā* and the appearance of phenomena 25
   - The *trimūrti* 31
   - The factor of *deva* and human interrelationships 34
   - *Buddhi* 42
   - The significance of Light 45

2. Mirrors, Time and the Logoic Mind 50
   - The *ālayavijñāna* and Logoic Mind 50
   - The Mirror-like Wisdom 53
   - Differentiation 58
   - The functioning of a mirror 63
   - Signposts of revelation and analogy 69
   - The factor of consciousness 75
   - *Śūnyatā* and *saṃsāra* 80

3. The Question of 'God' and the Physiological Key 85
   - The physiological key 85
   - The Anthropic Principle 94
   - Does a 'God', a 'Soul' or *ātman* exist? 96
   - The consideration of mind/Mind 106
   - The Creative process 108
   - Further considerations of Mind 115

4. Sources of the Causative Impulse Esoterically Considered 121
   - The mind/Mind and causation 121
   - The enlightenment process and causation 124
   - *Buddhi* and *ātma* 135
   - Planetary formation esoterically considered 142
   - Logoi as the creative Word 144

5. The Causal Body and the Monad 157
   - The planes of perception and the human constitution 157
   - The Sambhogakāya Flower 167

The factor of the *devas* 173
The Monad 178
The evolution of Logoi 184
Concluding statements concerning Logoic evolution 195

6. The Spiral of Consciousness, the Energy View 202
The general description 202
The spirals of consciousness 206
The time-space continuum 221
Chakras 237
Kuṇḍalinī and the serpent symbolism 245
The activity of the Heart centre 260
The ten stages of the evolutionary process 275
The evocation of *suṣumṇā* 287
The first seven verses of the book of Genesis 290

7. The Spiral of Consciousness, the Geometric View 296
Analysis of the zero 296
The geometry of causation 300
The function of the Dhyāni Buddhas 304
The foundational geometry 306
Einstein's theory of general relativity 313

8. The Nature of Bījas (atomic unities) and Causation 316
On *bījas* 316
The nature of *bījas* 319
Bījas as 'atoms' 326
The clairvoyant investigation of the Anu 334
The permanent atoms 349
The *devas* and the permanent atoms 362
Form building 365
The Logoic permanent atom 376
The Lipikas 385

9. The Atomic Universe, Esoterically Understood 397
The concept of ether 397
The problem of gravity 401
The four ethers 405
The Higgs field 408
Quarks and leptons 413
The Higgs field, dark matter and subjective space 422
The concept of gravitons 425

Gravity further considered....432
The thought field....434
The formation of the nucleus of an atom....440
Saṃskāras and cyclic time....443
The astrology of the subatomic particles....446
Dimensional perception....451

10. Meditation on the Electrical Nature of Mind....455
General esoteric considerations....455
The role of the Pleiades....463
The plasma and electric universe theories....482
Appendix One....505
The Heart is the Mind....505
Appendix Two....521
The Root Races....521
Appendix Three....545
A further note on figure five....545
Bibliography....550
Index....556

Tables

Table 1. The triune principles....111
Table 2. The triune aspect of Deity....115

Figures

Figure 1: The Trimūrti....33
Figure 2: The sun as a symbol....91
Figure 3. The Spiritual Triad....122
Figure 4. The relation between universal & empirical consciousness....123
Figure 5. The principles constituting a person....161
Figure 6: Summary of the planes of perception....165
Figure 7: Stages of expression of the Monadic Eye....181
Figure 8: The door to the cosmic Paths....181
Figure 9: The serpent of time....182

Figure 10. The spiral of consciousness....207
Figure 11. Spiral-cyclic motion....209
Figure 12: The geometry of spiral-cyclic motion....215
Figure 13. Fourth dimensional motion....217
Figure 14. The three types of motion....219
Figure 15. The time-space continuum....224
Figure 16: The movement of *bījas*....*226*
Figure 17: The expansion of thoughts....227
Figure 18: The wheel of the *dharma*....*228*
Figure 19: The time line....229
Figure 20: The swastika....231
Figure 21: The self-conscious individual....231
Figure 22: The lotus as an expression of energy and consciousness....238
Figure 23: The sun, earth, moon relationship....253
Figure 24: The inwards and outwards motion of consciousness....255
Figure 25: The three principal *nāḍis*....*256*
Figure 26: The symbolic horns of a ram....259
Figure 27: A cross-section of a *nāḍī*....261
Figure 28: Involution and evolution of consciousness....271
Figure 29: The face of the Waters....279
Figure 30: Formation of the Tau cross....282
Figure 31: The rectified Tau cross....283
Figure 32: The evocation of *suṣumṇā*....*288*
Figure 33: The hexagon....308
Figure 34: The ten principles....310
Figure 35: The spirillae of an atom....328
Figure 36: The ultimate physical atom....329
Figure 37: Babbit's view of the atom....342
Figure 38. Various views of combinations of Anu's....347
Figure 39: The hydrogen micro-psi atom....348
Figure 40: The cosmic law of Karma....381
Figure 41: The symbolism of Cancer the crab....391
Figure 42: Fermions and quarks....418
Figure 43. The six quarks....441
Figure 44. The Heart is the Mind....516
Figure 45: The evolution of the Root Races....539

Auu.....m! Cried the Bird (Kālahamsa).
Ahh.....h! Cried Prakṛti (the substance matter).
Huu.....m! Resounded the Universe.
(It has been accomplished.)

Throughout the Depths
of the fathomless cosmic Quietude
the slumbering universe stirs.
First a murmur,
the eternal Sound, Prāṇava – Aūṁ
resounds,
as Hiraṇyagarbha, the World-Egg
exuberates,
pulsating vibrantly,
crystallising
the dawn of a new cycle.

OṀ.
DEITY IS.
I AM
THAT.

YOU ARE
THE ETERNAL
INFINITUDE!

Arise matter!
Arise mind!
Arise Being!
Come and Know yourself.

Preface

Physics is in many ways an esoteric subject, at least to the lay person who does not have the mathematical skills to understand the fundamental derivations of the science, nor has had the time to learn the subtleties of many of the concepts and terms that are used. Most people are neither epistemologically or ontologically savvy and hence rely upon popular depictions and catch phrases of concepts describing new ground breaking discoveries and formulations. Many high Initiates and their disciples have incarnated into the scientific community in order to advance all fields of science. This is especially so for those along the fifth Ray of Scientific Reason and the seventh Ray of Ritualistic activity and Organisational Power. The endeavour of this book therefore is to incorporate the findings of these Initiates into a broader view that takes into account the nature of multidimensional space and the primacy of mind/Mind in the universe, and how it causes and organises phenomena. Consequently I wish to introduce a higher esoteric science that overlays or underlies what they empirically deem correct. This thesis however shall not delve into neurological science, of the mind-brain connection, which is a valid field of investigation for some scientists, but the information imparted may offer some further clues to these researchers.

In this work I shall diverge somewhat from the previous volumes in this series that deal mostly with Buddhist philosophy and its ramifications. This is mainly because Buddhist concepts concerning the 'creation' of things are either along the lines of a denial that such is possible (e.g., the Prāsaṇgika Mādhyamika position), or else are essentially derived from Hindu beliefs. The Western schools of thought on the other hand possess a wealth of concepts and semantics concerning the nature of how all things have come into existence. Much however that is promulgated along this vein is exoteric, i.e., materialistically biased, or interpreted empirically, with little or no comprehension of the nature of the multidimensional universe. The graduates of our universities are well fed with such fare. This has led to disquisitions or analysis which misconstrue some of the import, and are blinded to much else, of the religious doctrines some purport to be experts of.

Many scholars have never comprehended the main agenda of the exponents of the ancient schools of thought that were the living traditions of a gradual Initiation into the keynote of the Mysteries that were at the heart of the belief system or religion in question. Only the Initiated possessed the proper keys to unravelling the truths of the system, and when they died out those keys exoterically died with them. There is however an esoteric methodology of gaining enlightenment, known only to a very few that kept such knowledge alive. The propagation of the Mysteries is also veiled in the concept of *guruparamparā*, the 'ear whispered' tradition in Hinduism and Buddhism. Here the most esoteric teachings from the guru, the custodian of a particular lineage or tradition of Tantra are passed on to his/her spiritual 'son'. Consequently the exponents of the esoteric tradition have always existed throughout the ages possessing knowledge of the 'hidden Mysteries'.[1] They have reincarnated from cycle to cycle of activity to ensure that the Mystery tradition is kept alive.

Unbeknownst to them, some exponents of this tradition incarnated into the scientific community. Their purpose was to endeavour to explore and decipher the context of the mysteries hidden in the structure of matter and the constitution and evolution of the universe, with its many suns and planets. In this way the context of the ancient tradition was expanded to include the material domain in a way that was never before possible. Humanity has benefitted immensely as a consequence. Now is the time to unite the inner and outer Mysteries into a unity. My work is a step in this direction, however I have not the

1 Jesus for instance hinted at this when he stated to his apostles in *Mark 4:11:* 'And unto you it is given to know the mystery of the kingdom of God: but unto them that are without, all *these* things are done in parables'.

time to explore in depth all of the competing theories concerning the origination of the universe. Neither am I a physicist, with an extensive knowledge of mathematics and the intricacies of modern cosmological and subatomic observations, be that of the Big Bang theory, relativistic theory, or quantum electrodynamics. Myriad are the papers published by the scientific community on diverse subjects, such as whether the Hubble constant for the redshift, upon which the age of the universe is calculated, is indeed a constant. The published literature is vast and a researcher must wade through them the best he/she can along any line of intended research. Similarly with the alternate models, such as those that postulate no Big Bang, the incorporation of the Wolf Effect, plasma cosmology, or of the electric universe. To my mind, and for many others, the arguments concerning the current concept of the expanding universe that evolved from a Big Bang, verses concepts of the steady state theory (or versions thereof) have not been conclusively proven.

What I wish to present is the esoteric view of the causation of phenomena and cosmicogenesis, and to integrate it as much as possible with mainstream scientific thinking. The esoteric view (as known by enlightened Minds) presupposes a hylozoistic universe, where the appearance of all phenomena is governed by the laws pertaining to the application of mind/Mind, where mind refers to the reifying, dissecting, criticising empirical mind, and Mind to the enlightened, unifying all-embracive 'super-Mind' gained by enlightened Beings. The distinction between the two was adequately developed in my *A Treatise on Mind* series. Hence much that is written in this present volume finds its basis in the information given in that former series. Analysis of the laws of Mind presuppose a multidimensional universe, consisting of different planes of perception that is attainable by one who is mastering the vicissitudes of mind and developing the attributes of Mind.

The scientific community still needs to properly embrace this multidimensionality, not so much in terms of mathematical modelling, but through pure logic and the experiences derived via meditative awareness. Much that was taught by the ancient enlightened seers and their modern successors is obtained through direct visual perception, utilising extraordinary psyco-spiritual powers *(siddhis)* totally unknown to the empiricist. Direct psychic microscopic vision for instance, of what constitutes the atomic universe, should consequently not be automatically discounted by empiricists because they do not acknowledge the existence of such powers. When the scientific community as a rule will remove their blinkers, and consequently include such psychic evidence, by producing expanded Minds to solve their current empirical problems, then a new era of science will develop. A scientific revolution will occur that will take them far into a new epoch of 'miracle making', and so build the new era communities based on Mind fused with Love, where Love-Wisdom is the mainstay of all that ensues.

As the reader shall discover the esoteric view presents a grand universal theory of everything, where much that is still enigmatical to materialistic thinkers becomes meaningful. The esoteric view explains the seeming paradoxes in the subatomic world as part of the natural order of what is, being but extensions of the laws governing the subjective universe. Cosmologically, the esoteric view presupposes a vast Hierarchy of increasingly transcendent enlightened Minds that have evolved from solar systems and galaxies billions of years before our solar system was formed. They are collectively capable of causing the appearance of solar systems, galaxies and cosmoses, from out of primal ethereal 'substance',[2] by rightly organising and projecting the powers of MIND.[3] Everything manifests for a Purpose, is prearranged and impeccably formulated to produce the evolution of mind from out of that 'substance'. Nothing happens by chance. The (final) Anthropic Principle rules this universe.

Running throughout the former series, *A Treatise on Mind,* as well as this book, lies the gain of the millennial long deductions on causation and cosmology from both Hindu and Buddhist sources, aided by the rest of the world's mythological outpouring. There have always existed the enlightened, the Hierarchy of Enlightened Being, the Council of Bodhisattvas, who have cumulatively developed Mind, and so know the secrets of Nature. They have adequately demonstrated this knowledge in their monumental works, such as the mathematics encoded in ancient

2 What is meant by 'substance' here shall be explained later.

3 Such MINDS are termed Logoi in this book, a term explained in *A Treatise on Mind* Series.

monuments, such as the great Pyramid, and through the recorded demonstration of the *siddhis* they possessed. The scientific community still needs to properly acknowledge the manifestation of such phenomena, let alone do the appropriate experimentation, and so accordingly utilise these findings in the theorisations for the origination of things.

This present volume hopes to present some relevant information and concepts for those of that community to broaden the scope of their investigations and so incorporate a vaster, better understanding of what the universe actually is, and how it correlates to human consciousness. It is for such open-minded researchers and those interested in further developing wisdom that this book aims to assist. Similarly for those with esoteric inclinations, the aim is to give them a far deeper perception as to how the domains of perception within which they reside come into existence. Many aspects of this complex subject are analysed, which necessitates interrelating the most abstruse information in a logical, fresh and hopefully inspiring manner. This subject is viewed in terms of energy interrelationships, and the control of substance by Mind. The nature of the unfoldment of such Minds, hence the way of making prime Causative Agents, was explained in my series *A Treatise on Mind.*

It is not possible to overestimate the importance of studying the nature of Causation, for its process affects and effects everything seen around us in this material universe. It is hoped that the information presented will assist the reader to aspire to become a 'prime Causative Agent'. As this subject is interwoven with the entire enlightenment process, it is the objective of evolution itself.

We have reached the era where much that was formerly veiled can now be openly revealed for proper consumption of the worthy who will take the time to seriously study the presented material. This is indeed a fortunate provision for the future evolution of human consciousness. Accordingly, my works manifest as part of a well established agenda for the gradual revelation of the esoteric aspects of the world's spiritual literature. The first major expression thereto was that given to humanity under the auspices of Helena Blavatsky more than a century ago, followed by the many books penned by A.A. Bailey, the amanuensis of a highly enlightened Tibetan Rinpoche of the time, who telepathically dictated those books.

In many ways this present book is concerned with presenting the bridging information between the theories and findings of the scientific community and speculators in cosmology with what will later be expanded in depth in my future work on the Cosmicogenesis portion of H.P. Blavatsky's monumental work, *The Secret Doctrine.* Esoteric students will consequently better comprehend the advanced esoteric philosophy, partly explained by Blavatsky, veiled in the *Stanzas of Dzyan* (stanzas of meditation). Those coming from the scientific community that wish to comprehend the esoteric view will then have a far better basis to rationally critique a esoteric philosophy possessed by the enlightened that has up to now largely been anathema to scientists because of their materialistic bias. Through such bias this community have blinded themselves to much of what constitutes this universe. The veils of such ignorance need to be lifted, so that a proper marriage between the exoteric empirical sciences and the esoteric ageless wisdom tradition can propel human thinking far into a vast new renaissance of Revelation.

Upon study of the information presented the serious student will better comprehend the way that the *Heart that is the Mind* unfolds, hence will learn to reside consciously in that Heart forevermore. This is the hope of the author, and indeed, it is part of the series of Revelations that must be imparted to the world's disciples if the new era is to externalise and Shambhala be grounded by a *maṇḍala* of enlightened Knowers.[4] They must be well versed in the process that makes a Logos, be it a planetary, solar, and beyond...to THAT which embodies the universe—for each are prime Causative Agents within Their own spheres of Attainment.

Oṁ Tat Sat!

4 See *The Constitution of Shambhala,* volumes 7A and B of *A Treatise on Mind.*

The Great Evocation

From the point of light within the Mind
of the awakening one,
let light stream out to all celestial abodes,
let light fill all space.

From the point of Love within the Heart
of the aspiring one,
let the experience of Love fill the mind
with the rapture of Love's universal embrace.

From the focussed will of the meditating one,
let its purpose awaken the Fires of lighted resolve
of all seeking to escape from the *saṃsāric* morass.

From the places of abode of human minds
let them seek the way of liberation,
as taught by the custodians of
the Way of the Heart.

Let the Lords of Shambhala receive
each liberated pilgrim that passes through the Door
that opens to great Logoic Thrones,
and so to travel their cosmic Path.

Let all this, and more, come to pass,
as each person awakens to an enlightened stance.
May group purpose unfold as the multitudes
hearken to the compassionate call.

Love is the All that is the One that moves
and pulls us upon the cosmic Way
to the central Spiritual Sun.

Oṁ Maṇi Padme Hūṁ

Note that this evocation is a corollary to 'The Great Invocation' found in the books by Alice Bailey. A modernised form is explained in my *A Treatise on Mind*, volume 7B, 465-68.

1

Divine Causation, Preliminary Precepts

The primacy of Mind

The esoteric doctrine acknowledges the essential unreality and yet unity of all manifest forms, that the corporeal universe is an embodied expression of energy and is constantly modified by the interplay of various elementary and subjective forces. Next comes the concept that all forms manifest according to cyclic law, that everything thus reincarnates, consequently moving from subjective space into manifest objectivity at the appropriate cycle, and vice versa. The appearance and evolution of mind is the object of the cyclic coming and going of the manifest forms. Mind in fact is all there really is, in that it encompasses all that can be seen and known in the universe. Without the factor of mind (and its consequent evolution into Mind) the appearance and disappearance of universes would be absolutely meaningless. No matter how many such universes manifested and disappeared over countless aeons they will leave no record or factor of 'existence' without a mind/Mind to cognise the event. The imprint of phenomena upon mind/Mind hence makes the appearance of a universe in terms of what is experienced, cognised, substantially real. The evolution of mind/Mind hence is the logical objective of such an appearance. The corollary being therefore that one could say that a universe exists because Mind has evolved to comprehend, then encompass it. By utilising the creative aptitude of the Mind and the expression of its laws, an encompassing Entity can also logically cause a universe to come into being so as to incarnate into it. The universe therefore being the physical sheath that becomes the vehicle of expression of that Entity. In doing so it helps lesser mind/Minds to evolve increasingly vaster all-encompassing Mind-states. In this respect what the laws governing the expression of mind/Mind and how it manifests via multidimensional space to be causative of phenomena needs explication. The esotericist however rarely thinks in terms of such a vast field, much more important are considerations closer to home, our earth sphere, of the Logos that has incarnated into it as a vehicle of expression, of the solar Logos of whose Body of manifestation that Logos is part of, and the great Lives that inform the local part of our galaxy.

That human intelligence is one such factor that can recognise the phenomena of a universe is not an accident, rather it is a pre-planned forgone conclusion by vast Intelligences from former cycles of evolutionary attainment that have set the laws of what we understand as Life into motion. To understand the nature of mind, one must comprehend how it evolves and the attributes governing its ability to project thought-forms, and hence to create forms. It is a mistake to think that a mind needs a brain mechanism in order to function. Such a mechanism is needed for physical plane activity and cognition by means of the senses, but certainly the mind survives after the death of the form. Considerations concerning the nature of this subject is vast, and many books have been devoted to it. My series *A Treatise on Mind* presents the higher metaphysics, hence the information needs no repetition here.

Once we speak of survival after death then we are considering a multidimensional universe. This factor must then be included in our analysis of what the universe is and of how it came into existence. It is in this arena that the scientific community has yet much to discover, and without an understanding of the multidimensional nature of mind/Mind and its relation to physical phenomena many of their conclusions are short-sighted.

Energy follows thought, is directed by the Eye and projected by the will of the thinker. This entire world system is in fact an objectivised Thought-Form, in which we manifest the characteristics of actors undergoing a cyclic play, a play that is part of the creative Ideation of a primordial (Ādi) Buddha, a Logos, that embodies the Word for all time and space.[1] The Idea manifests in terms of an evolutionary pattern that

---

1 A similar concept is provided in the Vedas, R.L. Kashyap for instance in his *Hymns of Creation, Heaven & Ancient Fathers* (Sakshi, Bengaluru, 2011), 4, states

benefits all players on the world stage, as well as the divine Thinker. This present analysis is not however primarily concerned with the understanding of the purpose of that 'play', but rather with the method producing its manifestation.

All objectivised thought-forms progress through recognisable changes, of inception then maturation, as the expression of their intent is fulfilled. There is consequent old age and death when their inherent Life is withdrawn or modified to suit new manifesting environmental conditionings. This also happens when the will of the thinker is directed to another direction.

Inherent in the concept of 'Thought' are the Fires of Mind, and it is in the expression of Fiery substance wherein there is a confluence between the views of modern physicists and esotericists. The difference being that in the concept of physicist's view of the Big Bang the originating source of the energy from which the universe was formed is unaccountable. The esotericist has the impact of the energy of cosmic Mind to cause the manifestation of phenomena. One must also take into account that the cosmic Fire in question needed to produce the causation of all Logoic forms, even upon the stupendous scale when viewing a universe, comes from countless aeons of evolutionary development of Mind and ever-vaster arenas of creative expression of the powers of Mind, as such a ONE evolves in the ineffable cosmos. Intense is the pressure of Fire contained in the vastness of the sphere of containment of a Logoic Mind, the vaster the Logos, the more intense the energies concerned. With such energy then planetary, solar, galactic beginnings and that of a universes is possible.

The method of the appearance of phenomena shall be detailed in the last three chapters of this book. The preliminary chapters are devoted to laying the foundation for comprehending the esoteric lore and related terminology provided by our wise forebears, stemming from their meditative experiences concerning the subject of causation and the appearance of world spheres, and by logical extrapolation, the universe. The Sanskrit terminology introduced may be difficult

---

that: 'the *Rig Veda Samhitā* (10.72) in the next chapter 11 deals with the birth of the Gods. *Taittiriya Samhitā* (6.1.1.2) states that the world of Gods is interwoven with that of human beings. The two worlds function together. As TS (1.2.3.2) states, 'the Gods are mind-born yoked to the mind, have the blissful power of discrimination *(dakṣhiṇa)* and are the children of discernment'. Such a statement echoes later Buddhist statements concerning the nature of their Deities.

for many unaccustomed to the eastern philosophical systems, but if comprehended they hold the clues to many mysteries concerning the appearance of phenomena.

We must remember here that Buddhist and Hindu scholars took for granted the existence of the subjective dimensions of perception. They needed not exoteric proof that modern day materialistic scientists wish, because for them the subjective domains were simply part of their meditative world and directly experienced as such. The 'problem' for them was to explain how, precisely, the material domain appeared from the subjective domains, and what forces were involved in the process. The Sanskrit terms then relate to their form of scientific terminology explaining what, how, when and why. Similarly, the deities named in the process symbolise different types of forces and Mind-states that are brought to bear in the causation of things, a world or solar sphere, and the mode of the evolution of the units of Life, the bearers of consciousness and sentience, to a status where they are liberated from the phenomena scientists perceive to be the 'real world'. To the ancients, and esotericists alike, this material phenomena is the great illusion, and needs to be transcended by mind in order to experience the Real, namely, that pertaining to the domain of enlightened perception.

Thinking in terms of a planet, sun, galaxy or universe is but a matter of scale, as the same principles manifest throughout, but one must learn to think transcendently in order to comprehend the vaster perspectives of the Real, that manifest from within-without as far as what might constitute a Logoic Mind. Without comprehension of the meaning of the needed terminology, Sanskrit, or its equivalent in any sacred language, it is not possible to convey the teachings productive of revelations to the unenlightened.

That manifesting via the atomic world works via a similar, though reified paradigm. It is but a matter of applying the same principles upon the subatomic world as one would upon the macroscopic, but altering the view and terminology to fit. If a Logos at the level of embodying a solar system were to view the life of the human Soul on earth such a One would be faced with a similar problem, where the human unit would be a microscopic entity, an 'atom'. A Logoic Vision is so Vast that when looking to the human world the view will be of masses of human thoughts and the Soul-groups appearing upon the mental plane. Their vision can descend no further, because our mental plane represents the dense physical to them. To view into their 'sub-atomic world' represented by the activities of human units they must Peer through the Minds of great enlightened Beings, who have risen up the ranks of higher awakened perceptions and so are closer to the Logoi in Mind. The Minds of the enlightened ones thereby act as microscopes for the Logoi.

The great Symbol

All manifest Life is governed by cyclic laws. Cause produces effect, and all Lives eventually return to the emanatory cause. That cause is the essential reality upholding the rest of the play (manifesting as the *Great Symbol*[2] that is the *saṃsāra-śūnyatā*[3] interrelation*)*. The nature of the continued cyclic projection of the cause is dependent upon the expression and characteristics manifested by the play. A highly simplified example of cyclic projection can be viewed thus:

> A book plus fire becomes ash. Ash plus earth, plus seed and right environmental conditions, becomes a tree. A tree plus processing, plus ink and print becomes a book.

This cycle (as all others) obviously necessitates a mediatory manipulator—here the human hand, whilst the fulfilment of the instinct towards knowledge becomes the primal cause. As the information in the book becomes redundant, it is discarded to make way for another with more relevant information. The mediatory manipulators in Nature[4] are represented by the various members of the *deva* kingdom,[5] or by the will of a creative Thinker, a Logos. The law of *karma* becomes the leitmotiv or basis for all such action.

If we were to bring all cycles of cause and effect back to the original cause (and thus effect) to represent the sum total of all that is, we would have one 'infinite' cycle (or circle) of cause and effect emanating from a primal cause.

---

2 *Mahāmudrā,* explained in *A Treatise on Mind.*

3 *Saṃsāra-śūnyatā* is the nexus of enlightenment, *saṃsāra* being the transitory, illusory world and *śūnyatā* the place of liberation.

4 I capitalise this term to emphasise the inherent divinity of all Life.

5 The *devas* shall be explained later.

Before the originating primal cause existed, the universe must have been perfectly motionless, i.e., Void of all mental attributes (*śūnyatā*). This must have been its primordial state if naught existed then, there being no cause or effect of that cause. In Blavatsky's *The Secret Doctrine* this state of 'existence' is presented as 'DARKNESS ALONE FILLED THE BOUNDLESS ALL'.[6] The Rig Veda states that 'Darkness hidden by darkness in the beginning was this all'.[7] The first darkness is the darkness of *śūnyatā,* of that which exists beyond mind, and the second darkness is that of ignorance, of incarnation into the *māyā* (illusion) of *saṃsāra.* What this statement really means is that the mind to experience was not yet incarnate. Neither the darkness of the absolute, nor the darkness of the material domain was able to be cognised. Without the mind functioning everything is veiled. Once incarnate then the darkness of ignorance must be overcome so that the greater Darkness of the Real can be experienced.

Once we add an originating cause, seen as the primary motion of the thought of a Thinker, then its effects must eventually rebound upon itself. In effect, a sphere of action is produced that is viewed two dimensionally as a *circle.* This geometric form encompasses the maximum amount of space with the least surface area, and when seen three dimensionally it takes the form of a sphere. It can be viewed as the outer sheath or external boundary of the originating Thought encompassing the sum of the related qualities. In the microscopic world the sphere appears in the form of atomic unities.

Physical existence is in essence a reified expression of the conditionings existing upon subjective levels, (the higher planes of perception) and is cyclically made manifest for the purpose of the salvation of basic substance. Birth and death on the earth can been seen as an ever-progressive succession of an infinite number of finite cycles ('rebirths') that are always transmuted by a conscious mind/Mind[8] into

6 H.P. Blavatsky, *The Secret Doctrine,* Vol. 1, (Theosophical Publishing House, London, 1888, 2005), 27.

7 Kashyap, 14. He further states that 'This all was an ocean without mental consciousness *(apraketam)'.*

8 As stated, I use the term 'mind' for the empirical, concrete mind, (intelligence), and the term 'Mind' for the abstracted, liberated enlightened Mind. This dual aspect is deliniated as mind/Mind.

larger cycles of experience. These continually spiral into one boundless, 'infinite' cycle that represents the body of manifestation of an Ineffable Logos or Buddha of Meditation.

All the modifications of existence essentially stem from the symbol of the (cyclic) sphere, when its various qualities are metaphysically analysed. Such analysis was begun in my investigation of the nature of cellular consciousness in the fourth volume of *A Treatise on Mind.*[9]

The various patterns and symmetry observable in Nature, such as in the shape of leaves, flowers, animal bodies and many crystal forms in the mineral kingdom, are expressions or extensions of a blueprint derived from the geometrical properties of the sphere, circle, ovoid, a seed or ovum. Light is said (according to Einstein's mathematics) to traverse space in a curved path. Such a path will inevitably be productive of a spherical or ovoid shape.

Analysis of the properties of the *ovoid or sphere*, the world Egg of various mythologies, must thus form the basis for our concept of primeval causation. (Despite the fact that the limitations of our reasoning abilities may not allow us to see that far into the past.) In the Hindu philosophy we have for instance, the appearance of *hiraṇyagarbha,*[10] literally the 'golden womb, embryo' or 'golden Egg', the 'universal germ'. From it comes the universe or the world sphere. The term is a name of Brahmā, who was born from the radiant golden Egg (*hiraṇya*), or womb (*garba*), of the creative *(saguṇa)* Brahman. It is the womb of space-time, from which all evolution sprang.

---

9 *Maṇḍalas: Their Nature and Development.*

10 *Hiraṇyagarbha,* [from *hiraṇya* = imperishable substance, golden + *garbha* = womb, embryo, foetus, an interior]. Another rendering (provided by Kashyap, Ibid, 18) is that: *hi* is derived from *hita* = placed, hidden, and *raṇya* = delight, hence 'the hidden delight'. H.P. Blavatsky states in her *Theosophical Glossary* that this term refers esoterically to: 'the luminous "fire mist" or ethereal stuff from which the Universe was formed'. Brahmā is described in the *Rig-Veda* as born from a golden Egg formed out of the seed deposited in the waters when they were produced as the first *vikāra* (modification) of the self-existent (Brahman). According to *Manu* (1:9) this seed became a golden Egg, resplendent as the sun, in which Brahman, while remaining transcendent, evolved into Brahmā the Creator, who is therefore regarded as a manifestation of the Self-existent. Having stayed a 'year' (of *pralaya,* a period of dissolution, *nirvāṇa)* in the egg, Brahmā divided it into two parts by his thought, thereby forming the heavens and the earth; and in the middle he placed the sky, the eight regions, and the abode of the waters.

The ovoid or sphere is the veil of the Great Symbol, when fundamental reality is cognised. When the empirical universe is analysed it becomes the symbol of the Path. This Path must be pictured in terms of a spiral within the sphere because the nature of the evolution of consciousness must be added. Instinct, desire, and thought-form construction are factors necessarily taken into account by any causative agent. There are transcendent aspects to the concrete manifestation of divine impulse contained within the sphere.

The Great Symbol *(mahāmudrā)* is the Real, the cause and result of evolutionary being, of the nature of enlightened perception. It is the great Mystery, from which the other symbols embodying the lesser mysteries of all being/non-being emanate. Causation is effectively the vitalisation of the Great Symbol that delineates time and space as the Womb of being, and from which all the lesser evolutionary symbols embodied as manifest (and even imaginary) forms emanate. The analysis therefore of the mysteries of this and all related symbols will lead us to an understanding of the fundamental nature of being/non-being, and indeed, when rightly pursued, to full enlightenment and liberation from the realm of cause and effect. This then provides the ability to wield causative energy in the guise of a creative Deity.

Concerning the *mahāmudrā*, H.V. Guenther states:

> Mahāmudrā is the fact that all entities coincide with unoriginatedness, that the interpretative categories of subject and object do not obtain *per se,* that the veils of emotional instability and of primitive beliefs about reality have been torn, and the absolutely specific characteristics (of everything) are known as they are. Hence Mahāmudrā is said to be the immaculate effect. Its actuality is that (i) it has neither colour nor shape as all other determinate entities which have a beginning, a middle, and an end, that (ii) it is all-encompassing, that (iii) it is unchanging, and that (iv) it stretches across the whole of time. Therefore *mahāmudrā* is instantaneous awakening to Buddhahood, which means that the four time-situations and the four delight-intensities are not disrupted.[11]

---

11 H.V. Guenther, *The Life and Teachings of Naropa*, (Shambhala, Oxford, 1963), 222-224. The 'time-situations' and 'delight-intensities' are given in *fn.* 1, 222-223, as: 'variedness, maturation, absence of distinct characteristics, ferment; joyous excitement, ecstatic delight, co-mergence delight, recession of excitement'.

Guenther further states:

> Of particular significance, however, is the definition of Mahāmudrā as stretching across the whole of time. Mahāmudrā is not an event *in* time, it rather *is* time, not restricted to a particular now, but including the past and the future which we usually think of as non-existent.[12]

The Symbol is embodied in our temporal forms and manifests as all related qualities. The quickest way to understand its intrinsic nature therefore is to comprehend the nature of ourselves. This constitutes all of the qualities and energies, latent or empowered, that cause us to come to be and to manifest the complete potential of our evolutionary journeying. For, as all sacred books say, we are built in the image of the prime causative agent embodying our planetary sphere, who sustains all related manifestation. The Symbol however, is beyond causation and its resolution. It is an expression of the nexus between *śūnyatā* and *saṃsāra,* between the Buddha (of whatever description), who can express Himself no lower than *śūnyatā,* as the embodiment of divine Compassion; and his Consort, who characterises the attributes of *saṃsāra,* from which the wisdom principle *(prajñā)* is derived. Together they give birth to the Son, which is human consciousness encapsulated in a sphere of attainment (the *tathāgatagarbha).*[13] The best way to depict the nature of this nexus, the Great Symbol, is thus in the form of a sphere of containment, but what it contains is ultimately attributeless (when relegated to *śūnyatā*[14]*),* but its actual attributes that manifest upon the abstract mental plane are discerned by the enlightened Mind.

From the *tathāgatagarbha* inevitably a Ray of conscious awareness is projected downwards into the empirical domains to encapsulate a sphere of mental substance, then the astral and etheric, to finally manifest itself in the womb preparatory to being born in the physical domain. The mental

12 Ibid, 224.

13 Its qualities, in relation to the phrase 'the Sambhogakāya Flower', were explained in detail in volume 3 of *A Treatise on Mind.* It is also sometimes denoted the Causal body, the Ego, the human Soul. It is the cause for the reincarnation process, absorbing into its constitution the gain of the experiences whilst incarnate. It is group conscious, and is a sphere of Mind existing within a sea of Mind upon the higher mental plane.

14 The conventional concept is that *śūnyatā* is Void, attributeless, but it has characteristics, which I explain in Volume 3 of *A Treatise on Mind*, pages 259-63.

forces manifesting via the etheric double then uses the mechanism of the nervous system of the brain to interrelate with the transient world of material interrelationships and attachment. The purpose is to gain mastery of all-that-is by means of the evocation of the powers of the Mind out of mind.

The principle of incarnation for a Logos is similar, though of a far vaster scale and magnified scope and viewed in terms of transmuted correspondences.

The Symbol is thus the mode of containment at first of the principle of intelligence, which is conditioned by the factor of time, and thus of *karma*. An analysis of intelligence thus becomes fundamental to the understanding of the nature of being/non-being and of causative factors. Being the key tool with which we interpret and deduce the Symbol the factors that influence the rational mind hence need proper comprehension. Depending upon how the mind is controlled and directed, so the ability to think (and the related opinions) is accordingly swayed by means of the veils of imagery, and the use of symbols, such as the sphere. However, as it endeavours to comprehensively do so it must lift its imaginative faculty from out of the bounds of empirical constructs and thus into abstractions. Wisdom *(prajñā)* is thereby developed, wherein the enquiring one takes on the attributes of the Consort of the Buddha. The Buddha can then impregnate the Womb of the abstract Mind with enlightenment qualities that produce the liberation that *śūnyatā* represents. The nexus is breached and the nature of the Great Symbol stands revealed.

The above is an outline of the major part of the process that releases the energy and qualities that the Symbol embodies. Intelligence, as the foundation to the expression of the abstract Mind, is consequently an instrument needing to be perfected via self-mastery for the *mahāmudrā* to be comprehended. It must be developed before it can be transcended, otherwise there can be no 'tensity' of Mind substance that could withstand the potency of the thrusts of the impregnating Buddha. The Womb of the Consort would be incomplete, unable to process or hold his substance, yab-yum[15] cannot be thus maintained.

The factors influencing intelligence are given below. I have related them to their evolutionary purpose, thus the attributes of the Wisdoms of the Dhyāni Buddhas (Buddhas of Meditation).[16]

---

15 The posture of sexual union between Buddha and Consort.

16 The attributes of the Dhyāni Buddhas are of great importance in relation to the

a. Those upon the physical plane related to the evolution of the factors of instinct and of mind. Inevitable mastery leads to the All-accomplishing Wisdom of Amoghasiddhi.
b. Those related to the emotional and desire realms, the factor of desire (*kāma*), auras, desire-filled or imaginative thought forms. Inevitable mastery produces Ratnasambhava's Equalising Wisdom.
c. Those related to the mental body, the intellect and creation of thought forms. The factor of *manasic* (mental) input necessitates an understanding of the method of causation, mastery of which inevitably produces the Discriminating Inner Wisdom of Amitābha.
d. The factors arising from beyond the mental altogether, the liberated planes of causation. (The factor of being/non-beingness.) Receptivity thereto necessitates the inevitable expression of the Mirror-like Wisdom of Akṣobhya.
e. The interblending of the various Ray, astrological and cosmological energies, the factor of cosmos, *dharmakāya*[17] constituting the sum of human life. Here inevitably is awakened the Dharmadhātu Wisdom of Vairocana.

The concept of infinity (the boundless Space associated with the *mahāmudrā)* within our finite minds, translates as a vast though bounded sphere, circumscribed by the limits of our creative imaginations. It is an abstracted incomprehensibility that the intellect tries to grasp by means of some tangible form that comes within the parameters of its experiential possibilities based on former registered experiences. This necessitates the use of symbols for comprehension, primarily the Great Symbol and its derivations.

To make the process of the appearance of *saṃsāra* comprehensible we must relegate it to the realm of the mind, and therein this process takes the form of the profoundly metaphysical concept of the sphere

---

comprehension of the nature of Mind and have been thoroughly explained throughout *A Treatise on Mind,* with a summary presented in Volume 5A, to which the reader should refer for detail.

17 *Dharmakāya* means the fount or body of the *dharma,* the teachings of the laws of Life. *Dharmakāya* is the ultimate vehicle for the body of Truth. It is the primordial, eternally self-existing essentiality of *bodhi* (enlightenment) attained by liberated beings, and can be equated with cosmic Mind.

spiralling within itself and pushing out to fill absolute Space, which is motionless and dark (at least to our finite minds). The concept is represented on paper as a two dimensional circle.

In the various mythologies the circle ⊙ has always symbolised the full potential of Deity. It shows that the only aspects of Deity or a Logos that can be known are its effects; that the first result of Creation is the formation of a sphere, the cycle of time, the 'ring-pass-not' circumscribing manifest existence. This delineates the absolute time conceptualised as the *mahāmudrā* into finite sequences, known as *kalpas* and *yugas*. It also depicts the path of our solar Logos as He pursues His cyclic course in the Heavens, as well as delineating the potentiality of the solar disc to project light and Life. When a central dot is placed within the sphere then it represents the establishment of a Throne or Seat of Power of a Logos from whence the entire creative process of world-formation is enacted.

Darkness, *ākāśa* and *svabhāva*

It should be noted here that as the sun is the source of light to our entire solar system, so in a similar way there is a spiritual sun giving light to the path of *bodhicitta*,[18] the way of wisdom, resulting in enlightenment. The path of the sun and how it vitalises our planet is thereby the nature of the way of the gaining of enlightened perception.

Without light the darkness pertaining to ignorance cannot be overcome. Depending upon the degree of ignorance present, and how it manifests, so the qualification of light must appear in a skilful way to rightly counter it. Ignorance is affiliation to *saṃsāra*, the skilful means is *bodhicitta,* and the engendering of light is wisdom. The containment of this *bodhicitta* is the 'movement of the Heart', which is consciousness itself. Its appropriate symbol is the sphere, (as explained above) because it must work according to the limitations of time sequences that *saṃsāra* imposes upon it, if it is to be skilfully effective in eliminating the darkness of ignorance in all things. So also does the sun move in

18 *Bodhicitta,* the Heart's Mind, the Mind of enlightenment. The power or force productive of awakened realisations that emanates from the Heart centre. It is the compassionate force of the liberating Mind, the mind of pure perfection, the authentic nature of mind. This energy drives the entire Bodhisattva path.

the heavens above us. The skilful generation of *bodhicitta* will lead inevitably, in a far distant aeon, to one becoming a solar Logos; to enclose a vast domain of evolving mind-ful ones, plus the entire panoply of lesser sentient lives, towards liberation through the externalisation of the Great Symbol and the radiance of multidimensional light.

The originating ineffable Creative Potency has always been depicted in negative terms; such as Darkness or the Unknown. It is 'Darkness' because intelligence is trying to witness its own birthing. It can be known when one is able to fully identify with the vastness of abstract Space in its unmodified aspect. This book will obviously not focus upon the unmodified That aspect, but with the nature of manifest space and the evolution of consciousness within it.

Substance is but tangible energy in that it is expressed in a form that can be contacted and built into a body of manifestation, an instrument of experience, by an incarnate entity in any realm of existence. The person can utilise it at any time to contact a sphere of sensation and thus gather related experiences. The most intense form of energy in the material realm is but the grossest form of substance or sphere of sensation in another higher realm, and so forth.

Substance has always existed in one form or another and is ever becoming, ever changing, ever manifesting anew, as long as there is such a thing as imperfect being in the universe. As energy and matter are different modifications of the same thing, so we must learn to think in terms of energy-substance in motion if we are to understand the nature of being/non-being and how the universe came into existence.

If what is known as 'space' (*ākāśa*) pervades the whole universe, then a form of matter, however tenuous and subtle, must fill every iota of space. This is borne out by modern astronomical research, with its concept of background radiation, and also by the search of physicists for the one universal Law that logic tells them underlies all the others (the general Unified Field Theory that Einstein spent the latter part of his life trying to solve).

At present, there are four fundamental laws recognised by physicists: two long range forces, gravitation and electromagnetism, that integrate all aspects of universal space, and two short range forces, the 'strong' and the 'weak', that bind the nucleus of the atom into a coherent shape, as explained by modern quantum electrodynamic theory (QED).

That physical matter is a condensed form of a primeval Essence is a fundamental postulate of esoteric philosophy. Late in the nineteenth century scientists had a similar postulate concerning a substance they called *ether,* in their endeavour to explain some of the anomalies of the properties of space. Modern physicists have discredited such a theory, for to them Einstein's Theory of Relativity adequately explains such anomalies. A form of ether however does exist though it is far more refined than the substance of the phenomena that empirical scientists are investigating.

Consequentially there is a subjective medium through which light must also be conveyed, for the type of near vacuum[19] of space through which physicists think light travels in space is only a partial truth. There are subtler dimensions of existence which the scientific community has not investigated, or are only beginning to, being the threshold of their experiential zone of activity. The ether of space and the waves of light can effectively be viewed as virtually the same thing. We will see later that quarks, gravitons, electrons and photons emanate from the 'substance' of this ether, where the photons represent the consciousness-bearing factor in relation to the appearance of phenomena.

A negation of primeval substance would mean a negation of particulars, for naught could come into existence if the cohesive force that allows all to be sustained, integrated and interrelated, is not manifest. Essentially, energy is all there is, and consciousness consolidates and shapes that energy into the coherent forms (subjective and objective, visible and invisible) that we know and can come to know. Even what is viewed as 'formless' (*arūpa*) is only thus because the quality of the eye that sees is too gross to view finer, subtler forms. There are forms of substance and energy that exist beyond the parameters of the mechanism of normal vision, and which is responsible for the phenomena known as clairvoyance, and clairaudience in relation to hearing, to say nothing for all of the *siddhis* possessed by the *siddhas* (advanced *yogins)* mentioned in Hindu and Buddhist texts.

Old unstable states of matter, old forms, must eventually decay, disintegrate and give way to new forms. Causation results in changes

---

19 Nature certainly 'abhors a vacuum'.

of energy states and which manipulates them into set patterns, forms that have mass. When referring to the 'creation', or more specifically, 'condensation', of the tiny particles to form physical matter there will first be a concretion of primal essence. When galvanised into activity by the Will of the creative Thinker, (moulded and activated by an aspect of Mind) this essence becomes the inherent Life animating substance. It is the originating seed impulse cohering the essence into a form or atom possessing particular emanatory characteristics, a radiatory quality peculiar to it and to the class of interrelated atoms to which it belongs.

As all of manifest Nature adheres to the law of Economy, so the ovoid or atomic shape, being the most compact, becomes the preferred type of shape utilised by the Thinker to carry on the process of that One's evolving Thought Life. This ovoid contains the essence of the inherent Life of the evolving forms.

Cellular (sentient or conscious) Life[20] progresses in terms of spiral-cyclic motion, and the law of cycles manifests on a cosmic as well as a planetary scale. Thus there can be no absolute beginning of time, or of the causative Impulse, but rather cyclic recurrences of activity and obscuration of that activity, of day and night that are incrementally spiralling upon ever higher, or more refined cycles of expression. The intervals between universal or cosmic occurrences are, however, great enough to be called absolute or 'infinite' when compared to the human time scale.

The beginning of time, as far as our particular solar Incarnation is concerned, i.e., looking at our entire solar system as a cellular structure, can thus be likened to the awakening from a 'deep sleep' by the embodying Logos to herald a new Day by means of the thinking of a new Thought *(mahāmanvantara).*

The next act of the awakening Ineffable Mind could metaphysically be described as the 'Breathing of the Divine Thought'. *Mantra* is then emanated to effect the appearance of the desired phenomena. The *mantra* attracts to it the appropriate substance from the various planes of perception, according to the intrinsic energy qualifications built into it, and the force of the energy it possesses, as 'breathed out' by the divine

---

20 This concept is explained in *Maṇḍalas: Their Nature and Development.*

Thinker. This is the originating act that resulted in the causative *motion* that condensed the universal prima matrix (*mūlaprakṛti*) into the seed form of the solar system.[21] The Logos (an Ādi, or primordial Buddha) of this cellular structure 'awoke' with the Thought of a new Day, as the consequence of a meditation that resulted in the action of manifestation. In doing so the *saṃskāras* (tendencies, residual impressions from past lives) of an ancient Thought pattern was built into the new by means of activating the necessary *bījas* (seeds). All Thoughts pertaining to the future (emanation) necessitates the past (*karma*) to be reinstated into the new, and the resultant tendencies projected into the future with a Desire-to-Be, producing a new 'clothing' or incarnation. This is the wish-fulfilling gem[22] of what must eventually form, according to the Will of the Thinker.

The idea can be likened to a stone (symbolising thought-projected *saṃskāras*) that is thrown into a pond, thus causing numberless ripples that disturb the clarity of the water by forming geometrical patterns. This motion is an active expression of the Intelligent faculty of the abstract Creative Logos. The process manifests from above-down and from within-without and utilises material that is already existent, or else is caused to condense from its *bīja* form.

A Hindu concept of primal substance matter in the term *mūlaprakṛti* is utilised, as their philosophy is well suited to the conception of causation of phenomena by means of forces emanated by causative agencies (the 'Gods'). In this arena Buddhists have not won their debates with the Hindus, because the emphasis of the creative process is upon the formed nature of things, the philosophy substantiating the existence of *prakṛti*. Buddhists however tend to emphasise the nature of consciousness, the evocation of wisdom (*prajñā*), of the non true existence of 'Gods', and of the way that consciousness affects the form. In their case one could use

21 When extrapolated upon a cosmic scale we thereby have the explosion from the 'singularity' postulated by the deans of modern physics, from which it is asserted by them that the entire universe emanated. However, in the esoteric view the conception begins upon the domain of cosmic Mind. By the time the physical plane is reached there can be more than one locus for the descending energies, because the factor of the *chakras* must be taken into account.

22 *Cintāmaṇī,* the diamond-Mind, the wish-fulfilling gem embodying the intrinsic energy field of enlightenment.

the term *cittaprakṛti* (mind substance) as such primal substance matter, in which case we would look to mental *saṃskāras* being activated first. The Vaibhāśika concept of *dharmas* is not appropriate here because of the inherent flawed nature of the philosophy.[23]

When it comes to the appearance of phenomena Buddhists also often use the term *svabhāva,*[24] which is that which is self-becoming, self-existent, which develops its essential self from within outwardly by emanation or evolution. *Svabhāva* is the intrinsic energy, the universal world-forming substance that causes the existence of 'things'. It is an innate essence underlying phenomenal appearance. It can thus be equated with etheric substance, as viewed from a higher perspective, the fourth cosmic ether *(śūnyatā),* denominated as *buddhi.*[25] Thus it can also be viewed in terms of being *ākāśa. Svabhāva* is therefore the plastic *essence* of matter, both manifest and unmanifest. It remains in its own energy field upon the subjective domains as the cosmic reservoir of Being, therefore of consciousness, of intellectual light, of Life. *Svabhāva* may also be considered as *parabrahman-mūlaprakṛti,* the one underlying cosmic being or substance, the divine source; the self-existent.[26]

The term *ākāśa*[27] needs further explaining, as it is often viewed in terms of being the creative energy, the Aether of 'space', where space is that through which things must manifest in order to make a visible appearance. Through this space 'things' (consciousness-attributes) come into being, hence *ākāśa* can be considered the 'space of consciousness'. Literally the term means 'not visible', space, subtle and ethereal fluid pervading the totality of the phenomenon of the universe, and a vehicle of Life. The Purāṇas state that *ākāśa* has one attribute: sound. From this

---

23 One of the four main philosophical schools. Others are Sauntrāntrika, Cittamātra, and Mādhyamika. They accept *dharmas*, the distinct and independently existing atomic factors comprising the material worlds and the moments of time composing consciousness.

24 *Svabhāva,* from *sva* = self, plus the verbal root *bhū* = to become, to be.

25 The term *buddhi* will be further explained below.

26 See volume 1 of *A Treatise on Mind* for further explanation of *svabhāva* within the Buddhist context.

27 *Ākāśa* is derived from the roots, *ā* = towards, to, near, plus the verbal root *kāś* = to be visible, appear, shine, be brilliant.

perspective space is the *upādhi* (vehicle) of Thought. The *Chandogya Upanishad* (7:12:1-2) equates *ākāśa* with Brahman. It is akin to *prāṇsa*, as it is the higher correspondence to the *prāṅa* that vivifies our etheric forms. It is that subtle supersensuous essence that pervades the space of the four cosmic ethers, our higher planes of perception. Specifically it carries the 'electricity' associated with *buddhi*.

Ākāśa is the plastic essence that is the vehicle of a Creative Logos expressing itself as the formative forces conditioning the manifestation of all phenomena in systemic space. As such it is the transmuted correspondence of the five *prāṅas* manifesting in a human *nāḍī* system, hence there are five levels or degrees of *ākāśa* relating to the five higher planes of perception.[28] The plane *ādi* conveys the Aetheric aspect of *ākāśa, anupādaka* conveys the Airy aspect, *ātma* the Fiery aspect, *buddhi* the Watery aspect, and the abstract domain of the Mind, the Earthy. The Aetheric, Airy and Fiery aspects can be considered the emanations from the domain of cosmic Mind, the Watery conveys the *prāṅas* from the cosmic astral ocean and the Earthy aspect is intrinsic to the cosmic dense physical plane. From this perspective *ākāśa* therefore is conveyed in terms of the Wisdoms of the five Dhyāni Buddhas, or rather, the expression *(prajñā)* of their Consorts, where Vairocana's Consort embodies the *ākāśa* of *ādi*, and so forth to Amoghasiddhi's Consort conveying the *ākāśa* of the higher mental plane. *Ākāśa* consequently is the conveyor of the force of compassion *(bodhicitta)* emanating from the cosmic Waters (the cosmic astral plane), hence the sum of the zodiacal and planetary energies that modify space. Thus it is the vehicle of the enlightened Mind, from the third to the seventh Initiations.[29]

Skandhas and *saṃskāras*

Here should be added the information that summarises the Buddhist concept of causation with respect to the five Elements from Brown's book, *The Buddha Nature,* where in his commentary upon the *Ratnagotravibhāga* he states that the *'Ratnagotra* grounds the absolute

28 See figures 5 and 6 in chapter 5 for an explanation of these planes of perception.

29 See volume 6 of *A Treatise on Mind* for the explanation of the first five Initiations, and volume 7B for the higher cosmic Initiations.

nature of the Innate Mind through the standard parallel to the infinity of space' and continues with:

> The *śastra*[30] then proceeds to establish correspondence between the macro-and micro-phase of popular Buddhist cosmogony, where the earth is supported by water, water by air, and air by space which is itself, as the ultimate dimension, unsupported by anything. While the first three primary elements are themselves subject to appearance and disappearance, evolution and devolution, the omnipresent *ākāśa* transcends all causation and conditioning. In a similar manner, all the constituent factors of phenomenal existence, classified into five elementary groups *(skandhas),* eighteen component elements *(dhātus),* or twelve bases of cognition *(āyatanas),* are akin to earth. They, in turn, have their foundation upon the active force and defilements which resemble the expanse of water. As this latter was said to rest upon air, so *karman* and *kleśa* exist on the basis of the 'irrational thought' *(ayonimanaskāra).* This last, signifying the originative force of ignorance, is nevertheless grounded upon the space-like, firm, immovable, unoriginated, indestructible essence - the naturally radiant, Innate Pure Mind....
>
> The permanent, steadfast and eternal *Tathāgatagarbha,* beyond all that is caused, conditioned or compounded, is the supportive ground or base of the innumerable Buddha natures, which are inseparable and indivisible from it. At the same time, it is said to be the foundation of the 'defilement stores' which are however, separate from and extrinsic to it. This metaphysical formulation is translated more specifically into the problem of the simultaneity of an innately pure consciousness and a defilement on that consciousness. If the radiant purity of the mind cannot be touched by darkness, since it is nevertheless said that 'there is defilement and there is a defiled mind'?[31]

A few notes concerning this quote needs to be made, first the 'water' that the 'earth' is said to be supported by here is really what is termed *mūlaprakṛti* above. Only when humanity have appeared and have properly developed the emotions in combination with their creative

30 Commentary, sacred verse or book, treatise, as is the *Ratnagotravibhāga.*

31 Brian Edward Brown, *The Buddha Nature,* (Motilal Barnarsidass, Delhi, 2004), 108-109.

imaginations can the Element Water be considered to refer to the astral plane (the Watery domain), explained in *A Treatise on Mind.* Second, the 'air' here really refers to *prāṇa,* therefore to the *prāṇamayakośa* (the sheath or body of *prāṇa)* and the *nāḍīs* that support all of manifest Life. The *prāṇamayakośa* is an integral aspect of the etheric body, from whose substance this body of energies is derived and which supports its existence. The etheric body, plus the that of the material domain is Earthy in nature.

As the human body possesses a body of *prāṇa* so do all other incarnate Lives (Logoi) in the universe. The higher correspondence to *prāṇa* has already been dealt with above, and it can be inferred that all Logoic *prāṇamayakośas* are linked, as the universe is a unity, and energy, conditioned by the laws of Mind is all there is. It is via the *prāṇic* field that the energies manifest from the domain of Logoic Mind, in its five-fold qualification, the five *prāṇas (vayus),* to eventually produce the type of phenomena that our empirical scientists are presently investigating. Those engrossed in meditation practices are effectively endeavouring to rightly control and direct the *prāṇic* Winds of their minds and the *prāṇic* fields constituting their bodies, and in time the *prāṇic* forces external to them. Once all is mastered by Mind then the supramundane *siddhis* can be demonstrated. Vast is the esoteric lore gained related to such control, when the higher Initiations are to be obtained with respect to incorporating the *prāṇas (ākāśa)* from cosmos in meditation. The subject of esoteric astrology is concerned with gaining such knowledge. Indeed, esoteric astrology is the science of the high Initiates. It incorporates knowledge of cosmic law and of the energetic interrelation between various Logoi and the planetary bodies. Such interrelations produce the appearance of all phenomena known to us, and relates to the ability to travel the cosmic Paths at the end of one's servitude to our tiny earth sphere.

The *skandhas* are bundles or groups of attributes that together constitute the human personality and are responsible for the factor of consciousness. Exoterically, there are five *skandhas:* 1) form, or body, the sense organs, sense objects and interrelationships *(rūpa),* 2) perception or sensation, feelings and emotions *(vedanā),* 3) aggregates of action, or the motives to thus act *(saṃskāras),* 4) the faculty of discrimination (*samjñā*), 5) revelatory knowledge *(vijñāna)*. Effectively, all of these forms of activity are attributes of the *saṃskāras* that are carried through from life to life collectivised in their various groupings. Of the *skandhas, rūpa* represents the sense-consciousnesses, whilst *vedanā* and *samjñā* are together the *kāma-manasic* (desire or emotional-mind) aspects of consciousnesses. The *saṃskāras* are expressed in the form of the five different types of *prāṇas* conveyed throughout the *nāḍī* system. They are one's karmic accumulations that must be worked with in that life and are eventually transmuted into the seeds of enlightenment *(vijñāna)*.

The term *dhātu* means 'root or base', fundamental stratum, realm of being, constituent Element, or part of a world construct. It is a state of existence *(lokadhātu)*, thus there is *arūpadhātu*, the formless realms, *rūpadhātu*, the formed realms and *kāmadhātu*, the desire realm. Consequently the categories of classes of all manifested things are implicated in the meaning of the word *dhātu*. When eighteen *dhātus* are mentioned they refer to the six sense objects, six sense organs, and the six associated perceptions.[32] They are the subtle elements producing the generation of knowledge. When the number six is referred to then the mind (the intellect), viewed as a cataloguing and collating tool for the impressions from the five senses, is also thought of as a 'sense organ'. Thus we have the five sense organs and the brain as the 'sense objects', the 'six sense objects' relate to the experiences of the senses of touch, sight, etc, whilst the associated perceptions are depicted as the hearing sense-consciousness, sight sense-consciousness, etc. The number six is of importance here because there is a corollary to the six quarks and six leptons of the subatomic world, from which the entire material world (the Earthy Element) is constituted. What manifests in the atomic universe can be considered to be reified homologues to the world of human experiences. The macrocosm is reflected into the microcosm, and vice versa.

That which interrelates all *dhātus*, within the spaciousness of *ākāśa*, is the Fiery *cittaprakṛti*. We thus have the five Elements indicated here:

---

32 The *Abhidhammakośa* presents a list of the four main Elements (Earth, Water, Fire, and Air), the five sense faculties, the five objects of perception, the phenomena of sex, of the heart, the phenomena of life and of nutrition.

Earth, Water, Fire, Air (indirectly here, as that which is purveyed by the *nāḍīs*[33]) and Aether as *ākāśa.* When looking to *ākāśa,* with its five-fold subdivision, viewed in terms of the underlying energy from which all phenomena is derived, the corollary in the subatomic world are the five forces (bosons) that are responsible for the formation of atoms. They are the gluon, photon, the Z and W bosons, and the Higgs boson. Creative energy (Fire – *cittaprakṛti)* is that which interrelates them all.

The 'Innate Pure Mind' is said to support all of the *bindus* of irrational thought, 'the impurity of stains of desires, etc., which are of accident and produced by wrong conception'. Esoterically, this 'Pure Mind' is the constitution of the Sambhogakāya Flower *(tathāgatagarbha)* and its methodology of absorbing the accumulations of attributes of mind developed via the rebirthing process, and also the methodology of their transformation inevitably into what can be described as a Buddha-Mind. The methodology is provided in volume 3 of *A Treatise on Mind.* What needs to be at least logically hypothesised upon here is that Logoi also possess their versions of the Sambhogakāya Flower, from which, upon a vastly transcendent scale, all that is known and experienced comes into manifestation once a Logos chooses to incarnate. The transmuted correspondences of such mechanisms can be meditated upon, but the meditator needs to take into account the level of development of the Logos concerned, Logoic interrelationships, Ray Purpose and the scale of the nature of the Lives incorporated within the Domain of such Flowers. (Such Lives being the *saṃskāras* Breathed out into active manifestation at the appropriate time.)

By 'irrational thought' here we can look to (in terms of the attributes of the five Elements):

a. Thoughts that due to the input of ignorant assumptions produce wrong conclusions.
b. The emotional portion of any thought, which distorts clear thought in accordance to the desire of the thinker, subtle or gross.

33 The Element Air properly considered is the substance of the Clear Light of Mind and of its relation to *śūnyatā,* via which the Wisdoms of the Dhyāni Buddhas find expression. *Prāṇamayakośa* is the lowest reflex of the energy of these Wisdoms manifested into consciousness-space.

c. Circuitous thoughts that lead nowhere because of lack of proper energy or thought input to produce logical conclusions.

d. Thoughts that may be correct with respect to the perpetuation of *saṃsāric* conditionings but are not conducive to the production of ultimate truth, the eventuation of liberation from the wheels of limitation.

e. Thoughts that may be correct with respect to the concept of liberation, but not with respect to the *dharmakāyic* vision viewed by all advanced Bodhisattvas.[34]

Each form of irrational thought is a causative source and as such has its genesis, inbuilt limitations, evolving history, and eventual cessation.

The sum of a humanity that have yet to evolve to Buddhahood contain within them the *bījas* that project the *saṃskāras* that are the force of defiled mind, that must yet be comprehended and transformed into enlightenment qualities. A consequent liberation from *saṃsāra* will be gained. All *bījas* towards liberation are activated by means of correctly sequenced thought by That (the *tathāgatagarbha*) which utilises the *cittaprakṛti*. (This is another term for the *ālayavijñāna*,[35] when looking purely to the nature of the substance that embodies it.) The *tathāgatagarbha* works to annul the effects of the originating causative action that it manifested by projecting the stream of personal-I's (one's incarnations) into manifestation. These personalites produced the secondary causative actions that generated the forms of irrational thoughts (defilements, *saṃskāras* of *kleśas,* and *kliṣṭamanas)* that sustain their world. These defilements represent the *samalā tathatā* that must later be cleansed through the generation of the Jina[36] Wisdoms.

---

34 A Bodhisattva is an enlightened one who has made a vow never to cease striving until all sentient beings have reached 'the other shore' of Life *(nirvāṇa),* thus have been liberated from thralldom of *saṃsāra,* and then and only then will he/she also do so. There are ten stages of Bodhisattvaship in Buddhist teaching.

35 Universal storehouse or abode of consciousness, the mind as a basis for all in the Yogācāra philosophy. Such a 'store' in the subatomic world can be viewed in terms of a plasma field, the fourth state of matter, of which the universe is largely consisted of.

36 The term Jina refers specifically to the five Dhyāni Buddhas, though sometimes it can refer to a Buddha.

The production (causation) of irrational thoughts, in all of their versions, leads inevitably to friction, pain and suffering, sickness and disease. This is because they lack vital life, the consequential effects that are progressive in nature, thus inevitably life sustaining. 'Life' here refers to the 'residual' that sustains the liberation process after the limitations of the defiled thoughts have been eliminated. It then produces a new causative impulse upon a higher cycle of revelation, the cycle repeats upon ever transcending spirals of transmogrifying revelation and consequent bliss. Each new revelation produces a liberation from that which transpired previously, and so inherent sickness is averted or healed. The *saṃskāras* of irrational thought, of the defiled mind are then no longer generated, allowing the innate Mind to shine free from obscurations. The causative process has produced its fruit, but what is the gain of the cycle? The short answer is wisdom (*prajñā*), the long answer is Buddhahood.

This then summarises the process concerning the causation of phenomena, of all thought structures that will inevitably produce defilements and their evolution to the Clear Mind.

The universe consequently can be considered the expression of the rational Thought of a super-Mind. It obeys the laws of Thought expression,[37] but may support 'irrational' aspects of Thought carried through from previous *mahāmanvantaras*. That which is 'irrational' in this Thought structure represents the attributes (planets, stars, the humanity and lives they carry) that are not yet perfected, that must yet evolve, to liberate attributes that will appropriately harmonise the whole. Such irrationality is seen for instance in the way that humanity presently construct their desire based thought-forms, which sow the seeds of much chaotic activity, pain and suffering. The overall governing principle supporting the evolutionary process of this irrationality is clear, rational Logic, governed by immutable Laws that condition the manifestation of the all. Such are the laws of *karma*, of Economy, Attraction, Synthesis, and those discovered by physicists. Because a universe, or any of its component parts is incarnate, so it implies that there are 'irrational thoughts', imperfect chaotic substance, still to master and bring into rational Law by the higher evolving embodying Mind that is aspiring to ever higher states of attainment and so must transform that aspect of its equipment that prevents this.

---

37 Such laws are governed by, and are the expression of, the nature of the evolution of the Dhyāni Buddhas.

The *saṃskāras* manifesting via a planetary or solar sphere then represent the streams of sentient and conscious Lives, subhuman, human, *deva* and enlightened, that find further opportunity for evolutionary growth because of the existence of such spheres of containment provided by Logoi. The compassionate activity of each Logos works to cleanse the impurities from the *saṃskāras* and to so educate the informing Lives to develop the needed characteristics so that in time they too will be able to play roles as Logoi.

The *ālayavijñāna*, *śūnyatā* and the appearance of phenomena

The *ālayavijñāna* (store consciousness) can be viewed as the womb of all manifest being as it is the store of all attributes of mind and those of the abstract Mind. Thus all forms of defilements of mind as well as liberating, enlightened perceptions, are accommodated by it. It is the Sambhogakāya Flower's domain. The *ālayavijñāna* is organised according to the causative agendas of the entire community of Sambhogakāya Flower's that it supports as a base. They manifest the cumulative meditations that are causative of the civilisations that affect humanity at any time. The sum of the subsidiary impulses *(saṃskāras)* are stored in ordered fashion in the *ālayavijñāna.*

The causative moulding of the *ālayavijñāna* environment can therefore be considered a proper function of the Sambhogakāya Flower *(tathāgatagarbha).* Through the cumulative action of all Sambhogakāya Flower's the *ālayavijñāna* will also be ultimately cleansed of low order defilements, however the imprints of what has transpired will forever remain. From the base of the *ālayavijñāna* arises the awakened Mind. This is the basis to the evolution of the Wisdoms of the Dhyāni Buddhas. The *ālayavijñāna* is effectively the Buddha-field, the *nirmāṇakāya* of Amitābha, who rules the Element Fire. The *ālayavijñāna* can be considered a sea of Fire, a domain of pure energy, organised by the attributes of mind/Mind. When this idea is relegated to Logoic domains, then it is an important consideration in relation to the energy that can be drawn upon in order to effect the causation of material phenomena, from whatever level of expression that Logos hearkens. We can view an ocean of cosmic and systemic Mind that is utilised by the Dhyāni Buddhas (here viewed as Logoi) as their base for meditative activity.

There is a difference between the *ālayavijñāna* and *cittaprakṛti* in that the former relates more specifically to the conditionings of the sea of mind/Mind conditioning a humanity wherein resides the Sambhogakāya Flower, and the latter represents the sum of the substance of the mental plane, irrespective of whether it being an expression of human minds. The *devas* for instance utilise the *cittaprakṛti,* but are also established in the *ālayavijñāna,* for all is constituted of their substance.

With respect to causation it should be noted that the starting point of any causative agenda generally concerns the reactification of a seed, a paradigm that already exists *in situ,* and which contains the 'genetics', so as to speak, of all that is to follow. Such a seed is a *bīja,* and a collectivised integrated grouping of *bījas,* containing interrelated purpose, produces a *bindu.* Lama Anagarika Govinda states:

> If we speak of the space-experience in meditation, we are dealing with an entirely different dimension (in connexion with which our familiar 'third dimension' only serves as a simile or a starting-point). In this space-experience the temporal sequence is converted into a simultaneous co-existence, the side by side existence of things into a state of mutual interpenetration, and this again does not remain static but becomes a living continuum, in which time and space are integrated into that ultimate incommensurable 'point-like' unity, which in Tibetan is called *'thig-le'* (Skt.: *bindu).* This word, which has many meanings, like 'point, dot, zero, drop, germ, seed, semen', etc., occupies an important place in the terminology and practice of meditation. It signifies the concentrative starting-point in the unfoldment of 'inner space' in meditation, as also the last point of ultimate integration. It is the point from which inner and outer space have their origin and in which they become *one* again.[38]

There is another aspect of the Logos that must remain tranquil, motionless, aloof from the creative process and its attainment. This has been termed 'the real duration of being', represented as the sumtotal of all that is and is not. It is beyond 'the Creation', yet from it comes the thinker of the Thought, and that which is thought about. This is the

---

38 Lama Anagarika Govinda, *Foundations of Tibetan Mysticism,* (Samuel Weiser, New York, 1975), 116-117.

śūnyatā-saṃsāra nexus,[39] which was explained throughout *A Treatise on Mind*. *Śūnyatā* veils the Laws governing our natural environment, the universe we live in. Without *śūnyatā* as the stable base, *saṃsāra* would be chaotic, there would be no rhythm or regularity to its actions, no predictability, no relative stabilities. Because *śūnyatā* is stable, so *saṃsāra* has a solid foundation wherewith to move, to come from and to go to. This allows consciousness to evolve. Consciousness needs relative stability, it needs the rhythm and regularity of a flow of thoughts in order to be able to name, to classify, to arrange 'things'. It needs to come from and to proceed to somewhere, even if that 'somewhere' be a Void, to which it moves as the thoughts created become increasingly abstract, refined, sublime and then totally clarified of attributes of mind, yet sustains the universality of Mind.

Śūnyatā can be considered the foundation of the house that is the universe and everything that is contained in it. For this reason *śūnyatā* must be devoid of all the attributes that relate to the ephemera of the universe, otherwise the foundation itself could not last, and everything would collapse into a rubble of meaninglessness.

The substance constituting the vehicle of the mind/Mind undergoes the action and is eternally in the process of transformation. *Matter* is the end result when the Waters of space[40] was sent into action by the 'rock of motion', of the recollected Fiery mental *saṃskāras* that are projected forward into a new disguise by calling forth associated Watery *saṃskāras*. This is effected by the Intellect of an embodying Logos.[41]

---

39 This nexus can also be viewed in terms of the *apratiṣṭhita nirvāṇa, nirvāṇa* of no fixed abode, residing in neither *saṃsāra* or *nirvāṇa* (here taken as *śūnyatā*), Bodhisattvas do not remain in a static condition, as *bodhicitta* motivates them to continually serve. On the other hand the 'static', or 'fixed' *nirvāṇa* of the *arhants* of the Theravādin tradition implies that once *nirvāṇa* is attained, there is no further development ('movement'). In the *A Treatise on Mind* the *apratiṣṭhita nirvāṇa* also concerns a consideration of the attributes of the Sambhogakāya Flower.

40 Cosmic astral or systemic space. By 'systemic' is meant the substance that is enclosed within the confines of a planetary or solar sphere of activity.

41 This is but the transmuted correspondence of what transpires in a human mind in its act of thought construction. A Logos represents the cosmic Mind that embodies or emanates the Word *(mantra)* that is projected to activate the seed *bīja* whereby the complete *maṇḍala* (the geometric foundation of what is to be) comes into being.

Because of its inherent qualities the Intellect limits this omnipresence to a point or sphere in time and space. Action causes reaction, cause produces a tangible effect, and we have phenomena as the result. On a grand scale the material universe is similarly born.

'Phenomena' means that the originating Fiery energy is briefly clothed in the elementary matter of the mental and astral domains, incorporating thereby Watery *saṃskāras,* and the momentum thereof galvanises etheric substance to empower the appearance of phenomena. The mechanism of the appearance of such phenomena upon the physical domain from the 'quantum vacuum' (event horizon) shall be detailed in chapter nine. Physical phenomena is the result of the 'condensation' of the substance of subjective (multidimensional) space. In time the same phenomena will be distilled, refined, rarefied and abstracted back into the space from which it originated. It consequently dissociates as it passes through the event horizon. This alchemical process utilised by the power of the Fires of Mind is not yet understood by the scientific community, but once comprehended will produce a major step forwards in the advancement of their science.

A Mind can recognise within the lists of images that were formerly created those that appear as the signposts leading to the future of whatever is to be for it. Similarly the Fiery nature of the mind and its abstraction into *śūnyatā* produces the liberation of mind, but not its extinction, because it then becomes the foundation for the awakening of cosmic Mind, the *dharmakāya.* The energy of *śūnyatā* is needed for the transmutation of substance-matter because *śūnyatā* obliterates the foundation of its empirical construct, allowing a rearrangement of the fabric of substance to manifest via the four ethers.

The way of the development of the mind indicates the way macrocosmic space originates. From above-below, and from within-without, the forces of Mind precipitate what is, as well as incorporating primal substance-matter, the 'black dust' of space for the establishment of every new planetary or solar sphere. The black dust is the new that must be converted by means of the appearing substance from the past to follow the evolutionary impulse of the overriding Will of the presiding Logos. Out of this primal substance new 'man-plants' will eventually form, which is the objective of every new Logoic incarnation procedure. That which manifests from within, becomes the substance that appears without, bearing with a *saṃskāric* imprint[42] that will allow it to eventually transform the primal blackness *(mūlaprakṛti)* that universally exists into a body of intense luminescence, be that of an enlightened Mind, or a radiant sun. The process of transformation is an aeonic-long procedure, however, when mind does not exist, or rather 'sleeps', then timelessness is the essence of things. Primal 'black dust' is thereby attracted to the primary seed *bījas* and consequently are organised into forms,[43] bodies of manifestation, that move and interact, and as they do so, an evolutionary progression from the mineral to the higher kingdoms of Nature slowly ensues.

The appearance of manifest space is the result of the cooling down of the intensity of the highest possible (Fiery) energy state or modification of being/non-being. Every form has at its heart a *bīja,* a seed of Fiery energy. That energy is in essence but the expression of the Fiery *śūnyatā* (Earthy *ākāśa)* manifesting from the higher mental plane, which has been condensed by the driving Will of the originating Mind. (Upon the physical domain such Will is known as the strong force that integrates quarks into the atomic forms.)

Here is veiled the secret of the nature of *kuṇḍalinī* energy. Its Earthy aspect is what scientists liberate via nuclear explosions. In reality all phenomena can be considered crystallised mind/Mind that binds primal elementary substance into forms. The originating energy in Nature being the Fiery substance utilised by the five Dhyāni Buddhas (or from the primordial Ādi Buddha who embodies the five as part of His Mind) who meditated to form it all, coupled with the *devic* component of the Consort with whom the primordial Buddha 'copulates'. The articulated

42 They are termed 'the Elemental lives', Baskets of Nourishment' or 'Blinded Lives' in the chart on page 35 of *Esoteric Astrology* by Alice Bailey (Lucis Publishing, New York, 1975). Such substance is considered as units of Life, because everything is constituted from the bodies of manifestation of the *deva* kingdom. They are units of mind that are the basis to the eventual evolution of the sentience manifesting in the fields of life known as the kingdoms of Nature. The *devas* embody the domain of the Mother, and were explained in *A Treatise on Mind,* especially in volume 7B, as well as in Alice Bailey's *A Treatise on Cosmic Fire.*

43 The attractive potency is an effect of the inherent gravitational force carried by the *bījas,* which will be explained in chapter nine.

Fiery energy impregnates its purpose into the little units of sentience in Her Womb. They constitute the 'atoms' of the Void Elements[44] She embodies. Thus are the *bījas* formed of a new world-to-be based upon the Ādi Buddha's Meditation of what once was (the world through which He evolved), recalling the former transmuted *saṃskāras* now adapted as the bases of the form-to-Be. The *maṇḍala* of time and space is hence impregnated with Fiery noetic purpose and the little lives that are the *bījas* of all-that–is-to-Be.

The entire universe can hence be but considered as an enormous Thought-Form unfolding. Thus the first action of the originating Thought-motion (or Thought-moment) is the concretion of energy into geometrical (or mathematical) forms, according to the laws of Thought. Order (or form) comes from the concretion of an initial organising universal flux of energy. Once the process is begun, then the geometrisation happens with a rapidly increasing complexity of continually expanding forms until our physical realm is finally produced with the types of cosmological and planetary phenomena known to scientists. This realm thereby contains an immense number of different categories and patterns of things, which can be considered the reified aspects of Mind.

The creative Logos literally breathed the divine Thought in movement, metaphorically described as the universal ripple in the bosom of infinite Quiescence. This produced a reaction on all levels of being, allowing the condensation of virgin primal formless substance into/as formed space, impulsed (fecundated) by the seed idea of the causative Mind.

The appearance of concrete substance is the last part of the event of an all-embracive multifarious awakening of countless Conscious and sentient Lives. It is a product of the interrelated chain of solar, planetary, *deva,* human and subhuman atomic unities. All are brought into manifest activity, allowing the entire solar system (or universe) to thrill rhythmically to the pulsation of a new Life. Life concerns the incessant breathing in and out (expressing also the reincarnations) of myriads of Beings as part of a integral *maṇḍala* of active manifestation.

---

44 See my book: *The 'Self' or 'Non-Self' in Buddhism,* 187-189 for an explanation of the Void-Elements.

The inherent Life, light, instinct, intelligence and Love-Wisdom is thus actively expressed in countless different forms, engendered as part of the evolutionary purpose of great informing Lives.

The entire evolutionary process is governed by the laws of mind/Mind embodied by the *deva* kingdom, who are yet to be discovered and analysed by the scientific community.

The *trimūrti*

Though an absolute Logos exists in an omniscient, omnipresent, quiescent state, He also manifests the ever-flowing motion that is the emanation, sustenance and destruction of what can be considered a universe. These qualities from this Logos are a natural expression of each other, allowing the universe to always be in a perpetual state of flux in which the old constantly makes way for the new.

There are certain refinements of what constitutes a Logos yet to consider. We can see here that the concept of universe does not refer just to the cosmos delineated by scientists, but rather to a relative succession of such subjective and objective universes, of Consciousness-Spaces wherein sentient beings, and entities that are increasingly aware, enlightened, evolve. We can look to the relativity of local universes, such as that defined by our earth sphere, solar system, local part of the galaxy, the entire galaxy, and so forth. Each gradation of 'universe' has its own embodying Ādi Buddha that must be considered at increasingly subtler levels of transmuted correspondences. Space is far vaster than that of the cosmic space investigated by materialistic scientists, and has many gradations to it. The esoteric concept of evolution therefore, and indeed what it is that evolves, and where to, is far more refined, meaningful, detailed, and exquisitely purposeful than what has yet been comprehended by our materialistic scientifically minded brothers. In fact the best results of their investigations can be incorporated into this philosophy as part of a eschatological, ontological and teleological revelation, a grand unified field Theory of Everything.

The Buddha aptly emphasised that there is nothing static in the manifested universe, everything is in constant change, hence is transient. The motion that effects one thing must naturally be carried on by means of cause and effect to influence all else. The past, present and future are simultaneous expressions of the meditative process of a creative Logos. Energy is equable with motion, which must have a cause, it must be sustained during its course, and is destroyed (that is, transformed) when it reaches its destination. Every manifest thing must also go through this process. This is the idea behind the concept that forms the basis for the *triune aspects* of Deity (*trimūrti*), the Father, Son and Mother unity in the various world creation mythologies.

Thus in the Hindu theology we have: *Brahmā, Viṣṇu* and *Śiva*. They correspond to the Emanator, Sustainer and Destroyer, the triple aspects of the one Being (personified as Īśvara, but then abstracted into the neuter, Brahman), the one flow of events. This however only affects the manifest universe, for on the archetypal planes there can be no such thing as creation, sustaining, or destruction, for there is nothing to be created or destroyed, as we understand it. There is only intensified Life as all-pervasive omnipotence manifesting in an all-embracive ocean of potential, veiled by absolute quiescence.

Such *quiescence* can be somewhat understood if we imagine an energy wave intensifying its frequency to encompass 'infinite' motion. Its momentum must increase until it is at the point before the infinite, whence ultimate momentum is reached. Upon the attainment of the 'infinite', this ultimate momentum must be transcended, to reach a state beyond motion (or beyond mental activity); thus we have a *motionless*, quiescent state, akin to the *dhyāna* state of the meditation-Mind. It appears this way to the thinker engrossed in a 'finite infinite' sphere of activity which precedes a subsequent transcendent state.

This is a reason why the Mādhyamika philosophy specifically states that there is no such thing (in reality) as 'creation' and consequently sustaining or destruction, because all things are ultimately empty of all such attributes. How can there be a 'creation' they say if such a thing as what is described as 'quiescence' above is all there truly is? Clearly however, both *śūnyatā* ('quiescence') and *saṃsāra* (the activity of phenomenal appearance) exist relative to each other in terms of a co-dependency, and the nature of how one comes from the other is a great mystery, somewhat unravelled in the first three volumes of *A Treatise on Mind*. More technical information shall be provided in this volume. True understanding of the nexus between *śūnyatā* and *saṃsāra* lies in the mastering of the meditative process.

If energy is emanated, sustained and destroyed (converted) in the phenomenal world the vehicle that allows the process that causes the appearing substance *(prakṛti)* to manifest, in its totality, is the universal undifferentiated substance *(mūlaprakṛti)*. This is illustrated by the figure below, which utilises the concepts of the Hindu *trimūrti*.

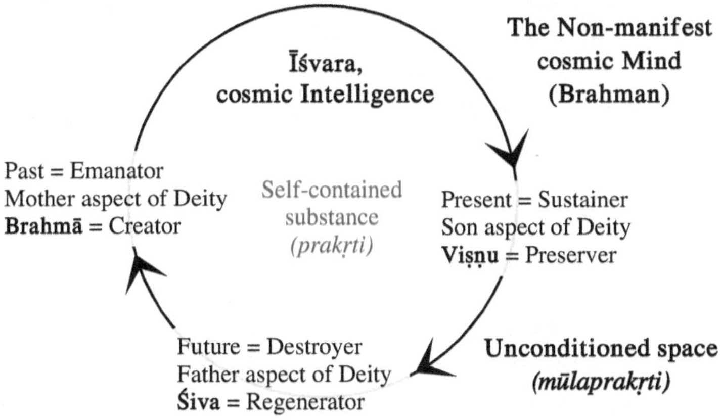

Figure 1: The Trimūrti

Brahman is considered the essence from which all things emanate and return, the immutable ground of the universe. There is also a concept of *parabrahman*, that which is 'beyond' *(para)*, the unconditioned absolute self-enduring space.

As energy and matter are dual attributes of the formed Universe, so the flow of energy that produces matter can also be considered to be the *assimilation of experience*. For as energy flows from point A to point B, it must experience the distance travelled, which is the process of the evolution of consciousness. Energy producing or tending to produce matter (the movement of a form which is inherently *manasic)* is the progress of experience. All is needed is a mind/Mind, to contain, accumulate and store this experience, and this we know to be human consciousness, and also the Sambhogakāya Flower. In the macrocosm (i.e., universal space) it can be represented as the universal storehouse of consciousness *(ālayavijñāna)*.

This view is obviously that of a hylozoistic universe, in that mind/Mind is incorporated into all aspect of what is observed and made known by means of the senses, hence sentient states condition all substance, whilst intelligence is a focal point of evolutionary attainment. This is a natural consequence of Mind causing all that is to come to be. Everything therefore is innately impregnated with primeval Mind, preconditioning the universe to evolve consciousness units that can comprehend that which is and which was established for their education. As earlier stated, without the appearance of such units that can consciously register what has appeared, then the entire process of the appearance and disappearance of universes is a meaningless exercise. Countless billions of universes could have come and gone in such a scenario to no avail, to no purpose, having produced nothing, being simply just a flash in the pan of nothingness. Once intelligence appears then it self-propagates and evolves to first comprehend, thence to embody the laws of physics. (Meaning the complete and utter control of all that has caused such a one to evolve in the first place.) This is the path that our present generation of scientists are upon, and as the process of their learning and concepts evolve over the centuries, so then we will have the appearance of great *mahāsiddhas*, creator 'Gods' to all intents and purposes. They will produce great strides of technological advances presently only articulated as science fiction. In a sense scientists already are Logoi, to the degree that they can manipulate energy and matter according to the level of technological advances they have achieved.

The factor of *deva* and human interrelationships

From the above we see that whatever embodies Mind sustains manifest being. During the entire evolutionary process the attributes of Mind are embodied by the functions of the *deva* kingdom, (the feminine principle in Nature, the Angelic lives), who are directed by the Logoic Will. The *devas* are embodied units of Intelligence, the creatively organising factors of a Logoic Mind. They are therefore aspects of the Mind of a Logos and manifest a basic duality, where the greater *devas* embody the Thought substance and the lesser ones correspond to embodying the substance of what might be considered the neuronal connection with the Thoughts of a Logos.

The Logoic Will organises *deva* substance, whilst the greater *devas* in their turn become creative potencies in accordance with the overall directives from that Will. On the path of evolutionary return human units form as a consequence of the evolved animal sentience aspiring towards the mental plane where this sentience interrelates with the solar Angels, the *devas* who build the Sambhogakāya Flower upon the higher mental plane. (The process is described in some detail in *A Treatise on Cosmic Fire*[45] and is called Individualisation.) The human units (Souls) Individualise from out of the animal group Soul, the purpose being the manifestation of an individuating will (hence is at first separative in action) that at first will evolve intelligence. The attributes of an enlightened creative Thinker manifests when the tendencies to separateness are countered through the development of group consciousness and Love-Wisdom.

The nature of the development of the will makes the human kingdom masculine to the overall feminine *deva* kingdom. The *devas* are feminine because they are receptive to directives that cause them to act as building potencies.

Upon the higher mental plane the principle of individuation manifests the individual Flowers, but the overall interrelation is that of group consciousness, group coherence (the principle of embodied Love). In order to develop the needed wisdom each Flower projects the seeds of a new personal-I into the feminine, receptive, matrix of substance that clothes the impregnating *saṃskāras* of the consciousness-stream of the 'I' with corporality. This substance is that of the lower, concrete, empirical mind, where the attributes of separateness is reified, hence the *ahaṃkāra* ('I am') principle becomes dominant. From the mental plane, the thought of the 'I' is then projected downwards through the astral plane wherein it is clothed in Watery substance, thence to the etheric sub-planes, where the *nāḍī* system is built, and finally into the womb of the mother to be, who was chosen because of karmic propensity or necessity. Thus a new human personality is born, to awaken consciousness and to manifest intelligently within the matrix of the substance into which that consciousness-stream now exists.

---

45 Alice A. Bailey, *A Treatise on Cosmic Fire*, (Lucis Publishing Company, New York). See also volume 3 of *A Treatise on Mind*.

The *māyā* of substance then engulfs the 'I', with which it identifies, and the 'I' utilises for its own inherent volitions to act out a role. This role is dictated by the force of *karma* (individual, group and national) and the guiding impulses from the Sambhogakāya Flower, coupled with the free will of the individual. This free will is heavily conditioned by the appearing *saṃskāras* from past actions that have been liberated from the originating *bindu* of that person's life projected forward through time.

The aim of all this at first concerns the development of the intelligent application of the mind with respect to all activities conducted in *saṃsāra*. Pleasurable experiences cyclically manifest, followed by pain, suffering and eventual dissatisfaction with phenomenal life, hence an eventual desire to be freed from the thrall of phenomena. For the human will to become a beneficent and not a destructive force wisdom and love must be developed, which become the focus of the evolutionary development of a human kingdom. For this repeated incarnations in dense form, the *māyā* of *saṃsāra* is needed. This then incorporates the sum of the philosophy presented in *A Treatise on Mind,* and other esoteric texts.

Wisdom is needed to rightly vivify incarnating forms with the qualities that lead to their salvation from material expression. Wisdom concerns the generation of intensified light directed by the will that becomes a *dominant factor* in evolutionary process along the upward way to liberation. Light is the pure expression of the intellect in action. The appearance of a *sun* therefore signifies the accumulation of a significant number of intelligent units, and their expression (conversion) in terms of the Light of Wisdom, by a supernal Lord of Life. Inevitably a solar Logos appears as a result of the accumulation of the Light of Wisdom, and further trials, Initiation experiences in certain Schools of Learning in cosmos. The Logos, through Wisdom and command of the *devas* that will embody what is to be known can then build a Body of Light, a solar sphere, and incorporate within the bounds of that circumscribed sphere of activity the planetoids that are the externalisation of the Logoic *chakras*. Therein will evolve the streams of the Lives who in their turn must learn the way of generating Light and Love.

Developed Love-Wisdom and Will is the principle that sustains Life in our solar system via manifest Activity. The engendered Light is collected specifically by a plant kingdom to sustain the Life of all the kingdoms of Nature higher to it. The plant kingdom absorbs the light and contains it as the vitality (vitamins, nutrients, etc.) and the *prāṇas* needed to be absorbed by all. For this reason our *nāḍīs* and *chakras* are esoterically part of the plant kingdom. They similarly collect and store the *prāṇas* from interrelations upon the earth, the subjective domains, and eventually from the Heart of the spiritual Sun, which sustains the life of all vital Bodhisattvic activity. The purpose of such *prāṇic* empowerment is to gain the perfection of consciousness as a consequence, also of the evocation of *bodhicitta,* evoked from the Sambhogakāya Flower *(tathāgatagarbha)* via the Heart centre on the Bodhisattva path and eventual liberation from the trammels of *saṃsāra.*

The *masculine aspect* of Nature thus inherently expresses Love, which is slowly converted to Love-Wisdom as a consequence of repeated incarnation into the *devic* feminine substance that denotes inherent intelligence. Love-Wisdom then is the evolutionary gain of material evolution, of the fecundation of the Mother by the Father (the originating Thinker). This process is symbolised by the images of the union of a Buddha and His Consort, wherein the non-dual aspect of all Life, is emphasised.

The Father aspect sustains the manifest evolving being (the personal-I, and later the *nirmāṇakāya,* the phenomenal appearance of a liberated being[46]) through the potency of Thought, founded on the accumulation of the magnetic principle of Love that has been evoked from former rounds of experience as a human. It further evolves when that human becomes a Bodhisattva, thence a Buddha. The Buddha continues evolving within the domains of space that is cosmos. Because of compassionate reasons, plus the residual *karma* that must be played out on a vast scale, the evolved Buddha later manifests as a Creative Logos (or else plays a role in another Jina's world-play[47]). A Buddha is compelled so because the *saṃskāras,* the karmic threads that originally tied such a One to a earth (in a former solar system) have never been annihilated. Annihilation of such threads is not an objective[48] because

46 This idea can be extended, through thinking in terms of transmuted correspondences, to a planetary or solar system, the Logos of a constellation, of a galaxy, a cluster of galaxies and even to a universe.

47 As all cosmic paths do not necessary lead to one becoming a Creative Logos, but nevertheless the accumulation of the united *karma* is thereby cleansed.

48 Even threads linked to those that revert to dark brotherhood methodology.

all sentient lives and consciousness streams with which he formerly related have continued evolving, progressing towards Buddhahood. As they aspire upwards to where he has gone the law of Love demands that he reciprocates, to draw them to him as part of the vaster *maṇḍala* of being/non-being that he is involved with.[49] The form of *karma* that now manifests conditions the realms of enlightenment.

Also, as the sentient lives that once constituted the *nirmāṇakāya* of a Buddha (and of the stream of his former lives) evolve to become a humanity and the corresponding *devas,* so the (evolved) Buddha works for their salvation. He[50] has become the Logos embodying their entire evolutionary journey. He is bound thus by *karma,* though being infinitely vaster than they, as generally this *karma* is the residual when they constituted the cellular forms of his earlier *nirmāṇakāyas.*[51] In His *dharmakāya* state the sum of this intelligent substrate of substance becomes His Consort, with which He is embraced. This allows the impregnation of Her Womb with the seeds of evolutionary Life, the *tathāgatagarbha* aspect of a human kingdom. They are thereby predisposed to aspire towards the light-filled domain in which He resides.

Such union, the embrace with the feminine counterpart of an Ādi Buddha, lies at the Heart of all being/non-being and conditions the quality of *bodhicitta* that emanates through the entire substance of Her Womb (with the basic overriding Ray quality of the presiding Jina). This makes travelling of the Bodhisattva path possible for all who are to evolve out of that Womb through receptivity to *bodhicitta.* This necessitates the accumulation of experience, which can eventually be utilised practically as wisdom by an evolving human kingdom.[52]

Compassionate grounds and long distance *karma* keeps a link, unless the evolving Logos becomes far too exalted (removed in Initiation Stature) to being able to embody the world sphere wherein the dark one resides.

49 The continuation of the Bodhisattva vow upon a cosmic scale.

50 I use the masculine gender here because of convention, however, such a Logos can equally be feminine in nature.

51 Technically the entire *maṇḍalic* structure of which the Jina is the central dynamic powerhouse must be added to this concept, to complete the picture of the entire genesis of a complete universe. Each unitary component of that *maṇḍala* is similarly engaged as the central Jina, according to their function as part of the *maṇḍala.*

52 It should be noted that when analysing such concepts as the enlightenment of a

When one utilises such terms as Father-Mother-Son one should take care not to specifically anthropomorphise the subject, because it is an error to concretise ideas in terms of the human personality. Rather, one is analysing the nature of energy qualifications and other seminal processes that can be likened to the functioning of these three attributes in the human kingdom.

Everything in this manifest universe reincarnates, including the universe. All reincarnating entities obey similar laws of cause and effect. Everything consequently is transient, thus death is but a process of change from one state to the next, of an incarnating principle going from a lower to a higher enlightenment level. One dies to the old state of awareness and is progressively reborn into the new one. So also is the evolutionary process of all in Nature.

Even in the most concretised of the kingdoms, the mineral, many rounds of evolutionary change will transform rocks to metallic, then crystalline form, with a final incarnation into a radioactive substance, allowing the Life within the form to be released. It escapes in an expanded form of expression after entering the mineral *śūnyatā*. In a new emanatory cycle it becomes a participant of the evolutionary attainment of a higher kingdom of Nature, the plant kingdom (via a transitional algae stage), which eventually evolves into the animal, and finally a human kingdom.

The plant kingdom gains its nourishment directly from the mineral kingdom through sending its roots into it to directly incorporate mineral nutrients. It also displays its leaves into the air to capture sunlight and substance (carbon dioxide), and by means of photosynthesis to convert it into the cellular constituency, the cellulose and starches that are the mainstay of its form. This massed utilisation of mineral substance by the plant kingdom is part of the process whereby the mineral kingdom gains plant-like sentience, allowing the Blinded Lives to eventually make a leap forward into a new awareness state for them.

Light results when the dynamic positive Life energy of the Father aspect reacts with the passive inherently 'intelligent' physical substance

---

Buddha, or even what constitutes their own enlightenment, Buddhists need to think more universally than they do. They must therefore begin to utilise the expanded form of their philosophy as presented in my books, to better comprehend the nature of the enlightenment they seek. Similarly for the Hindu practitioners of yoga-meditation.

of the Mother. The reaction (or resistance) of the physical substance to the energy of the Father aspect results in friction, which in turn produces heat and light.[53] When the two meet and react in the field of human consciousness the light of the Son eventually blazes forth as the light of illumination, the path of wisdom. Experience (light) is born through friction, when sluggish forms of awareness must be overcome. In Buddhism that 'friction' (resistance to change) is called ignorance, because it resists evolutionary progress, continuing therefore to reside in states of darkness. Ignorance and the resultant attachment to transient substance is the cause of pain and suffering. Overcoming ignorance is the engendering of light.[54]

The manifestation of light can hence be seen in terms of the unfoldment of consciousness in a vehicle of reception. The evolution of the various kingdoms of Nature and planes of perception can be viewed clairvoyantly by using the supramundane spiritual perceptions and categorised in terms of different hues and states of vibrancy. Light manifests with ever-increasing brilliancy until the vibrant light of the Sambhogakāya Flower stands revealed. Solar light is the expression of a Son in incarnation and reflects into the physical domain the innate light of a kingdom of Sambhogakāya Flowers. A 'Son in incarnation' exists for the purpose of disseminating the light of consciousness throughout Nature's domain. Each solar Logos has incarnated into the dense form of a solar system precisely for this purpose. Because there appears to be an apparent lack of physical evidence of what is purported to be the conditions for 'Life' in any associated planetary sphere does not mean that such does not exist there upon subjective realms.

When one looks to the various physical forms of fire, as sources of light; by friction, ignition of flammable materials, chemical, electrical,

---

53 Here is veiled the terminology of electrical interplay, of a positive and negative factor interrelating via an electrically resistant medium to produce heat and light.

54 In this idea lies the Buddhist concept of *pratītyasamutpāda*, the twelve links *(nidānas)* of dependent arising, dependent origination. One treads the wheel of rebirth thereby. First is ignorance *(avidyā)*, producing predispositions *(saṃskāras)*, consciousness producing name and form *(nāma-rūpa)*, the six fields and five sense consciousness along with mind *(saḍāyatana)*, contact *(sparśa)*, feeling *(vedanā)*, craving *(taṇha)*, attachment *(upādāna)*, becoming to be *(bhāva)*, rebirth *(jāti)* an old age and death *(jarā maraṇa)*. There are four characteristics to each: its objectivity or goal, the necessary appearance, the foundational nature and conditional attributes.

geological and nuclear, then there appears little to connect them to the domain of the mind, except by the hand of humans. However here one must look to the factor of the *devas*. The *devas* are the essence of Fire, it is their Element, hence they embody every aspect of this phenomena. When the scientific community finally discover the factor of the *devas* in Nature and the evolutionary process it will entirely change the sciences, producing a new revolutionary technology hitherto only dreamed of in science fiction, or in occult texts. Scientific cooperation between the *devas* via an open-minded scientific community will allow revelation of the subjective nature of human evolution and the laws conditioning Nature's kingdoms previously veiled, except to esotericists.

The actual dense appearance of a sun indicates that the *kuṇḍalinī* Fires of the Son[55] have been ignited to produce the necessary light and warmth needed to sustain the sum of the incarnation processes of the myriads of Lives evolving throughout that form. Preconditions have been built into that form that will flower into zones or schools of experience whereby human units (i.e., the bearers of consciousness) can develop aspects of consciousness in any of the subjective states therein. The concept is similar to the processes of developing and transmutation of *saṃskāras* in a human unit, however the *'saṃskāras'* now *are* human units coursing through the *nāḍīs* of the Logos concerned.[56] Collectively they undergo similar processes of manifestation, transformation and transmutation as happens within a individual human unit. A particular solar system can also be viewed as one or other of the *chakras* within the Body of an even greater embodying Logos.

A Logos is the Emanator, Sustainer, and Transmuter-Terminator of a planetary, solar or other cosmic form, Who therefore fecundates the universal storehouse of energy, consciousness and of substance to bring into incarnation a world that can be known by conscious units. This trinity can also be viewed as Life (energy), Quality (consciousness), and Appearance (substance-form).

---

55 The 'Son' here represents the integrated Creative Hierarchies of liberated Lives that together work to embody subjective Space, and assist in the engendering and evolution of objective space.

56 Logoic *skandhas* and *saṃskāras* represent the Lives of the various kingdoms of Nature that evolve during the evolutionary process of the system.

The energy flux underlying the manifestation of all phenomena conveys the seven Rays of light that exist in various grades. At first substance-energy is viewed in terms of the five Void Elements from whence are derived the five types of *prāṇas*. They can also be considered the emanations of 'space' *(ākāśa)* from which all is derived. These Void Elements are the seeds *(bījas)* of what is to transpire in the phenomenal domains. When they are acted upon to build the forms of things then there is a manifestation of *karma*, and the *karma* of phenomenal appearance is cleansed when they are resolved back into their Void state. The emanation of the five types of substance are permeated with the seven types of light, and of their subrays.

Buddhi

When looking to the combination of the effect of the energy of the Rays of light within the domains of liberation, I use the term *buddhi*,[57] The term *buddhi* is sometimes translated as 'intelligence', but it is far more than that and can be best considered as supramundane pristine cognition, pure reason, intuition beyond thoughts. *Buddhi* is the Clear Light of the enlightened Mind, which transmits the energies of the seven cosmic Rays that govern the sum of the conscious evolution of all planetary Life. Light is the direct expression of consciousness. *Buddhi* is Airy in nature in that like the air the intuition is not contained by anything, and manifests like lightning (which passes through air) to influence consciousness.

The tendency to name is an aspect of the empirical mind, it sees something and instantly mental formations occur, based upon language and patterns of things known from the past, which allow classification and thence naming. The faculty of inner knowingness that I term *buddhi* here does not classify in this way, it simply knows the essence of what the thing represents, it instantly sees that thing as part of a composite of a complete whole and sees its place therein.

The fourth plane of perception[58] *(buddhi)* is also equated with

57 *Buddhi*, from the verbal root *budh*, to awaken, enlighten, or to know. I use this term in preference to *pratyakṣa*, which is also defined as 'direct perception, intuition, spontaneity, without conceptual processes'. The Buddhist term *dharmatā* can also effectively summarise the qualities of *buddhi*.

58 See Table 1 on page xx of volumes 7A and 7B of *A Treatise on Mind*, which depicts these planes and figure 6 of this present book.

śūnyatā, but *śūnyatā* is more specifically the fourth sub-plane of the fourth, and acts as a mirror reflecting the attributes of the higher cosmic etheric sub-planes *(dharmakāya)* into *saṃsāra* via *buddhi.* This is the empowering source of the Intuition of the awakened Ones. A lower perspective relates to intuitive perception emanating from the Sambhogakāya Flower *(tathāgatagarbha)* via the Śūnyatā Eye. Here is revealed the true nature of the *śūnyatā-saṁsāra* nexus. Its evocation allows one to see all things in a flash of Revelation, 'at-oned', embraced within the Heart of Life.

Buddhi can be equated with the 'blood' or 'ambrosial wine' of the Christ. It is 'the blood of all the prophets, which was shed from the foundation of the world' *(Matt. 23:29-39).* The lightning flash symbolises this quality, and the spotless mirror, freed from all discernible attributes manifests its function. This energy is the expression of the Soul of the World (the *anima-mundi*), and can only become known as one dies completely to the vicissitudes of the empirical mind. It is the Voice of Silence that speaks within the silent recesses of the Heart, producing the fourth Initiation when continuously listened to. It is the Wind blowing away the dust and stains of the mind. The word 'Buddha' is derived from the same root term as *buddhi. Buddhi* thus implies the evocation of the wisdom that is the heritage of the Buddha. The Mahāyāna dispensation, emphasising the doctrine of the Void *(śūnyatā),* brings one to this revelation, which produces liberation from limiting empirical concepts. *Buddhi* is the faculty that delineates the channel of divine inspiration from higher sources to the Sambhogakāya Flower and to the Mind/mind. Its effects are instant comprehension, right discrimination, intuition, boundless love, and consequent universal compassion *(bodhicitta).*

The phrase the 'supramundane pristine cognition', (which I have equated with the term *buddhi)* is derived from *The Laṇkāvatāra Sūtra.* Therein there are said to be three types of pristine cognition (mundane, supramundane, and most supramundane):

> Now these three kinds [of pristine cognition respectively] generate the realisation of individual and general characteristics, the realisation of that which is created and destroyed and the realisation of that which is neither created or ceases. The mundane pristine cognition is that of the extremists who manifestly cling to theses of being or non-being and of all ordinary childish persons. The supramundane pristine cognition is

that of all pious attendants and self-centred buddhas who openly cling to thoughts which fall into individual and general characteristics. The most supermundane pristine cognition is the analytical insight of the buddhas and bodhisattvas into apparitionless reality. It is seen to be without creation or cessation, for they comprehend the selfless level of the Tathāgata who is free from theses concerning being and non-being.

Furthermore, Mahāmati, that which is characterised as unattached is pristine cognition, and that which is characteristically attached to various objects is consciousness. And again, Mahāmati, that which is characterised as being produced from the triple combination [of subject, object and their interaction] is consciousness and that characterised as the essential nature which is not so produced is pristine cognition. Then again, Mahāmati, that which is characterised as not to be attained is pristine cognition, since each one's own sublime pristine cognition does not emerge as a perceptual object of realisation, [but is present] in the manner of the moon's reflection in water.[59]

Now, the 'mundane pristine cognition' can be interpreted in terms of the exoteric rendering of *buddhi* as 'intelligence'. The 'supramundane pristine cognition' can be interpreted in terms of the word *pratyakṣa* (perception as valid immediate means of knowledge). As such, 'direct perception without conceptual process' can be considered the basis to the experience of the *pratekyabuddha* ('self-centred buddhas'). This can be considered to be a general rendering of the term *buddhi.* The 'most supramundane pristine cognition' is therefore the esoteric rendering of *buddhi.* Once the 'intelligence' has been cleansed of its attachments to *saṃsāra*, and thereby residing in its natural state, it can act as a receptive tool wherewith the *tathatā* that is the *dharmakāya* can manifest in the Mind's Eye with a view to being expressed in *saṃsāra.*[60]

Buddhi consequently is the gain of the evolutionary process, and its effect can be thought of as the light of intensified Fire (i.e., of the

---

59 Dudjom Rinpoche, *The Nyingma School of Tibetan Buddhism* (Wisdom, Boston, 1991), 180-1. For the *sūtra*, see D.T. Suzuki, (trans.) *The Lankāvatāra Sūtra,* (Routledge and Kegan Paul, London, 1932).

60 The term that can here also be used is *dharmatā,* defined as 'actual reality, ultimate truth of phenomenon, *śūnyatā*'. *Dharmatā* however manifests as the natural force of things, inherent nature, the essence of existence. *Dharmatā* is the force that projects the *dharmakāya* into manifestation via the spaciousness of the abstracted Mind. *Bodhicitta* acts in a similar manner, but in relation to the liberating energy of compassion.

Fires of the Mind), hence the radiance of a sun can be considered the physical plane effect of this energy carried through the atomic world. The paradoxes of the particle and wave-like attributes of the photon, the carrier of light, can be considered to be the homologue of *buddhi*.

I have explained the term *buddhi* at length because the fourth plane of perception is governed by *buddhi* and this is the higher correspondence of the fourth etheric sub-plane, which can be considered the atomic 'event horizon' or quantum vacuum. The properties of this 'event horizon' needs to be appropriately comprehended if the true nature of the appearance of phenomena is to be comprehended. As there is a triad of characteristics denoting this higher correspondence to the fourth ether, so we will discover a triad of entities that are responsible for the appearance of mass, and hence the phenomenal universe. They move in and out of the quantum vacuum and are the quarks, leptons and the bosons, as will be discussed in chapter 9. Similarly consciousness of an enlightened one moves into *buddhi* when absorbed in deep meditation *(dhyāna),* and from it into mental space when it needs to deal with any aspect of *saṃsāra,* utilising the gain of what was experienced in *dhyāna*.

The significance of Light

The significance of light is well known to physicists, as Lee Smolin states:

> But, of course, light is the ultimate source of life. Without the light coming from the sun, there would be no life here on earth. Light is not only our medium of contact with the world; in a very real sense, it is the basis of our existence. If the difference between us and dead matter is organization, it is sunlight that provides the energy and the impetus for the self-organization of matter into life, and on every scale, from the individual cell to the life of the whole planet and from my morning awakening to the whole history of evolution.
>
> We will never know completely who we are until we understand why the universe is constructed in such a way that it contains living beings. To comprehend that, the first thing we need to know is why we live in a universe that is filled with light...Why is the universe filled with stars?[61]

61 Lee Smolin, *The Life of the Cosmos,* (Universities Press, Hyderabad, 1999), 27-28. This quote is part of Part One of his book, which is titled: THE CRISIS IN FUNDAMENTAL PHYSICS. *Why is the universe hospitable to life? Why is it full of stars?*

Because of the appearance of solar light phenomena can be perceived, hence the mind can develop and evolve. This is important, because without the existence of the photon the universe would be in darkness and consciousness could not evolve. Suns appear, to shed their light into the darkness of space, to convert that darkness into knowledgeable units of Life via the agglomeration of that dark substance into planetary spheres (incarnations) embodied by informing Logoi. This then presupposes a form of evolution for the substance of a planetary sphere to eventually incarnate as a radiant sun. The substance evolves to shed light into the darkness.

A similar effect occurs with that elementary substance that is incorporated as the thought structure of human minds. One needs to consider an involutionary process of the descent of primal mental substance, the elementary black dust, to the physical, to be integrated with the elemental particles therein. This is effected by the Will of the empowering Logos working via a *deva* hierarchy, who 'impregnate' that substance with an innate Fiery essence, producing the hylozoism inherent in the universe, and which is the basis for the Anthropic Principle debated by scientists. Nature's laws are fine-tuned so that the inevitable effect is the evolution of consciousness, nothing happens by chance. There is consequently an evolutionary conversion of that 'dust' into lighted substance.

An innumerable number of units of Mind/mind *(devas)* condition the all, regulating the appearing *karma* of the manifestation of phenomena that is expressed as their bodies of manifestation. *Karma* here is the effect of the laws set in motion by the originating Thinker, producing the phenomena we experience in the external universe. Because all exists within that Mind, so all laws discovered by scientists are but the expression of the laws of Mind/mind, hence they can use their minds to try to comprehend it all. Such laws cannot be discovered, for instance, through irrational (emotional) thinking.

When an Initiate can consciously reside in *buddhi/śūnyatā* then his/her consciousness can (instantaneously) travel to any part of space (be it in the solar system or the local cosmos), because we are now viewing that energy substratum that is common to all such Logoic 'spaces'. The enlightened consciousness is not bound by the constrains of the formed spheres. It resides in 'that which is neither created or ceases' therefore it has the capacity to reach far out into cosmic space. Speculation re what exists 'out there in cosmic space' no longer manifests as the perceiver is already 'there' and Knows. When scientists begin to function via the 'most supramundane pristine cognition', wherein intelligence is cleansed of attachments to *saṃsāra,* then the true era of 'space travel' can begin, wherein the secrets of the far bounds of cosmos can be properly revealed. One need not leave the square metre of space upon which one sits to visualise what actually is, but certainly then ships of supramundane Light could also be constructed to convey incarnate forms to many interstellar destinations.

The law of correspondences, 'as above, so below, that which is within is also manifest without', applies just as well to physical plane phenomena and its relation to the cosmos, as it does to human physiology and how it relates to the cellular consciousness of the externalised Logos. This axiom must always be applied to help in the understanding of the nature of causation, be that of humanity or of the universe. The microcosm being a reflection of the macrocosm is an ancient truth already well documented and can be proven by modern physicists in their investigations into the nature of substance.

All that is known is an expression of the *One Universal Mind*, as symbolised by solar light and its interrelation with all other stellar spheres in our galaxy and universe. The evolution of the Bodhisattva-like consciousness of the entire human kingdom will, in the aeons to come, transform itself into that of a blazing sun, shedding light into the darkness of abstract space. Such is the way when all have become Buddhas and Bodhisattvas. One can then say that the 'frictional fire' governing the interrelation between forms has died down and solar fire becomes all-consuming.

The Logoic form constitutes:

a.	The negative energy of matter—motion on the material realms, seen as inertia. It is the path of concretion of energy (or primal matter) which produces physical substance. **The Phenomenal Appearance.**	Mother Creator
b.	Solar Fire (light) produced through the union of the evolutionary 'sparks' of Life, the merging of primal atoms into forms (coherent organisms), and	Son Preserver

by the reaction between these forms. This concerns the evolution of consciousness and that which acts as a store of that consciousness. **Consciousness.**

c. The positive dynamic energy (cosmic electricity) which breaks up forms and releases the Life inherent in matter. **Life.** — Father Destroyer

The *sun* is part of the dense body of expression of a Heavenly Man who is incarnate as a solar Logos. The *maṇḍalas* associated with the constitution of Shambhala, explained in volumes 7A and 7B of *A Treatise on Mind,* presents the nature of a planetary Logos that exists as part of the body of manifestation of the solar Logos. This was explained (in part) as an extended philosophy of the expression of the Dhyāni Buddhas and of the overriding Ādi Buddha. Buddhists should now be able to expand their philosophy of the Dhyāni Buddhas, to conceive them as part of a Logoic expression incorporated within a grand scheme of cosmos. Such a conception can be considered the higher expression of what a human being is, for as there are five of these Buddhas, so we have five attributes to our form, two arms and hands, two legs and feet, torso and head, an integration of five main *chakras* and *prāṇas*. They have five respective Wisdoms, and we have the five sense-consciousnesses. There are five Elements to account for and the five *skandhas*, and they have five Consorts. As such, the human family represents atomic cellular unities in the body of the meditation Minds of Logoi, that were 'thus gone' ones of former aeons of evolution in earlier solar systems.

Everything in cosmos evolves together as part of the cumulative expression of such Dhyānis (Lords of Meditation) constituting Hierarchies of Logoi for each galaxy, and for their superclusters extending throughout the far reaches of cosmos. Vast must be the reckoning in terms of consideration of transmuted correspondences if one is to arrive at any meaningful accounting of the source or cause of all that is in this hylozoistic universe.

The ancients perceived the light body that represents the physical emanation of the combined Rays of an incarnate Logos, and this is one of the principal reasons why the ancients worshipped the sun as an Absolute Deity. He is the centre of all Life and Light, giving all that is needed for the well-being of the constituency of His incarnate form. The seers amongst ancient humanity knew that the sun was the exterior symbol of the human Soul. They also knew that the solar Deity was a cellular unity or Son of another more embracive cosmic Heavenly Man, and so forth. The esoteric vision was always an accomplishment of the training given to the candidates of the ancient Mystery Schools. The higher Mysteries they safeguarded through secrecy, but they encoded the result of their visions into the myths that have been handed down throughout successive generations of humans. When decoded these myths give plenty of evidence of the true cosmic understanding gained by these ancient scientists/seers. Though they did not manifest their concepts in the terms of modern scientific opinion, it does not mean that the results of their observations were not a valid explanation of what actually is re the cosmos and its secrets.

The scornful disdain of such esotericism by materialistic scientists must yet change into a smile of revelatory understanding when they finally appropriately analyse the collective wisdom of our forebears without the prejudice of blind narrow-mindedness. They must eliminate their presumption that the only way to comprehension of the nature of things is through what they have described as 'scientific methodology'. (Where only what is most material and concretely evident is taken to be real.) The physical plane universe which they have analysed thereby is only a tiny portion of the picture of what constitutes cosmos. When they comprehend this fact then we shall see a new era of revelatory bliss and outstanding observations enter human minds. It will involve the marriage between the alchemical, meditative, ritualistic and magical activities of the ancient seers with that of unbiased modern scientific methodology. Hearts and minds will then be united in a common accolade of beauty and truth.

2

Mirrors, Time and the Logoic Mind

The *ālayavijñāna* and Logoic Mind

The doctrine of the *ālayavijñāna* (the universal store of consciousness) in the Yogācāra philosophy postulates that everything can be viewed to be an aspect of consciousness, that this store is:

> A stream of consciousness. It is neither permanent, nor impermanent. Vasubandhu says that 'it is in perpetual evolution like a violent torrent'. From the beginningless time this consciousness has evolved in a homogeneous series without interruption. As a cause it is perished and as an effect it is born from one moment to another. Because the effects are born, it is not impermanent. Because the causes are perished, it is not permanent.[1]

It is thus safe to say that consciousness exists within evolution, because from 'the beginningless time this consciousness has evolved'. It however may be more expedient to say, from the point of view of the philosophy I have presented, that the originator of the evolutionary push is Consciousness (or rather, Mind). As consciousness evolves we can ask 'into what?' The quick answer would be 'into enlightenment, wisdom'. Another answer is that the consciousness-moments go on to produce the ineffable Mind of a Logos or 'God', the Ādi Buddha encompassing a world system.

---

1 Swati Ganguly, *Treatise in Thirty Verses on Mere-Consciousness,* (Motilal Barnasidass, Delhi, 1992), 42.

To properly comprehend such an entity and the nature of the formation of a world sphere one must look to the transcended extension of the philosophy of the *ālayavijñāna*. The foundational basis is the *dharmakāya*, from which emanates that which is 'neither permanent, nor impermanent'. Much has been discussed in volumes 1 and 2 of *A Treatise on Mind* concerning 'a homogeneous series' in the context of the *ālayavijñāna*, especially with respect to the rebirthing principle. Such a series may exist in the *ālayavijñāna,* but heterogeneous streams of thought are far more likely, thus thinking in terms of the expression of 'a violent torrent' of energy is possible in the domains of causation. From this perspective such streams exist in the lower strata of the mental plane. If we looked into the Mind of a Buddha such 'a violent torrent' cannot exist. A 'homogeneous series' of meditative sequences would however more correctly describe the way that such a Mind manifests. There are no sharp, incongruous thought structures, no tempestuous, impulsive thoughts or ideas therein. Everything is correctly sequenced in serene contemplative poise and has its basis in the Void. Cause and effect are merged into a non-dual expression, where neither *saṃsāra* or *śūnyatā* exist as independent entities. Meditative Mind-sequences are thus formulated, but remain in their domain of intrinsic unadulterated purity. What is it then that is causative of the phenomena we see around us?

This question is important if we wish to extrapolate considerations concerning the formation of world and solar spheres, as they may exist upon the domain of the Mind of a Creative Logos. The processes between that which causes a human mind to formulate ideas and to carry them through to actual conclusion as physical plane artefacts, which humans deem necessary or useful, and Logoic Considerations re the formation of world spheres are similar. We must extrapolate information from the domain of the known in order to arrive at some consideration of what the unknown might be. Everything starts from the domain of the mind/Mind. Just from an observation of our present solar system, with all of its component parts consisting of an astounding diversity of planets, comets, etc, we can see that heterogeneity is very much the context of a Logoic Thought structure with respect to the causation of things.

At the level of the appearance of a solar sphere however, the argument would tend to favour homogeneity, as all suns are spherical sources of fiery light, created by the same processes. Some stars may be larger more intense luminaries than others. Others may be young stars, still to properly ignite their furnaces. There however is a relatively ordered sequence of stellar evolution, to its eventual demise according to astronomers as a super nova, neutron star, or even a black hole, depending upon such considerations as initial stellar mass. Such is the most corporeal aspect of the dense body of manifestation of a solar Logos. We see therefore that both heterogeneity and homogeneity have their roles in the thought processes relating to the manifestation of 'things'.

The manifestation of phenomena happens via the nexus between *saṃsāra* and *śūnyatā.* At the level of this present enquiry we will see that this nexus is conditioned by the Wisdoms of the five Dhyāni Buddhas. They embody the seed conditionings that flavour the appearance of manifesting phenomena with the attributes of inherent intelligence, the principle of sentience that will eventually flower into the birthing of human units. Upon a universal scale such consideration is the basis for the inevitable manifestation of the Anthropic Principle, allowing the appearance of Life in a universe. The Dhyāni Buddhas factor into the appearing phenomena an inherent 'five-ness', the attributes of the five Elements, and their development in time and space.[2] When the energies of the primal Ādi Buddha and Consort are included then a septenary is formed.

All comes into manifestation via the seed Idea, the *bindu* that is activated as part of the *dhyāna* of the Ādi Buddha. The primal Thought produces *saṃsāric* permutations that are instantaneously elaborated in terms of the *jñānas* (Wisdoms) of the Dhyāni Buddhas. A world-play can then be effected once such elaborations are seeded into the Wombs of the Consorts of the Jinas. *Saṃsāra* then manifests, which is held within the embrace of the meditative concentration of the Consorts. They work via their agents, the *devas* and *ḍākinīs,* personified as the gods of the creation myths. Such gods are personified forces of Nature, as well as symbolising the great Ones constituting a planetary or solar Head lotus (from whence the factors of Logoic Thought emanates), as described in the two volumes of *A Constitution of Shambhala*. The entire creative process known to us

---

2 Upon such a vast scale as a universe one must look to the homologue of the Dhyāni Buddhas.

is thereby produced, from initial beginnings to the path of active return, wherein *saṃskāras* are in the process of being terminated or transmuted by a great One in conjunction with the cooperative humanity that have evolved within such a One's body of manifestation.

Primarily, the *bindu* containing the seeds of all that is to transpire in the realms of manifestation is reflected into the lower strata of being via the Mirror-like Wisdom of Akṣobhya. The *ālayavijñāna* environment (of the higher mental plane) is then conditioned according to the primordial patterning of the Logoic Thought Impulse originating in the *dharmakāya*. The Impulse of this pristine *maṇḍala* of the Thought projected then enters the Mind-space of Amitābha, who governs the *ālayavijñāna* environment. This environment is then organised according to the karmic pattern of what is to be, and Lipikas[3] are called forth to circumscribe the boundaries of the new manifestation. They delineate the parameters of the *maṇḍala* and set the *karma* into place so that all will manifest according to the Plan of the originating Logoic Thinker. Once the 'blueprint' of the *maṇḍala* has been established upon the higher mental plane, then the Lords of Flame (embodied aspects of the Logoic Mind) can 'rush in' to fill the appointed places. The associated *devas* and all of the other stream of Lives then move through the *maṇḍala* and so the Shambhala of a new world or solar sphere is established. Physical plane phenomena can be then appropriately effected.

The Mirror-like Wisdom

One way of comprehending Akṣobhya's Mirror-like Wisdom is that things are reflected from a higher source into manifestation in accordance with the validity of their purpose. Consciousness also manifests a degree of reflection between what is external to the consciousness and that which is internal to it. If one says that there is nothing outside of the internalised consciousness, then confusion manifests regarding the nature of what actually exists, because if there was nothing external to the consciousness then the sense-perceptors could not gain any impressions at all, there would be nothing they could contact. In this scenario there would be

3 Scribes, Lords of *karma*. See *A Constitution of Shambhala,* volume 7B for their description. They shall also be somewhat explained later in this volume.

little or no consciousness-growth, as it would have nothing but itself to experience and no other input.

The internal and external always exist relative to each other, which presents a relative understanding of the 'I', the world, and universe. Every level of comprehension can be considered a *ring-pass-not*[4] that delineate consciousness limits, within which consciousness works and which it must fill with consciousness-moments and *maṇḍalas* of complete ideas. They create forms of three dimensional physicality within consciousness-spaces. Each ring-pass-not can be considered a 'bubble of space'[5] that represents a consciousness unit, leading to a completed fully integrated mind/Mind space of a personal-I, the Sambhogakāya Flower, or Logos. Such 'bubbles' of containment can thus manifest in the form of a Logoic sphere, such as a solar or planetary system, a galaxy or extended to the confines of the universe. Logoi view each other through such spheres of Consciousness-attainment that represents their manifesting Personalities. Such Personalities are conditioned by Ray lines, the *saṃskāras* of past karmic activities (the manifesting streams of human and *deva* units), the developing sense-contacts established by the Dhyāni Buddhas (governing the kingdoms in Nature), group interrelationships with their Brethren, and an overriding Purpose.

The nature of the manifestation of mirrors and this particular Wisdom was treated throughout *A Treatise on Mind,* especially in relation to the functioning of *śūnyatā*, to which supplementary information can now be added. *Śūnyatā* is now to be taken to mean specifically the fourth sub-plane of the fourth plane of perception *(buddhi),* which reflects the higher cosmic domains into the planes of manifestation. This is possible also because *buddhi* embodies the fourth cosmic ether, the higher correspondence of the fourth ether, which separates the subjective universe from the three dense planes whereon our physical senses derive their sense-contacts. Similarly a Logos experiences 'sense-contact' with what represents dense physical phenomena to Him (the mental, astral and physical domains) via *buddhi.*

---

4 This phrase is derived from H.P. Blavatsky's *The Secret Doctrine.* Stanza V, part 6 of 'Cosmogenesis, which simply relates to a circumscribed sphere of limitation into which a Logos incarnates and is bound to until its purpose is accomplished.

5 See volume 4 of *A Treatise on Mind* for an examination of this subject.

In this sense *buddhi* and the fourth ether reflect the attributes of each other. What manifests via *buddhi* then are the attributes of the Logoic 'sense-perceptors', which are the energies of the Consorts (the *deva* kingdom) of the five Dhyāni Buddhas (Jinas). The gain in terms of 'sense-responses' acknowledged by the Logos concerned are the attributes of the developing Jina Wisdoms in Nature.

As stated, the 'fiveness' of Logoic dense incarnation is the reason why our bodies are shaped like a pentacle, two hands and two legs attached to a torso, and also a head, which is ovoid, a Logoic form. There are also five fingers and toes to each hand and foot. The mental plane, from whence our intellects are derived is also the fifth plane of perception.

Once enlightened beings appear in a human kingdom then an enlightened decision consists in using the *reflected forces* of the Dhyāni Buddhas, as well as the inherent wisdom developed from the entire evolutionary process. These forces manifest via the *buddhi* mirror to impress upon the Mind of the enlightened one the gain of those that had previously mastered their evolutionary Journeying, and hence are much higher Initiates than the one enquiring. The reflected forces carry the Mind-spaces of Logoi manifesting in cosmos.

The forces that manifest become formed because they are reflected from the spacious and non-defined universality of Logoic Mind, and then expressed in a Consciousness that defines and categorises. (This reflection process is then utilised to help build the structure of 'things'.) The forces of reflection come through the side of the mirror that forms and creates Images of the Divine for enlightened Minds. Those that have not yet built the pathways (*antaḥkaraṇas*) via the manifesting Logoic senses cannot comprehend what exists on the other side of the mirror, it represents a darkened surface for them.

If the mirror reflects perfectly then there is no difference between the inner and the outer manifestation. Consequently, on the other side of the mirror some type of form would exist. That beyond *śūnyatā* will therefore be 'formed' with respect to its own level of interrelationships. (Here the *śūnyatā-saṃsāra* nexus manifests as the mirror, from whence we get the Mirror-like Wisdom.) If one contends that the reflection is not as it appears, which is generally the case with the uncomprehending reifying empirical mind, then one side of the mirror modifies what it reflects, so that those with form (existing in the *rūpa lokas*, the formed realms) can comprehend. The mind then acts as a creative principle, destroying the function of a mirror. In Nature such an effect is seen in the modifying action of Amitābha, Ratnasambhava and Amoghasiddhi. These Jinas are responsible for the phenomena manifesting in the three planes of human livingness (the mental, astral and physical). Everything however is relative, consequently what is considered *arūpa* (formless) is only so with respect to the nature of the consciousnesses existing in the *rūpa lokas*. The functionality of this mirror exists at the junction between the *arūpa* (lighted) and *rūpa* (darkened) universe.[6]

If we observe the subatomic world then a similar dichotomy appears between the formless, the wave front, and the appearance of particles of substance. The plane of transition is the fourth ether, here the mirror between the sensory (experiential) and subjective domains.

When consciousness is transcended and formlessness is achieved, then a new criteria is utilised (as perceived by the enlightened) and a transcended form comes into perception. The *sambhogakāya* domain also bespeaks of this at its own level of expression. The Sambhogakāya Flower is but an 'atom' when viewed from a Logoic perspective. From the viewpoint of an incarnate human it is formless to the empirical mind, but is also a form to the awakened seer, consisting of lines (waves) of energy crossing each other, producing nine whorls of petals.

Can we question a reflection if the reflected elements are the same as what is on the other side of the mirror? Are the particles forming the reflection in a different phase or state from what is reflected? Is there a difference to consciousness as to what it perceives in the reflection compared to that which is reflected? The answer to these questions is generally taken to be yes, because the image in the mirror is viewed to be illusional, faithfully depicting something that is taken to be real, tangible. But are not both the reflection and the 'real' but attributes of consciousness? It is the way that consciousness views things that makes things real to it. What is an accepted reality to one may not necessarily be viewed so by another. However the concept of a mirror here is but a metaphor for processes that cause the appearance of phenomena. These processes are therefore what I wish to explore.

---

6 The terms 'lightened' and 'darkened' here are relative and do not relate to the physical plane situation, rather to the fields of consciousness.

That which is reflected in the Mirror-like Wisdom is not *śūnyatā* (because *śūnyatā* is empty of all that can be reflected). Indeed, it is necessarily incorporated as the (substance of) the mirror itself. If the *śūnyatā* mirror does not distort or add to the image formed then it is true, it reflects truth as it is. Therefore we can only conclude that what exists on either side is form, but that the form of one side, i.e., expressing the *dharmakāya*, is the Real, non-illusionary, and the *saṃsāra* side, is unreal, the illusory image of the Real.

This means that if the mirror reflects truthfully, then the essential substantiality of *saṃsāra* must be in the image of what is contained in *nirvāṇa*. *Nirvāṇa* here does not refer to *śūnyatā*, but to the *dharmakāya*. The inescapable conclusion is that *saṃsāra* is fashioned in its essential reified delineations by what already exists in the *dharmakāya*, though this *dharmakāya* exists upon a transmuted level of being/non-being. This is the level wherein Buddhas are born as members of a cosmic Hierarchy, and to which high-level Bodhisattvas frequent. The mirroring process allows *dharmakāya* to be the realm of the Creative Logoi that cause world spheres, such as that of our earth sphere. All are part of a Mind-structure of a grand Incorporating Logos, which for sake of any better term is called THAT Logos, or as depicted in *A Treatise on Cosmic Fire* as the 'One about Whom Naught may be Said'.

The emphasis here is upon the term Naught, which does not mean 'nothing', though in the minds of the great majority this is indeed what exists in relation to comprehension of such a One. Rather, it means that which is veiled by *śūnyatā,* manifesting as a sphere, a 'nought', the cypher zero (O), the circumscribed sphere of attainment via which a Logos builds His/Her Body of Manifestation. The Logos in question here is that which embodies all other Logoi constituting the stellar spheres and constellations in our galaxy observable with the naked eye. Such a One is a member of a cosmic Humanity consisting of similar Ones whose Bodies are constituted of *Chakras* that are stars and constellations of stars. As we all live within the Body of such a One, the true characteristics of what lies beyond such a Body can only be surmised rather than directly known, even by the highest Dhyān Chohan (liberated being) on our planet. From this perspective naught can be directly said about this Domain, but the enlightened have surmised many avenues of veridical Knowledge. There are limits to what can be comprehended in the Minds of non 'thus-gone'[7] Ones.

In Advaita (non-dual) Vedānta philosophy *nirguṇa*[8] Brahman can be equated with *śūnyatā*. It is Brahman (the universal principle of being/non-being) without attributes, other than residing in a self-contained state of *sat* (eternal being), *chit* (inherent consciousness) and *ānanda* (bliss). From it, by means of the power of *māyā,* the illusion-creating force, emanates *saguṇa* Brahman, Brahman with (formed) attributes, the three *guṇas: sattva* (truth, rhythm), *rajas* (kingly activity) and *tamas* (darkness, inertia). *Sattva,* rhythm or balance, is what must be realised; *tamas* opposes this realisation; and *rajas* is the appropriate activity that overcomes *tamas. Sattva* manifests as peace and serenity, from which truth can derive. *Rajas* establishes knowledge of how to overcome, and *tamas* represents laziness, sluggish activity, lack of interest and ignorance. *Saguṇa* Brahman embodies these attributes in the form of a Logos (Īśvara), and the *guṇas* become personified in terms of the forms of activity of the deities Brahmā, Viṣṇu and Śiva. They are the creator, preserver and destroyer of whatever manifests in the form of *māyā*. Despite the illusional appearance of phenomena all is in reality One – Brahman. The illusion is the separative individuality of the 'things' we take to be real. All this then really relates to the phenomena that appears in our minds. The ramifications of the three *guṇas* manifest all of the way into the subatomic world, as shall later be explained.

Differentiation

Both *saṃsāra* and *nirvāṇa* contain self-willed individualities that can alter the phenomena of their respective realms. However one form of this divide is deemed illusional because short-lived, impermanent, and the other the Real because relatively so. The forms of phenomena existing there are vast, transcendent, and last for enormous durations of time, but also obey similar laws of change with respect to their own durations of expression.

Differentiation however happens in the process of reflection from the liberated domain to the phenomenal world wherein we reside, (i.e.,

7 This phrase is a translation of the Sanskrit term for a Buddha, Tathāgata.

8 *Nirguṇa* meaning no form, without attributes *(guṇa).*

there is something different about what the mirror reflects or what we perceive) but this is not because the mirror reflects imperfectly. The problem lies in the organs of reception or translation of what it is that is viewed in the mirror. The mechanism of reception (the human mind) distorts the image, because of an incapacity to view it in its pristine purity or intensity. Therefore fogs and mists of incomprehension take the place of the Real.

The *ālayavijñāna* environment can however experience the image of this cosmic sphere without undue distortion, because it is patterned according to its paradigm. Therein for instance, resides the *tathāgatagarbha,* which is the reflection of a far greater cosmic Entity. All is then perceptive to the Mind that is held steady in the Clear Light. Attachments *(kleśas)* produce modifications of the mind so that instantly distortions appear in terms of images known from the empirical world, or as created by human minds and attributed to the *sambhogakāya* state.[9] The images determine what people think aught to be, rather than what actually is and then people act as if the fabrication was real. We then get a host of conflicting opinions.

There is a truth associated with the *saṃsāra-śūnyatā* nexus that helps to clarify the confusion relating to the nature of the mirror and what is mirrored into *saṃsāra.* As already stated, this truth is: *'that which is within is also without, that which without is also within, as above, so below'.* This ancient adage is the golden key to help fulfil revelation of what this mirror truly reflects. It means that a proper study of *saṃsāra* is a means to the revelation of *nirvāṇa*, because *saṃsāra* is fashioned in the image of the Real. All who wish enlightenment accede to comprehension of this adage. From this perspective we see that the most wondrous human form has its exalted prototypes in *dharmakāya.* The liberated ones, the Mahābodhisattvas and Buddhas that inhabit cosmic space, are therefore rightfully depicted in human form, with all of the attributes of fully enlightened ones when viewed in terms of radiance and transmuted correspondence of mind. Forms however can change according the mode of the expression of the mind/Mind to do so.

9 The body of sublime vision, the ecstatic transformation body. The second of the three vestures *(trikāya)* of a Buddha or fully liberated Being. The form of the great ones depicted in Buddhist art.

From this perspective we can also deduce that just as no two humans are identical, so consequently Buddhas are not identical. Their fundamental base awareness may be identical, because *śūnyatā* and the *ālayavijñāna* are the two foundations, or grounds, upon which Buddhas stand. Those who have mastered the *ālayavijñāna* environment (the five attributes of mind) find that these attributes become an instinct to them, having been transformed into the Jina Wisdoms.

Ordinary people however have their emotions and intellects as their base conditionings. All humans have the sense-consciousnesses and their minds to analyse things with, but their personalities are different, different modifications of *saṃskāras* manifesting as their personality structures, as well as the nature of the *kliṣṭamanas* (emotionally afflicted, defiled mind) they incorporate. Buddhas have similarly developed higher correspondences of the *saṃskāras,* whilst instead of *kliṣṭamanas* there are the permutations of the various types of Wisdom in which they have specialised, of which Love-Wisdom is the key. Different attributes have also developed because of the inherited characteristics from the world periods wherein they evolved to became a Buddha. As humans evolve, so also do Buddhas on their level. The Buddhas become the major components of the *maṇḍalas* of Logoi of planets and solar systems. They have evolved to incorporate arenas of far greater spiritual power and thus of service capacity than was ever possible upon an earth sphere. From this perspective we see that there are perceptible problems in the type of question and answers presented in a commentary by Tsongkhapa:

> Is there an empowering condition *(adhipatipratyayaya)* for mental consciousness *(Tib. yid shes)* at the Buddha level?
>
> [In response] it asks in the *Guhyārtha-vyākhyā* if mirror-like transcendental wisdom (*ādarśana-jñāna*) is perfumed with residual impressions or not? If it is perfumed then earlier Buddhas would [absurdly] be more powerful because the perfuming of the residual impressions would have been there for a longer time, while later Buddhas would [absurdly] would be less powerful because theirs would have been there for less time. Also, it would no longer be a hard and fast rule that the location of perfumes [i.e., the consciousness carrying the seeds] is not a subject of moralizing [because transcendental wisdom, unlike the neutral *ālaya-vijñāna,* is always virtuous]. If [on

> the other hand, transcendental wisdom] is not perfumed then there will be no cause for the infinite good qualities such as [the Buddha's ten] powers *(bala)* and [four] fearlessness *(vaiśāradya)*, etc.[10]

The main problem with the answer here is that it derives from the point of view of those accustomed to thinking in terms of the mentalistic conditionings derived from their earth experience, of the types of 'perfumes' (the consciousness carrying the seeds) that they are accustomed to. (These are *saṃskāras*, with their inherent forms of defiled mind, *kliṣṭamanas*.) The answer has not accounted for the process of the sublimation, transmutation and transcendence of these 'perfumes' when a high level Bodhisattva reaches out to cosmos and begins to partake of awareness states not derived from earth conditionings. They therefore develop new *saṃskāras* that betoken of a complete awakening in *dharmakāya*. In Tsongkhapa's time the foundational teachings to explain what existed in and as cosmos did not exist, therefore he was severely hampered in what he could say. Only generalisations were permissible, such as '[the Buddha's ten] powers *(bala)* and [four] fearlessness *(vaiśāradya)*, etc'. The Buddha's cosmic links and relative attainment are nowhere described in such generalisations. Therefore it was not possible to depict how a Buddha actually manifests in the true cosmic perspective. Therein the earth evolution is but a relative grain of sand within a magnificent omnipresent astoundingly vast form of Kalachakra Maṇḍala, with its five transcendental levels of expression, embodied by Deities, Bodhisattvas, and Buddhas of differing capacities. They are situated and empower higher or lower strata of the *maṇḍala* according to capacity and evolved functioning.

The result of an evolutionary process associated with a particulate grain of the entire *maṇḍala* is not equal to the expression of the whole, even though an evolved Buddha may have the paradigms within of the entire structure, and contain the *bindu* for the growth of a similar structure aeons into the future. Similarly a cell of a foetus in a womb, possesses the *maṇḍala* of a human body, but is not yet that (human) body. It will grow into that form given the right conditions and sufficient

10 Gareth Sparham (trans), *Ocean of Eloquence, Tsong kha pa's Commentary on the Yogācārya Doctrine of Mind*, (Sri Satguru, Delhi, 1995), 117.

time. Similarly the sentience of the lesser kingdoms of Nature will over aeons of time grow into that which in a later world sphere will be called 'human', as such a world sphere will provide the conditions for such an eventuation. Hence the *maṇḍala* for the eventuation of a human form is already pre-existent in the sentient entity, but this eventuation will take a far vaster time scale than for the sentient unit constituting a cell in a mother's womb to evolve into the human form. That cell already bears the *saṃskāras* of aeons of evolution in the lesser kingdoms of Nature. Buddhas similarly must yet develop the appropriate *saṃskāras,* attributes, in cosmos to eventually become Logoi.

We see therefore that there is nothing 'absurd' about what would be perfumed by earlier Buddhas to be 'more powerful because the perfuming of the residual impressions would have been there for a longer time, while later Buddhas would [absurdly] be less powerful because theirs would have been there for less time'. At any rate, 'time' is itself but a perception of the human mind accustomed to think in terms of its own livingness on the earth. Esoterically one thinks in terms of cycles of accomplishment. Also, how does one otherwise properly account for the appearance of the macrocosm, of such planets as the earth, with their immense bio-diversity, and of the guiding Mind behind the evolution of the sentient entities of an entire solar system, thence of galaxies, and clusters of them containing trillions of solar evolutions? A thing not understood by Buddhists when they consider the nature of a Buddha is that he has evolved, has grown in Bodhisattva stature and never really stops being a Bodhisattva. It simply stops with regard to earth evolution, for he has outgrown that limited mode of expression, discarding it, similar to one discarding one's clothing before going to sleep at night, or body at death. Once again the aphorism of 'as above so below' finds its truism.

There are more lives in the universe to enlighten than just those of the earth. Also, the growth of a Bodhisattva is inclusive of all those that he has personal identifications with, that become his disciples. They grow with him, for the enlightenment-Mind that he has developed is inclusive of theirs, consequently they all evolve together. Those *arhants* that walked with Gautama 2,500 years ago still have karmic links with him. They are part of the Buddha's auric field and are viewed as part of an organism, of which the Buddha is at the centre. He has simply left the earth to build a zone of combined activity for them elsewhere, wherein they will together preside over the evolution of a similar planetary evolution to the earth in another solar system. They await a time when a humanity such as ours will appear there. The great Ones[11] will then become Lords of the Shambhala that will then be established, allowing the Buddha to preside over the sum of that evolutionary paean. This is the outline concerning the making of a planetary Logos. The two volumes of *A Constitution of Shambhala* provide much more information concerning the nature of such a kingdom.[12]

The functioning of a mirror

How does a mirror work? A mirror normally is a sheet of glass, silvered on one side that allows reflection of rays of light coming onto it. Science tells us that that the angle of incidence equals the angle of reflection. The rays of light from an object come to and from the same side of the mirror to hit the eye, which then sees the 'mirror image' of the object because a mirror reverses the image. However, in the esoteric scenario the reflection comes from 'something' above the mirror to 'something' below it. Also, though it may be static, it appears as if this mirror is actually a turning mirror, to account for the gyrations in the movement of the continuum of consciousness (the 'something below' that perceives) through the time needed for its growth. This also depicts the functioning of a *chakra*. The nature and functioning of *chakras* must then become the proper esoteric study for this subject of causation.

The functioning of mirrors can also be viewed in the way that the planes of perception are structured. These planes of perception (the Earthy, Watery, etc.) manifest in the form of septenaries, with the central *buddhic* level functioning as a mirror. Looking to the overall septenary we see that there are five exoteric planes expressive of the five Elements that are Mind-borne and two are esoteric, the realms

11 Each being Initiates of a higher status than the Buddha was when he gained his *parinirvāṇa*.

12 See also A.A. Bailey, *A Treatise on Cosmic Fire* (Lucis Press, London) for the basic groundwork of this immense subject.

of synthesis, of Love-Wisdom and Will. The central plane *buddhi* is the mirror reflecting the abstract attributes from the higher planes to the lower ones. This function manifests as the nexus that is the *saṃsāra-śūnyatā* fusion. This fusion represents the mirror (and its silvered coating). Here *śūnyatā* represents the darkened silvered layer which reflects the higher to the lower, and vice versa. It is darkened to the mind because it must die to reside in the Void. *Buddhi* represents the mechanism of reflection from both sides of this mirror, hence it manifests as illumined consciousness.

The higher three realms (*ādi, anupādaka* and *ātma*) represent the three levels of the *dharmakāya* which is reflected downwards, whilst here the abstracted mental plane represents the actual substance of the mirror. From another perspective the substance on either side of this mirror is darkened. One side is so because it is steeped in the darkness of ignorance, of materiality, consisting of the substance of the lower (concrete) mental plane. (The Watery astral domain, and the dense Earthy field of expression being zones of reflection.) The other side is darkened because those in *saṃsāra* cannot see into it with non-awakened Eyes.

Buddhi illuminates the bright shiny substance of consciousness, the luminosity of the Clear Mind (the higher abstracted Mind). This Mind is thus a thinly veiled *māyā,* the substance of the Clear Light of Mind of a fully enlightened Being. A small degree of distortion of the *dharmakāya* is thereby possible when it is to be expressed in terms of the empirical patterns of mind when active in the domains of form. If this reflection process is reified through the desire-mind then aberrations occur according to the nature and strength of the desire of the thinker. The wish-fulfilling gem of the imagination sets in to make what was originally pristinely perceived into the image of what one believes aught to be. In this way many types of philosophy considered to be inspired have come into being, be this theistic, philosophic, scientific, occult, Buddhistic, or Tantric.

We can also question whether what is viewed as a reflection can be that of a vast Entity, a Logos; whose Eye gazes upon the Mind(s) of the meditator(s) at the other side of the mirror. What is seen are Images in the Mind of a Being whose full extent is incomprehensible to the little mind/Mind of the meditating one, because he is viewing it through the mechanism of a mirror, and thus only sees that part of the greater Mind that is reflected and upon which he focuses his gaze. This gaze can focus only on a small part of the greater whole that the mirror captures or can capture. The visioning is similar to looking through the eye of a telescope. (Whereas a Logos would have the 'telescope' reversed, i.e., He is peering through a proverbial microscope.) Also, what part of the reflection upon the mirror is the meditating Mind actually viewing? The complete extent of the impression may be immense, thus what is being focused on may only be a part of the rest. Aspects also may be veiled to the enlightened seer because of his/her incapacities, or because the Logos, in His infinite Wisdom may deem it premature to unveil more than a few Images in His Mind. This then indicates a part of the reality of *dharmakāya* into which a Bodhisattva is progressively initiated as he/she gradually awakens to the higher *bhūmis*.

The organ of vision in each case is the Eye. Its function and symbolism here is therefore of great importance. The mechanism of the eye must thus be analysed to gain an understanding of the nature of Logoic Vision.

Does this stare or gaze of a Logos reflect His total Reality, i.e., revealing the complete nature of His body of manifestation, of the extent of the Consciousness-attributes that constitutes the component parts of such a Body? Can one travel through this Eye in the *dharmakāyic* sphere to see the reality of what is veiled by That 'corporeal' Body, or be able to view the vast extent of such a One's Mind? Does the Eye serve sufficiently as a mirror for the view the meditator is aspiring to reach? Is it possible to fully comprehend the vast expanse of what is contacted and seen, thus are there limits to the extent of one's enquiry? Such questions represent ways of approach of the enlightened to cosmos. Obviously the reach of the Mind's Eye depends upon the Initiation standing of the meditating One.

When we view the corporeal Logoic Body we are no longer concerned with the function of the mirror, but with what it reveals, therefore with the attributes of the three Dhyāni Buddhas into which the *dharmakāya* is reflected by Akṣobhya.

Amitābha expresses the Wisdom of the Clear Light, the purified higher mental substance that allows perception of the differentiations in the Logoic Mind, thus the constitution, purpose, status and lesser integrated Mind-spheres of the unities (Individualities) constituting that Buddha sphere (or domain) in cosmos. Amitābha allows one to perceive the Thought-structures (Minds) of the rest of the Deities or Buddhas that form the *maṇḍala* of the presiding Logos. The ability to converse with any of the enlightened Minds chosen, or to receive Thought-impressions from them, is then possible. The View through the Logoic Throat centre then stands revealed.

Ratnasambhava embodies the projection of the integrated *nāḍī* system of the entire *maṇḍala*. Through him is seen the reticulation of the various channels of energy and lines of communication, the evolutionary progression of all within the confines of the Logoic whole. Thus the Way of the Logoic Heart is seen clear, but this is viewed via the Door of experience of the Logoic Solar Plexus centre. All *chakras* reveal their potency and level of awakening in the pure Light of his Equalising Wisdom.

Amoghasiddhi allows one to delineate the forms of all constituent beings embodying the *maṇḍala*. One's gaze can therefore travel to the far reaches of *saṃsāra* for that Logoic expression, to analyse the extent of the progression towards liberation of any aspect of that Logoic structure. Consequently the note or tone of the cries of suffering of the ones enchained to *saṃsāra* can be heard and the Logoic disposition analysed. This the All-accomplishing Wisdom will reveal what needs to be accomplished to produce the evolutionary goal. The Logoic Sacral-Base of Spine centres become objects of specialised discernment.

Akṣobhya helps establish the inner quietude of the meditator, linking the substance of the meditative stance with that of the Logoic Heart centre, as expressed in His Eye. All meditating ones perceive images through this *chakra* to experience whatever is possible through the Void. (The *śūnyatā* conveyed by the Heart does not discriminate, evaluate, or reject, thus allowing faithful observation.) The Heart centre links one to the omnipresent All. When there is a reciprocity of energy qualification, a harmony of meditative expression, then the images can be viewed or projected from one to the other. The Mirror-like Wisdom views the impress without distortion and contemplates upon the expression.

Vairocana represents the focal point of the Eye (generally the *ājña chakra*) that with pin-point accuracy selects the totality that is to be experienced, the image to be reflected as experiential phenomena. The Eye projects the meditative purpose to its target in the utter stillness of unobtrusive *dhyāna,* which allows linkage to the Logoic Mind (Head lotus). The *dharmakāya* is then revealed in accord with the objective of the meditative focus of the Dharmadhātu Wisdom.

At first the mirror of the Mind spins to encompass the entire view. The meditator then selects the particular facet or petal that will accord him/her the specific vision, contact, or revelation desired. This happens automatically as one focuses one's gaze. The cosmic landscape focused upon is then reflected into the meditator's Mind-space. The reflected vision interacts with the entire organ of reception, which then:

a. Spins at an accelerated rate to accommodate the intensity of the energy that has been contacted.
b. Expands to the extent that it can record the entirety of the impress.
c. It instantaneously sends currents of energy to awaken all associated centres (*chakras*) that must share in the completeness of the vision. Thus the totality of the *maṇḍala* of the Dhyāni Buddhas becomes utilised.
d. Everything then becomes merged, as the meditator and the object of meditation become unified in expression. Eye to Eye there is a simultaneity of Vision and experience. Transmission of *dharma* has ensued from Logos to mediator in the *dharmakāya,* with the meditator being the worthy supplicant.

The separation of the process from a point on our earth space out to the whole, and the impress of revelatory expanse going in from the whole equals expansion and contraction, inbreathing and breathing out. Observing the two simultaneously (the holding of breath) equals *synthesis.* Synthesis implies that there is a co-existence of expansion and contraction at the same time, descending and ascending *antaḥkaraṇas,* from Logos to meditator and vice versa. The person is integrated with the whole and the whole is everything the person is and beyond.

Thoughts that pertain to the small concreted processes (images) manifest arrow-like from the unit to the whole via the mirror. The process of integration of information is apparent after absorption into *śūnyatā* and we have emerged again into the *ālayavijñāna.* Here non-comprehension of certain aspects of the vision may be possible, especially when couched in terms of language. To avert this possibility one must stay in the Clear Mind. The bridges of comprehension must be built for other minds if the complete *maṇḍala* that the meditator experiences is to be vivified. The universal then becomes actualised in the particular via the laws of group evolution.[13] Such laws govern the interrelation and evolution of the stars in the night sky.

One's sense-consciousnesses notice the small variations of movement, the elements or changes in the environment one is in. The intuition of those vivified similarly demonstrate the perception of changes in their Mind-scapes, of the impressed images that have been directed thereto. The images appear as seed ideas that their Minds then transform into relevant purposeful or meaningful expression in their lives, so that they can play out their proper service role within the entire *maṇḍala*. *Bodhicitta* is thus enacted. The *dharmakāyic* vision can then be particularlised into the service arenas of others who have not yet developed the capacity to view the entire panorama, but where relative comprehension is possible. What is necessary to effect the potential of the fabric of that part of the *maṇḍala* that the recipient is an aspect can come to that Mind. Smaller thoughts then occur within the womb of the overall paradigm, which when actualised and bearing fruit in *saṃsāra,* prove to be a mechanism of travel to comprehend the nature of what feeds the entire *maṇḍala*. This allows the necessary expansion to bear the next level in the overall scheme.

The way of perceiving the changes produces constant realisation of the larger concept, hence an awakening of the Mind. Every small revelatory experience in the path of action broadens one's mind, because one has redefined the boundaries of one's thinking. One becomes linked to the greater Thought-domain, which impresses the smaller thought-structure with something greater, more enlightening than the current *modus operandi*. In doing so one utilises the past definitions of what one conceived to help define what constitutes the whole. The grand scheme of enlightenment comes to view and is now dwelt in. With every passing moment the past life experiences and revelatory awareness can be accessed, producing perception of what needs changing, the

13 See volumes 6 and 7B of *A Treatise on Mind* for a detailed explanation of these laws.

fleeting phenomena, *saṃskāras* to be cleansed. The horizon can then be broadened to encompass the all. Every little change of attitude enables the smaller thoughts to tap into the higher thought-structure. They produce memory cues to keep one thinking universally, to keep consciousness focused upon the broadest picture possible. Thus the true nature of cosmos comes to view as consciousness expands to inevitably include the content of a Logoic Mind in its embrace.

The unit aspires upward to gain greater clarity of Vision because the *chakras* can now spin with heightened vibrancy, the Heart can expand to reflect a greater portion of what the mirror of *śūnyatā* conveys. The cosmic *nāḍīs* are accessed, which pour forth vibrant invigorated *prāṇas* of transmuted *saṃskāras* that bear a higher potency of energy, because *buddhic* input is achieved without distortion of the Vision incurred. The *dhyāna* activates the mirror of Revelation that the *dharmakāya* portends. The group-conscious unit manifests the meditation that enables him/her to see Eye to Eye with the Source of all possible revelatory attainment. What constitutes a Logos, or of cosmic beginnings, then no longer presents a mystery. The rebirthing process of it all is seen clearly in the diamond-Mind of the Mirror-like Wisdom. Group-conscious Visioning expands to include the Eyes of Perception of great Logoic Minds, the interrelated stellar spheres of our local cosmos. What They Know can then come into View, to be Known and pathways of Revelation accessed allowing the meditator to be in the Mind-Space of a Lord of a star system with the speed of Thought, which lies outside the bounds of the limitations of the speed of light. The mirror is thereby surpassed.

Signposts of revelation and analogy

Analogy is the key to comprehension. The smaller unit of consciousness resides in a smaller but similar structural universe than the overriding presiding One. Therefore to reach out to that which is beyond it the smaller unit must build an image of what it is yet to comprehend, based upon the signposts on the way it has already observed that has given it the right clues. The signposts are important, those created by former events are part of the organisational structure of the all-encompassing consciousness.

Signs happen in consciousness from a number of sources and it is important to recognise the quality of the light, or clarity of impression. What should specifically be noted are those from the Sambhogakāya Flower, or from the Master within presenting symbolic imagery that can be rightly interpreted, prescient instructions through an inner Voice speaking. They are often coupled with quick impressions of what is or must be. The signposts manifest as the conscience, a knowingness of what should be done, of how one should act, or even impression of where to go or to be.

The fact that particles, *i.e.*, human individualities, can choose to go with or against these signposts is what creates individual *karma*.[14] Such *karma* is created by the force of free will, as to which way to drive within a set of circumstances. Gradually one can move out of these circumstances and become a causative principle for the all that exists around oneself, the various appearing forms of sentience and embodiments of awareness. A person then manifests with the higher or greater streams of cause and effect, as produced by universal Consciousnesses, and responds in unison with the greater streams. Thus Logoi are eventually born. Such beings act as mechanisms whereby greater universal *saṃskāras* can be cleansed.

The images created by self-willed entities generally come in the form of what exists in the mind, but generally evolve into much more than itself, as the mind/Mind aspires to view broader panoramas than what it knows. The image then comes to represent a vaster expansive horizon of thinking and which serves to help embody that which the smaller unit aspires to. It becomes the signpost of the future. Thus we have the images of Deities, (peaceful and wrathful) Bodhisattvas, Buddhas, created that help serve as guides on the way to revelation and liberation. From a historical perspective the nature of the images (the iconography of religions) have progressed as the respective advanced members of any society have developed and transmuted the *saṃskāras* governed by a particular *chakra*. Historically, from an esoteric perspective, the evolution of civilisations can be traced according to the nature of the *chakras* that those of any particular epoch were involved in awakening.

Ancient civilisations, such as the Egyptian and the Vedic, explored the potency of the *maṇipūra chakra* and therefore the entire Inner Round

14 See chapter eight ('Signposts of Consciousness') of volume 2 of *A Treatise on Mind*, for further detail concerning the nature of manifesting signs.

of *chakras*. These potencies became symbolised in their various Gods. Buddhism evolved the way of the Throat and Heart centres, with the Tantric Vajrayāna view exploring the relation of the Head lotus to the Base of Spine centre. Christianity should have developed the way of the Heart centre, but this hoped for advent was largely captured by the methodology of the dark brotherhood. Now the Dharmakāya Way is on offer, which can explore the sum total of the *maṇḍala* in relation to what is expressed in the external universe.

Analogy concerns refocusing the eye away from what it knows to ascertain the bigger picture. It moves from a myopic concern with small detail to include an analysis of the paradigm of a vast scale Vision encompassing Images formed in Mind-structures of immense proportion. A human unit but mimics the visioning process of a Logos, who reverses the process by looking from the greater Ken to a microscopic universe.

Using the concept of analogy as the point of reference one can reference the smaller points within the greater unit, especially when the smaller points are those that bring one to the higher analogy.

- The first point of reference is the observations made about the body and its consciousness.
- The second point of reference is the analogy made when the observation is extended to the outer universe. (Projecting the will out into space.) This point of reference exposes far more detail than the first point ever could about itself. The picture that is built upon by this process is the *maṇḍala* of vast understanding, and it helps develops wisdom.
- The third point of reference concerns establishing an understanding of the subtle body, constituting of the *chakra* and *nāḍī* system. This allows comprehension of the nature of the *prāṇas* and the role that *saṃskāras* play in one's consciousness, and later, in the external universe.
- The fourth point of reference concerns establishing the purpose of Life, the nature of the future and of the subtle influences playing upon and directing consciousness. Thus such concepts as *bodhicitta*, the Sambhogakāya Flower, *śūnyatā*. Bodhisattvas and Buddhas are pondered upon. Links are established thereto via the way of meditation.

- The fifth point of reference concerns Seeing the exalted nature of the *dharmakāya*, Buddhas, Deities, Logoi, and their placing in the scheme of things in cosmos. By utilising the concept of analogy, and via direct experience, their attributes can be Identified within *dhyāna* (meditative absorption).

When one anthropomorphises, one uses the human body, subtle or gross, and its consciousness as a reference for the vaster streams of Being/non-being. Expansion is then possible through comprehension of the bodily consciousness and its identification with the universal consciousness in all of its diversified attributes. Vision can then extend to more than just the gross physical form, to include the five points above in the ontology of one's scope of revelatory analysis.

When the *chakras* in the stars are seen, as well as being constituted as groupings of stars, then a better understanding is obtained of what a human unit really represents and evolves to become. A better time concept is also developed, of the law of cycles, of the duration of Being/non-being in terms of absolute time, as these *chakras* unfold at a vaster scale and rate of unfoldment than those in a human unit. Time has been elongated with respect to a human life-span. When there is more time to view things then one can obtain an intricately detailed picture as to the nature and composition of the *maṇḍala* being observed.

The Mind of the vaster entity dealt with perceives vast cosmic arenas of activity, expanded 'time sequences' through which to view the phenomena of the universe. Happenings in the microscopic human world manifests at lightning speed to such an Entity. Such a One's available 'time' is vast and even a 'seconds' observation from His domain upon the human world would account for many millennia of human activity. He therefore can 'take the time' to make many sweeps across human history along the lines of *saṃskāras* developed by human civilisations, and quickly make the appropriate decisions re desirable outcomes. (Such a sweep would take but a tiny portion of the 'time' of such an entity, being little more than a causal 'glance'.) This produces the conception of omniscience and omnipresence attributed to a Logoic Mind. Energy can then flow from His Eye through the increasingly vaster successions of Eyes/eyes below, allowing the effective changes to be made. Omnipotence thus enters our conceptualisation with respect to the effect of such a Logoic Mind, as our eyes have limited scope and vision.

Motion happens at an apparent slower rate at a Logoic level, allowing detailed observations to be made. Separations in space can be thoroughly analysed and the innate essences *(prāṇas)* directed within the macrocosmic Body to arenas of need. Here the karmic play is completely comprehended. When our minds are enlarged to the size of an ocean, or to become a stellar unity in cosmos, one can then better observe the nature of what the future holds for all. *Saṃskāras* will also be seen in all their glory in a vaster scale in the sky. The smallest details existing in human consciousnesses can be elaborated from a transmuted perspective in the cosmic landscape. The *saṃskāras* will be noticed for what they are in terms of creative and formative forces. The procreative process relates to the formation of stars (cosmic Ideas) and galaxies, and we will see also a version of *pratītyasamutpāda* (dependent origination) in action. We will see well Reasoned elaborations of processes akin to human mental-emotional volitions. Small details are defined there but become huge and transmuted in the telescope of universal thinking and the appearance of things.

Everything comes under karmic impress, the scenery of interplay, of action and reaction relegated to the law of cycles and governed by the expressions of the vicissitudes of mind/Mind, as already analysed in *A Treatise on Mind.* Observations however must be made with reference to the scale of the vision and to a proper understanding of time sequences.

Time is created as consciousness moves. Because consciousness moves, so evolution exists. Another way of viewing this concept is that because consciousness moves progressively so human evolution *is* a factor in Nature. Within the magnitude of Nature we see that the evolutionary process pushes the All through to higher sentient states and conscious awareness. This process is but the expression of a vast Consciousness moving it all. Its effects may be viewed in terms of Sheldrake's 'morphogenetic fields' governing biological processes,[15] or may even manifest in the form of Vasubandhu's 'violent torrents'. Here the perspective is of viewing the *prāṇas* moving in the *nāḍīs* of the lower strata of Mind on a planetary or solar scale. The torrents create the

15 Rupert Sheldrake, *A New Science of Life,* (Blond & Briggs, London, 1981).

ever-changing scenario seen when aspects of them are crystallised into physical manifestation by an organising Mind. The 'torrents' however need not be violent, but rather can simply be peaceful expressions of streams of *prāṇic* forces within a Consciousness-stream on a vast scale, where the Mind in question is that of a Logos. They appear 'violent' because of the intensity of the *ātmic* energies perceived from the domain of the empirical mind. Streams of energies in a Logoic Mind then sequences events upon the time horizon, effecting the moving line of evolutionary progression.

The existence of a greater Mind-continuum is what many would posit as a 'God', especially when coupled to the factor of Individuality on a vast scale. Evolutionary progress in Nature is seen to be the effect of such a Mind working out a Thought process to completion, of streams of energies within spheres of containment.

We know that everything that is conscious has an external and internal to it. To be consciously aware one must be aware of something 'other'. That 'other' may be reflected in consciousness, but consciousness may view within itself created images, ideas, derived from that existing external to it. Such images create ring-pass-nots, limitations of possible expression, the bounds of any type of consciousness-experience. If such Thought-Bubbles of containment did not exist then the original cause and also the subsequent effect that was created would expand to nothingness. They would be inchoate and cease to be. Everything has to be properly defined if it is to be cognised and reasoned out to be a 'something'. Ring-pass-nots are similar to the skin of a human unit that defines the shape or context of the form it covers.

If the effects generated by a Logoic Thinker are not totally controlled then the structure impelled by the originating cause could easily change and an indeterminate process produced that may not delineate an evolutionary progression. Rightly ordered sequences of events may vanish, as the attributes of Nature that appear will not be brought together in a meaningful way, or interrelated in an appropriate manner. Rapacious individualities will evolve that fight against the structured order of things, to build for themselves power bases that resist evolutionary change and which grow only as an expression of the power of the self will. They will destroy the common good in order to build for themselves extensions of that power in competition with other individualities thus empowered. Inevitably the most powerful will fight to control all the others and whip them into obedient slaves of their overriding will. Time, as we conceive it, would change because there would be no true continuum of progressive cause and effect of things. Cause and effect will exist, but will be manipulated by the dominant will, producing a stagnation, a truly cancerous growth of a separative thought-structure that will destroy the fabric of the body of manifestation sustaining it. Such is the way of the evolution of the dark brotherhood.

Inchoate thoughts are thus obviously not produced by Logoi, because then we would have disorder and indeterminacy manifesting everywhere, rather than the precise laws so evidently obvious. Compassionate laws sustaining the evolution of consciousness (the Anthropic Principle) manifest via Logoic Thought because of the aeonic long training in the art of meditation that the Logoi have undergone. They have evolved past the stage of being a Buddha, whose meditative process has gotten him 'there'. All things and the laws that interrelate them, are collated in the Consciousness of such Meditation-Minds. Thoughts within such Minds consequently are precisely regulated and bounded by an emanatory Clause, sustained by a duration of Thought, and eventually terminated when the consequences of that Thought have reached their conclusion. Thus a world sphere or universe evolves.

The existence of all things necessitate a ring-pass-not of their manifesting form, even of the forms existing upon the *arūpa* levels. Existences upon such levels possess their own types of delineations, or 'skins' of energy fields, if one can style it so.

The factor of consciousness

To be conscious one needs to notice the appearing sequences of things manifesting a patterning of events. One needs an appearing sequence of images and to interact with them through being aware of their existence. A point of contact with a sense-perceptor correlating with the mind and its continuation creates a sense of experience. This is because it manifests as myriads of little 'dots', photons of light, or atoms of substance (taking a 'atom'[16] here as to mean the smallest possible thing) manifesting in a pattern that contacts and excites our senses. Experiences necessitate the relationship of these photons (or atoms) to our senses and the mind. Time exists because the points of contact are manifested and experienced in a series. Those points are the phenomena appearing from fifth dimensional space[17] (wherein the intellect functions) interrelating with the third dimensional objectivity, thereby correlating the appearance of objects, flashes of cognition, and the series of events collated by the mind.

Our minds thus represent fifth dimensional tensor fields cognising three dimensional space and time by means of a neurological system that manifests upon a one or two dimensional framework. (Interrelating points of experiential activity and sheets of visual illumination that are correlated by the mind into three-dimensional images.) The three-dimensionality represents that which is outside of consciousness coming into the domain of the consciousness. Consciousness then links up the images via the factor of time. A consciousnesses however can step out of the bounds of time when it simultaneously links the past to the future in one moment of timeless revelation. It has then instantaneously visioned the panorama of the all and has taken from that panorama what is needed to build the new. Such instantaneous linking procedure (of all three dimensional images) and creating a complete picture from them if need be is the nature of fifth dimensional thinking.

The phenomena persists, consciousness persists, as long as the ring-pass-not of the purpose of the incarnating consciousness exists, and the boundaries of this universe of forms persists. That outside the consciousness represents the not manifest to that which is inside. *Śūnyatā* can also be considered the not manifested, as consciousness does not recognise it. Only when consciousness is transcended can *śūnyatā* be experienced for what it is. This is because consciousness consists of moving 'things', moving images, and *śūnyatā* is freed from all such.

Śūnyatā represents zero dimensional space, there are five dimensions of expression below it (including that of the abstract Mind), as explained above. We can also consider five transcendent dimensions of expression

16 I use metaphorical language here, as we know atoms to be composed of smaller particles.

17 'Fifth dimensional' if we take time to represent the fourth dimension.

above it in the *dharmakāya* universe. From this cosmic perspective the *ātmic* plane represents the first dimensional space, because from this domain emanates the forces pertaining to the manifestation of the world of appearances, the world of phenomena, known as *saṃsāra.* From here are seeded the atomic unities, the human Souls, and hence ultimately the 'I' of the personality. The Monadic domain presents the cosmic second dimensional space, because here the members of the cosmic 'plant kingdom', human Monads,[18] are esoterically 'planted' in the soil of the cosmic dense physical plane. Plants can be considered to 'think' two-dimensionally in that they grow upwards and downwards, with their leaves, like sheets of paper receiving the energies of the nourishing central spiritual sun.

The plane *ādi* then represents the third dimensional Thought Life in cosmos, in that this plane is directly energised from the cosmic mental plane, which allows the Initiates upon *ādi* to Think properly from a cosmic perspective, hence to see appropriately the depths of the cosmic scenery that confronts them at the border between the cosmic physical and cosmic astral domains. The cosmic astral plane signifies the attributes of the fourth dimension of time, because here the Logoi that are to, or who have incarnated into, planetary or solar spheres delineate the durations of such sojourns, *manvantara, pralaya,* and the appearance of the various *yugas.* The cosmic mental plane, from which all proceeds then manifests fifth Dimensional Ideation.

When the fifth dimensional space of the higher perceptions becomes limited by the phenomena of three-dimensional appearances the perceptions become particularlised, manifesting a patterning of things spread out through space.

There are fields of moving images consisting of spheres of Logoic domains presided over by Logoi and Their compliments all linked by Mind. A vast cosmic panorama unfolds of the integrated Lives, the colourings (Rays) of their Thought structures and lines of communication. All simultaneously work to assist the Other in a grand harmony of organised Mind, within constellations of stars that are part of galaxies revolving. This internal organisation reflects in the laws governing the unfoldment of the universe observed by astronomers.

---

18 The principle of Life, the Spirit within, which will be explained later.

Consequently there cannot be an expansion of the universe into infinity. Though expansion manifests according to the mode of the Mind to seek new Knowledge, however the bounds of the universe, the ring-pass-not, has been set for that and every other cosmic Incarnation. This produces a dynamic bounded universe, with a cosmological constant that holds all into unity set by the Originating Mind. Within that space there is a moving of Thought Forms in relation to each other, represented by stars, constellations and galaxies. There are set bounds, limits to the expansion of a universe in space for any particular Incarnation, in a similar fashion that there is a limit, a heliosphere, to the extent of a solar system's existence. What can be considered as Space therefore manifests far past the extent of the duration of that cosmic Incarnation. The rebirthing process necessitates something that an entity is born into and can experience by means of that activity. The Thoughts are moving to prearranged destinations, they were born, and hence shall also die, to be abstracted back into their Source. All Lives manifest according to a pre-existing Plan for that Incarnation.

The interrelated Fires of Mind produce a universal background warmth to the sum of the substance of the universe, which is reflected in the cosmic microwave background. As all Logoi are working to produce a similar purpose, (though variegated accordingly to accommodate their different degrees of evolution and individual challenges) and are integrated into the pattern of the originating Mind, so it produces an overall homogeneity of activity, to fulfil the purpose of the originating Meditation of the One MIND. The structure of Thought contains the patterns of what is to be for any Incarnation.

The progress of time manifests in terms of being part of an organism evolving increasing sophistication. Atoms (as the building blocks of forms) meld into planetary spheres revolving around suns. The higher supersensory perceptions allow the true impressions of Logoi to appear in consciousness. Phenomena appears as planned, according to the mathematics and geometrical reasoning of universal Consciousness in the case of a Logos. This concept is highly stylised in the *maṇḍalic* diagrams of the universe. It can have further visual dimension in the extended vistas of our minds locked in the embrace of extended time.

Human units are similarly learning to plan the minutiae of the course of their lives, if they are to be successful in their chosen professions. In the right handling of their individual time lines and of the appearing resources as they progress, so they are learning the elementary arts of eventually being a Logos. Most humans however cannot factor in the appearance of *karma* in their lives, and also are sometimes careless with the way they handle their resources (intellectual, emotional and physical). Logoi, on the other hand have mastered all such considerations, as also those that have become enlightened. We see therefore that attaining enlightenment represents the educational process preparing one to eventually demonstrate the perfect coordinative power of a Logos in maintaining a world sphere throughout its evolutionary epochs.

Time represents the domain of phenomena, and there are aspects of that phenomena, units of inertia (hence *tamasic)* resisting change, working to perpetuate, or stretch out the duration of a cycle of accomplishment infinitely, if possible. They represent the unregenerate aspects of one's *saṃskāras,* the darkened *prāṇas* within consciousness, which need converting into arenas of light. Such *saṃskāras* can represent people's sensuality, attachments to various forms of desires and emotional attributes. They can also represent the willful manifestation of hatred and spite, separateness and forceful opposition to progress, to accumulate materialistic power, control over the forces of Nature, and of human society for the empire of 'self'. Such are the members of the dark brotherhood, and they will work fervently to 'stretch time' for as long as they can. They will the conditionings of the material domain wherein they can breed their power for as long as possible. Such dark forces can exist in huge numbers and with vast power in cosmos. Consequently not all Logoic incarnations are successful, especially in the earlier stages of a planned series of Lives, and many stars and constellations have been consumed with the cancerous Blackness.

All incarnating Logoi must calculate the probable effect of such Tamasic opposition (the inviolate factor of individual free will of units of consciousness) within each duration of any cycle of time. Consequently cycles of accomplishment are normally of vast duration at the beginning, and become increasingly briefer as the Tamasic residue is overcome. What cannot be mastered in one cycle is recycled for a later cycle to complete. The higher echelons of the dark brotherhood have their means of escape from the ring-pass-not of a completing cycle, but not from the eventual prison house of their *karma.* A cosmic line of evolution can therefore persist long after its Planned time of delivery to liberating Light because of the karmic pull of the Tamasic ones. Nevertheless the overall force of the liberation of the successful Logoi pulls the entire universe towards evolutionary gain. Human groupings, planets, suns, constellations, That Logoi, galaxies, galactic clusters and universes manifest incarnation after incarnation together as they spiral ever onwards and upwards to supernal heights.

In making calculations concerning the spiritual age of Entities we must factor in the time element and realise that the vaster the Entity observed, the further back into time one is traveling. A vast Entity, such as a galactic Lord takes far greater durations of time, hence meaning a slower pace of activity, movement of events, than say a humanity on an earth sphere needs to accomplish what might be called its liberating Purpose. Hence when looking at vast Entities, one travels backwards in time, not in terms of the events of the proceedings of some postulated Big Bang, but in terms of the fields of Consciousness, of the development of cosmic Mind. Consequently one travels towards arenas of dark space. One must also note that in observing, say galactic evolution, not all are at equal development. Some are at the comparative level of low grade human units and others may be at the level of enlightened Beings in terms of their overall Lemurian (third Root Race) environment.

Such vast domains however represent enormous pools of potential energy, consequently when such a One Thinks in order to initiate a movement, such as the 'creation' of a solar sphere, an intense amount of energy can be focused to do so. Hence though the amount of energy available to such a One in say per cubic metre of space may be far smaller than to an enlightened being (a *siddha*[19]) upon earth, cumulatively what can be accomplished (i.e., the manifestation of *siddhis)* is far greater.

Śūnyatā and *saṃsāra*

Through comparison with what we humans demonstrate and embody in the expression of the senses on a physical level we can deduce how *skandhas* and *saṃskāras* play out on a universal level, for the Eye that sees can reflect in terms of transmuted correspondences. One can thus

19 A *siddha* is a *yogin* that has developed demonstrable psychic powers *(siddhi).*

look at the universe and observe the effects of energy exchange, hence Consciousness interplay in each body of manifestation, be it of stellar spheres or larger amalgamations thereof. Such understanding was the basis for the creation myths of millennia ago. The relationships are ordered in the geometry of space and are clearly delineated, frozen in time (because of apparent slowness of movement with respect to our earth forms of activity). They can be properly comprehended with reference to the blue-print sequence of the *chakra* and *nāḍī* system of various *maṇḍalas* of evolving expression. (This concept will be explored throughout my cosmological writings.)

Similar processes that hold the human body into a unity also holds the universe intact, because everything is interrelated and interdependent. There is no true separation between the small body and the stellar spheres seen at night. If one is to analyse the human unit (a minute 'stellar sphere' in its own right) then one can also look to the stars and observe similar evolutionary processes. The process of universal integration is innate in humanity, helping to push their march onwards to evolutionary perfection, their total enlightenment as a human species, rather than just as integral unities. The collective consciousness of humanity looks outwards to the universal space around them (which is the sum on their environment) in order to find their place therein and thence to project the human world thereto. For this purpose also the integrated Minds of Bodhisattvas work, being the way they demonstrate their compassion for the all.

Humanity however must look inwards as well as outwards to be able to journey thus, as the inner and outer universes are also an integral unity. Accordingly, all meditators, Buddhists or otherwise, should not separate themselves from the human trend by endeavouring to only explore the inner space. Knowledge of the macrocosm is also needed to gain the enlightened stance. If one only queries inwardly in this modern empirical epoch then one is effectively saying that one is separate from the external universe and its awareness-scapes, thus only a partial enlightenment is possible. This was not so in the time of the sages of the past millennia because the *iḍā nāḍī* stream (of scientific understanding) in human civilisation was only partially developed, hence those that gained their enlightenment earlier have had to undergo compensatory development, incarnations to gain the necessary Knowledge. The

Buddha has also had to 'complete' his education in a star (Regulus) in the constellation of Leo the lion, wherein cosmically the *iḍā nāḍī* is exemplified within the Love-Wisdom *(piṇgalā)* stream.[20]

Those that look only inwards may say that the outer phenomena is an illusion, but they actually personify themselves in this statement, consequently that with which they meditate with will be an illusion as well. There may be a seed of truth here, but where can they grow further, unless they see that their expansion into the realms of meaning, of being/non-being that is inclusive of *saṃsāra?* Without *saṃsāra* they have no mechanism of travel in consciousness, no purpose for the enlightenment to follow. Complete knowledge of sum of the processes associated with the *saṃsāra* in which they reside, extended into the further reaches of cosmos, is indeed their ultimate purpose. It is that for which they have incarnated into this 'precious human body' to learn from. The attainment of *śūnyatā* is but an addendum to the main fare of their evolutionary process, but a step on the Way to Knowing what *saṃsāra* is truly about. Buddhists must learn the Ways of cosmos, from the starting point of *śūnyatā,* but then travel on backwards and forwards to other *saṃsāras* on a vast universal scale. For to be Buddhas is to achieve no less. New mantras of power are to be gained, new cosmic acquaintances, unimaginable orders of Being/Non-Being at whose Feet one must yet learn to prostrate as one travels through the various echelons of *dharmakāya.*

If one grows inside then one must grow outside as well. Some Buddhists may also say that they grow neither outside and inside, neither not outside nor inside, both inside and outside. The middle way may espouse such logic, relegated to concepts of *śūnyatā.* The middle way however cannot be found by a separative mind that tries to define in absolutes by negating purpose, order, any start, beginning or any ending. There is no 'middle way' if there was no beginning or ending, or any polarities of any sort. The polarities needed are those that combine the inner and the outer, where the microcosm of the earth and the macrocosm of the universe become unified in meditation, wherein is found the nexus of Being/non-Being.

---

20 This esoteric fact can be ascertained by those that have developed the Eyes to See, and until then can be accepted as a logical hypothesis by others. Leo governs self consciousness, the 'I am' principle.

There are reflections within reflections, of smaller patterns within the greater ones. The simile is that of interconnected spheres (of consciousness) gradually expanding and spiraling within greater spheres (*maṇḍalas*) of similar substance, where the surface of each bubble can be viewed to function as a mirror. The mirror is a moving expression of Akṣobhya's Wisdom. The unmanifest is on the outside and as it makes contact with consciousness its expression is mirrored inwardly to the field of consciousness. This surface is not meaningless, it has expansionary properties, and receives new instructions to maintain the existence of the cellular structure. This surface consciousness or skin-consciousness that interrelates with the external universe awakens inevitably the higher attributes of Mind. Such a Mind can pierce the boundaries of each surface consciousness it apprehends, to see beyond their boundaries and hence register that which exists as the *maṇḍala* of each sphere of conscious attainment.

If it is said that *śūnyatā* and *saṃsāra* are identical then one need not focus exclusively upon *śūnyatā* as an object of attainment, analysis of *saṃsāra* will get one there too, as explained above. *Saṃsāra* is a way of bringing one to Truth. To master *saṃsāra* one must come to know it in its completeness as an expression of the Mind-scape of the universe, and therein will be found *śūnyatā*. Conversely, if one experiences *śūnyatā* then the fundamental nature of *saṃsāra* is no mystery.

If one desires to gain an understanding of the nature of causation however then the experience of *śūnyatā* alone will deny this, as in *śūnyatā* there is no such thing. This is the reason why Buddhists (the Mādhyamika specifically) have denied that there is causation of anything. Their focus has been only upon one factor (albeit an important one) in the schema of being/non being. They have thereby not properly comprehended the functioning of consciousness and the way it interrelates with *śūnyatā* via this mirroring process, of the true nature of the relation between Vairocana and Amitābha via Akṣobhya. They have therefore not perceived why indeed there *is* such a relation, thus why all Buddhas are *not* identical. They function differently, sequenced in space, to produce different fruitions, but have one ultimate goal, the liberation of all *saṃsāric* being.

Saṃsāra-śūnyatā are the basis of each other's expression, like the yin-yang (male/*śūnyatā*-female/*saṃsāra*). Such duality exists in a manifest consciousness that is capable of bringing one to the other. Consciousness therefore is the bridge or mirror between *śūnyatā* and *saṃsāra*, or rather between *dharmakāya* and *saṃsāra,* where *śūnyatā* is the equalising mediator. Here the shiny part of this mirror, the reflective capacity, is made up of the intensified light of clarified Consciousness, of illumination. This illumination is the integration of *dharmakāya, śūnyatā* and Consciousness. Behind this is a backdrop of darkened space, of varying degrees of *ignorance,* where consciousness is lacking from many perspectives. This is the actual substance of this mirror, which must in time be consumed by light, and that which illuminates is the shiny reflective surface. The spiral tunnel of time-space is thus coated and so constitutes the *nāḍīs* conveying the *saṃskāras* of consciousness-volitions from here to There. We travel through increasingly larger 'tunnels', and so onwards to a greater comprehension of cosmos, as the darkness transforms into the ever greater Light of Comprehension.

3

The Question of 'God' and the Physiological Key

The physiological key

The abundance of Life is far greater than has yet been imagined by the speculations of the average philosopher or scientist, yet its Mysteries can be wrought from the base metal of *saṃsāra,* by everyone willing to undertake the necessary meditative journeying into Consciousness. If one wishes an insight into the nature of Life and the relation between Logos and humanity, or the other kingdoms of Nature, the *physiological key* is virtually indispensable. A major postulate from which full understanding of all relationships can be derived can now be considered.

Just as a cell is a unified, coherent, functioning entity composed of many diverse parts, so a large number of cells, possessing many diverse attributes together constitute a functioning entity, an organ (with specialised qualities) in the body of a person. The totality of many organs with a multiplicity of functions, when working rhythmically and harmoniously together, constitute the physical form of a person, an integral entity. When the analogy is extended, then we will see that a person, constituting of a form, possessing consciousness and embodied by Life would then form a cell (or atom) in the Body of a sensed but unseen greater entity, which we can call a Logos of the earth. That Logos manifests similarly with respect to the Logos of a solar system. Literally we have *maṇḍalas* within *maṇḍalas* of expression to consider, all obeying similar though transmuted correspondences of attributes and functions. They obey similar laws of evolutionary growth and are tied by the same laws of *karma* and that of cycles.

To rephrase this concept, we see that a Logos constitutes a large number of self-conscious units (human and *deva* 'cells') all interrelated in some way to form a coherent unity. From this postulate we see that just as there are unseen messengers (hormones), channels of supply and elimination (blood vessels) and of energy reception (nerves), between the various organs in the body of a person, so there must be in the body of that Logos similar channels of energy flows. These can be divided into three categories:

A. The dense material world, the flesh and bones of That Body.

B. The Watery world, the circulatory and digestive systems.

C. The Airy and Fiery worlds, the breathing and thinking mechanisms.

This tabulation can be exemplified thus:

A. *The dense material world.* The flesh and bones of a Logoic Body.

This can be seen and touched, forming the body of action, the corporeal form of both a person and a Logos. It can be noted in the analogies between the observable dense physical universe and human physiology. It is that which provides the ability to move, contact our surroundings and to make and manipulate things.[1]

For a Logos the dense physical world can be considered to be the realm of the collectivised human empirical minds incorporating the substance of the four lower mental sub-planes of the mental plane, and their mode of contact with the physical plane. This mode of contact constitutes the three worlds of human livingness. The concrete realms of mind are the correspondence to the terrestrial Airy sphere, the Watery astral plane corresponds to the domain of the oceans, etc. The purely Earthy sphere is our dense physical plane. The Fiery Element can be considered a universal ingredient. It is inherently manifest in all things, as all are integrated into the ring-pass-not of the creative Will of the Mind of the Logos.

---

1 See also the sections in my book *Maṇḍalas: Their Nature and Development* entitled 'Logoi and the organisation of *citta*', and 'The structure of cellular units', pages 75 – 101.

The three sub-planes of the higher mental plane can be considered a type of 'event horizon' integrating the cosmic world and the purely phenomenal sub-planes of the cosmic dense physical *(saṃsāra)*. It is a transitional zone between the intensity of the energies of the higher domains and the world of phenomena. Hence here is experienced the *ālayavijñāna enlightenment* explained in the earlier volumes of *A Treatise on Mind.*

Counting from above-down the sub-planes of the higher mental plane can be depicted as:

a. The first mental sub-plane, the Clear Light of Mind, the sea of rarefied Fire, the most rarefied aspect of the Fiery Element, an extension of *buddhi.*

b. The second mental sub-plane, wherein most Sambhogakāya Flowers reside. We have Fiery Love, the reflex expression of Logoic Love into manifestation via human consciousness, the energy of *bodhicitta.*

c. The third mental sub-plane, the abstract Mind, radiant Fire. The pure Fire that integrates the three worlds of human livingness into a coherent unity. If the four concrete mental sub-planes below it are considered then the abstract Mind represents the Aetheric aspect of their relative Earthy expression manifesting in the form of the four lower Elements as attributes of Fire.

Śūnyatā (Emptiness, the Void) is the fourth sub-plane of the fourth plane of perception (*buddhi*). The three buddhic sub-planes below it plus the three higher mental sub-planes represent a distillation unit for the aspects of Mind, wherein Mind becomes increasingly rarefied, until at the level of the seventh distillate, *śūnyatā,* what is left is *sat,* the essence of (Logoic) Life, *chit,* the essence of Mind as a compassionate force *(bodhicitta),* and *ānanda,* the pure bliss, which is the experience of intensified energies. The three higher sub-planes of the buddhic plane represent the places of impact of the *prāṇas* from cosmos. The highest of these transmits the Fires from the cosmic mental plane. The next projects the impact from the cosmic astral plane, and the third the effect of the sum of the *prāṇas* of the cosmic dense physical. The energies from the cosmic mental plane manifests as *sat,* the eternal Life aspect sustaining all via *śūnyatā.* Those from the cosmic astral plane effect the experiences of *chit,* as the force of compassion via *śūnyatā.* Finally those of the cosmic ethers produce the experience of bliss. The force of compassion represents the energies from the various Logoi ensconced upon the cosmic astral plane that impact our solar system according to the right cyclic expression of their *Manasic* considerations that express the nature of these energies to the awakened Ones upon earth. From this perspective *śūnyatā* represents Emptiness as far as the empirical mind is concerned, but becomes the pure conduit for the downward expression of cosmic Mind in the form of *sat-chit-ānanda.* This then represents the nature of the *śūnyatā enlightenment.*

Logoically, the four higher sub-planes of the cosmic dense physical plane, *ādi, anupādaka, ātma* and *buddhi* are the four cosmic etheric sub-planes.

B. *The Watery world,* the circulatory and digestive systems of a Logoic Body.

The Logoic Watery world is the cosmic astral plane. Upon this plane exists the general interrelation between Logoi and the *nirvāṇees* that have evolved from various solar systems. However the 'circulatory and digestive systems' are an expression of the establishment of a Logoic Seat of Power upon the second and third of the cosmic ethers (*anupādaka, ātma*). Shambhala is such a Seat of Power for our earth. Shambhala represents the Head centre of the planet and synthesises the *prāṇas* from the lower *chakras.* The 'circulatory and nervous system' is governed by the *deva* kingdom enthroned upon the plane *ātma* and then the lower planes of perception they embody. They regulate the energy exchanges between various orders of entities within the Body and the karmic interrelations between them all.

The functions of the 'digestive system' is governed by the human kingdom. They digest various streams of information derived from interrelation with *saṃsāra,* and evolve what is digested into higher units of Consciousness-bits. The principle of enlightenment *(prajñā)* ultimately evolves, the wisdom that is the real food of the Body of manifestation of the Logos.

C. The *Fiery and Airy world* is the domain of the higher strata of *dharmakāya* reflecting the expression of Logoic Mind upon the cosmic mental plane.

Comprehension of the nature of the Logoic Head centre and the Way it regulates the law of cycles for all within the Body of manifestation must be taken into account here. This represents the *dharmakāya* enlightenment. Every Thought sequence projects into manifestation streams of *deva* and liberated *nirvāṇees* in order to ascertain the expression of the Logoic Thought, and so complete another aspect of the *maṇḍala* of expression that is the Logoic Ideation or world-sphere.

There are basically *seven parallelisms* between the physical life of a person and that of the solar system:

1. The physical sun is the dense heart of the solar system, to which there is an approximate twenty-two year cycle of sunspot activity (areas with intense magnetic fields). Their numbers increase every eleven years or so and then diminish. Often accompanying them are solar prominences, which discharge an immense amount of energy to the solar system. This is analogous to the cyclic coursing of blood through our veins, which conveys energy to all parts of our bodies.
2. The solar system travels on its interstellar path and is influenced by the energies of other stars and constellations. Thus it responds to external stimuli, similar to the way that a person travels on his/her path in life and is influenced by friends. All such interactions have considerable effects upon those experiencing them. For the Logos of the solar system there is therefore the higher transmuted correspondences of people's emotions and mental impulses. *Saṃskāras* drawn from past interrelations are enacted upon and eventually transmuted, though on a vast scale and concerning immense time periods in the immeasurable domain of cosmos. The entire science of esoteric astrology was founded by ancient seers to analyse the effects of such influences and those from the planetary Regents in our solar system. They endeavoured to prognosticate future events by understanding the nature of the energy qualifications from such sources that may influence any civilisation and the related environment.

3. The solar system also breathes, with respect to each planet, which expands or contracts, depending upon whether the planet is in aphelion (the point furthest from the sun in the orbit of a planet) or perihelion (the closest point to the sun). The 'breathing process' affects the nature of the terrestrial Life of the planet concerned. In a similar manner the chest cavity of a person expands and contracts as he/she breathes in and out. Our lives can also be viewed in terms of expansions and contractions as we undertake a course of action that provide a set of experiences (an expansion), then the contractions of those actions when the experiences serve their purposes. We are thus born and die to each new set of experiences, and to our lives as well, as they are also cyclically breathed into and out of existence.
4. The solar system was born, it undertakes many evolutionary stages, and shall eventually die, as also does a person. By such methodology both entities gather the experiences or qualities they need to further their evolutionary progress. (The sun does this by absorbing stellar energies from those cosmic entities with which it has relationships during its interstellar journeying.)
5. Just as:
    a) Every atom has a nucleus (the positive centre) and surrounding electrons (the negative particles) that are not inseparably bound to the centre; similarly every cell has a nucleus and surrounding protoplasm.
    b) Also, as the solar system has a nucleus (the sun) and surrounding planets ('electrons'); so people also respond to a central nucleus.

    In one who is purely sensual, the *organs of reproduction* form the positive nucleus.

    In one who is emotionally polarised, the *solar plexus* area is positive and the remainder of the body is negative.

    In the case of one that is mentally polarised, the *head* is the 'positive nucleus', and the body constitutes the 'negative particles'.

    In the spiritual person, the *heart* is positive and the body is negative.

Every personality gravitates around a central nucleus, such as their home circle, social group, professional or business group, ideological or religious faction. It holds their focus of attention and is the central sun, the source of their emotional, idealistic and/or mental life.

c) Each nucleus or centre of Life for the unit becomes an 'electron' to the nucleus of a greater Life of which it is a part.

6. Each atom or unit of Life is basically spherical or else evolves toward the spherical or ovoid shape. Though the human body can be likened to a pentagram consisting of a head, two arms and two legs, the head, the seat of that principle which distinguishes us from animals is spherical in shape. The start of all Life is a cell, and the ultimate form is spheroid in shape, the *tathāgatagarbha*-Sambhogakāya Flower.

   Most ancient religions had a ideogram of the sun thus: ☉ as the symbol of its supreme Deity.

   There are many levels of interpretation to this symbol, some examples being:

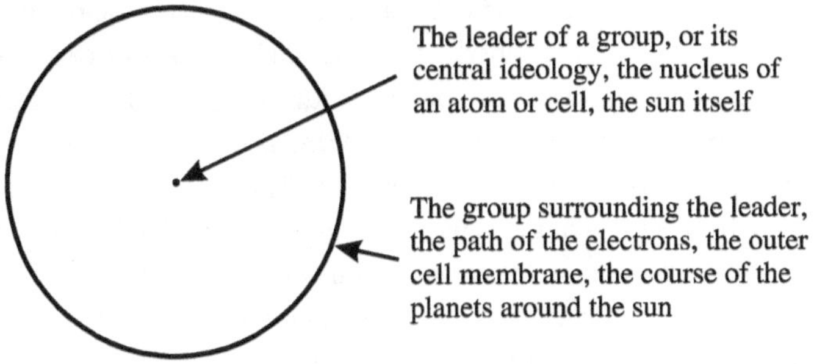

Figure 2: The sun as a symbol

7. Ultimately form is transcended, and formlessness, the energy at the heart of every atom or sun is released (experienced), freed from bondage. The atom becomes radioactive and disintegrates, a sun often becomes a nova, similar to the eventual destiny of the Sambhogakāya Flower.

Presently people largely limit and confine themselves to forms in their thought processes, to the use of words, circumscribed ideas, ideals, and picture symbols (bits of information) that relate to the process of living. Those upon the path to enlightenment transcend the need for such tools to think with, as spontaneous revelation, universality and abstract thoughtless thought becomes the norm. Enlightenment has been likened to a perfectly controlled atomic explosion in the realms of consciousness, though slowly brought to the critical stage. Then the mind is no longer self-centred, but focussed upon the greater whole, the infinite multi-dimensional universe, which is a boundless and therefore a formless sphere. This is a seeming paradox, but nevertheless imparts an idea of the nature of the enlightened Mind.

The entire life process must be viewed in terms of relationships, seen entirely from the angle of the viewer, and his/her relation to the object of perception. Just as the consciousness of a person is ineffable, boundless, as related to that of an animal, so the Awareness of a Logos (or any liberated being) is likewise ineffable, boundless or formless, compared to that of a person. The enlightened Mind relates the one to the other.

A state of consciousness is ineffable to the sentience below it, yet all are bound by form, a sphere of contact and a means of expansion (that is limited by natural laws or self-made activity). All such forms have an objective purpose. The greater Mind is inclusive of the lesser, which can therefore be viewed as part of its constitution.

Though the above parallelisms in Life are generalised, not necessarily true in detail, nevertheless, the same general principles and laws governing the evolution of all forms of Life can be observed. This can be properly comprehended if we acknowledge that every atom or form has at least latent sentience and is evolving to and beyond the human states of consciousness. Humanity has evolved the capacity to reason and think and must progress to stages of evolution where the ability of clear deductive reason becomes instinctive, below the threshold of consciousness.

All that our minds can cognise in the immediate environment can been said to be the result of the unfolding Thought process of a great One in Whom all live and move and have their being. Such a One is incorporated in a succession of greater Ones, until That ONE that moves the entire universe is reached.

The nature of the embrace of the Mind of a Buddha cannot be comprehended by the empirical mind. People can only think of limited infinities, or a series of successive embodiments of lesser units of consciousness. A Logos can be defined as that Life or Mind that embodies and abstracts the collective-consciousness of all lesser unities that are included in that Life's sphere of activity as Its subtle, psychic, and gross constitution.

There are many differing conceptions concerning the question of what constitutes a 'God' or Logos. Some have risen because of theological assertions, many of which are self-contradictory. Other philosophies, such as the Buddhist, negate such a concept. In systems such as the Hindu, there is a seemingly confusing plethora of gods. We also have scientific materialism and the atheistic assertions of thinkers arising from its schools. Because of the widely differing conceptions many people have reacted in a negative way whenever the subject is mentioned. They have refused to even begin to rationally tackle the problem, saying categorically that it is impossible. The question is however not an incommunicable enigma. It can be answered by the enlightened Mind. All enlightened beings testify to this, and by means of the Initiation process veridical Knowledge will eventually become the common heritage of humanity.

It should be here emphasised that when referring to a Logos, or 'All Creating King'[2] it does not imply that there is a personalised anthropomorphic God, such as has been created by the imaginative minds of many Christians, Jews and Muslims. Neither need one look to the many idealised Hindi images of the gods, or of other religions, past or present. No such being exists, except what has been created by the mental-emotions of people out of astral substance. There is however a Creative principle, an embodied Mind, that has caused all that is known to come to be, that sustains it, and which will also call for its eventual dissolution, thence consequent rebirth.

A Creative Mind is needed for us to logically describe the appearance of the phenomenal universe and its laws, without falling into philosophical and logical quagmires of thought, and speculative

---

2 In Buddhism this term refers to Vajrasattva. See the section entitled 'Does a God exist', in my book *Considerations of Mind: A Buddhist Enquiry*, pages 195 – 203.

wish-fulfilling absurdities. One may also rationalise the way of the Advaita Vedānta of a dispassionate non-dual Brahman, that existed 'alone', and which somehow interrelated with the principle of illusion, *māyā,* to form *saguṇa* Brahman, or Brahman with attributes, hence the appearance of Īśvara, the Lord of Creation. This Lord is but an illusory appearance, from whence came Hiraṇyagarbha, 'the golden womb' or world Egg (Prajāpati, the Lord of Beings), from which the subtle bodies of all that is seen has emanated.

The Anthropic Principle

Blind chance, the throw of the dice, is not the mechanism for the appearance of anything in the universe. Everything obeys the careful consideration of Mind in application. If scientists properly considered the fact that intelligence and its precursor, sentience, is inherent throughout Nature, that everything is governed by the laws of mind/Mind in action in a hylozoistic universe, then many of their present mysteries will become self-explanatory. The Anthropic Principle, where all the laws in the universe have contributed 'miraculously' to be productive of human consciousness alone, should be sufficient to convince most of this point. Currently there are four versions of this Principle, summarised by James Gardner thus:

> Systemic analysis of the anthropic cosmological principle has, over time, revealed that the basic concept—that the universe is life friendly—actually encompasses four separate but related subprinciples:
>
> 1. The "weak anthropic principle," which merely asserts in tautological fashion that the universe we inhabit must perforce be life-friendly since it happens to be inhabited by living observers like ourselves
> 2. The "strong anthropic principle," which states that the eventual emergence of life and intelligence in the universe is actually predestined by the laws and constants of inanimate nature
> 3. The counterintuitive "participatory anthropic principle," which hypothesizes, on the strength of the Copenhagen interpretation of quantum mechanics, that observer-participancy is necessary to summon the universe into existence and to give it structure
> 4. The "final anthropic principle," which advances the extraordinary claim that once life has arisen anywhere in this or any other

> universe, its sophistication and pervasiveness will expand inexorably and exponentially until life's domain is coterminous with the boundaries of the cosmos itself[3]

The version of the Anthropic principle that is here advocated is a fifth point, or adaptation of the 'final anthropic principle' with the adjunct that we live in a hylozoistic universe, that the principle of Life, absolute Intelligence, is what commanded the Universe and all its phenomena to manifest in the first place. The purpose is to produce experiences stemming from whatever transformation of base substance (black primal dust) that has been incorporated into the boundaries of the Primordial Logos is possible during that Incarnation. The Incarnation has been designed such a way that the principle of intelligence evolves from out of the elementary substance and works at the refinement of the elements of mind so that a Logoic super-Mind is the eventuation. This must happen in accordance with the exigencies of the law of *karma*, of predisposing causes seeded in a previous existence of that Intelligence.

This presupposes the existence of something that must be converted, an originating substance, or a form of dark matter or dark space, and an Agent that incorporates that substance so that it can be converted. Such dark space then must be considered universal, primordial, pre-existent, rather than coming as a consequence of a Big Bang. An explosive expansion from a seed *(bīja)* there might have been, like the exponential growth of the cells of an egg in a womb, which starts as a unicellular form before fertilisation, and rapidly divides and diversifies, according to a prearranged genetic code until the time for the parturition of the child from the womb. However such expansion originated from the domain of Logoic Mind. The manifestation of the physical domain is an effect of this expansion, which shall be further explained in the later chapters.

Nothing comes from nothing, there is no *creatio ex-nihilo,*[4] neither from a Creator God, or from a physicist's conception of a incomprehensible singularity preceding the Big Bang. What initiated that Big Bang in the first place is left unanswered in the annuals of

---

3 Gardner, James N. *Biocosm,* (New Age International (P) Ltd., New Delhi, 2006), 38-9.

4 'Creation out of nothing', St Augustine's and then Thomas Aquinas's formulation of the Roman Catholic doctrine of how the universe came into being through the Grace of God.

physicists, it simply 'happened' is their dogmatic assertion, and one can only presume the circular argument that 'it happened' because we are here to observe the results.

If one assumes the pre-existence of substance, then there is a form of steady state universe via which reincarnating Entities manifest, and a Universe also follows the same laws concerning the rebirthing process, though on a far vaster scale than a human unit. Significant problems in the data concerning redshifts, from which the Big Bang hypothesis has been extrapolated has been discovered by the work of Halton Arp in the 1960's and the problem with quasars. He documented instances showing the inter connectedness between some low redshift and high redshift galaxies, which demonstrated that their redshift signified something other than recessional velocity. This work is well documented in those that advocate the electric universe hypothesis, which though not yet fully explored, has some merit, because of the electrical nature of the attributes of consciousness. We also have the Wolf Effect. I shall deal with these subjects later.

Because of Theistic paradigms concerning the use of this word 'God', I have largely omitted its use, rather using the more appropriate term Logos, which means embodied Word, which is closer to the Buddhist thought-life. The term implicates the Truth of a Creative Sound, an emanatory vibration *(spanda*[5]*)* that sweeps substance and coheres, agglomerates them into the forms seen in the universe depicted as filamentous strings of galaxies, their super-clusters and all of the stars, etc., contained within them. Emanatory Sound admits the appropriate weaving in the purpose of the Dhyāni Buddhas (or their cosmological correspondences) into the edifice of what must be.

Does a 'God', a 'Soul' or *ātman* exist?

The Theistic religionists have the right idea concerning the existence of what is considered as 'God', but do not have the correct sophisticated reasoning, subtleties of argument to properly explain the qualities of such an entity. Anthropomorphic absurdities have arisen, such as the

5 *Spanda,* vibration, throb, self-movement, creative pulsation, apparent motion within the motionless, yet serves as a cause for all other motions. A central doctrine of Kashmiri Śaivism.

unique 'Son of God', and the elevation of Mary as 'God', *creatio ex-nihilo* and many quandaries, such as whether 'He' created evil or not. We also have concepts of a 'personal God' that can be logically refuted.

Buddhists have the subtlety of the philosophy concerning the background of how such a Being can have arisen, as I have explained in *A Treatise on Mind.* However they have been too fearful in their ontological system to appropriately elaborate upon the nature of a Buddha's continued evolution in cosmos *(parinirvāṇa),* because they deemed it necessary to sustain the thesis that there is no 'God'. Neither did they deem it necessary to properly extend the doctrine of the five Dhyāni Buddhas or of the Ādi Buddha to cosmological principles. They consequently developed no ontological, teleological or cosmological argument for the existence of such a Being. Hence let us now weave a true Madhyamaka philosophy, a true 'middle way', between the two extreme opinions and hence comprehend what the Buddha's intention was when he refused to speculate as to whether there is a 'God' or not.

The Buddha rightly denies the existence of 'a creator-god', the nature of which has been anthropomorphised in the Hindu setting, but he was also rightfully silent as to whether a 'God' exists, as Narada points out:

> On several occasions the Buddha denied the existence of a permanent soul (attā). As to the denial of a creator-god, there are only a few references. The Buddha never admitted the existence of a creator whether in the form of a force or a being.
>
> Despite the fact that the Buddha placed no supernatural god over man, some scholars assert that the Buddha was characteristically silent on this important controversial question.[6]

Many esoteric teachings had to be veiled from the Buddha's time till the present. Only now have the foundational concepts and developed language of the esoteric doctrine been able to be presented that will allow comprehension of what was formerly not possible to explain. The development of empirical science in the field of cosmology and of comprehension of the nature of energy is one major reason why this

6 Mahathera Narada, *The Buddha and His Teachings, Gradual Sayings,* I, (Buddhist Publications Society, Kandy, Sri Lanka. 1988), 229. See also the teaching concerning the Buddha's silence over the existence of a Soul (chapter 5, 'The Buddha and the Soul Concept') in volume 3 of *A Treatise on Mind.*

is now possible. The development of a sophisticated thought life by a large number of people is another.

The problem with the asserting of the existence of a (personalised) Creator God, seen for instance as an Īśvara or Brahmā, is mainly that of correct interpretation of the associated symbolism and anthropomorphism. The Hindu Deities do have symbolic truth concerning certain forces in Nature related to cosmological and creative potencies. Thus they cannot be arbitrarily denied, even when anthropomorphised, as long as one does not accept the existence of such Deities as literal truth. We also have the myths which have a 'God' interfering in the activities of humans. If a 'God' had 'favourites', then there could be no true justice in the universe, and the law of *karma* could not function as it does. Clearly also no liberation can come from mere worship and devotional ritual.

The Buddhists objected to the existence of something as 'permanent' (i.e., a 'self' or a 'God'). Logically the concept of such a 'God' is but an extension of the nature of a concept of a 'self', an *ātman,*[7] because a Creator God is really but a version of an elevated or developed *ātman* that includes humanity as its mode of personal domain, which it can influence accordingly. The *ātman* concept is fraught with problems, which the Buddhists early on managed to elaborate and give the counter philosophy of *nairātmya,* (non-self).

The Advaita Vedānta concept of the *ātman* is thus:

> The ātman is the innermost essence of a man back of the bodily sheaths known as the *annamaya, prānamaya, manomaya, vijñānamaya* and *ānandamaya kośas,* being respectively the material, the vital, the mental, the cognitive and the blissful sheaths. It is not to be identified with the sense organs...The ātman is not the breath of the individual or his mind or his intellect or his inner senses, his *prāṇa, manas, buddhi* and his *antahkaraṇa.* It has no size or sex or caste, or status. It neither acts nor enjoys. All these, size, sex etc. and action and enjoyment pertain only to the body and manas. The ātman is ever pure, untainted by any blemish. In fact, no quantity can be attributed to it. It is mere

7 *Ātman,* (T. 'dag nyig) self-identity, the innermost essence of a person as well as the universal 'Self'. The probable roots, *an* = to breathe, *at* = to go or eternal movement, *ah* = to pervade and connected with *aham* = 'I'; *avātman* from roots *av* and *vā* to satisfy one's self.

> Intelligence or *kevala cit*. It is the witness or *sākśī* of the three states of *jāgrat, swapna* and *suṣupti*.[8] It is that by which the eye sees, the ear hears, the mind thinks, the understanding and the ego function. Being incorporeal, it is not to be identified with anything physical. It remains by itself in its lone luminous character of intelligence...
>
> This ātman, understood as pure intelligence or *kevala cit,* is the same in all individuals or jīvas. The difference between one jīva and another is only in respect of the body and the bodily features encasing the ātman, in which the ātman indwells. The ātman by its nature is unlimited, (vibhu) but it appears as if limited within the confines of the body. These physical limitations are known as *upādhis* constricted into which the unlimited ātman is identified with those limitations. In fact, it is not the actual limitation by the upādhis so much as the wrong view of the ātman as thus limited. It is the ajñāna[9] of the individual that is responsible for the seeming individuation of the ātman within the upādhis, and that leads to the impression of the distinction of one person from another. In metaphysical fact, no person is different from any other *in his character as the ātman.* The difference pertain to the upādhis or limiting adjuncts encasing the ātman...If one is asked: When did this constriction of the unlimited ātman into the limitations of the upādhis begin, no answer can be given...Not only should one realise the ātman as different and apart from the bodily upādhis; one should also realise the individuated ātman in each jīva is the same as the Supreme Ātman or Paramātman or Brahman...the individualised ātman in several jīvas is of the same essence as the Supreme Ātman or Brahman. Both are of the nature of *Sat, Cit* and *Ānanda.* The jīvātman is only an abridged edition, as it were, of the Paramātman. Both are of the nature of pure Intelligence, which makes all cognition through the physical senses possible.[10]

The concept of *ātman* has often been confused with that of a Soul in the minds of many thinkers. Though he had much to say about the non-existence of a personal *ātman*, the Buddha however never denied the existence of a Soul. Later Buddhist metaphysicians however, took

8 Waking consciousness, dreaming or trance state, and dreamless sleep.

9 Ignorance, nescience.

10 P. Sankaranarayanan, *What is Advaita?* (Bharatiya Vidya Bhavan, Mumbai, 1999), 61-64.

the teaching of *nairātmya* so much to heart that they found repugnant a concept of a soul as the expression of divinity within the human persona that was capable of being responsible for the rebirthing process.

Edward Conze aptly summarised the Buddhist and Hindu ideas concerning this:

> we cannot be quite sure what notions of an ***ātman*** were envisaged by the early Buddhists when they so emphatically denied it. I personally believe that these notions were of two kinds, i.e. (1) the ideas implied in the use of 'I' and 'mine' by ordinary people, and (2) the philosophic opinion, held by the **Sāṃkhya** and **Vaiśeṣika**, that a continuing substratum acts as an agent which outlasts the different actions of a person, abides for one or more existences, "and acts as a 'support' to the activities of the **individual**. It is, however, doubtful and a mater of much dispute among experts, whether the Upanishadic doctrine of the ***ātman*** had any influence upon early Buddhism." 'What in general is suggested by Soul, Self, Ego, or to use the Sanskrit expression ***Ātman***, is that in man there is a permanent, everlasting and absolute entity, which is the unchanging substance behind the changing **phenomenal world**.'....The Buddha never taught that the self 'is not', but only that 'it cannot be apprehended'.[11]

W.Y. Evans-Wentz presents a similar viewpoint:

> Expositions of the Buddhist doctrine of non-self, or non-soul, frequently exhibit looseness of thinking and misleading argumentation, sometimes by Buddhists themselves. The Buddha did not teach that there is no self, or soul; He taught that there is no self, or soul, that is real, non-transitory, or possessed of unique and eternally separate existence. In Buddhism, salvation is not of a self, or soul; it is entirely dependent upon what the Buddha declared to be the deliverance of the mind from the *sangsāric* bondage imposed by Ignorance (Skt. *Avidyā*), from the erroneous belief that appearances are real and that there are individualized immortal selves, or souls.
>
> When there is no longer a clinging to selfhood, when all the external play of *sangsāric* energies is allowed to subside, because there is no longer attachment to any of them, then there is that state

11 Edward Conze, *Buddhist Thought in India,* (George Allen & Unwin, London, 1983), 38-39.

> of absolute quiescence of mental activities which our text refers to as the natural state of the mind.[12]

Of the fourteen questions *(avyākṛta-vastūni)* that the Buddha decided to not answer for various reasons, the last two are: 'Is the self *(jīva)* identical or different from the body?'[13]

I have reminded my readers about this ancient debate because I have aptly demonstrated the existence of a Soul-form in volume three of *A Treatise on Mind,* describing it in terms of the Sambhogakāya Flower (*tathāgatagarbha*). The Sambhogakāya Flower functions for an individual human unit, though it's full potential may yet be in the process of unfolding. The personality that it embodies is its thought form, the phenomenal appearance, that has been made manifest for a set purpose. The Sambhogakāya Flower unfolds its qualities (and thus evolves, contrary to the concept of an *ātman*) by means of a series of successive embodiments of personalities, but as all forms, it must at some time reach the bounds of its possible attainment, and thus die. This information was detailed in the above book but what needs to be elaborated here is that we can similarly posit a more universal over-Soul existing in *dharmakāya,* embodying the sum of all human Flowers, and the attributes of the categories of Nature. Such a concept has been called the Dharmakāya Flower in volume three, chapter 6 of *A Treatise on Mind,* where some of its properties have been described.[14] Such a Flower, when personalised in terms of the united functioning of the five Dhyāni Buddhas and the constitution of Shambhala, can be equated with the term Logos in my works.

Regarding the question of 'God', H.V. Guenther states that the term:

> may be used in the deistic sense, deism being the doctrine that "there is a certain part of the Universe which is not existentially dependent

---

12 W.Y. Evans-Wentz, *The Tibetan Book of the Great Liberation,* (Oxford University Press, London, 1971), 76-77.

13 See Alex Wayman, *Untying the Knots in Buddhism,* (Motilal Banarsidass, Delhi, 1997) 183-84.

14 The detail concerning the nature of this Soul-form presented in my book indicates why the Buddha was so reluctant to delve into the subject. It was not the right time for its explication.

> upon anything else, that all the rest of the Universe is existentially dependent upon this part of it". Another use is to apply the term 'God' to the whole Universe as having certain characteristics from which all others necessarily follow. This is a kind of pantheism. Lastly the word 'God' is used to denote those features of the Universe which actually belong to it and are not mere distortions or illusory appearances. On this view the Universe is in reality purely mental matter, space and motion are distorted appearances of this mind. There is nothing to show that Buddhism falls in with any of these three views. It eschews a First Cause as well as the mentalistic premise that the Universe in its totality is a mind or society of minds. And it also rejects the thesis that the Universe in all its aspects is God. Therefore, before we speak of God or Gods in Buddhism and jump to the conclusion that it is something polytheistic or that a theistic element has been introduced into it at a later phase of its development we had better find out what the label 'God' *(lha, deva)* means in Buddhism.[15]

Another important thing to realise concerning this question is that its answer necessitates comprehending the nature of energy, of relativity and of meditation. Everything is in a state of flux and can be seen as quanta of energy in dynamic motion and interrelation, in which nothing is permanent or remains unchanged in time and space. The idea exemplified in the Buddhist doctrine is that no matter how the idea of 'God' is presented, it suggests permanency, and thus something that is static or lifeless. In his consideration of the Buddhist 'pantheon', Guenther further states:

> Another objection that is likely to be raised is that all these gods are doing temporal acts. If it is really true, as theologians claim, that the Divine is non-temporal, how can we ascribe temporal processes and qualities to that which is non-temporal without becoming involved in endless contradictions? This difficulty does not exist for a Buddhist, as he does not think in terms of 'things' and their 'qualities', but in terms of dynamic processes which, by virtue of their dynamics and variability, are vivid and therefore 'divine'. The gods are functions and their formulations in concrete forms are symbols for the inner experiences that attend man's spiritual growth.[16]

---

15 H.V. Guenther, *Treasures on the Tibetan Middle Way,* (Shambhala, Berkeley, 1976), 24.

16 Ibid,. 32.

The idea of looking at Being in terms of 'dynamic processes' is a valid viewpoint if we focus our attention upon the substance behind the form, to the energy fields that cause the appearance and fleeting existence of all manifest things in the phenomenal universe. It dispenses with the notion of 'God' in that such a concept is superfluous, a thing, which in terms of its essential nature and interrelatedness is non-existent. Such things nevertheless *do exist,* even if only illusively and temporarily so, for this is what is presented to our cognitive faculties in the world around us. Things also exist as ideas when limited to the noumenous level where ideas do have a lasting permanence, for their purpose and concepts can be carried through from life to life, such as for instance, the concept of *bodhicitta. Bodhicitta* may be an interesting idea at first, but later becomes a driving force, driving the being onwards through to liberation. How much more real do ideations become when pertaining to the realms of the *dharmakāya* and beyond?

Within *dharmakāya* one can conceive of the existence of Deities that are but the embodiment of the driving force of *bodhicitta.* They have ridden its ever-progressively expansive energy to the far reaches of the universe and have gained an expanded Buddha-Mind that has further evolved in cosmos. The attributes developed by such a Being is so far removed from what the general population of humanity has developed, or even the enlightened ones on earth, the Council of Bodhisattvas, that They manifest as ineffable Personalities. Such Personalities will possess transcendent, virtually omnipresent ever-lasting qualities, when related to the phenomena manifesting upon our world. They might possess illusory, 'Dynamic Variability' on their own level of expression, but such Variability is well nigh incomprehensible by all but the most enlightened on earth.

In the *Devadaha Sutta* the Buddha says:

> So, then, owing to the creation of a supreme deity men will become murderers, thieves, unchaste, liars, slanderers, abusive babblers, covetous, malicious, and perverse in view. Thus for those who fall back on the creation of a god as the essential reason, there is neither desire nor effort nor necessity to do this deed or abstain from that deed. (*Majjhima Nikāya*, II.)[17]

---

17 Narada, 229.

Here we can see that the Buddha was concerned with the concept of a personal 'God', one that would absolve people from the consequence of their actions. Such actions imply concepts of attachment and direct concern to humans existing within particular groupings such as being part of a culture or religion, etc. Group concepts inherited from past cycles of activity of the teachers, the sages they ascribe to, have their limitations and must later be expanded upon. All limitations that warp the naturalness of the progress of the entire evolutionary sequence must later be transcended, which necessitates meditative concentration via absolute impersonality, via which maximum salvifical wisdom can be gleaned. A Buddha-Mind is the expression of such qualities.

The real question is, 'what does such a Mind do after the *parinirvāṇa* of a Buddha?', which is a question the Buddha himself refused to answer, but which can now be answered, here, and in my earlier books wherein the subject of the Initiation process and the group Laws are discussed (volumes 6 and 7 of *A Treatise on Mind).* This teaching should be applied here.

Having discussed the evolution of such an entity as a 'God' one can then ask how then can such a 'God' manifest in relation to the causative process? First we must define what exactly is meant by a 'God' or Deity, and the properties that can be expected of such an entity. The concept consists in the first place in understanding what mind/Mind is and how it functions. The Mind is what embodies or en-Souls the essential Life in such a way that it can relate to the material world. It en-Souls the activities of a person because it is the Eye of the Divine, and is the mediator or mirror between the abstract and the phenomenal. It embodies the essence of the past and is the seed of future Divinity.

In terms of consciousness, and limiting our vision to humanity's tiny scale, then first we see the divinity that relates to the Sambhogakāya Flower. This Flower incorporates the evolving personality in an ideal form as a continuum in time and space throughout the births and deaths of successive generations of such entities. It is an embodied flux of conscious receptivity incorporated into a form that can relate the transcendental to the corporeal, wherein the ocean of being/non-being and the unit of consciousness (that is the incarnate person) can interrelate without the destruction of the latter and the abnegation of the purposes of existence. (A purpose that emanates from the *tathāgatagarbha,* which is a true cellular unit in the ocean of being/non-being.) The *tathāgatagarbha*-Soul is essentially a radiant Sun that resides in the heart of all manifest Being, a 'divine body of perfect endowment'. It manifests as the *sambhogakāya* form (subtle body) of a Buddha (to be). Here the Buddha nature inherent in everyone is referred to. The evolution of the *tathāgatagarbha* is humanity's destined fulfilment.

Such a Soul-form has a seeming permanence in relation to the life span of a person. In its simplest connotation *the Soul* can be defined as that which is the mediator between the undefinable, the archetypal or dynamically omnipresent all-encompassing aspect, and the phenomenal form. It is the Mind nature relegated entirely to the numinous level as a self-contained sphere of activity.

The Buddhist idea of the three bodies of a Buddha *(trikāya)* find their application here. That which I have termed *the Divine* can be considered to be the *dharmakāya,*[18] the body or vehicle of the *dharma*. It is the ultimate nature, body of Truth, the primordial, eternally self-existing essentiality of *bodhi* (enlightenment), the highest of the three-fold bodies (*trikāya*) of a Buddha, or of any Initiate of the fifth degree or greater. *Dharma* is the fount of the (spiritual) Law, and *kāya* is its vehicle. The *dharmakāya* as a body of expression will later be explained in terms of being the Monad, which is the true cosmic traveller, the awakened Buddha within.

The corporeal form, when perfectly expressed and embodied (by an enlightened being), becomes a *nirmāṇakāya*, the Incarnation body or appearance of the Buddha (or a great Bodhisattva's) form. The second of the three vestures *(trikāya)* of such a One is the *sambhogakāya,* the Bliss body, that of sublime vision, the ecstatic transformation body found in the heavenly world (higher mental plane). This is the form of the great ones depicted in Buddhist art. It is the Sambhogakāya Flower when related to a normal human unit that has not yet gained liberation from the trammels of *saṃsāra*.

The reincarnating principle of a Logos is similar to that of a human unit, wherein in the case of Logoic activity the Sambhogakāya Flower is of vast scale. Stored in its petals and in the constitution of the permanent atoms are the *bīja* forms (*saṃskāras*) of the forms of Life (the Lives) that are to incarnate throughout the *mahāmanvantara*. The nature of the entities stored depend upon the level of expression of the Logos one is considering.

18 *Dharmakāya,* from *dharma* law, continuance, from the verbal root *dhṛ* to support, carry, continue + *kāya,* body.

By this is meant whether one is envisioning a planetary Logos, solar Logos, that of a constellation, a That Logos, etc. The appearance *(nirmāṇakāya)* is what the scientific community is actually trying to investigate, being completely oblivious of the subjective processes whereby a *nirmāṇakāya* manifests. Consequently much more needs to be understood concerning the process of the incarnation of an entity.

The consideration of mind/Mind

As previously stated, the mind is the bridge between the involutionary (sub-human) and evolutionary (para-human) states of awareness, and is therefore a combination or product of the characteristics of *instinct, feeling, desire, imagination, intellect, pure reason (intuition),* and *the Ineffable or Universal Mind,* which were explained in *A Treatise on Mind.*[19]

The key to the revelation of the nature of the evolutionary process and the 'creation' of the universe is found in the laws governing thought-form creation in the mind/Mind. The fleeting, phenomenal appearance of things, the entire physical world that people live in and are involved with, can be likened to the images produced in the imaginative Mind of a great presiding Logos. The 'cells' in the brain of that Being can be envisioned as being great angelic beings *(devas).* They are the feminine principle (*śaktis*[20]) acting as Consorts to the Buddhas and fashion the images out of the substance of their own forms. Lesser units of consciousness within the bodies of such Beings embody the various diversified aspects of the material world. All substance is consequently feminine, including that which is termed *manasic* (mental), whilst the principle of *bodhicitta* imbued into the Sambhogakāya Flowers makes humanity masculine, inherently loving. Thus we have the establishment of the polarities in Nature.

The purpose upon the path of liberation via Tantric ritual and meditation is to resolve the dualities by integrating them into unity. Such activity happens upon a vast scale via a Logoic Mind as it works to transform the appearing *nirmāṇakāya.* Streams of Lives (the categories of the kingdoms of Nature) are sent into expression at their

19 See volume 6, pages 10-13.

20 From *śak*, 'being able to be done'; feminine power manifesting in the form of *siddhis*, where the potency of a Buddha or Bodhisattva can come to manifest phenomenally.

appropriate cycles *(yugas)* to play their appointed role according to the vicissitudes of *karma* and the Logoic Purpose for the Lives informing the *nirmāṇakāya*. The evolution of the species, and the marvellous biodiversity then happens as planned, according to the wills of the lesser enlightened Minds that are aspects of the greater Mind. All is the effect of planned activity, by Logoic design, and so the species of Life evolve. Nothing is left to chance, all is designed to manifest according to Plan over the millions of years of evolutionary time.

The entire universe is created out of the Thought-Substance of such a Logoic One, an Ādi Buddha, to fully express a yet unresolved possibility, which is the objective of such a One's 'Desire'.[21] The use of such terminology as 'Desire' is problematic to an Entity that is as inexplicable to the present human mind as humanity's developed consciousness would be an amoeba. A better term would be Purpose, and better still, Directive Ideation, however the Logoic quality expressed transmutatively equates with what humans consider as 'desire'. In considering such terms the difficulties in talking about the process in a non-anthropomorphic way come to the fore. There are correspondences to the human qualities that can be utilised, but these concepts have to be properly transmuted in the thought process. The understanding is that all Logoic attributes have *śūnyatā* as their base. *Śūnyatā* acts as a mechanism of transmission, rather than being a mere enigmatical 'Empty' zone. It is simply empty of the vicissitudes of human activities and mind.

The nature of a Logos will always remain inexplicable unless there is comprehension that a human unit is built in the image of the Divine. Analogy is the major tool towards understanding. 'As above so below' the ancient adage reads, and so it is. Though this tool is an invaluable aid for interpretation, one must take great care in not deriving too literal a meaning in one's deductions. The concepts must be transmuted by utilising the Clear Light of the meditation-Mind's analysis and the revelations accessed without distortions by the empirical mind. This is necessary, for our objective (as for all of human endeavour) is to *know* all things by developing a Buddha-Mind. Therein such knowledge provides comprehension of what 'God' is and/or is not, which will make

---

21 Which collects and collates incorporeal substance from subjective domains that is then externalised into the appearing phenomena.

us 'sons' of or 'joint-heirs' thereof.[22] The nature of such revelation can then be disseminated, being part of the entire enlightenment-process, a reverie in the *dharmakāya.* Though such concepts are a necessary theme of this book it is possible to only give general, broad analogies, the thought form or seed from which direct awareness may develop.

The creative process can be likened to that of meditation, in which the conception of a thought takes place and then becomes a definite form as the meditator imbues it with the vitality of the Fiery lives (atoms) of the mind/Mind. All aspects of being/non-being can then be summoned and incorporated into the form to be.

Because all who meditate are limited by the qualities of the substance (properties of their mental constitution) that they must work with in their meditative endeavours, therefore it can be inferred that the Logos is similarly limited at the exalted level upon which such a One works. All substance (human, *deva,* or atomic) can be considered 'tainted', for it represents calling forth the *saṃskāras,* the residue of past action, on a vast planetary, solar, or cosmic scale. It has its inherent karmic imprint, and can be carried into further (future) manifestation that involves a panoply of entities, once the necessary *bījas* are activated. A Buddha's compassion carries with him the evolutionary impetus of an entire milieu of evolving and informed lives. If this is clearly analysed, then one will comprehend that this is the natural result of a Bodhisattva's vow to 'never cease striving until all sentient beings have been released from suffering'. A Buddha never relinquishes his Bodhisattva vows, he just carries them on to new vaster spaces than ever before. Consequently Buddhists need to think deeper into all of the implications of the nature of a Bodhisattva and his vow, because they have not yet done more than scratch the surface of what it is that truly constitutes a Buddha-Mind into which a Bodhisattva evolves.

The Creative process

The universe was not created out of nothing, unless that 'nothing' be energy that is ever-present and omnipotent, and which from another angle of expression, is substance. (For energy and matter are interchangeable.)

22 *Romans 8:16-17.* 'The Spirit itself beareth witness with our spirit, that we are the children of God: And if children, then we are heirs, heirs of God, and joint-heirs with Christ; if so be that we suffer with him, that we be glorified together'.

Because all of the entities on our earth-system are evolving, so they are imperfect, and if so, then the energy, or substance, that the Logos had to use must also have been imperfect, that is, limited, for cause ever precedes effect, the future is the effect of the past as it unfolds through the present. The past is imperfect, whilst the present imperfection is evolving to a future perfection. A Logos (or a 'God') can thus be said to be evolving to what represents a point of perfection. This is a logical deduction, utilising the principle of analogy, and is true if the concept of 'man' being built in the image of Deity is valid.

The objective of evolution by means of the evolved principle of *bodhicitta* is to allow Deity to fully embody the lowest as well as the highest realm of perception (types of substance). The objective is to 'lift' or transmute that substance by infusing it with the highest spiritual energies. The three worlds of human livingness represent the unregenerate *saṃskāras* yet to be mastered by the Logos. It is the lower nature to be controlled and fully infused with *bodhicitta*. The evolutionary purpose will then be completed, the Bodhisattva vow serving its purpose.

In summary it can be said that in deep Meditation the Logos formulated a sphere, a limit of possible attainment for that cycle and imbued it with the necessary Life, the rightly activated *bījas* and *saṃskāras* for the purpose of the manifestation of that sphere of activity. The Meditation produces a sphere of limitation, considered an Incarnation. Within that limitation the roles of the entire action-reaction effects of energy interplay that compose all lives in a world-sphere are played out. This proceeds towards the perfected fulfilment that already exists in That Mind, concurrent with the Meditation process, but which is not yet fully manifest in form.

From this we can gather that by learning the arts of meditation then one in reality is gaining the credentials that will in the far distant future (at the attainment of the ninth Initiation[23]) allow one to manifest Logoic mentation for an entire world-sphere. The art of precipitation of thought forms from the mental plane to effect physical plane phenomena by the enlightened has yet to be demonstrated to the scientific community. Once comprehended then that community will better understand the nature of the appearance of all phenomena in our universe. An objective of the practice of meditation therefore is to provide the necessary qualities and needed training that will eventually allow one to become

---

23 Explained in *The Constitution of Shambhala*, part B.

a Creative Deity concerned with the evolution of an immense number of entities. Such entities are the higher correspondences of a person's corporeality, his/her body nature and thought-engendering equipment.

The path of becoming a Logos is via that of the Bodhisattva, who creatively manifests thought sequences and activities in such a way that all sentient beings can be helped to gain enlightenment (evolutionary perfection). If this is true for a Bodhisattva, how much more so for a Buddha that has developed *bodhicitta* far further than any Bodhisattva? Compassionate activity *(bodhicitta)* is ingrained in all Logoic activity, being but the continuation of the force that is the drive to liberation gained whilst practicing Bodhisattva virtues upon an earth sphere. *Bodhicitta* consequently is the driving force behind the manifestation of all Logoic spheres. Consequently the primal substance originally incorporated within the originating Logoic ring-pass-not is driven towards evolutionary perfection via the stage of developing intelligence by a human kingdom that has evolved out of that substance.

On the abstract level of Mind the three times (past, present and the future) are an expression of present timelessness encompassed by the duration of the existence that is the Mind of a Logos. All is played out in the realms of Mind. Even astronomers have likened our expanding universe to the effects of an enormous thought process.

The Spirit, or Monad, that manifests in the form of *tathatā* (suchness, thusness, thatness, the fundamental Buddha nature, the 'God' within each entity) can be considered unchanging, unformed, unbounded, eternal in duration. (The *tathatā* is considered so in relation to the fleeting life-span of a human unit.) The *tathatā* is the heart of the Sambhogakāya Flower and becomes the mechanism of containment of the *dharma* in the form of the *tathāgatagarbha*. The existence of the Monad[24] is gauged on a cosmic time scale (on the patterning of aeonic universal cyclic duration) in contradistinction to that associated with our planetary life. It can also however be considered 'That which must evolve', for it exists and thus has a purpose for that Existence. When viewed in terms of words many of its properties become abstractions. Its energetic potency would produce the annihilation of the mind nature that would try to experience it, hence is contained within the Sambhogakāya Flower, whose form is destroyed as one moves to experience the Thatness in consciousness. This necessitates the

24 Its properties will be further examined later.

refinement of consciousness to such an extent so as to evolve it into Mind, allowing it to withstand the Monadic potency. In the process the attributes of mind are completely transformed into what is phrased as 'Clear Light'.

Buddhism	Equivalent Concepts	Theistic Concept
Dharmakāya The Real	The Divine/Monad Spirit	The Abstracted Deity (The Father)
	Expression = Will or Power	
Sambhogakāya Subtle body of a great One	Consciousness Soul	The Meditating Deity The Christ (The Son)
	Expression = Love-Wisdom	
Nirmāṇakāya The appearance	Phenomenal Appearance Personality	The Divine Mother The world of forms
	Expression = Activity	

Table 1. The triune principles

This process is similar to the mechanisms that physicists need to build in order to contain plasma fields and produce nuclear transformations. They construct cyclotrons and nuclear reactors to contain the intense energies generated, whilst the human mind must develop similar safeguards to safely handle the potencies from cosmos. This is done via passing the testings upon the Initiation path and the transformation, etherealisation, of the substance constituting the physical form. The entire bodily nature is transformed via the increasingly refined *prāṇas* flowing via the *nāḍī* system, wherein the *chakras* become the organs of transformation, producing the magnetic fields that can contain the energies received. First the lower *siddhis* are developed and later the supramundane ones.[25] The way of awakening *siddhis* manifests as the antithesis of corporeal form of physicality, even that of the mind, as we understand the term. The *nirmāṇakāya* produced becomes but

25 This process is explained in volumes 5A and B of *A Treatise on Mind, An Esoteric Exposition of the Bardo Thödol.*

an extension of Mind and can appear and 'disappear' at will. Such is the process of the arc of return to the higher domains from whence phenomena originally appeared. All is governed by law, but this law is that of the forces pertaining to Mind.

It is possible to postulate therefore that the essential embodied form of Deity which incorporates the sum total of the *maṇḍala* of any manifest being is also in a state of dynamic unfoldment, though on a scale or state of perception that is far beyond anything and of vaster duration that can be cognized by the intelligent person. Basically, Deity can be defined as a great liberated enlightened Being that has creatively built a Body of Activity that incorporates the collective-consciousness of all kingdoms of Nature (including the human and *deva* kingdoms), giving them a purpose for existence and coherent unity. (These kingdoms must yet evolve to the liberated 'God' state and for this purpose have been appropriated into the body of expression of Deity.) Such a 'God' is thus the originating Cause, the embodying Thought, and That which is indicative of the future of all being/non-being. It can also be considered to be a 'causeless' cause in the sense that everything is continually spiralling onwards and upwards together. Everything in the universe acts as a cause to everything else, there is no true beginning or ending because everything acts in relation to something else, though there are transient beginnings and endings of smaller cycles of attainment within the greater whole.

Deity is thus an evolved Buddha that has long ago passed the need for corporeal evolution, yet sustains a world sphere or universe and the related evolving entities for a specific purpose. This purpose originates in that Buddha's Compassionate Understanding or Meditative Awareness of need, on a cosmic, an inter-planetary or inter-solar scale. It should therefore be understood that the terms Deity, 'God', or Logos here refer normally to That which can be viewed in relation to humanity and its evolution in this world, (unless otherwise stated). It is thus that great Being that incorporates humanity as part of 'His' body of manifestation.[26]

A world sphere is made manifest by that quality termed the *Will,* is sustained by the expression of *Love-Wisdom,* and incorporates what is

26 This then elaborates the concept of the world sphere of a Buddha, as symbolically described in such texts as The *Saddharma-Pundarika, The Lotus of the True Law.*

fundamentally *Activity* in its dynamic interrelatedness. These are the three aspects of Deity that have been anthropomorphised above as the *Father, Son* and *Mother,* the Trinity as One, symbolised thus:

The corporeal personality of humans sums up the qualities plus the evolutionary purpose of the lesser kingdoms of Nature (including the atomic lives composing our forms). In turn the Logos, the Word made corporeal for our planetary sphere, embodies the sum of the qualities of both human and *deva* evolutions. These evolutions thus represent bodily functions or aspects to such a Being. They demonstrate the sum of that Entity's *nirmāṇakāya,* the manifest Personality. The corporeality of such a Personality is what can be known through the use of analogy and the deductive reasoning obtained by means of the mind. One must however transcend the limitations of mind to comprehend what lies beyond. Meditation leads to the portals of the *dharmakāya* wherein revelation is accorded as to the nature of the organisational structure and constitution of the Mind of the incorporating Deity, the Logos. The pathways thereto necessitates being aware of the emanatory mantras, the causative Words of the Logos, which one rides upwards along the line of descent to the primal Source. In doing so one learns the Secret Mantra *(mantrika śakti*[27]*)* that liberates one from *saṃsāra,* thereby manifesting in the guise of a creative Logos for the sum of the embodied form. Thus are *mahāsiddhas* born.

There is an attribute of Deity that corresponds to the archetypal, the unmanifest, the Soul of a person. This cannot be known or understood by those involved as personalities as part of the constitution of the divine Personality. To this the ancient Hindu philosophers have given the name SAT, TAT, THAT, the Immutable, the Unknown, and refused to explain 'THAT' because of the impossibility of the finite mind to comprehend.

27 The *śakti* or psychic power of these mantras, and their numerological ordering and arrangement of the sound patterns. This is done according to the esoteric knowledge, directed will, and psychic purity of the person or group sending forth the Sound. Mantras empower the formative, vibrational, creative power of sound. The vibration of every sound has its own numerical keynote, and this numerological ordering and arrangement of the sound patterns are manifested to order and alter formed space.

The Buddhist philosophy is effectively engaged in the analysis of the qualities relating to, or which will be productive of THAT, thus they emphasise the essential non-reality of Deity. Theistic religions on the other hand, do not concern themselves with the qualities of THAT, but with that which it incorporates as a means of expression, and which has been termed 'God', or the 'Gods' and their powers.

From this idea arises the concept of the 'personal God' of the various theistic religions. As a consequence there is a validity to such a concept, but not to the point that such an entity interferes with personal *karma*, or to it being concerned with individual suffering. Rather, an overview is seen of the way humanity is evolving, and by working through mediators, Bodhisattvas, the Initiates of various degrees, that the salvation of all can be attained. This is possible via those that are the dynamically active lovers of all, as *bodhicitta* then is the force that allows them to travel the way of either the creative or liberating mantra.

Manifesting in the guise of a deity a person must be considered in terms of a fusion of three aspects: Father-Son-Mother, Spirit-Soul-personality. A person manifests therefore in the form of the triune bodies of a Buddha, though yet to be fully expressed as such without defilements. Each human Soul can be considered a cell in the body of Deity. The fusion of a large number of such integrated cells embodied by a similar Ray quality expresses a common embodied quality that constitutes an organ in the body of Deity. Many diverse organs constitute the sum total of the internal equipment and emanatory characteristics of the divine Personality, as far as the conscious awareness of such a 'corporeal' body is concerned. As the human personality is a composite of dense form, emotional and mental body, all of which are emanations of the Sambhogakāya Flower existing upon the abstracted levels of Mind, so also the Personality of a Logos would have a similar composition. From such a Soul-form all manifest Life proceeds.

Once this basic paradigm is comprehended then we can apply the logic to ever vaster, more inclusive Logoi, where the 'cells' constituting such Ones are Logoi, such as the one that embodies our earth sphere, and so forth, until the Logos embodying our universe is revealed. As we travel to increasingly vaster Mind-Spaces of the greater transcendent Logoi, so we are travelling back in evolutionary time, towards darker, and more sluggish, though vaster Mind-Spaces. As we travel to the origination of things cosmically from this perspective we travel towards darkness, hence to levels of comparative Ignorance on a Logoic scale. (The more such a One needs to KNOW.) The level of the Light of the cosmic bearer of Mind becomes dimmer as the vastness of the enclosed space increases. However, because of the vastness of the Space encompassed, the specific gravity, or 'weight' (force) that can be applied through the resonated *mantra* becomes exponentially greater. Such resonance effects the domain of physical space to form the strings of galaxies observed by astronomers. Upon the physical domain, because of the force applied, the effects of the mantras produce the observed brightness of the galaxies.

Further considerations of Mind

What is corporeal to us is effectively non-existent to Deity, being below the threshold of awareness for such a One. It is a part of His/Her internal equipment, but not that which 'His' Mind normally reaches. (Similar to the constituency of the organs in our bodies.) The essential Divinity in us manifests as part of the Conscious Body of expression of the Logos, thus it is the *dharmakāya*. Yet to humanity such characteristics are unknowable, unimaginable, until people have consciously evolved the enlightened awarenesses that are the characteristics of the Sense-perceptions of the Logos. We hence have:

	Aspects of Humanity	**Aspects of Deity**
1	*The Spirit/Monad* The Transcendent	An aspect of the Brain structure Mind of Deity
	This is above the threshold of human consciousness	
2	*The Soul* The abstract Mind	Cellular structure of Deity Sense-perception
3	*The Personality* The empirical mind	Below the threshold of Awareness Dense physical substance

Table 2. The triune aspect of Deity

The abstract Mind is the common denominator that allows an interrelation between humanity and Deity.

All energies and factors of the Bodily expression of a Logos are in a constant state of mutable Activity, in a similar way that the qualities of the personal-I are governed by thought and desire processes. When speaking in terms of manifest Life one is concerned with the resultant expression of the Logoic Mind. Collectively the Sambhogakāya Flowers of humanity and the triads of the *deva* kingdom represent factors of that Mind. They automatically respond to the Thoughts (Energies) coming via that Mind. This manifests as divine Law effecting the corporeality of the Logoic Form via the activity of the *devas* (the *iḍā nāḍī),* and the human Souls (the *piṅgalā nāḍī).* The *suṣumṇā nāḍī* aspect is represented by the liberated Initiates, who form the constitution of Shambhala. Consequently Logoic Mind is the energy conditioning all manifest being.

When H.V. Guenther stated that Buddhism 'eschews....the mentalistic premise that the universe in its totality is a mind or a society of minds', he was stating a view that is not consistent with Buddhism as a whole, except as far as it relates to the idea of a 'God'. For instance W.Y. Evans-Wentz states:

> In its totality, the Universal Essence is the One Mind, manifested through the multitudinous myriads of minds throughout all states of *sangsāric* existence. It is called 'The Essence of the Buddhas', 'The Great Symbol', 'The Sole Seed', 'The Potentiality of Truth', 'The All-Foundation'. As our text teaches, it is the Source of all bliss of *Nirvāṇa* and all sorrow of the *Sangsāra'.*[28]

This is an exemplification of the Yogācāra-Vijñānavādin doctrine, which stresses the importance of meditation as a means to liberation. Mind is considered in terms of both transcendent thought and empirical thought, it is all there is. Regarding 'transcendent thought', this doctrine teaches in terms of the creation of phenomena similar to what is given in this passage by Alexandra David-Neel:

> Void, we have already said, does not mean "nothingness". On the other hand, this term does not belong to a cosmogenic system beginning

28 W.Y. Evans-Wentz, *The Tibetan Book of the Great Liberation,* (Oxford University Press, London, 1971), 4.

with a declaration analogous to "In the beginning was the Void".

Nevertheless, some people have thought to find an explanation of the origin of the Universe in the classic statement *"Ji ka dag... Tsal len dup"* which means "In the originally pure base, an energy arose by itself".

The Secret Teachings expounded in the sayings attributed to Srong bstan Gampo contradict this opinion. According to them, the original Void *(ji ka dag)* is the inconceivable form of the Mind existing before an autogenous energy *(tsal len dup)* caused the *saṁskāras* (mental composition) to arise in it, creators of the images which constitute our world. It is in this void of the mind, comparable to the special void, that are born, act and disappear all the phenomena perceived by the senses.[29]

In its universal aspect as the *ālayavijñāna,* the universal storehouse of consciousness, mind/Mind is said to be the cause of all things. How this 'autogenous energy' caused the *saṁskāras* to arise is not really explained in Buddhism, for that one needs the doctrine of the Sambhogakāya Flower which I have detailed in volume 3 of *A Treatise on Mind.* David-Neel further states that:

"Ideas of continuity or discontinuity cannot be applied to the mind; it escapes them, just as in the case of space one cannot conceive it either as limited or as infinite.

" It is impossible to discover a place where the mind is born, a place where it dwells afterwards, a place where it ceases to exist. Like space, the mind is void in the three times; past, present and future.[30]

This has a remarkable similarity to the words of *Jacob Boehme* (1575-1674), a Christian mystic, who states:

"Within the groundlessness (that by which some writers is called the 'Non-Being' - a term without any meaning) there is nothing but eternal tranquillity, an eternal rest without beginning and without end. It is true that even there God has a will, but this will can be no object for our investigation, as to attempt to investigate it would merely produce

---

29 A. David-Neel & Lama Yongden, *The Secret Oral Teachings in Tibetan Buddhist Sects,* (City Lights Books, San Francisco, 1967), 126-27.

30 Ibid., 127.

> a confusion in our mind. We conceive of this will as constituting the foundation of the Godhead. It has no origin, but conceives itself within itself." *(Menschwerdung, xxi. 1.)*
> "Divine Intelligence is a free will. It never originated from or by the power of anything. It is in itself, and resides only and solely within itself, unaffected by anything, because there is nothing outside or previous to it." *(Mysterium, xxix. 1.)*[31]

Whether one wishes to equate Mind with 'perfect Buddhahood', or 'God' or abstract it in terms of 'space' is a matter of personal predilection, for it is all these things, as they are only symbols of the inexplicable. The conception of 'God' used here refers to the statement 'God is Mind'[32] for it has a validity in relation to the knowable universe. Mind is the only facet of Deity that can be comprehended by our minds, being reflections of That Mind, moulded according to the patterning of divine Mentation.

It is also easy to posit that the community of Bodhisattvas are indeed a society of Minds working towards a common purpose, that of the liberation of all sentient beings, to say nothing of the community of Buddhists and all other meditators all over the planet, that are similarly engrossed with respect to themselves. Do not all share a similar thinking process, and can not this collective thought process be called thus a 'society of minds'? They represent but one grouping of humans, but what about the rest? Then there are the *ḍākinīs (devas)* and liberated beings to account for, past and present, throughout all of the dimensions of time. When we look to the domain of the Sambhogakāya Flower, the concept of 'society of minds' begins to become quite staggering. Certainly it is well worthy of consideration when extrapolating the concept of the evolution of mind/Mind into cosmos.

Upon becoming liberated and thus a Buddha one will then have the faculty to identify with the expression of a Buddha-Mind that aeons ago had gained release from identification with even the most

---

31 Franz Hartmann, M.D., *The Life and Doctrines of Jacob Boehme,* (Kegan Paul, Trench, Trubner & Co., London, 1891), 60.

32 Compare the statement in *Deutronomy 4:24, Hebrews 12:29*, which states in relation to the 'God' of the Israelites: 'God is a consuming fire', Fire being the Element of the Mind.

subtle types of substance that constitute the dimensions of perception associated with human evolution. This Buddha-Mind incorporates it as the Sambhogakāya Flower incorporates the personality for each succeeding incarnation, but is not bound by it in any way.

With reference to humanity in particular, a liberated Buddha ('God') is here considered a planetary Logos. Such a Logos incorporates the lives of a planetary scheme and the related kingdoms as 'His' body of manifestation. It is inclusive of these kingdoms, for He/She had thoroughly evolved through and transcended the stages of evolution that the various beings that constitute those kingdoms must yet unfold. The compassionate Identification of the Logos is so complete that the Logos is befitted with the qualities needed for planetary liberation by utilising the substance and lives of those kingdoms as the subjective and objective sheaths of 'His/Her' incarnation. An entire world sphere is thus the *nirmāṇakāya* of the Logos. It is possible therefore, for an enlightened one possessing the type of attainment as Maitreya or Christ to attain a specific communication, or covenant, in a seeming 'personal manner' with a Logos, but not those immersed in *saṃsāric* mire. A planetary Logos has projected and embodies the Word for a planetary system and is thus 'God' to that system.

A planetary Logos is an integral aspect of the constitution of a solar Logos in a similar sense that humanity is an integral aspect of the former. By the utilisation of analogy, a solar Logos constitutes a 'Cell' in the Body of an ineffable cosmic Deity, about 'Whom Naught may be Said'.[33] The nature of the Mind or Purpose of such an Entity is so stupendously vast and supernal that even the most enlightened Mind on our earth will find great difficulty to Identify with.

The authors of the Buddhist texts, in accordance with the religious presentation of the Buddha, focussed the framework of their teachings most pragmatically to that which was *directly related* to a person's liberation and consequent enlightenment. Their focus was hence (as previously stated) upon the THAT, denoted as being/non-being, which predisposed the entire framework of their doctrines. The theistic religions on the other hand were little concerned with the idea

---

33 Derived from *A Treatise on Cosmic Fire* by Alice A. Bailey, (Lucis Press, London 1977), 148, 553-54, etc.

of 'liberation' (except with respect to the *yoga* tradition of India), but rather with a conscious receptivity to, mergence into, or union with, that which they saw as the cause of all Being and which interceded in their lives in some way. It produced thereby a holistic ontology with an efficacy concerning a transcendent, imminent, relationship with the Divine Personality. A 'God', or the 'Gods', thus became real to them.

Both methods of realisation will eventually produce similar experiences, for as the goal of each approach is reached, then the ongoing development of the other's method that up till then was not developed becomes the Path. (For the sum of the human life experience must become known *in toto* before liberation is possible.) The Buddhist form of internal meditative capabilities and abstract reasoning merges into the spaciousness that is the Heart of the Divine. By means of an outward going contemplation and mystical reasoning the Theistic type is absorbed into the personified vicissitudes of the space that is Mind. One must inevitably learn to perceive the Heart that is the Mind and thereby come to know *dharmakāya*.

There are recorded statements of those who could hold a direct covenant with Deity in religious scripture. However, as the modern era has shown seers who could truly do this (without distortion from an 'I' centred desire, or ego involved) are scarce. So the many conflicting theological arguments as to the nature of Deity has developed. In fact, the many conflicting arguments extant at his time concerning this question caused the Buddha to frame the philosophy that he did. It was the middle way between all opposing views, as it avoided the rhetoric of religious speculation. It concentrated on what was of immediate practical concern, the way to attain the enlightenment that could enable one to Know.

Also, during his time it was virtually impossible to explain the real nature of Deity, because it concerned an understanding of what *energy* was, of which his contemporaries knew very little. Neither was there any sound physiological knowledge to relate to. The premise here is that as an expression of his wisdom therefore, he kept silent on the subject, for he knew that a time would come when humanity would evolve the necessary terminology and experiences to understand, and only then could such teachings be given.

4

Sources of the Causative Impulse Esoterically Considered

The mind/Mind and causation

With respect to the human kingdom the sources of the causative impulse from an esoteric perspective are three in number.

A. The mind/Mind.

B. *Buddhi* and *ātma*.

C. A Logos as the Source of Mind.

They shall be analysed in turn.

When considering the sources of the causative impulse we must look at the planes or realms of enlightened being and the related conditionings. These are the realms of reality that reflect their energies into the world of the great illusion (*saṃsāra*). This world constitutes the arenas of livingness of the human personality—the physical world, the astral, and lower mental, of which much has already been presented.

The lowest of the planes of causation from the point of view of Logoic Activity causing the appearance of manifesting phenomena is the domain of the abstract Mind (the *arūpa* domain of *manas*). Here resides pure consciousness, the Son in incarnation, the *tathāgatagarbha* (Sambhogakāya Flower). Therein the forces that will impulse the illusional form (the personality) are gathered together. The reflected expression, the personal-I, becomes a causative agent in turn, albeit one completely enmeshed in the realms of form. The personal-I is totally conditioned by the karmic waves of ephemeral phenomena and perpetual incident that are generally projected by the desire-mind.

The lesser spheres of mind move completely within the precincts of the Bosom of the greater ocean of Mind, the lowest level of which is the collective consciousness of all Sambhogakāya Flowers. As a people evolve so they must inevitably look to the *dharmakāya* and cosmos as the source of all phenomenological appearance. Mind is the cause and result of emanatory being; whilst mind is the 'great slayer of the real', so *The Book of the Golden Precepts*[1] informs us. Mind must be used to 'slay the slayer'.

When looking to the physical plane, then the causative activities therein are effects of the forces manifesting via the domain of mind. A person thinks as he/she acts to produce or move things. Rarely nowadays do creative activities stem *purely* from the plane of desire and emotions. The first produces attachments to the objects desired, and the second generally manifests in a violent, destructive manner. One could hypostasize on actions that stem from purely physical volitions, where no mind or emotions are involved at all. It is difficult to conceive if such actions could be considered 'creative', as the physical body then would not be in control by any wilful conscious entity.

The realm of the mind (the mental plane) is dual, subdivided into an archetypal and concrete portion. The nature of this division was explained in *my previous books* and can be further elaborated in the contexts of the causative process by means of the figures below:

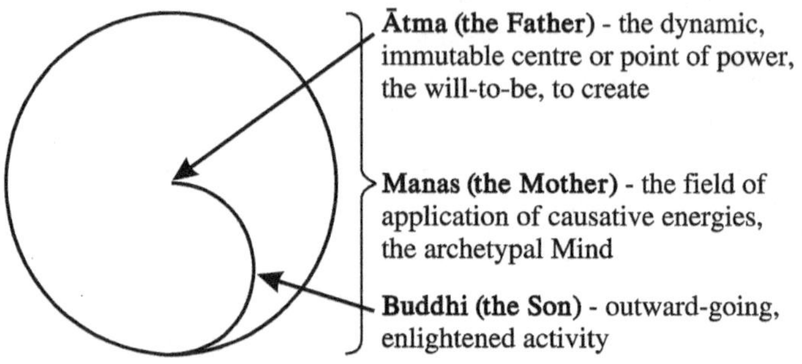

Figure 3. The Spiritual Triad

1 Translated as *The Voice of the Silence* by H.P. Blavatsky, (Theosophical Publishing House, 1998), 13.

Manas, the universal consciousness (being a reflected attribute of *ātma)* then acts as a Mother, to give birth to empirical consciousness. Here we see that the action of the 'Son' *(buddhi),* the principle of enlightenment in figure 3, carries with it the energies of the four cosmic *prāṇas (ākāśa)* into manifestation via the higher Mind. This becomes expressed as *prāṇa* in figure 4.

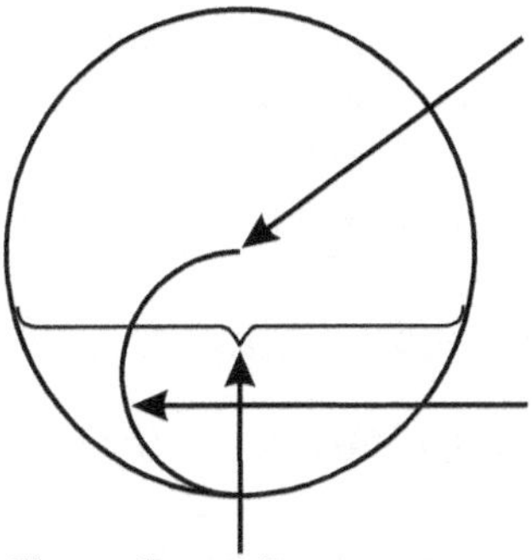
Point of contact as physical perception
The Father-Mother (abstract Mind). The resultant experience is seen as the Son, for when expressed on the path of return it finds its application as enlightened consciousness *(buddhi).* Causative action can now find its most concrete expression
Underlying energy (prāṇa)
The five *prāṇas* first convey desire, resulting in sensual experience and then illumined activity, when at-oned with the mind
The resultant gain, manas
The daughter, underlying cause of phenomenal existence, expressive of the energies of the Father-Mother. It is receptive to the impulses emanating from below and directed from above

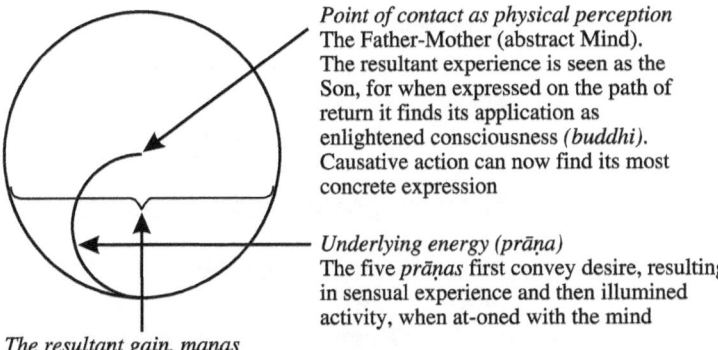

Figure 4. The relation between universal and empirical consciousness

Here empirical consciousness can be understood as the activity of the Mother undergoing rounds of experience, to eventually birth enlightened consciousness (the Son), the result of such experience. The focal point of the activity shown in figure 4 is *manas,* vivified by energy *(prāṇa)* and motivated by desire, which manifests upon the physical plane as sensation via the experiences derived by means of the five sense-perceptors. Satiating the desire for sensation eventually produces understanding of the object contacted via the five sense perceptions. The mind is thereby formed, which registers, collates, stores, analyses and divulges such information. Understanding then produces desire for further (or new types of) experiences. So the wheel ever turns.

When universal consciousness merges with or embraces empirical consciousness then the thinker is born. Such a one can not only consciously classify information, but that can also form ideas, dream, envision and aspire to greater heights of revelation. The empirical evolves towards universality (the Ineffable Mind), whilst universality is inclusive of empirical consciousness. The only field of action whereby all this is carried out is the mind/Mind *(manas)* of humanity, or its transmuted correspondence in Deity. The Mind is universal and Real; whilst the mind (the intellect) slays the Real by limiting the Mind to a sphere of action, moving focal points upon the screen of time and space. It imagines itself to be separate, distinct from all other similar points (or 'atoms') that exist as tangible images in the mind-scape. Such distinction however only manifests in terms of the mind's own particular self-focused field of sensation. This becomes the focus of the personality life. Sensation is ephemeral, constantly changing, and thus is considered unreal *(māyā).*

The enlightenment process and causation

If one desires something strongly then one either objectively or subjectively sets the process in motion that will allow desire to be crystallised as a concrete fact. (This is the working basis to the power of prayer.) Money, for instance, is really crystallised desire energy, a form of *prāṇa*. Logoi manifest a similar Desire upon their own realms at the right cycle for incarnate expression. In their case such Desire is an emanation of Love-Wisdom which expresses itself as a combination of the Will-to-manifest and Desire-to-liberate the sentient and conscious streams of Lives that come into manifest activity, by guiding them to higher levels of attainment.

The process of becoming enlightened concerns knowing the nature of mind by controlling mental substance. One must be able to consciously utilise it constructively and to simultaneously learn to think with the Heart. It means gradually mastering the sea of thought-forms that constantly cloud and envelop us, until one stands fully in the presence of the Spiritual Sun and merges with it. The thought-forms have then transmogrified into fields of radiance. Light can then be brought into the world to dispel people's glamour and illusions. Each of us must learn to do this as the light within us grows.

Also, once humanity rapidly dispels and disperses the fogs and clouds of its ancient thought-forms and massed glamour, so group enlightenment, the 'light of the world' will manifest amongst us. Consequently humanity is the immanent Christ, and this the imminent

Christ (the Head of the Council of Bodhisattvas) referred to when He stated in relation to one that believes in the Christ: 'Greater works than these, shall he do'.[2]

Again the importance of meditation needs emphasis. Only through the control and wise utilisation of the mind can one stand free and work consciously as an enlightened causative agent. Such an enlightened one works via his/her point of contact with the physical plane—the physical permanent atom.[3] The Mind is the essence of being, whilst the mind (or desire-mind, *kāma-manas*) is the battlefield wherein the results of life's experiences are expressed. The Mind can, through mind, express its adamantine potential to manifest all things in form and space. The ability to identify with the Mind that fully controls the potency of the mind implicates an Initiate of the third degree, or greater. Such a one is an enlightened builder in the three worlds, wherein karmic streams of energy that form the basis to the standing of one's Soul-group can be directed.

The Initiate is part of the collective meditative activity of the Hierarchy of Bodhisattvas that rightly colour and direct creative mental substance so as to produce beneficent effects in the world. Initiates are in tune with the streams of thought that underlie all that is manifest. They can contact any aspect of those streams in the consideration of the various projects, the service work to be accomplished. All manifests within the context of the total picture constituting the world-view as related to the particular group (*maṇḍala*) the Initiate is part of. Such a *maṇḍala* is part of a greater one in the scheme of things. The hierarchy of *maṇḍalas* become ever vaster in scope, and so one inevitably reaches the level wherein exists the Logoic Thinker. (Who is but an Initiate of very high attainment.) An Initiate thus becomes a co-worker and co-planner, co-adjudicating world *karma*, resources and human goodwill for the benefit of all. Such work is coloured by the governing Ray conditionings and the world period involved.

Those that have undertaken the third or higher Initiations act collectively with the Council of Bodhisattvas as mediators channelling and adapting primal Causative energy emanating from the greater Logoic Mind. Consequently long planned beneficent results manifest

---

2 *John 14:12.*

3 The permanent atoms shall be explained in chapter eight.

in the world during the evolutionary period ahead. The rolling cycles of such events becomes for them the eternal Now.

General humanity also function as a causative agent because of their ability to utilise and mould mental energies coupled with desire of various grades. They build the images, concrete forms, appliances, artefacts, etc., on the mental, astral, and physical planes wherein they consciously reside. Their creative building is however relatively transient. It is generally related to themselves, thought of as separate parts of the whole, and conditioned by extraneous forces over which they have little or no control. Natural catastrophe, man-made disasters, chaos, war, and obsolescence can at any time reduce those forms into rubble.

Such creative building is an effect of the ocean of conditionings and karmic impulses that govern all manifestation. The directive originating streams of Causative impulse is however seeded by the Initiated. Such Bodhisattvas direct the subjective streams of Causative energy from Logoic sources by way of the Mind through the Heart's purpose (*bodhicitta*). Humanity relegates aspects of those streams as vortices of energy that produce the external aspects and qualifications of their civilisations and societies. One group works consciously with full knowledge of the process involved, and the other (humanity) largely unconsciously and sometimes destructively, against the stream of the good and wholesome, the future beneficence of the all. Such destructive, aberrant, self-focussed potency of humanity is also part of the Logoic Plan, because it is a necessary stage of the process, through free will, that will eventually teach people (through karmic activity) to become appropriate, conscientious creative thinkers. Humanity will in time learn to think with their Hearts.

Those that are actively striving to aid humanity in some way form the personnel of the Council of Bodhisattvas. They differ only in their degree of realisation, their sensitivity to the primary Will emanating from Shambhala (the Plan of the presiding Logos), and therefore of their ability to express the conditioning forces determining what is to be. This manifests according to the inherent Ray conditionings governing the Initiate concerned. This is their part of the task of world service for the salvation of the tainted streams of human activity, and to rightly direct evolutionary progression of all Lives. The 'degree of sensitivity' is determined by their Initiation status within the Council of Bodhisattvas.

The many sacred books and esoteric texts that have mentioned the executive members of this Hierarchy, have called them the council of the holy Sages, the Seven Rishis, the four and twenty elders, the great Adepts, Masters of Wisdom, Avatars, and so forth. The liberated Ones heading the Hierarchy are esoterically given the titles Manu, the Christ, Mahāchohan and the Chohans of the Rays. They actively work with the principle of the Shambhalic Will (the 'kingdom of God'), as explained in the two volumes of *The Constitution of Shambhala,* as well as in the writings of Alice A. Bailey.

The members of the forces of evil (dark brotherhood) also work in highly organised groupings. They are a hierarchy of focused, wilful units that embody the concrete mind *in toto.* They have followed the path of self will maximally to serve their own ends. They destroy and retard the functioning of the common good through building separative empires of selfish concern around themselves with the power of manipulative thought. Thus they fight the forces of Love and Light, everything truthful and wholesome, for such empire building cannot abide in a potency of evolutionary progression. Lying deceit and the methodology of war are their *modus operandi.* They therefore work to retard the flow of the effects of time, to make things go backwards to the environment of dark spaces of desire and mind that is their home. Their line of evolution is towards becoming Logoi of black destructive and cancerous intent. Otherwise they work directly with the grey lines, typified by the mind of the Anubis, explained in *The Constitution of Shambhala,* part A.

Initiates of the *fifth degree* (Masters of Wisdom) have the ability to contact the Plan as it exists in the *dharmakāya* and direct it to the material realms as primal Causative energy. Such energy emanates from the *ātmic* plane and must be 'toned down' and directed in such a way that it can be utilised by intuitive, inventive and inspired thinkers. They can use those ideas as ideals to accomplish their self-initiated tasks in any of the departments of Life. Thereby much is produced that happens in the material world.

The *ātmic* plane, from which these Masters of Wisdom work, is the plane of causative *karma,* from which emanates world, group, and national *karma*. It is the plane upon which the impact of the Fiery *dharmakāyic* substance from the Dhyāni Buddhas (depicted here as causative *karma*) finds its objective externalisation in our solar system. *Ātma* is but the reflex of the energies manifesting from the Logoic Mind (the cosmic mental plane). Masters are therefore but emanations or aspects of the Mind of a Creative Logos. The Masters rightly wield and direct this causative *karma* so that all involved in *saṃsāra* will progress according to the Shambhalic Plan. They are the recipients of the greater Will, thus are the custodians of the Plan that is the Purpose during any evolutionary epoch. This Plan is then particularised into the minutiae needed for the appropriate evolution of the streams of consciousness in the world via the *maṇḍalas* representing the Ray Ashrams of each Master. Masters are thus agents that allow a Logos to focus His/Her Purpose to produce necessary effects in *saṃsāra*. Masters consequently are agents of directed Purpose, units of the Logoic Will. Those who aspire toward the manifestation of new age ideals, or improving any aspect of Life on the planet, will be directly energised from such sources.

The course of events in the material world, and the other kingdoms in Nature are also effected in a similar manner by members of the upper echelons of the angelic hosts, the *deva* equivalents of Adepts and Masters. All work via the agency of greater and lesser *deva* units via the laws of sound, ceremonial order and colour.

Everyone must walk every step of the path to enlightenment through the complete development of their own capabilities and right discrimination. By mastering all of Life's qualities one becomes enlightened, thus a prime causative agent. This is the first major step after Individualisation into the human from the animal kingdom and subsequent development of intelligence. Eventually such a one becomes a Logoic Creative Will.

Guidance may be given by the enlightened to all seekers, but such is not a dictum or command, for each task that a disciple takes is self-initiated. It is the result of meditation upon the cause of a particular aspect of suffering, and upon how best to serve. It is not imposed from above, though impressions may certainly come thusly. *Siddhis* are developed through the generation of the compassionate will. *Siddhis* betoken of a causative potency unknown to the unenlightened.

Intensity of feeling (mistaken as love) however, is what the world's disciples still mostly express. Generally they will ardently follow any teaching disguised as 'Words from God', conventional renderings of *sūtras* and religious texts, that intensify devotion or aspiration. Intensity of feeling, even ecstatic vision, has a *selfish* connotation because it is intrinsically related to, or focussed upon, the personal 'I'. This helps build the great prison house of personality (the desire-mind) and of separative attitudes—such as promoting an exclusive religious doctrine. The divinity within one, the causative function, is then relegated to a menial and limiting task, the building of the concepts of mind affecting the most material and limiting of all bodies of manifestation (the material world), or to produce glamoured, astral images containing the 'waters from which it', the often zealous aspirational mind 'drinks'.[4]

Such attitudes intensify the conditionings surrounding and containing the 'speck' of unreality that is the human personality, (the 'atomic unity' within a solar sphere) whereas Initiates are concerned with directing the causative streams of entire groupings of such 'specks' into the realms of reality. This necessitates the application of methods that will allow one to defocus the Eye (the directive agent) away from the 'I', so that it can become more inclusive, to view the entire panorama of which the 'I' is but a minute part. Expansive inclusive perception is the way of developing a Logoic Vision.

Logoi play a similar part in their greater schema. Therein energies from other stars, entire constellations of stars, and from the signs of the zodiac, must be conveyed and integrated into their Bodies of manifestation so that the various streams of Lives therein will benefit. Such energies must be directed during the appropriate cycles in accordance to what is needed to vitalise those they envision within the ring-pass-not they embody.

Inclusiveness, not separateness is the way forward. The great heresy of separateness is one of the major causative factors promulgated by the forces of evil. It produces much of the evil seen in human societies. The effect of this (and its base of massed selfishness) is obvious to all who are learning to rightly vision.

4 See appendix one.

The keynote of all that comes from the mouth or pen of a Master of Wisdom is always *inclusiveness* stemming from far sighted Vision. Such teachings can always be noted by their meaningful quality, forthrightness of imagery and dynamism of presented truths. They work to feed the Heart that is the Mind, to awaken the intuition, rather than feeling-perceptions or devotional tendencies, though these qualities may also be inadvertently kindled in the aspiring ones.

Fully enlightened teachers will be found by those that are awakening the powers of the Heart via the intelligent application of the task at hand, through consecrating their lives to serve. Being liberated Souls the enlightened ones work directly through people's Sambhogakāya Flowers, which are also prime divinely guided causative agents upon our earth. These Souls possess no sense of separateness. People thus come to know the Master of Wisdom via contact with their own higher selves. Often messengers are sent, the direct disciples of an enlightened one, and how well they are received by the aspirant indicates the readiness of the aspirant for the way of approach to the spheres of causation.

There is consequently no need to actively seek for the Master or guru in the world by means of desire or aspiration, for the Master will appear only when the karmic underlay for right action has been accomplished. The disciple will then already be well established as a prime causative agent in the field of service.

The time for changes in the world of religious opinion is nigh, through the appearance of the *Avatar,* as explained in *The Constitution of Shambhala.* This Avatar will work to cultivate the causative seeds sown previously by his predecessors such as the Buddha and Jesus, according to the Plan. He reaps the gains of the past, which are then moulded into the form that will be causative of the new World Religion. He will be able to do this because he is the embodied externalised power of the *ātmic* plane, (the third level of the *dharmakāya*) which is the recipient of the Mind of the planetary Logos. Initiating new revelations world-wide, in the face of immense prejudice and mental-emotional reactions may necessitate miracles indeed. Enlightened ones however do not work in phenomenal realms to amuse, bewilder, or convert the gullible, or to meet the demands for sensationalism by the aspirant and beginner on the Path. They reveal the new ways of thought and proffer the testings that will allow disciples to overcome limitations of mind. All must learn the way of approach to Logoic domains, all must in time become Shambhalic recipients. Eventually they will become Logoi themselves, members of the stellar community seen in the night sky by a future evolving humanity.

Drastic action is needed in this epoch of concretised mental-emotions of humanity, if the cycle of Love-Wisdom is to become prevalent. The Plan calls for a world-wide externalisation of a similar process that happens within the Initiate's consciousness as he/she awakens to enlightened perception. Similar causative functions that were used to awaken the nature of the Mind within must be also utilised to awaken the minds of the intelligentsia. Such subjective experience must now project itself as observable and even miraculous happenings on the world stage. Many will be swept by waves of reactionary sensationalism and bigotry, but the lower mind structure (the separative 'I') of humanity will inevitably be transformed in a way never before possible on such a wide scale. The march of time will make this possible. Decades, centuries and millennia go into Hierarchical planning in accord with the way of manifestation of the law of cycles.

Much that was previously hidden may now be revealed in such a way that humanity can quickly progress towards the Light and learn the ways of the higher Creativity. The Laws of causation necessitate such development, if the primal Logoic atoms that are the Sambhogakāya Flowers of humanity can be stimulated to the vibrancy of the next level of expression ahead of them. The ways of Love must begin to dominate world affairs, for this Love is but Mind actively beating out more embracive cycles of endeavour, as fostered by the Lords of Life.

All that is needed on the Path will unfailingly be given to one that walks the way of loving, wise service (at the time one needs to utilise the gifts), for service is but the directed application of the power of a causative agent. Such agents must transcend commonplace thinking and belief systems, they must comprehend the nature of the law of *karma,* thus of the ways that the Lipika Lords[5] play their hands. Right concepts

---

5 Lipika, meaning 'scribes' or 'recorders'. They are karmic agents, the geometricians of the universe, circumscribing the *maṇḍalic* patterns that become the blueprint of

of *karma* must be appropriately understood, hence much that transpired in the past is visioned if one is to be a divinely inspired causative agent. No Logos can work without taking *karma* into account. Nearly all that happens in the world around us is because of the necessary *karma* that has transpired from past cycles of endeavour. People must comprehend why that *karma* manifested and not react so ignorantly and mentally-emotional, or be so reactionary when they or others must pay back their karmic dues because of the ill that was sown in past cycles. The vicissitudes of the way *karma* works must eventually become a facet of common discussion and well-reasoned topics in many books. One must also comprehend that not everything is the result of past *karma,* because people are quite capable of manifesting new volitions, producing karmic consequences to be paid in a future cycle, especially through mental-emotional reactions.

Lives of learning mathematical reasoning within the field of science may eventually allow one to evolve into a Lipika, responsible for building the lines of activity within any Logoic *maṇḍala* and to interrelate that *maṇḍala* within those expressive of the greater whole constituting the universe of 'things'. Many are the cosmic Paths that enlightened ones are preparing to travel. If enlightened beings do not evolve into Logoi, then they will play roles, such as that of the Lipika in relation to a Logoic Incarnation.

Right action and sincere endeavour to become enlightened will by its very nature evoke a response by those that guide human evolution. All that is hid will then be revealed, with right compassionate action that will evoke the revelation. Fanatical adherence to a doctrine, wishful thinking, or knowledge of occult subjects alone will not do it. Neither will a meditative path that is fundamentally concerned with one's own self and its problems. The divinely embodied Teacher will manifest when the disciple can adequately handle the energy transmission of which the Master is the custodian. The potency of the established *maṇḍala* (Ashram) of such a one must become the experience of the disciple, who then must integrate the lines of reason to view the integrated

---

space through which the Word of the Logos can sound out and attract the Builders that build the substance of the forms. See chapter 8, and *The Constitution of Shambhala* for further information.

maṇḍalas of the Logoic whole. The individual disciple is but an aspect of the world disciple. The world disciple is but an attribute of the Logoic Mind that is endeavouring to turn about its Seat of Consciousness away from material considerations (the world of human affairs) upwards to cosmos. In doing so the world disciple is inevitably projected upon its cosmic journeying in the stellar domain that is cosmic astral space.

The precondition for each disciple is the process of battling his/her way through *saṃsāric* turmoil to the mountain top of realisation upon the Initiation path. The way of Initiation must be thoroughly comprehended by the disciple in the new era because little progress upon the higher Way can be made without such a consideration. The world religions teach the elementary steps upon the way, and those that expound meditation teachings provide higher revelations as to that way, however, for enlightenment to ensue the topic of Initiation into the Mysteries of the 'kingdom of God' (Shambhala) must be understood and the Initiation path demanded by the Lord of the world (Sanat Kumāra) followed.

Complications arise when an aspirant upon the Initiation path looks to Causative agents functioning from planes higher than the mental, for they must use the mind to understand what transcends the mental process altogether. By the use of analogy however they can obtain some comprehension, and progressively more so as the foundational philosophy presented in the series of books I have written is understood.

Here the reader should also refer to the information presented in chapter 7 of *Meditation and the Initiation Process,* entitled 'The Beatitudes: The way of evolution of Prime Causative Agents'. I begin that chapter with a definition of causative agents, which I shall quote below:

> By 'causative' is meant that which produces a fundamental change in the structure or existence of something, or in the environment in which they reside. The concept generally implies the existence of directive wills (also seen in terms of the expression of natural laws) that impinge or act upon that which exists to change it. These agents can be categorised into five main groups.
>
> 1. Unconscious agents that are capable of manipulating the various categories of substance without the factor of intelligence involved. The actions come as a consequence of automated responses to stimuli manifesting in the environment concerned, through instinct, desire, or emotional projections of force. They do not

comprehend the processes and laws governing their actions, they simply act in accordance to inherent predispositions.

2. Semi-conscious agents that work with the impulse of desire and mental-emotional actions. Here the actions are largely the expression of massed conditionings, or inherent tendencies, instinctive urges, brooding sentiments, fears and the like. The emotions generally sweep over the conscious factor, impelling it to action. Little thought is actually used to meld the controlling forces into a preconceived direction, however, what is to be desired is comprehended.
3. Intelligent agents. Here the thought processes are involved. The mind clearly formulates what it desires to accomplish, analyses the requirements necessary, the tools and materials available and the longevity of the construct. The causative activity then proceeds according to plan.
4. A prime causative agent. This is one that meditatively comprehends and can utilise all of the laws and energies constituting the multidimensional universe to create the forms desired at will. Such a one can also consciously use forces, materials and substance other than that which is purely material. Planning ensues from the abstract Mind or the *dharmakāya* and the *maṇḍala* that ensues evidences a multidimensionality and far-ranged purpose not possible by means of the use of intelligence alone.
5. Logoic creators. They start the entire gamut of the evolutionary process happening from first principles, directing that which is originally static and inert into motion by means of Mind. Inevitably a myriad minds capable of becoming prime causative agents will evolve from the originating impetus.

The way of evolution and form of action of prime causative agents has by now been somewhat understood by the world's esotericists. It constitutes the evolution of Bodhisattvas and of their various grades of expression to Buddhahood. This concerns the nature of the making of a *mahāsiddha,* the treading of the higher Tantras, stripped of their complex layers of congealing veils. This necessitates the path of Initiation, the making of a Master of Wisdom, and eventually the evolution of a Logos.

Buddhi and *ātma*

The state of perception directly beyond the mental is termed *buddhi*. *Buddhi* represents the fourth or middle of the seven planes of perception. Though I have explained *buddhi* in some detail in chapter one, here a little more information can be provided. This fourth plane is the true home of the human kingdom, whereas the mental plane serves a similar function for the *devas*. *Buddhi* is the mirror that reflects the archetypal into the concrete realms. It reflects *ātmic* perception (which embodies the archetypes of the Causative Impulses from the *dharmakāya*, the Mind of Deity) into manifestation as the underlying quality of the mental realm. Omniscience is reflected into the physical realm when the mind of the receiver is free from emotional and mental blemishes, the qualities that might distort the Image.

A physical mirror free from blemishes perfectly reflects the colourings and forms of a scenery in terms of visible light. *Buddhi* has a similar function with regard to the light and energy qualifications of the enlightened realms. The potency of *buddhi* is such that only a group demonstrating a group service, or an Initiate of high degree embodying Love-Wisdom, (who is but the Heart of such a group) can adequately channel, direct, or express this energy. It was explained in terms of the *śūnyatā-saṃsāra* nexus in *A Treatise on Mind.* Therein I explained that nexus as representing *buddhi* from one perspective, from another it is the Sambhogakāya Flower. When abstracted into *śūnyatā* it represents the heart of the Mahāyāna Buddhist dispensation.

Buddhi is that realm to which one aspires after one's consciousness has turned about because of no longer being controlled by impressions from the empirical world. It is the Intuitive realm from whence emanates enlightened perception in the form of universality (comprehensive Inclusiveness), the fount of all compassionate undertakings. This then is a translation of the Buddhist term *bodhicitta*.

Buddhi is the vehicle fusing all dualities into at-onement. It is the attribute of the fourth cosmic ether that grounds the energy emanating from the Heart of a Logos, hence demonstrates as the Life found at the heart of each manifest atom. It interrelates all forms into a unity, because these 'atoms' are the corpuscles (*anu,* atomic unities) pumped by the one cosmic Heart by means of the Love of the Lord of All.

Buddhic perception thus becomes the embodiment of compassion, for by means of direct Realisation of what it mirrors one identifies with the interrelatedness of all being. The vision is inclusive of where disease and disharmony exists in any body of manifestation, and consequently what is needed for the rectification of the disease. This 'mirror' allows one to be fully cognisant of That which is the basis of all that is, and which floods one's entire being with the will-to-good for the united whole.

Living consciously in this realm constitutes attaining the fourth Initiation, thereby becoming a Lord of Compassion, capable of shedding esoteric tears of Blood (Love) for the suffering of all in *saṃsāra*. This compassionate one has mounted the fixed cross of the heavens and will stay thereon until those that are incorporated within that one's vision have responded to the emanating flood of Love, causing them to tread the path to Light.

Here is portrayed the concept of a *Bodhisattva*—who has vowed never to cease striving until all sentient beings have been delivered to the other shore of *saṃsāra*. He/she is an embodied expression of the enlightened Mind. A similar concept is indicated in the New Testament by the phrase in *Matthew 20:16* (and *19:30):* 'so the last shall be first, and the first last: for many be called, but few chosen', and in *Mark 9:35:* 'If any man desires to be first, the same shall be last of all, and servant of all'.

This well known Bodhisattva vow is in fact fulfilled upon the path of becoming a Logos and its aftermath. Having vowed to serve all upon an earth sphere such a one eventually gains complete release from *saṃsāric* involvement as a Buddha. Cosmos is then entered. Here the Buddha works upon developing further qualifications that will allow him to later embody a planetary sphere (or become a major functionary of a solar system) whereby younger Initiates and the members of humanity that did not pass the 'grade' at the appointed cycle for evolutionary expression of the planet from which the Buddha evolved can continue to develop the attributes needed upon the Initiation path.

Being the nexus that governs the middle ground of the evolutionary *milieu*, relating the One to 'the other', *buddhi* is the veil of abstracted Space, the awesome vastness of cosmos, the full omnipotent potential of the All. As a mirror one who is absorbed in the *buddhic* state of revelation can instantly perceive several aspects of fundamental Truth as it exists in the *dharmakāya*, or the sum of those aspects, but not the full 360 degree panorama. (In a similar sense a mirror reflects the scenery immediately before it, not that behind or around it.) The concern here is with Truth in the absolute sense, comprising entire cycles of relationships and *karma* associated with the planetary and solar Schemes, of cosmic evolution, and not just with that associated with our relatively insignificant little planet.

The next highest plane of perception (*ātma*) on the other hand allows the Mind of the Creative Deity to be objectivised. It allows the laws and related energies of the associated creative Impulse to be projected and expressed according to the Initiate's vision of the Plan.

Alaya Avalokiteśvara (the 'downward looking one'), the bearer of the Lotus (*chakra*) that supports all manifest Life, embodies the qualities associated with *buddhi*. He is the prototype Bodhisattva (compassion personified) who is said to have originated the mantra Oṁ Maṇi Padme Hūṁ. The symbolic one thousand arms to his *sambhogakāya* form are said to touch the hearts of every being with compassionate benevolence. *Krishna* embodies this function in Hinduism. He plays the song of *māyā* on His flute to His consorts (*Gopis),* who represent aspects of the Mind nature fully in love and in ecstatic embrace with the inner and outer Divinity.

The *Christ* is said to energise this realm of perception in his role of world Saviour and Server, for the Hierarchy of Enlightened Being, the Council of Bodhisattvas, (to whom the Christ is the heart or Master of) esoterically reside there.

As *buddhi* as the fourth etheric sub-plane of the cosmic physical plane, so the physical, astral and mental realms (the three planes of human livingness) have an analogous function to our solar and planetary Logoi as the dense, liquid and gaseous sub-planes of the physical realm have to us. They represent the substance of the dense body of manifestation built into the personality aspects of the Logoi. This solar Incarnation is said to be tainted with the gross energy, the remaindered substance and *karma*, from the past solar Incarnation. The objective of this solar Incarnation is to transmute these effects.

The esoteric fact that *buddhi* is the lowest of the four cosmic ethers, and therefore the home of the *chakras* that relate to the Personality life of the solar Logos, automatically makes it a plane of at-onement. Here all the causative energies that produce tangible results on the lower three sub-planes of the cosmic physical plane find their point of emanation. On the buddhic plane therefore, the Life that animates our world and its relationship to solar and cosmic Intelligences, the ineffable etheric web constituting the Body of the One in whom our Solar Logos resides, is first truly known for what it is. Here one comes into conscious contact with the various aspects composing the omniscient Mind of Deity, and That which it embodies. Those that have attained buddhic perception have therefore attained a measure of 'cosmic consciousness'. (To use a term that is nowadays much used by the spiritually minded, but so little understood by them.) They have become an integral part of the Heart centre of the planetary Logos.

Complete participation in the cosmic energies expressed by these *chakras* and the related laws confronts the Initiate of the fourth degree after the Sambhogakāya Flower has died. There can be no separative consciousness for such a being, only the isolated unity of group consciousness, and at-onement in its purest connotation with all that Is. The Initiate enters into a multidimensional universe veiled by the energies of Space.

The Initiate of the *fourth degree* has mastered the qualities of the buddhic plane, whilst the fifth degree Initiate (a Master) can reside consciously on the next highest plane of perception, the *ātmic* plane. This term is derived from the Sanskrit roots *āt-ma*, meaning 'to breathe, eternal movement, to pervade, to go', or even 'the self'. *A* is the first letter of the Sanskrit alphabet and connotes the primordial Sound from whence all manifestation came. It is also That into which all knowledge or wisdom can be condensed. In one connotation *ma* means 'mother', from this viewpoint the *ātmic* plane can be said to symbolise the primal Mother of all being. The Mother is that which emanates from the Mind of the creative deity, the activity aspect of the incarnate solar or planetary Logos. Being in conscious contact with this Mind the Master thus becomes the prime agent of transmission for these Causative energies.

The Dharmadhātu Wisdom, as governed by Vairocana, is an emanation of the *ātmic* plane. This, the third *dharmakāyic* level, is given the property of the Element *Aether*, implying ethereal, primal, the highest of the five Elements, not knowable by means of sense perception.

Planetary *karma* (the crystallised expression of the Logoic Will) emanates from the *ātmic* plane, whilst all is resolved back into this plane at the end of any major cycle of Logoic activity. The in and out-breathing of everything associated with the *tathāgatagarbha* (Sambhogakāya Flower) is eventuated here, as are the subjective experiences and *karma* of incarnate humans.

As the *ātmic plane* is the higher correspondence of the mental plane, the intellect becomes an automaton for those on this realm. The substance of the three lower realms is directly controlled by the will by those absorbed in the *dharmakāya*. They have become creative agents for the dispensation of Logoic *karma* to the human kingdom. To the at-onement experienced on the *buddhic* plane is added the quality of the *will*, expressed as a potent force, such as the will-to-good, the will-to-create, according to the revealed Plan. This Plan emanates from Shambhala, within which the *dharmakāya*-absorbed one resides.

The gain of the karmic process, the effect of all evolutionary progression and rightly focussed meditation adds the *ātmic* Mind's will to the divine Love earlier developed as a Bodhisattva. This allows the returning *jīvas* (life-forms) to be inevitably abstracted into the second plane of perception, *anupādaka,* hence into their Monads, the Sun/Son in manifestation for each human unit.[6] This necessitated the process of transforming the structure of the empirical mind so that it can embrace the awesome extent of the universality of Love. Love is essentially the magnetic field that sustains and contains All as a coherent unity. Mind (*ātma*) instantaneously discerns the interrelatedness between the component parts of that Unity and can particularise them if need be. It immediately discerns the *karma* of what must be and what must remain.

One must be careful here not to conflate what is termed *ātma* here with the *ātman* of the Vedānta and Hindu philosophy. The Buddha was right to apportion erroneous thinking there, because the way that *ātman* is actually 'enclosed' so that it could serve an individualsed human unit was not appropriately thought out. The main error is by omission, and

6 The attainment of a sixth degree Initiate.

the limitation of teachings concerning *cit* (consciousness) and how it affects human intelligence without being affected itself.

I had earlier provided a definition of the *ātman* from the Advaita Vedānta viewpoint (the other Hindu sects hold similar views here) because belief in a *ātman* by Hindus is a fundamental distinction between Hindu and Buddhist philosophies. Buddhists have no such belief. The Hindu view however is correct to a degree, but conflates the attributes of the Monad (which here can be considered to represent the doctrine of Brahman) and that of the Sambhogakāya Flower, which can be considered 'pure intelligence or *kevala cit'*. The Buddha correctly focussed upon the teachings related to gaining the attributes of *śūnyatā*, which was the correct focus for his time period, and was the evolutionary progression for the revelation of the *dharma* then. The doctrine of *śūnyatā* could eliminate the need for manifesting forms, hence complete liberation from *saṃsāra*. The *buddhadharma* however just brought one to the midpoint of the entire evolutionary milieu, and barely penetrated in any meaningful way into the nature of what happens after liberation. Here the teachings of a Monad (or an *ātman* properly considered), a unit of cosmic consciousness and 'container' of the liberated Mind comes into view. The Monad is the form attained by a Buddha in *parinirvāṇa* when such a One travels to stellar fields.

Human consciousness has now sufficiently progressed to begin to comprehend the next level of the revelation of *dharma*, being the teachings of what *śūnyatā* veils, hence the nature of the higher planes of perception, of the constitution of Shambhala and the true nature of cosmos. This includes whatever is possible to reveal in words concerning Monadic Identification with the All.[7]

The attributes of the Sambhogakāya Flower have been adequately explained in my former books, especially volume 3 of *A Treatise on Mind*. A little further information concerning this subject, plus an exposition of the Monad, shall be explained in chapter five.

Mind possesses an inherent separateness that allows it to particularise between this and that. Love is an inclusiveness that knows no bounds.

---

7 It should be obvious that as we move to the higher dimensions of perception, and to cosmos, words and concepts become a limiting adjunct. They fall far short in depicting what must be experienced in order to be understood.

Both exist, allowing the interrelation between the One and the null character that is phenomena, being with non-being, within the domain of consciousness wherein perpetual dynamic growth is possible. Their fusion is the *bodhicitta* that is the true nature of enlightenment.

The creative Impulse emanating from *ātma* (the Mother) is the means that allows expansive growth. Consciousness manifests as a product of *ātma's* sphere of activity, until such a time that all particulars within that sphere are fully mastered. This process involves the evolution and evocation of the five instincts (self-preservation, sex, group or herd, towards knowledge and self-assertion). We consequently have an expansive approach, via the sex instinct to incorporate the 'other' (sphere of activity), from which comes the propagation of the species, hence the chance to eventually develop attributes of consciousness derived from the instincts, coupled with the developed sense-consciousnesses. Consciousness then transforms into Mind once it includes the ineffable Vastness of that greater 'Self' through 'touching', signifying the commingling of substance.

The attributes of Love are born through the unfoldment of the sense of Touch (which is the prime tool of any causative agent) on a vast, most inclusive scale. The within-without motion to contact another similar sphere of action draws the other to it, until the other is fully absorbed in the Heart of its involvement, and vice versa. Thus we perceive fourth dimensional motion. It ruptures the ring-pass-not of self-involvement and identification, and a greater inclusiveness is produced. The way of the Heart thus unfolds, for myriad are the expansions of consciousness that thus take place, making the Heart to be the Mind. Such a Mind then evolves into the universal Mind that is expressed by a Logos, Who can manifest a world sphere of activity by means of the ability of the Mind to individuate particulars.

Touch has inherent within it friction and pain, when propelled by mind, with its separative and self-willed attitudes; or bliss, when lubricated by the waves of Love. Love is the energy that facilitates the associated motion. The energy of Love as we know it is but an expression of cosmic astral substance. All primary Causative agents utilise this Watery substance to make malleable the clay and dust of the cosmic dense physical plane (the systemic mental) so that it can be moulded into the shapes desired by the Logoic Thinker, and the 'man-plants' can be adequately nourished.

Planetary formation esoterically considered

The basis to the more esoteric understanding of divine causation lies in the energy exchanges between the planetary *chakras* found on the *buddhic* plane. Each *chakra* (which can be given an astrological name) is the repository of a particular Ray force, a specific type of energy or quality. The Logoi of the sacred planets can thus be considered the Lords of the seven Rays, as expressed in the solar system.

The earth sphere upon which we reside is but a petal of a Flower *(chakra)* viewed as a planetary Scheme. Our globe is one of seven globes, six of which exist on the subtler planes of perception. Together they form a *Chain*, which constitute the seven sheaths of the body of a planetary Logos. In a similar manner humans possess seven sheaths of increasingly rarefied substance. We inhabit the dense physical sheath whilst incarnate, and the etheric, astral or mental sheaths when disincarnate. Similarly the Logos incarnates into whatever globe is the focus of attention in any particular cycle, (thus vivifying it with Life energy) through which primal Causative energy will then flow. This Causative energy carries forth with it all the streams of Lives of the twelve Creative Hierarchies (the embodied *prāṇic* streams flowing from the twelve petals of the Heart centre of the solar Logos) that will find their field of application upon that globe.

These *prāṇic* streams of Causative energy flowing through the *nāḍīs* of the planetary system take the form of a primal Atom. They are directed by the streams of application of Ideation by the Logos. They become the Plan as seen through time. Different intonations and hues of the Ray Lives are made to effect arenas of application of the various *chakras* constituting that body of manifestation. The energies flowing through these *chakras* condition the interplay of entire groups of organs and the associated Lives within that manifest expression. They are directed by the conscious mediators between the Logoic Mind and the unconscious streams of Lives. The mediators are Chohans of the Rays, Masters of Wisdom, plus their *deva* (angelic) correspondences. There is much overlapping of cycles, whilst the three modes of energy expression (the three *guṇas)* find application on each level of unfoldment of these Causative streams.

There are seven Chains (of Incarnations) to one planetary Scheme. Each can be seen as a psychic centre, a *chakra* in the body of a solar Logos. (Making 49 Planetary Incarnations to one solar Incarnation.) There are ten (7 + 3) such Schemes composing the *chakras* of the Logos of our solar system, according to esoteric lore. These are the major sources of Causation within the embrace of the triune Logos that govern our solar system. These sources draw their potency (the will-to-be, the will-to sustain the Impulse, and the will-to-resolve the All into the One) directly from the Seven Rishis of the Great Bear via the Seven Sisters (the Pleiades[8]). The central Plan, the modifying karmic adjudication, stems from Sirius, the central Heart of all evolutionary journeying for those in our solar sphere.

All constellations within the Body of THAT Logos[9] Live within a reticulation of Causative energies that become the major lines of force that draw them all into the one united Organism. All constellations are related and revolve round the central hub of the great wheel of Life, of the Law. *Draco,* the Fiery Dragon of Life, embodies the furnace (cosmic *kuṇḍalinī*), that vivifies or sustains the Fiery Breaths of all Causative streams. We must however look to the twelve great zodiacal constellations to understand the Way of unfoldment of the Logoic Heart that is the originating Mind of All. Logoic vitality is drawn from *Orion's Belt,* from which all creative Impulses are sustained, and to which the rejected streams of *prāṇic* Lives must go. Orion is the Heart, but Taurus the bull leads the Way, for His Eye (Aldebaran) pierces the veil of all that is to be born in the Womb of the Great Mother, wherein these Causative energies must play their roles. The Pleiades, assisted

---

8 The role of the Pleiades with respect to the formation of a solar sphere and the forces governing the conditionings of the cosmic dense physical plane was given in *The Constitution of Shambhala,* part B, and will be elaborated with respect to the causative process in chapter nine.

9 The great presiding Logos of which the 88 constellations observable in the night sky represent the *chakras.* Note that there are also an unspecified number of constellations that are also presently disincarnate existing in cosmic etheric space. Logically, in terms of the organisation of the *chakras,* there should be 96 constellations, hence we should look for eight etheric constellations, whose energy fields should be detectable in space with the right instruments.

by all constellations and stars with a feminine name, such as Virgo and Coma Berenices, are the Mothers of all that we see. They were named thus by enlightened Seers long ago because of this role. The Pleiades build the forms of the planetary spheres from the available substance existing in nebulae of cosmic dust in the manner described in chapter nine.

Logoi as the creative Word

In the section below I shall provide some information concerning the executives governing the evolution of our planet. All other planetary and solar spheres, Lords of constellations, etc., have a similar governing Council directing the evolutionary process of that which they embody.

The incarnate representative or Personality of the Logos that informs our planetary Scheme (for the two are one in actuality and effect) is called Melchisedec in the Bible.[10] The equivalent term utilised in my book *The Constitution of Shambhala,* part A is Sanat Kumāra, the Lord (King) of Shambhala. He is a cosmic Avatar, embodying (anchoring) the functions of the triune Logos of this planet, (Father-Son-Mother). I shall repeat below somewhat what was previously stated in that book.

Sanat Kumāra is the embodied manifestation of the combined function of the Avatar of Synthesis, the Spirit of Peace and the Mother of the World. Some of His appellations are: 'the Ancient of Days', 'the Eternal Youth', 'the Great Sacrifice'.

From the Buddhist perspective the best term that depicts the attributes of this great one is the Ādi Buddha, Samantabhadra. The term means the one who is all good, ever perfect, manifesting universal goodness or joy. It is a synonym for the *dharmakāya,* manifest in the form of a primordial Buddha, as recognised by the Nyingma tradition of Tibet. Samantabhadra can be viewed in terms of the natural or spontaneous luminosity that is the 'masculine' attribute of *dharmakāya*, secondly as the 'feminine' form (Samantabhadrī), which represents Emptiness (*śūnyatā*). We also have the non-dual attribute (yab-yum) that coalescences the appearances of *saṃsāra* with emptiness. He is also one of the eight great Bodhisattvas, and it is in this fusion in the

10 King James version. Also spelt Melchizedek

qualities of primordial Buddha and eternally manifest *mahābodhisattva* we look to the basic quality of Sanat Kumāra. Other terms utilised for the Ādi Buddha are Vajradhara and Vajrasattva.

The Kumāras are said to be the 'mind born' sons of Brahmā, 'virgin youths', who refused to procreate and thus remain *yogins*. There are seven of these, three esoteric and four exoteric. They embody the substance of Mind for a planetary Scheme, and are thus responsible for the dissemination of the patterns of the Mind of a creative Logos. They are Sanat Kumāra's representatives for the sum of the evolving lives within embodied space.

Sanat Kumāra can be considered the awakened Personality of the Planetary Logos. (Who embodies the Sambhogakāya Flower aspect of our planetary Scheme.) For this reason he can be represented as the 'God' of the Theistic religions, for we 'live and move and have our being' in His body of manifestation. The symbolism presented in the Bible relating to Melchisedec can also be utilised to gain an added dimension as to the nature of such a Being. The Planetary Logos is an abstracted Trinity, whilst Sanat Kumāra represents a manifest quaternary with the three Buddhas of Activity, which were described in *The Constitution of Shambhala*. There is a Divine Mystery here as to why this relationship exists. This is hid in the significance of the earth being considered a non-sacred planet, and happenings upon the former moon Chain of activity.

In utilising the Biblical terminology we find that the Christ is a direct representative of this Regent, as Paul explained:

> For every high priest taken from among men is ordained for men in things *pertaining* to God, that he may offer both gifts and sacrifices for sins[11]...So also Christ glorified not himself to be made an high priest; but he that said to him, Thou art my Son, today have I begotten thee. As he saith also in another *place,* Thou *art* a priest forever after the order of Melchisedec[12]... And being made perfect, he became the author of eternal salvation unto all them that obey him; Called of God an high priest after the order of Melchisedec. Of whom we have many things to say, and hard to be uttered, seeing ye are dull of hearing.[13]

11 *Hebrews 5:1.*

12 *Hebrews 5:5-6.*

13 *Hebrews 5:9-11.*

The Christ's role of 'high priest' here betokens his placing as the head of the Hierarchy of Light. If one takes Melchisedec to represent the planetary Logos, then 'the order of Melchisedec' implies that the Christ is following in the 'footsteps' of this One, by being 'made perfect'. The phrase 'we have many things to say' implies that 'we', the liberated beings upon earth, would speak much concerning this Melchisedec, however the information to be presented is very difficult to convey ('hard to be uttered') because of the high esoteric nature of its content, whilst the listeners are 'dull of hearing', meaning that the spiritual beings, the disciples who would try to listen have not yet developed the inner faculties (inner hearing) to be able to comprehend the nature of the teachings to be presented. Such it has always been when concerning Shambhalic matters and the planes of Causation whereby the Logos resides.

In Chapter Seven Paul explains the nature of Melchisedec:

> For this Melchisedec, king of Salem, priest of the most high God, who met Abraham returning from the slaughter of the kings[14] and blessed him; To whom also Abraham gave a tenth part of all; first being by interpretation King of righteousness, and after that also King of Salem, which is, King of peace. Without father, without mother, without descent, having neither beginning of days, nor end of life; but made like unto the Son of God; abideth a priest continually. Now consider how great this man was, unto whom even the patriarch Abraham gave the tenth of the spoils.[15]

'Salem' is generally taken to be an abbreviated version of Jerusalem, the 'holy city' or 'city of God', the term meaning 'possession of peace' or 'foundation of peace'.[16]

The title given to Melchisedec is 'King of Peace', but that normally given to Jesus and to the Jewish Messiah is 'Prince of Peace' (*Isaiah 9:6*). A prince is subservient to a king, but the quality of 'Peace' is

---

14 This story is told in *Genesis* chapter 14.

15 *Hebrews 7:1-5.*

16 See *Easton's Bible Dictionary* (Thomas Nelson, 1897), under the index heading for Jerusalem. There it is stated that 'It stands on the edge of one of the highest table-lands in Palestine, and is surrounded on the south-eastern, the southern, and the western sides by deep and precipitous ravines', which symbolises the exalted heights of Shambhala, of which Jerusalem is the earthly representative.

that which qualifies them both and is fundamental to their beings. In addition the Christian eschatology makes of Christ-Jesus the (unique) 'Son of God', then it is easy to transpose Melchisedec as an even more exalted 'Son', hence the Logos above Jesus-Christ, and to whom Christ submits as a supplicant under the general rubric of 'Peace', which is a major attribute of Shambhala.

The energy of peace, or of tranquillity, manifests via the buddhic plane. It can also be equated with *śūnyatā,* that which is Void of space-time as we understand it. Peace implies a state of tranquillity in a person's mind and entire body nature, allowing the highest spiritual energies to manifest and be unequivocally recognised for what they are. This energy is then conveyed to the entire Hierarchy of Enlightened Being, those that bear this energy in such a way that its attribute can manifest upon a planetary scale. That such an effect is not immediately evident upon the planet is due to the fact that humans have the free will to counter that which is broadcast for their beneficence. Nevertheless the energy of peace will inevitably travail upon the earth as the *karma* of human failings is rectified and the peace bringers increasingly manifest to produce the Logoic agenda.

Both abovementioned beings are 'high priests', worshippers and receivers of the Word of God in a direct and unequivocal manner, though the one who qualifies this particular priesthood and gives it its particular characteristics is Melchisedec. (Even Jesus was 'made an high priest forever after the order of Melchisedec'—*Heb. 6:20.*) As an embodiment of the planetary Logos, Melchisedec is also 'high priest' to the 'the most high God', which can be considered the solar Logos (the One above Him).

As the Christ-Jesus was anointed a high priest 'after the order of Melchisedec', he can thus dispense the Word of our Logos to all sentient beings who are receptive to the influence of this priesthood. This Word is an expression of that magnetic 'pull' that is Causative of our paths to the realms of Light Eternal. Christ-Jesus here stands as the head of the hierarchy of Bodhisattvas who have vowed to perpetually demonstrate that Word in such a way that eventually all humanity will indefatigably be drawn to the zone of peace and serenity wherein these Lords of Love reside. As 'priests' they administer the attributes of the Logoic Word to the general world at large. They maintain the mantras of the Logoic Command so that the planetary purpose can manifest as planned. They deal with the fine-print and minutiae of the dissemination of the Word, thus ensuring its effectiveness amongst the human population, taking into account all of the laws that dictate the evolutionary process.

All enlightened beings who choose to remain on earth with our humanity therefore automatically become subservient to the Will of this great presiding Lord (Ādi Buddha), which manifests as the energy of Peace working to overcome the strife and disharmony that is prevalent amongst humanity. All must give a symbolic tithe to support the 'clergy' of Melchisedec's order, as did Abraham, the Father of the 'chosen of God', or else become 'priests', as did Jesus. This tithe[17] concerns taking upon oneself part of the burden of the *karma* of a planetary Logos. In doing so one serves the magnetic potency of the Logoic Creative Will as agents of rightly directed evolutionary change. The Will manifests as an expression of the compassionate undertaking of an Initiate of high degree, a Bodhisattva. For though the Bodhisattva may be relatively *karma*-less as far as his/her own material plane *karma* goes in the sphere of human suffering, he/she appropriates some of the burden of the world and walks in the 'fellowship of the Christ's sufferings'.[18] This 'fellowship' represents the sum of the Ashrams of the Masters of Wisdom, which constitutes the Ashram of Melchisedec, headed by His chosen representative, the Christ (Maitreya),[19] the Master of Masters.

Other Bodhisattvas may go to become 'priests' of the order of other planetary Regents (Ray Lords), or even leave the solar system altogether after they have become Buddhas. A priest serves at the altar of a Lord (of sacrificial Love-Wisdom), dispensing His benediction to those that lift their faces upwards for nourishment. The priest serves to draw the streams of evolving lives back to the ultimate source of Causative power via the teaching dispensation. They are predominantly along the second Ray. (The entire Hierarchy in fact embodies the various

17 A tithe is traditionally one tenth of one's earnings, where numerologically the number one relates to the energy of the will, and the number ten to that which makes perfect.

18 *Phil. 3:10*, 'That I may know him, and the power of his resurrection, and the fellowship of his sufferings, being made conformable unto his death'.

19 Note that there is a difference between the Christ and Maitreya, as explained in *The Constitution of Shambhala*, part A, though the two bear the same energy of compassion.

Ray and subray attributes of the second Ray.) The priest dispenses the benefits of Love-Wisdom to everyone, assisting them to eventually become primal agents of causation.

The *Jerusalem* to which Jesus 'steadfastly set his face to go to'[20] is the earthly embodiment of the 'city of the great King' (of righteousness). It is the Biblical correspondence to Shambhala, the city of the great King of Asiatic mythology and Western esoteric tradition. David made this city both the royal and religious capital of the Jewish nation, in which the house of the Lord was finally built by Solomon. (As described in the first book of *Kings*.) Shambhala is said to exist in etheric substance somewhere in the Gobi desert. It represents the Head centre of our planetary system, the kingdom of 'God', presided over by Sanat Kumāra, the 'Eternal Youth'. From His council chamber emanate the Commands (words of Power) that effect all major changes and cycles of our planetary Life and related civilisations.

Dynamic Peace (*śūnyatā*) is indeed a description of the nature of the quality of the power that is at the base or heart of the Causative impulse. Only via such a zone of quiescence can emanatory streams of Causative energy flow, for immovable and immutable Peace is the foundation of being/non-being, around which there is perpetual change and incident, the evolving transient forms.[21] Peace is the foundational substance, the basic building material of the City of this great Lord. The emanating Causative energy however stems from a higher *dharmakāyic* zone of the expression of Life, viewed in terms of cosmic astral and mental energies.

The key to understanding the nature of Melchisedec is the statement in *Hebrews 7:3:*

> Without father, without mother, without descent, having neither beginning of days, nor end of life, but made like unto the Son of God; abideth a priest continually.

Here Paul speaks of an eternal Man who was not 'created' as humanity was said to have been, who had no line of descent, and who

---

20 *Luke 9:51.*

21 In the subatomic world this quality has its analogue in the ground or rest state associated with the quantum vacuum.

existed in a timeless zone, that of the eternal Now. He was however made 'like unto a Son of god' that is, made to look like, or take the form of a Christ or human unit, for both are 'Sons of God'. In Buddhist terminology His *nirmāṇakāya* would thus take the shape of a fully enlightened Buddha-like human form, a causeless, eternal, king of the *dharma*. (As the Buddha is generally represented.)

The fact that this persona 'abideth a priest continually' means that he continually lives in the capacity of one who Initiates humanity into the Mysteries of the kingdom of 'God'. He does this forever, for he has 'no beginning of days nor end of life, without father, without mother'. He is therefore 'unborn', similar to the Buddhist and Hindu state of Mind that a Buddha (Tathāgata) or fully enlightened being continually resides in—*nirvāṇa*, *mokśa*, the *dharmakāya*, etc. These words, which mean 'unborn', 'unmodified', 'unmade', 'the fount of the Law', relate to states of being liberated, supra-human. Hence this verse describes the attributes of the 'Son' or Love-Wisdom attributes of Deity, a liberated Christ, who promulgates the *dharma* of the 'most high God'.

Esoterically Melchisedec is therefore *The One Initiator*, 'The Great Sacrifice', Who is said to stand steadfast for aeons over our planetary life, ever watchful over the destinies and activities of humanity. When people pass the necessary tests of Initiation, he then wields the Rod of Power that Initiates them into the Mysteries and energies related to the source of their being. He manifests the appropriate beneficence befitting the subjects of his domain that serve him and his Kingdom well. He is the Father to whom the Christ, the 'Prince of Peace' *(Isaiah 9:6)*, must go and 'sit at the right hand of', eventually ascending to his throne as a rightful heir. (For Melchisedec must ascend after His aeonic servitude to this planet to greater spheres of Being and thus unfold other roles in this great 'Day of Brahmā', the *mahāmanvantara.)* Melchisedec is the epitome of the act of sacrifice,[22] for he receives no personal gain from his ever-watchfulness, the protecting and nurturing of all the kingdoms on earth, and the sanctification of all streams of manifest Life.

It was stated that in the Hindu philosophy Melchisedec's correspondence is called *Sanat Kumāra*.[23] Sanat Kumāra is said to be

22 *Sacrifice* meaning 'to make holy or sacred'.

23 Meaning 'the eternal Virgin Youth', as explained in *The Secret Doctrine* by H.P.

one of the four 'mind born' sons of Brahmā-Rudra, and is the oldest of the progenitors of mankind. Esoterically, he and his three brothers form a quaternary or Seat of Power, a Base of Spine or *mūlādhāra chakra*. They embody the sum of the Life streams of the mineral, plant, animal, and human kingdoms. As Melchisedec is the oldest, the human kingdom is his direct concern.

The lesser three Kumāra are also called Buddhas of Activity, as they are attained Buddhas from past aeons of unfoldment who are in continuous active creative dispensation. (As far as the realms of form are concerned.) They particularise the potency of the triune Word embodied by Sanat Kumāra, and dispense it to the sum of manifest Space constituting the three kingdoms of Nature (animal, plant and mineral) below that of the human. Consequently they are causative of all progressive cycles of opportunity for those kingdoms. They stand firm on their high zone of Peace, but continually resonate the mantras of spiralling mutable activity to the streams of lives unfolding in the formed realms.

Sanat Kumāra sounds the Word of liberation, of abstraction for all groupings of lives when these lives master each particular cycle of endeavour. Two of the Kumāras are said to have 'fallen', i.e., directly come to embody the substance of the sheaths of the kingdom of Souls and of a certain portion of the angelic kingdom. They have thereby incarnated into cosmic dense substance, and are accordingly Causative of all base (primal) conditionings thereon.

The four Kumāras and their integration with a fifth (the Mother of the World, explained in *The Constitution of Shambhala),* correspond to the five Dhyāni Buddhas. Sanat Kumāra, the 'One Initiator', is the Head of the various Hierarchies and streams of Life on this planetary system. With the three Buddhas of Activity, the Lipikas, plus the symbolic 'four and twenty elders',[24] etc., He embodies the Head Centre (the city of Shambhala) within the body of the overshadowing planetary Logos

---

Blavatsky Vol. 1, 176-179, sixth edition. The four Mind-born sons of Brahmā who refused to create progeny remained therefore ever pure 'virgins'. Their names in Hinduism were: Sanat Kumāra, Sanānanda, Sanāka and Sanātana. Sometimes a fifth was added, Ribhu.

24 *St. John's Revelation,* Chapter four.

(the Sambhogakāya Flower aspect). This Centre anchors His active will, which in turn becomes the prime Causative agent for all on earth, distributing the Fiery *dharmakāyic* energies from the Mind of the greater Logos. This manifests via the mediatorship of the Masters of Wisdom (necessarily functioning from *ātmic* levels) and then to their respective Ashrams.

The Hierarchy of Enlightened Being embodies the Heart Centre of Deity, manifesting active Love-Wisdom for the planet. The focal point is the Christ, who expresses the energies of the Son aspect, the second or great Teaching Ray for our planetary system. Together with the Lords of Shambhala they build a great protective aura around humanity, helping to prevent the intrusion of the destructive potency of (cosmic) evil, so that purposeful causative evolution of the evolving Lives can come to a meaningful conclusion.[25] To assist in this task this great Bodhisattva works to establish the dispensation of the various Schools of divine Wisdom and philosophic groups in the world. The Christ is the world Teacher[26] that slowly but surely educates humanity (manifesting throughout the lower centres of our planet) to become causative agents, to create with the Light of the seven Rays, by becoming Initiated into Hierarchical modes of expression.

This work is assisted by that of the *Manu*, who embodies the energy of the first Ray of Will or Power for humanity. He rightly assesses and regulates the sum of their initiatory wills, be they creative or destructive. (Essentially however, his purpose is the direction of all of the streams of Life upon our planetary system.[27]) It should be noted here that *karma* is created once the energy of the will is imposed to change or manipulate anything, for *karma* is but causative energy set in motion by the will, and which must later be rectified by a similar, but opposing force. Desire is but the lowest reflex of the will, working out via the personal-I. The Manu and his agents are responsible for the

25 In a similar way the earth is protected from harmful cosmic rays that would be destructive to all life on this planet by the magnetosphere.

26 In *The Constitution of Shambhala* the office of the Christ's department is described and the historical evolution of three Christs is explained.

27 See *The Constitution of Shambhala* for detail concerning this and the other planetary executives, of their modes of interrelationship.

cyclic evolutionary progress of the various races and groups of humanity, of the cyclic and terrestrial changes of all upon the earth, as well as the various movements within the political and international arena of human activity. His is the destroying energy that must rectify the imbalances of old crystallised forms and forces in human societies so that the forces of the Christ can build the New. His is the transforming energy that prepares the way for the coming of Light.

The third executive member of this Hierarchy is termed the *Mahāchohan.* He embodies the third Ray of Activity and is responsible for the projection of the activities producing right progress of the economic, artistic, scientific and social developments within our civilisation. Under him work the Chohans of the five Rays of Mind (third, fourth, fifth, sixth and seventh). Then come the Ray Ashrams of the Masters of Wisdom that are responsible for the scientific, devotional, etc., development of humanity. The mahāchohan's department represents that of the Mother for our planetary civilisation.

This is a very basic outline of the planetary executives who guide humanity, but they were explained in some detail in *The Constitution of Shambhala.* Sufficient information was then given for the student to gain appropriate information concerning their expression as causative agents under Ray impetus for the cycle immediately ahead of us. Much detail has also been presented in the works of Alice A. Bailey, therefore no further comment need be added here.

The intelligentsia of humanity embody attributes of Sanat Kumāra's Throat centre. They are the repositories of the Creative Word, for from this centre the Fires of Mind find their application. They are the creative intermediaries between 'his Mind' and the rest of the body of manifestation that is the human *māyāvirūpic* cesspool.

Saturn, the Lord of *karma,* rules this centre astrologically. Humanity thus embodies the manifested effect of the contact and associations of our planetary Logos with extra-systemic (cosmic) entities. Humanity represent the focal point or means whereby the Lord of Saturn helps to rightly resolve the effects of human thoughts (with their emotional responses) that distort the Bodhisattvic Impressions arising from Shambhalic realms. Via Shambhala *karma* works to cleanse the effects of slovenly thought structures that keep people trapped in *saṃsāra.*

Human *saṃskāras* from the long distant past are brought to the fore to be reinvigorated upon a higher more vibrant manner. New ideals and images can then be projected into the manifest, concretised human mind-space which will produce future intent for them. This will manifest through the development of human self-will in such a way that they can in time learn to become directive agents of these elevating karmic streams. They will then ride the waves of causative energy that are expressions of the Will of the Logoic Thinker. To bear that Will they must learn to 'think with the Heart', for only the Heart can withstand its potency without disruption of the refined structure of consciousness.

Much of the suffering associated with the human form, of the nature of pain viewed in terms of it being a major tool of evolutionary impulse, finds its source in the effect of human response to the incoming waves of causative energy. For as humanity struggles with the pressure of what the future holds in store, the resistance of their oft unwieldy, concretised, ungainly emotional and reactionary desire-minds then produces its inevitable shower of friction and pain. Humanity's extreme resistance to changing for the better, to the new impulses and ideas to which they must continuously adjust produces its inevitable reward in the associated societal upheavals. History records these struggles in terms of wars, revolutions, the 'clash of civilisations', the 'march of science', secularism, and so forth.

As Melchisedec pursues His Meditations on cosmic realms and contacts increasingly more refined primal sources of causative Revelation, so the effects pass down to the human domain via a concerted effort by the mediating Masters that are responsible for the sum of human education. The continuous need for change and progressive development towards Mind is a reason why human Monads are esoterically called 'Lords of Sacrifice'. (Those who choose to suffer and die for the purpose of the advancement of the greater whole.) They carry with them the mode of evolution of many lesser kingdoms of Nature. They are the active agents for cleansing the karmic impediments that prevent the manifestation of more vibrant states of being attaining the way of ascension.

The Hierarchy of Enlightened Being embody the Heart centre of the Logos via the Airy realm, the *buddhic* plane upon which they largely consciously reside. The entire Life process is directed via the Heart centre beating out its directives via the Hierarchy that project the Airy *prāṇas* of the planet into manifestation. These *prāṇas* contain attributes from cosmic sources that manifest as the 'nutrients' and 'hormones' that vitalise the planetary whole, the entire Logoic Body. All is sustained by the rhythmic pulsations of the Heart. It influences every aspect of the embodied form via the *nāḍī* system. Obviously disease and debilitation can take their toll, as do the forces of evil, but the Heart (Hierarchy) works feverishly to counter the negative aspects of such *karma* engendered by humanity. From this perspective they are agents of the Lord of Saturn, but the true method of karmic adjudication manifests via the *devas.*

The at-onement that serenity and peace bring is the common effect of all that work from the Heart. Air sustains the Fires of the world, in a similar sense as *buddhi* sustains the *manasic* principle. Vapour or moisture condenses from air to form the Waters of Life, both subjectively and objectively. From this realm therefore, emanate the causative streams of energy that breathe Life into all manifest things, as was also 'breathed' into the Adam of clay by the 'God' of the book of Genesis.

Self-conscious *deva* units that are the correspondences to members of humanity basically constitute the forces pertaining to the Solar Plexus centre of the Logos. They are the evolutionary emanations of the Logoic Mind that are galvanised into action by means of the creative Word and energy directives sent from above. Here it should be noted that the Solar Plexus is the seat of the personality will, of all self-focussed activity. From it upon a planet-wide scale manifests the many wills (the higher orders of the *deva* kingdom) that cause the manifestation of all the diversifications seen in Nature, the myriad individual entities and species that have evolved throughout the epochs.

The lower mental-emotional wills of humanity govern the turbulent energies that need to be pacified and transformed by the forces working via this centre. The *devas,* on the other hand, are the forces that regulate the evolutionary development of all aspects of the various kingdoms of Nature. They are the Creative Intelligences designing all of the vicissitudes of every animal, plant and mineral entity. They direct all species to a pre-planned conclusion. They can also be found as the primal, 'unruly' *deva* forces manifesting the forces of Nature in its most 'passionate' stages of expression—storms, volcanic eruptions, cataclysmic continental changes, plagues, and the like.

The clashing arenas of tension or focus of the human personal will must become subdued if this will is to be directed by the Mind that is the active Heart. The Solar Plexus is also the centre of the fluid Watery emotional body that must eventually be directed by the prime transformative principle of the Heart. The Watery substance clothes, gives a body of manifestation, to the causative streams of the creative agent, be it the Logoic, Hierarchical, or human thinker.

It would be erroneous to think that humanity embodies only the Throat centre, or the *devas* embody only the Solar Plexus, for the self-conscious *deva* and human units can be found in all the *chakras* in the body of the Logos, whilst the subhuman kingdoms form the organs and substance of that body.

5

The Causal Body and the Monad

The planes of perception and the human constitution

This chapter is mainly concerned with presenting some necessary background information concerning the constitution of the subjective planes of perception. The reader can then comprehend better the mode of the causation of things and visualise clearer the attributes of what constitutes a Logos, as well as where such a One fits within this multidimensional universe. Consequently much esotericism and its associated terminology is introduced that is alien to modern empirical thinkers educated in the fare of the scientific community. Those unfamiliar with the subjects introduced should keep an open mind and take what is presented as a working hypothesis until they have developed the inner vision to clearly see for themselves the nature of internal reality.

In the section below I shall present two figures. Figure 5 gives a basic view of the interrelation between empirical and universal consciousness, with the focus being the illusory appearance of the phenomenal form. It depicts the principles governing the average person ensconced in *saṃsāra*. This appearance is depicted in the form of a square. Figure 6 summarises the planes of perception and places the Sambhogakāya Flower (Soul) as the focal point for the evolution of the various kingdoms of Nature. In the juxtapositioning of the two figures we see that the Sambhogakāya Flower resides within *manas,* the third aspect of the triad of universal consciousness shown in figure 5.

This information will allow the reader to better visualise the esoteric schema concerning the attributes of a human unit. This is needed because when considering the causation of phenomena, either as a product of a Logoic Mind or a human mind, one needs to comprehend the subjective constitution underlying the appearance of things. From this perspective we see that the phenomenal appearance, though important, is not all there is to consider when trying to comprehend what a universe is and the forces that condition it. In reality the subjective constitution is more important than the objective, however the objective universe is the place for the appearance, transformation, transmutation and elevation of its substance-matter to the domain of mind. That which resides upon the subjective domains works to achieve this end, for which reason great Lives incarnate, at whatever level of expression one wishes to observe the process, human, planetary, solar or beyond.

Figure 5, which summarises the information presented so far concerning the principles constituting a human unit, shows the differentiation between our physical, psychic, and spiritual constitution. It presents the mechanism for deed, word and thought for the human personality. The square that is depicted symbolises what is related to the material world, the sum of a human personality, as dominated by the desire or emotional mind. It shows that the intellect, astral body, etheric body, and dense physical form can function as a separate unit without guidance from a higher self.

Prāṇa is the binding energy that coherently interrelates these principles and vitalises their expression. It is the vehicle of transmission of the force vortices directed by the causative agent. It is that with which one builds if the originating thought is to be objectivised in concrete substance. *Prāṇa* conveys the *saṃskāras,* the mental-emotional attributes developed through past cycles of activity. The *saṃskāras* manifest into the psyche of the individual as strong tendencies pushing the mental-emotions to manifest in a certain way, depending upon the strength of the force by which the *saṃskāras* were generated.

The higher correspondence of *prāṇa* is termed *ākāśa*, of which *buddhi* is the medium of transmission, in a similar way that the fourth ether is the transmitter of *prāṇa*. The word *ākāśa* derives from the root *'kas',* meaning to shine, radiate. It is the radiant cosmic electrical energy that causes the appearance of solar light when contacting dense substance. (Here of the higher mental plane.) Hence it is the energy that sustains the activity of the Causal form of the Sambhogakāya Flower, and is projected through its central jewel (see figure 6). The zodiacal and stellar energies are conveyed to our system in the form of *ākāśa*. *Ākāśa* has the same relationship to this Flower as *prāṇa* has to our corporeal form, consequently *ākāśa* is the vehicle of transmission of the causative principle of an enlightened Being. It is thus *space*, the subtle and ethereal fluid pervading the totality of the universe, and is a vehicle of Life.

Ākaśa then can be considered the primordial substance of systemic (manifested) space. It can be considered to represent the four cosmic ethers shown in figure 6. This etheric substratum of the cosmic dense physical plane therefore conveys the five cosmic *prāṇas*. *Ākaśa* can also be considered the elementary (plastic) substance of the *dharmakāya* wielded by a Buddha-Mind to convey the expression of his wisdom as the *dharma*. Without the *ākaśa* the spontaneity of the enlightened Mind could not be expressed. It conveys the energies governing the manifestation of the elements, atoms, and attributes of phenomenal life. It is the medium that allows the *saṃsāra/śūnyatā* nexus[1] to convey the wisdoms of the five Dhyāni Buddhas. *Ākaśa* (in this five-fold division) therefore also has veiled in it Mahat (cosmic Mind), *āpaḥ* (cosmic Waters), which makes a sevenfold differentiation, that represents the seven-headed serpent upon which Nārāyaṇa[2] rests as he floats upon the Waters of cosmic space. When *ākaśa* impacts upon consciousness the potency of its energy is toned down and is conveyed by *prāṇa*, the streams of energy qualifications that vivify consciousness and which consciousness qualifies by the action of its own volitions via the activity of the sense-consciousnesses.

The life of a human personality, shown as the 'square' in figure 5, is dominated by whatever principle rules. Presently for most this is the desire or emotional mind. Very few break free from the tyranny of the desire-emotions, to live almost exclusively in their mental body.

---

1 Explained in the first three volumes of *A Treatise on Mind*.

2 The 'God' in humanity, the one who moves on the Waters of space, the ocean of consciousness. (The ability of a Logos to travel through the cosmic astral Ocean.) It is the god Viṣṇu, who reclines upon Nāra the primeval (milky) Waters upon Śeṣa (seven-headed serpent) as his couch. This signifies The primal manifestation of the consciousness aspect of the Deity that spreads outwards to encompass absolute space.

(The world's intelligentsia are mostly focussed that way.) Such focus then determines the quality of expression of a person manifesting as a causative agent. The triangle pointing upwards in figure 5 represents the *Spiritual Triad,* of *ātma-buddhi-manas,* and signifies the principles gained by an enlightened being, once such a one has transcended the limitations of the 'square'. The attributes of this triad is generally summarised by the term Mind in my writings, but we can see here that the term is generic and veils various differentiations of the states of Awareness obtained by enlightened Ones. It can refer to the abstract Mind, the Clear Light of Mind, the consciousness of the Sambhogakāya Flower, the Heart that is Mind *(buddhi), ātmic* perception, and the various stages of development of cosmic Mind.

The triangle in the circle above the Spiritual Triad symbolises the Monad, whose qualities will be explained below, or else the attributes of the triune Deity, within Whose body of manifestation we all reside. The dotted line stemming from the Monad is the *sūtrātmā,* the Life line that is projected into the formed realms to cause the appearance of the embodied form. For the Monad this form is the Sambhogakāya Flower, and for that Flower it is the human personality. The life of the form persists as long as the higher entity maintains this link. When the link is withdrawn the form dies, as it is no longer vitalised in any way. For humans the *sūtrātmā* is anchored in the Heart centre, and from it radiates out the entire *nāḍī* and *chakra* system.

The dotted triangle pointing downwards in figure 5 is the symbol of the smallest unit that can function as a separate entity which can be created as a causative agent. Below that, the unit disintegrates into the elements that composed it. When this triangular unit functions as a separate entity from the higher Mind, then it is occultly termed *the dweller on the threshold.*

The 'dweller' represents the *prāṇic*, emotional, and mental elements incorporated within all our constitutions. It is a bundle of aggregates consisting of the built up and reinforced desires, wishes, phobias, etc., that have been amassed throughout the aeons and which constitute a karmic heirloom. It is the 'devil within', and there eventually comes a time when all beings must battle with their 'dweller on the threshold' (of enlightenment) before real spiritual progress can be made.

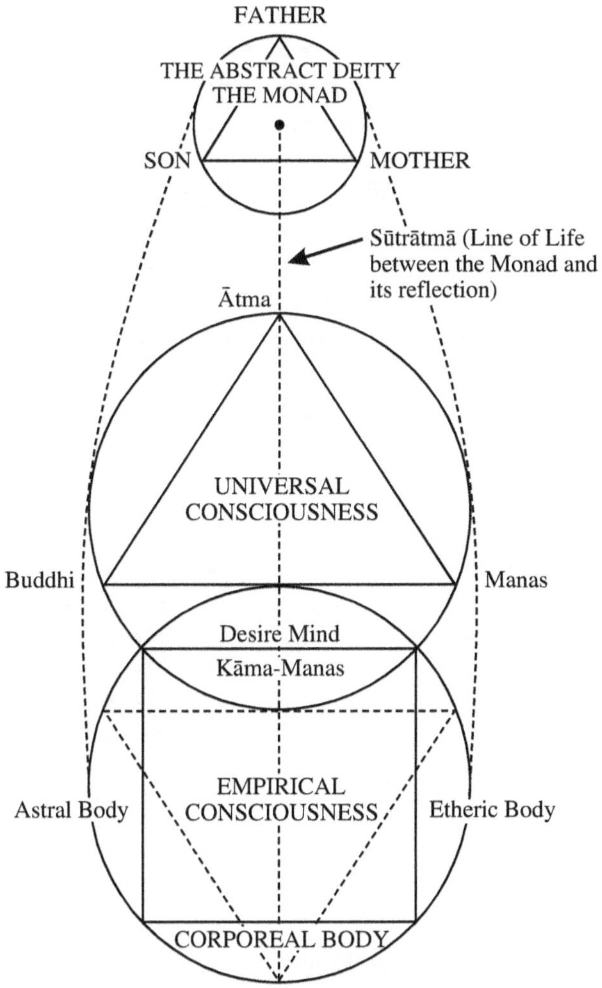

Figure 5. The principles constituting a person

Humanity have built an immense cumulative 'dweller' by means of their misplaced thought-forms and self-focussed causative streams of interrelationship. It is constituted of the myriads of glamorised images and thoughts generated by them, of everything constituting the *māyā* wherein people reside and which prevents their gaining enlightenment. It can be considered 'evil' whenever it influences the general population (as it inevitably does), preventing their right evolutionary attainment. It is effectively a thick, darkened cloud of repetitive, cancerous images of consumer items, desired ideations, commercial slogans, religious and cultural images, hallucinations concerning the world of sports and movie stars, sexual imagery of all types, political sloganism, avariciously looming wealth generating schemes, and malicious broodings surrounding us all. When it is destroyed through right thought, the demonstration of light, applied Love, and goodwill to all, then humanity will receive its full endowment of divinity.[3]

The *astral body*, a subtle ephemeral form, has sometimes mistakenly been called the Soul. In Egyptian mythology this principle was called the *ka*. The information presented in the popular book by Oakes and Gahlin explains some aspects of this body from the Egyptian perspective, which is erroneous in relation to the astral body needing physical food. The real food for this body being the *prāṇas* of one's mental-emotional *saṃskāras*:

> The *ka* was thought to come into being at the birth of an individual. Dying was sometimes described as 'joining one's *ka*'. The *ka* was intimately linked with the physical body, which was regarded as the vessel for the *ka* after death. This explains the belief in the need for the survival of the body, and the measures taken to preserve it whenever possible. *Ka* is often translated as 'spirit' or 'vital force', as in the creative life force of an individual that enabled the generations to continue through the ages. It was believed that the ka required food and drink, so offerings were made to it for as long as possible after death. In fact the word *ka* sometimes means 'sustenance', depending upon the context.[4]

A new astral form is formed with each new incarnation. It embodies the nature of the growing emotional and desire body of the human unit. It leaves the physical form when one is asleep, allowing travel in the world's astral space whilst the personality slumbers. The *ka*/astral body is then attached via a 'silver cord', known as an *antaḥkaraṇa* ('consciousness link') that links to the personality via the naval area, allowing the individual to meet with loved ones anywhere in the world, visit the heavenly and hell

3 See *Glamour A World Problem* by A.A. Bailey for an elaboration of this subject.

4 Oakes, Lorna & Lucia Gahlin, *The Mysteries of Ancient Egypt* (Hermes House, London, 2005), 393.

states of the deceased, and bring back the experiences if need be to the awakened consciousness. These are often revealed as lucid dreams. Upon death the *antaḥkaraṇa* is severed and the astral body is freed completely, but often the astral form is 'earth bound' for a while, frequenting the places of habitation, friends and family known whilst 'alive'. This fact then in part is what possessed the ancient Egyptians to offer the *ka* foodstuffs, etc., as objects of oblation and devotion, to try to keep the *ka* as close to what was considered good in life as possible. The ignorant masses were taught the most reified aspects of the doctrines, thus learnt to think in terms of physicality of the deceased life as part of the objective of their religious devotions. There however was a truism here, for the astral plane (the Egyptian *Amenti,* the 'underworld') is constructed by, and as a consequence of, the collective desire-minds of humanity, thus manifests a close parallel to what people were most familiar with whilst incarnate. We thus have astral houses, lakes, animals, food, music and forms of pleasure, plus experience of the *devas* that are indigenous to that realm. All activity being conditioned by the laws applicable to the astral plane.

When a person dies his/her vital (etheric) body is intimately integrated with the astral form and takes some time to dissipate, which keeps the astral body 'earthbound' during this period. This interconnectedness between the vital and astral body caused the confusion of terms, allowing the definition "*Ka* is often translated as 'spirit' or 'vital force', as in the creative Life force of an individual" to be meaningful, and also of the connection of this body to the vital *prāṇas* emanating from the food offerings to it.

The physical component of a human persona is dual and incorporates the dense corporeal form, about which the researches of modern science has told us much, and the etheric portion, constituting of the *nāḍī* and *chakra* system, which is explained in Eastern yoga and meditation treatises, and much further information has been added in my books.

Figure 6 summarises all the above, and attempts to show how each hierarchy of lives evolves to the next.[5] The action of each hierarchy has a reciprocal effect on those near it, and is specifically responsible for the evolution of the hierarchy below it. This is essentially seen in our relationship to domestic animals and is an immutable law in this solar system, whose major characteristic is Love-Wisdom. The substance

---

5 See Appendix 3 for further information concerning figure 5.

of the planes of perception are governed by Deva Lords (Rāja Lords). The plane *ādi* embodies the Father aspect of the cosmic physical plane, *anupādaka* expresses the Son aspect, and *ātma* the Mother aspect.

The *antaḥkaraṇa* is the link of increasing sentience and then consciousness that must be built by a lower kingdom (or entity) in order to be born into the next kingdom.

In the *mineral kingdom* the *antaḥkaraṇa* is built by means of the transmutative process of alchemical Fire. This process is found within the plant kingdom, when it incorporates mineral substance in the form of its nutrients. Transmutation can happen when plant tissue draws the mineral elemental into the etheric domain, wherein rearrangement of atomic structure is possible.[6] We also have the process of radioactivity, which releases the inherent mineral lives. The instinct of self-preservation has its genesis in this kingdom, which relates essentially to the tendency of mineral atoms and compounds retaining their structure by means of which they can be identified as a unique entity, often lasting for vast periods of time. The sex instinct also has its genesis here via the laws governing chemical reactions. (The limit of the 'sensory' contact of the entities composing this plane incorporates the ethers.)

In the *vegetable kingdom* (whose embrace encompasses the lower astral sub-planes) the *antaḥkaraṇa* is built by means of the transformation of the substance of light and the earth into colour, perfume, and geometric design. They eventually gain their liberation through emanatory perfume (scent) and magnetic attraction to units from the next higher kingdom. They constitute the food of the animal kingdom because of this attraction. They sacrifice themselves for the kingdom above so that they can evolve animal-like qualities, whilst the animal kingdom is thereby sustained and nourished. This is an expression of the law of Sacrifice (a first Ray quality) that sustains and underlies the entire evolutionary urge of the solar Logos. In this kingdom the qualities of feeling and sense response are born and the sexual instinct is thoroughly developed via the many colourations, smells and shapes of flowers, mainly aimed to lure the insects that will carry genetic material from flower to flower.

---

6 See Kervran, C. Louis, *Biological Transmutations,* (Crosby Lockwood, London, 1972), also Peter Tompkins and Christopher Bird, *The Secret Life of Plants* (Harper Collins India, New Delhi, 2000) for detail.

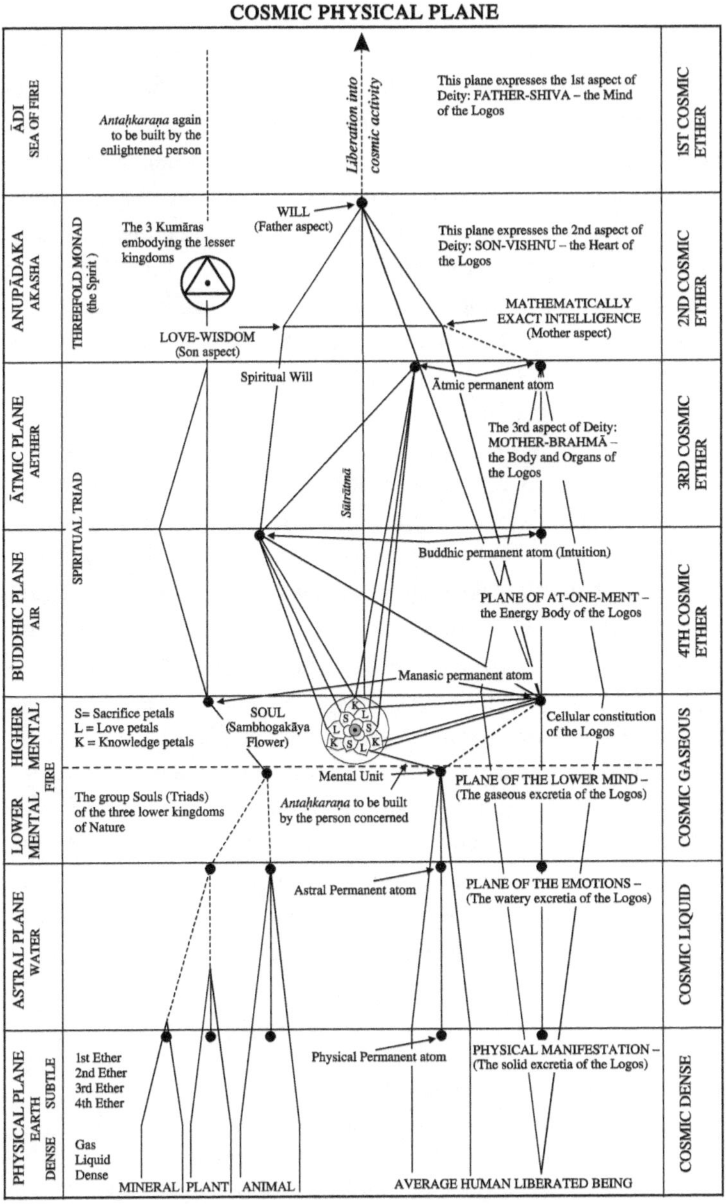

Figure 6: Summary of the planes of perception

In the *animal kingdom* the *antaḥkaraṇa* is built by means of an experimental adaptability to the environment and the translation of the animal's experience into rudimentary thought. They obtain their liberation and eventual birth into the human kingdom mainly through the development of thought-aspiration (the feline line), devotion (the canine line) or general obedience to human intelligent directions (horses). Here lies the spiritual purpose behind the domestication of animals, whereby the basic qualities of devotion and intelligence are born. Thought-aspiration, devotion and general *manasic* obedience relate to the first three Ray aspects for this kingdom.[7] The group or herd instinct is exemplified in this kingdom.

The *human kingdom* develops intuition via their ability to image the Real, and then to creatively express the related qualities in the world in such a way that beneficence is achieved. As they do so they evolve the qualities of a causative agent and achieve liberation from form by means of creative illumined aspiration. They thus control the illusional *māyāvirūpa*[8] constituting the material domain by eliminating the tendencies to be controlled by it. People thereby come to stand at the heart of being/non-being, the *saṃsāra-śūnyatā* nexus. Illumination implies light energy, the ability to express the causative energy from the Sambhogakāya Flower, whilst aspiration is a quality of the personality. Aspiration is the lower reflection of the will. When the will is developed and properly expressed it allows the *antaḥkaraṇa* to be consciously built for the first time. Aspiration calls for a descent of light from the higher planes. This necessitates contact with the Sambhogakāya Flower, and then the energies of the Spiritual Triad, the permutations of the transmutative Fires that liberate as they burn. With the attainment of

7 It is difficult to adequately describe these character traits because they are commingled with clairvoyance and clairaudience. Also obviously members of these three groups can somewhat develop each other's traits.

8 A temporary vehicle created for experience of one kind or another. It is thus literally the body (*rūpa*) of illusion (*māyā*), the body of ever changing desire and mentalistic patterns. It is the sheath of substance and of materiality through which we must all live and labour. As we start to control the related *saṃskāras*, so we begin to become rulers of our destinies. We then live in the substance of the centres (*chakras*) above the diaphragm and not be subjugated to the effects of those below the diaphragm. *Māyāvirūpa* can also be viewed as the etheric body, through which *māyā* finds its expression.

the third Initiation the person gains full participation as a member of the fifth kingdom of Nature, the Hierarchy of Enlightened Being.

The *liberated being* develops receptivity to divine Will, and the Initiate's service eventually releases him/her from the restrictions of the almost omniscient womb that is our solar system, into an ineffable cosmos. The instinct towards knowledge, governing animal development, is exemplified in the human kingdom, though it is the main driving force of the entire evolutionary milieu. The instinct of self-assertion that has its true genesis in the human kingdom is fully developed by the Sambhogakāya Flower in order to impel the reincarnation process. It makes the Flower a prime causative agent.

The Sambhogakāya Flower

The expression of *the Soul* falls below the domain of the Spiritual Triad, which consists of the *ātmic, buddhic* and higher manasic principles. (The triangle pointing upwards in figure 5, labelled Universal Consciousness.) It is constituted of the substance of the sub-planes of the higher mental plane and has also been termed the Causal body, Soul or Ego.[9] I have used the terms Sambhogakāya Flower[10] and *tathāgatagarbha* in *A Treatise on Mind,* wherein much detail has been supplied (to which the reader should refer), especially in relation to my interpretation of certain passages of the text *The Uttaratantra of Maitreya.* Because its corporeal expression resides in the higher mental plane, so the Spiritual Triad has a similar relationship to the Causal body as the etheric, astral and mental bodies have to the incarnate personality.

The Causal body is formed from a single causative Ray (which is inherently triune) from the Monad, which embeds itself on the mental plane in response to definite aspiration, or primal projection of the thought process by an animal kingdom that is on the verge of *Individualisation.* Individualisation refers to the formation of the Sambhogakāya Flower,

9 Specifically in the writings of A.A. Bailey.

10 The term *sambhogakāya* means 'body of enjoyment, delightful participation', and is the 'bliss body', the second of the three bodies of a Buddha or Bodhisattva when incarnate. The ideal representation of the Buddhas and Bodhisattvas exist upon the abstract mental plane, hence I have adapted the term as a descriptive illustration of the way that the Causal body looks and what it embodies for each human unit.

which becomes a major distinguishing feature between humans and animals. The Sambhogakāya Flower needs to exist in order to collect and collate the experiences *(karma* and *saṃskāras)* gained by a human unit, and to direct the rebirths of the consciousness-stream of the rebirthing ones. At Individualisation there is a welling-up of devotional intensity, or an embyronic thought impulse from the target members of the animal kingdom that is met by a reciprocal action from above by the associated *deva* agents and the Monad. These energies interrelate to form a reticulation of energy lines, making nine major whorls (petals of a lotus) and three inner bud-like whorls. The three bud-like whorls hide a point of (Monadic) Light that becomes the jewel in the heart of the Lotus.

As the Causal form slowly obtains the capacity to hold the related consciousness and energies by attracting to it the appropriate mental matter, so in time the interrelation between the Monadic and mental energies shines forth as the *Solar Fire* or Robe of Glory, the Augoeides.[11] The intensity of luminosity of this Fire indicates the 'agedness' of the Soul in terms of the quality of the substance of the higher mental plane that is utilised. As the animal aspires, so then the causative spark of Divinity descends and the union that eventuates forms the nexus between *śūnyatā* and *saṃsāra,* as well as the means for the consciousness that arises to find in the Causal body that comes into existence a stable placement for further expansive growth. This Causal body, the human Soul or Sambhogakāya Flower, is then the Buddha Womb *(tathāgatagarbha),* the seed for future development of the 'spark of mind' that was the embryonic animal-intelligence. The Flower then works to elevate the developing consciousness from the domains of desire and emotional mind to the abstracted levels of Mind. Therein it will eventually experience what the *śūnyatā-saṃsāra* nexus represents, and so gain eventual liberation of Mind from the trammels of form. Thus is the story of human evolution told, the how and why humanity have come into existence.

Sambhogakāya Flowers are sacrificial units of love (governed by *bodhicitta*) that have collectively come into manifestation for the purpose of liberating a quanta of sentient substance that has evolved slowly from the originating pool of substance (*mūlaprakṛti*), the basis of the *māyā* (illusion) governing us.

---

11 A Gnostic term for this sheath of substance.

The Causal Body can then be considered the manifest encasement for the Spirit-Monadic aspect in the field of the Logoic dense physical substance. This has its correspondence in the formation and appearance of the Life of an atom from out of the ethers, that collectively are the substance of the physical domain, which we, as comparative Logoi, experience. By utilising this substance as our experience base the transmutation of all the *saṃskāras* developed by incarnating into the corporeal form is effected. This concerns the alchemicalisation of substance and its elevation into the domain of the mind.

The entire philosophy presented in *A Treatise on Mind* lays the foundation for comprehension of the true nature of this process. Eventually the life of the personal-I becomes consubstantiated with the greater Life of the Monad over the aeonic course of evolutionary time, hence at the attainment of the sixth Initiation the personal-I and the Monad are One. What is known as Mind becomes the vehicle of the Monadic Mind, and so a Buddha appears in the material domains. It should be noted that contrary to Buddhistic thinking not all appearing Buddhas are *mānuśi* or teaching Buddhas, most sixth degree Initiates appear upon different Ray lines than the second, and so play a different role. Hence Buddhas are not as rare as Buddhists make out, as they are focussed upon only one aspect of their manifestation. There is placing upon this planet for only a comparative few such beings, thus most enter their cosmic journeying.

The Monad can only manifest Itself on the lower planes by means of a Soul, or by a consciousness that has purified its gross *saṃskāras*, has detached itself from the allurements of *saṃsāra* and attuned to Monadic energies through yogic direct perception.

On the higher mental plane the Sambhogakāya Flower expresses the sum of empirical attainment, and yet is the lowest level of the universality reflected by *buddhi*. It thereby becomes a universal storehouse of consciousness *(ālayavijñāna)* for a human personality. It is a fit vehicle to transmit what emanates from the higher planes to the personality, and also to receive the developing consciousness-attributes from it. Only what is most virtuous, ideal, untainted by desire, selfishness or pride, plus that what is inspirational or ideates from the Spirit, can be absorbed into the Sambhogakāya Flower. The disharmonious, destructive, or gross forms of qualified energies will exist as latent *karma* within the genetics of the permanent atoms.[12] The *karma* in the form of *saṃskāras* manifest as inherited tendencies which the succeeding personalities must tackle and seek to transmute. Eventually a Soul-personality develops which becomes a fit transmitter of powerful subtle energies, allowing it to be a prime causative agent. Our modern technological civilisation is testimony to that. There are however more subtle, direct forms of causation than discovered by the scions of science and technology. They directly emanate as a consequence of the higher psychic abilities *(siddhis),* which humanity must yet properly discover and so incorporate such abilities as the mainstay of a new civilisation.

The process of Individualisation is a mass happening that transmits planetary and group energies. All Souls share a collective consciousness which can consequently be considered an Over-Soul. This Over-Soul constitutes a collective *maṇḍala* of the groupings of human Sambhogakāya Flowers organised in a similar fashion as the whorls of petals of an individual Flower. Each grouping of Flowers then embodies a particular Ray quality that manifests its expression within the context of human civilisation. Such expressions can be considered as *saṃskāras* manifesting from the higher dimensions.

The energy of Life, transmitting interplanetary and cosmic forces into our planetary womb, flows through the buddhic and higher manasic channels via the organism of the Over-Soul. Each distinct Soul (a cellular unit in that organism) knows its relationship to the other cells and of the work to be done to ensure the unfoldment of a human civilisation.

The relationship between the Sambhogakāya Flower and the personality explains why, for instance, that *yogins* take such great pains and many years to acquire enlightenment *(satori, bodhi).* This is because the thought pattern and energies of the mind and the entire bodily nature of the personality, has to align and function in a harmonious rhythm with the Flower/*tathāgatagarbha* and its cycles, in terms of the Sambhogakāya Flower's Thought sequence.[13] This cannot

---

12 They shall be explained in chapter 8.

13 I am considering the *ālayavijñāna* enlightenment here (the third Initiation). One could also consider the *śūnyatā* enlightenment (concerning the death of the Flower's form, (attained at the fourth Initiation) and the *dharmakāya* enlightenment, attained at the fifth Initiation.

be accomplished in one or two meditations, but requires a long process of assimilation, of right understanding, stabilisation, and alignment of body, speech, and mind (the physical, astral and mental bodies).

The symbolic alchemical transmutation of the leaden corporeal nature to spiritual 'gold' is then accomplished. Every 'atom' (Anu)[14] in the body is then transformed into a dynamo of energy, a transmitter of radiant golden energy. When all the spirillae in the various permanent atoms have been fully vivified, then the evolutionary goal is reached, and the person is able to control the potent forces of the corporeal planes through them.

The dotted line that descends vertically down the median in figure 6 symbolises the *sūtrātmā,* (literally thread-self) the thread of Life energy between the Monad and the personality. Upon the path of active return it symbolises the *antaḥkaraṇa,* the 'rainbow bridge' leading to the symbolic Valhalla of the Norsemen, or the path to the Mt. Olympus of the Greeks.[15] The *antaḥkaraṇa* is a bridge in consciousness built consciously by the illumined one via meditative practice. This necessitates the focussed will, and represents the prime line of unfoldment of the thoughtful aspiration of a causative agent. Its projection utilises the *creative imagination* and an extension of the threefold thread of consciousness (*iḍā, piṇgalā,* and *suṣumṇā nāḍīs*) through a Fiery sea of consciousness, from the *mental unit*[16] to the *manasic* permanent atom. This link is thus from the lower to the higher Mind.

When completed, it directly unites the Spiritual Triad to the personality, making the undertaking of the fourth Initiation possible. The *antaḥkaraṇa* is a stream of living light projected across a gap in the consciousness of a person existing between the lower and higher minds, and also in the etheric body. There are webs of etheric substance in the spinal column between the major *chakras,* which must be sublimated through right yogic and living style practices. The webs of etheric substance prevent clairvoyance and other *siddhis* from arising prematurely in the normal person. This is a necessary adjunct otherwise the normal selfish individual would quickly fall into the trap of becoming a black magician.[17]

14 See chapter 8 for the explanation of an Anu.

15 These terms relate to Shambhala, the planetary Head centre, wherein resides the Monad.

16 The *mental unit* and the *manasic* permanent atom shall be explained in chapter 8.

17 This theme could be significantly enlarged here, but falls outside the scope of this book.

The *antaḥkaraṇa* is symbolised by the neck which separates the head (the governing or spiritual sphere of action) from the body. A person who can cross this bridge and stabilise his consciousness upon the higher mental plane is *'born again'* into 'the spirit', and thus into the 'kingdom of God'.[18] When intuitive beings briefly traverse this bridge, they get flashes of insight, inspiration or genius, that they do not normally possess.

Literally speaking the *antaḥkaraṇa* is the path that starts at the Base of the Spine and leads to the Head centre and beyond. It is the path that starts with the fully integrated personality and leads to Monadic Identification. It is the razor-edged path to enlightenment and the higher Initiations. Humanity must eventually take this path from the earth to traverse to the Heart of the universe.

Figure 6 depicts the Sambhogakāya Flower as a nine petalled lotus, that is sustained by the interplay of the triune forces emanating from the *ātmic, buddhic*, and *manasic* permanent atoms.

The energy from the Monad resides in the 'jewel at the heart of the Lotus' which is shielded by three bud petals. Nine spokes (vibrations) or petals of energy are emanated from this jewel, in conjunction or interaction with the evolutionary *devas* from the higher mental plane *(Agnishvattas)* and the *lunar pitris* ('fathers' of the form, or lunar nature, thus they embody the substance of our personality structures) of the lower mental plane. Their combined activity produced the spheroid form of the Causal Body. As this Body is composed of *deva* substance and denotes that a 'Son' is in incarnation, it is also called *the solar Angel*—the 'angel' that wields the forces of the second, the Son or Love-Wisdom aspect in Nature. The Sambhogakāya Flower can be seen as a multi-hued, dynamic, opalescent splendour, the tone and colouring of which depends upon the stage of evolution of the particular Flower. Each petal is vivified by a different combination of hues and colourings as the gain of the incarnation process gradually acquires the characteristics needed. The general colouring is *orange* for the Knowledge petals, *rose* for the Love petals and *yellow* for the Sacrifice petals.

The bulk of the Sambhogakāya Flower's evolutionary time is spent unfolding the Knowledge petals. Contained within them is the essence

---

18 *John 3:3-9.*

of the total knowledge gleaned from all the past lives. Therefore these are generally the first to unfold. The average person has one or two of the Knowledge petals vivified and maybe one of the Love petals, but rarely, except in the advanced stage of discipleship are any of the Sacrifice petals seen unfolded.[19]

The factor of the *devas*

Be it concrete or subtle, all matter is really *deva substance*. All embodied forms are an expression of their corporeal bodies. The *devas* are feminine in nature, being products of the Mother aspect of Deity, and thus express aspects of innate intelligence, joyous dedication, and activity. They are receptive to the line of Commands, the emanatory Word from liberated Beings, thus they follow the line of least resistance to the evolutionary Will. They therefore automatically unfold the Way of the Love-Wisdom of those that direct their purpose.

The human Soul-group is correspondingly masculine, hence their rebirthing aspects, humans, evolve through the path of greatest resistance, involving friction, discord and strife.[20] By this means they evolve the first Ray attribute of the directive will, at first via desire-emotion, then adding intelligence, and inevitably Love-Wisdom, which is the true objective of their evolutionary journeying. They will inevitably learn to wisely and lovingly direct their will (in accordance with the evolutionary Plan from Logoic sources) so that it is constructive and cooperative in nature and not war-like and destructive. This process involves considerable evolutionary time and involves the *karma* of much pain and suffering, which becomes a prime educational tool for humanity. The *devas* know not this suffering, though they embody its substance, however they have a responsive intelligent concern for the arenas of disharmony and discord humans sow through their callous selfish and self-willed actions.

---

19 For those who wish further detail concerning the nature and functions of the Sambhogakāya Flower they can consult volume 3 of *A Treatise on Mind,* and *A Treatise on Cosmic Fire* by A.A. Bailey.

20 From an energy perspective, humans manifest a positive polarity, and the *devas* a negative one.

Despite the above, the Love-Wisdom principle is innate within the human kingdom, being the major characteristic of the Sambhogakāya Flower, and also for most Monads in this Incarnation of the solar Logos. Because of the developing will, the human kingdom normally evolves at a comparatively faster pace than that of the *deva* evolution.[21] Up to this era intelligence has been the major focus for evolutionary attainment by humanity. Love-Wisdom, the true heirloom of humanity, can awaken upon a large scale from the deep recesses of people's Hearts, once the evil machinations of those governing the corridors of power in our societies have been overcome. For this to happen first Ray energy must be directed rightly by humanity to constructively liberate rather than for destructive avaricious activity.

A major function of the appearance of the Christ/Maitreya will be to facilitate the outpouring of the combined energies of Love-Wisdom and Will by humanity. When humanity and the associated *devas* have perfected their goal for evolution and have fused in a common expression of Love-Wisdom and Divine Intelligence as active service, then the Son aspect of Deity will be manifest in the full flower of its expression.

Each plane of perception has its presiding Deva Lord, who through His subordinates, determines the way of expression and *karma* of the matter on that plane. Thus in the Sanskrit terminology:

The Deva Lord Kśiti rules the physical plane.
The Deva Lord Varuna rules the astral plane.
The Deva Lord Agni rules the mental plane.
The Deva Lord Indra rules the buddhic plane.

Kśiti, Varuna, Agni and Indra, are the sum total of the lives of these planes. Each wields a triple force that has a sevenfold application. Each sub-plane has its lesser Deva Lord presiding over it, and so forth. Agni thus rules the Element Fire, be it solar, electrical, mental, chemical or frictional. It can be inert, latent or active, gross or subtle. Pre-eminently however, He rules the mental plane and the mental sub-planes of each

21 This is true, unless a human unit follows the ways of the forces of evil, the dark brotherhood. 'Brotherhood', because they work collectively for persistent selfish and separative mode, using developed psychic power, great intelligence and cunning in all their activities. They are 'dark' because that is the colour of the auras they generate.

of the other planes. The subject of the *devas* was explained in volume 7B of *A Treatise on Mind,* as well as in Alice Bailey's *A Treatise on Cosmic Fire.* It is outlined here for the sake of consistency of the imparted information, for without the agency or assistance of the *devas* no causative agent could function, because they embody the substance with which the causative agent must work. Enlightened Knowers have utilised Yantras, Tantras, and mantric sound to rightly command the *devas* through the developed Love-Wisdom principle for all magical or creative endeavour. An enlightened causative agent must have certain knowledge as to the order and quality of the *deva* substance with which he/she works, as it also constitutes the corporeal form that is one's bodily appearance.

Because of this last point much danger is accrued by the unwise endeavouring to work with *devas* before they have developed the right motives and can channel the right energies. The *devas* can run amuck in the body of a 'sorcerer's apprentice'. For this reason the training provided by the enlightened preceptors of yoga-meditation for their selected students is long and involves the testings for Initiation explained in volumes 6 and 7B of *A Treatise on Mind.*

The basic ordering of the various classes of *devas* is given below:

- *The Fire elementals*—the minute Fiery *prāṇic*, essences on the involutionary arc that can permeate the bodies of all forms.
- *The Fire spirits*—are latent in all focal points of heat; they are internal warmth.
- *The Salamanders*—they manifest in any flame, be it deep in the earth or a lighted taper.
- *The Agnichaitans*—on the evolutionary arc. They embody physical plane fires when viewed on a large scale.
- *The Agnisuryans*—their outer expression is an astral body, and they are also concerned with buddhic Fires. They form the emotional and desire bodies of a human or Logos.
- *The Agnishvattas*—the mental *devas*. They embody the sum of the substance of the minds of all thinkers.[22]

22 These terms are adapted from Bailey's book *A Treatise on Cosmic Fire.*

The *devas* see sound and hear colours, which is the reverse for humanity. Because *devas* are primarily active intelligence (and not expressed wisdom) their basic means to evolution is to be responsive to enlightened directions, the Words of Power spoken from the awakened Heart. Elemental *deva* lives form the substance of unevolved and intelligent human desires and thoughts. As the human attributes become more refined, so also the quality of the *deva* lives that incarnate to embody those attributes. When one progresses upon the path and the more base Elemental lives are to be expelled, then a period of sickness often occurs at the places of exit of that substance.

Contraction, or lessening the sphere of influence of their base substance, is the means of evolution for *devas,* as they do so their energies become more intense and radiatory. Expansion of consciousness is the key phrase explaining the evolution of the human kingdom. Because the *devas* are impersonal in their actions, so the lesser *devas* and elementals automatically obey the sound patterning of the creative will or desire of the controlling entity, no matter what the purpose is. This then forms the basis to the expression of the law of *karma,* as the *devas* inevitably return to their original conditioning, returning the force of impression, the energies of conversion, upon those that have appropriated them for whatever purpose manifested.

Due to the dangers involved in the premature release of knowledge that will indicate the means of control of *deva* substance, much occult information is esoteric, hidden, 'ear whispered truths', given only to pledged disciples of long standing by an enlightened master of meditation. Both black and white magicians are able to control the *devas* with whose substance they use to create. They learn the right mantras that will invoke or evoke the type of *devas* needed for their purpose. A black magician, however, uses the force of personal self-will to dominate and manipulate the elementary lives, consequently works against the processes of *deva* evolution. He has no influence whatsoever over those of the higher orders, of purified vibrant colourings. The white magician ever works cooperatively with the *devas* to build what is needed according to the evolutionary plan for all. He/she works with vibrant substance and the emanation of love, whilst the black magician works with the sombre hues and darkened shades, the elementary lives of the shadows, utilising that which is disintegrating and the forces of death. The secrets to the occult control of the forces of Nature are consequently well guarded, but will be slowly given to humanity when they develop the needed loving sensitivity, wisdom and harmlessness to be able to use them properly. As stated in *Mark 4:22-24* :

> For there is nothing hid, which shall not be manifested; neither was any thing kept secret, but that it should come abroad.
> If any man have ears to hear, let him hear.
> And he said unto them, Take heed what ye hear; with what measure ye mete, it shall be measured to you; and unto you that hear shall more be given.

In this coming Aquarian age some of this knowledge will be scientifically discovered, especially in relation to the etheric body. Thus certain of the etheric (violet) *devas* will be contacted and their energy constructively used in a working relationship with humanity. Many violet *devas* embody the force of healing, as they constitute all etheric forms through which healing vitality or the *prāṇas* of sickness and disease must manifest. Contact with them will be possible by means of ceremonial magic. (Two keywords for the Aquarian age.) Much of the immense success of such books as *Lord of the Rings* by J.R.R. Tolkien, adult fairy tales, and books on occultism, are a prelude to this eventual contact.

Human sheaths are largely controlled during the early stages of our evolution by:

1. The *lesser elementals* that pamper our sensual nature. They represent primeval Life still on the involutionary arc.
2. The *lunar pitris (pitṛ)*, the astral (or *kāma-manasic*) entities that come under the domination of the energies from the moon.[23]

Consequently our evolutionary and spiritual development essentially concerns the ability to lessen the control of such *devas* over our threefold constitution, to master the related instinctual desires, lethargic energies and *karma* of the past. We must learn to control, dominate, then

---

23 The meaning of this phrase will become clearer when the history of the moon Chain is understood. See *The Constitution of Shambhala* 7A. Chapter 13 of *St. John's Revelation* also presents a wealth of information, when properly analysed.

transmute the substance of our threefold bodies, to lift it up to higher states of perception, if we are to become true divine causative agents, to be masters and not that which is controlled.

The *lunar pitris* embody the essence of the past solar system, or world period. We must metaphorically learn to live eternally in the present, as the Son aspect of Deity. Thus, we must come under the embrace of the *solar Lords*, the great *devas* who embody the sum of Logoic vitality for all evolving Life. We must learn to live within the light of the Spiritual Sun, and not in the shadows of the lesser lives *(pitris)* that hinder and veil all that is; who live by the reflected light of the moon and embody the substance of all forms. They qualify the light of *saṃsāra* through which people obtain their world-view. Their energy is that of *māyā*, the realm of illusion.

Devas are consequently the great Mother, embodying the forms of all Nature.

The Monad

The nature of the Monad (from *mono:* one, indivisible) and Monadic evolution is barely understood by anyone at all. The term for this, the highest aspect of the human persona, is often generalised as 'Spirit', *paramātman,* Brahman, or 'the God within'. Esotericists following the Blavatsky and Bailey traditions have an extended concept, which still barely skims the surface concerning comprehension of this principle. That relating to an understanding of the nature of the Sambhogakāya Flower is difficult enough, as well as the concepts of *śūnyatā* and *dharmakāya,* though Buddhists have had more than two millennia to theorise upon these subjects. Then we have empirical minded people who find anything esoteric hard to accept, and rarely delve honestly into such subjects at all without their materialistic bias demanding concrete proofs, 'evidence', according to parameters they set.

Only upon the Initiation path is it possible to gain proof of the existence of such attributes by means of direct experience, an Identification with the principle. Attainment of the third Initiation produces Identification with the Sambhogakāya Flower, though the first two Initiations can produce intuitions and moments of experiential awareness. The fourth Initiation produces Identification with *śūnyatā,* whilst the fifth Initiation provides access to Monadic Impression for the Master. At the attainment of the sixth Initiation the Chohan is fully Identified with the Monad. Such a One is a Dhyān Chohan, a divine being of meditation substance who has integrated both the human and *deva* components within. The Chohan is a Buddha preparing to undertake one of the cosmic Paths.

Having stated the above it should be obvious that I can only provide here a little more of this subject than has hitherto been elucidated. People should then have a better view of this esoteric subject, to help prepare them for that time in the future when they too are ready to make their Decision as to which of the cosmic Paths they are to follow. Hence it stands that the Monad is a cosmic voyager, only 'briefly' tied to our earth sphere of attainment.

The glyph for a Monad is similar to that of a Logos, in that both are represented by a circle of abstraction with a central dot. The Deity is essentially triune, as is the internal structure of the Monad. The Monad is the One from whence all manifestation associated with the personality has emanated. It resides on a plane beyond the five normally associated with human evolution in the domains of *karma.* It wields cosmic forces, bringing them into concrete manifestation via the Sambhogakāya Flower.

The Monad empowers the Light that is the *maṇi,* the diamond-like jewel from whence stems the Lotus of being, and from which is extended the *śūnyatā-saṃsāra* nexus. The Monad is the Father of the entire evolutionary journeying of a human Soul in the realms of form, and can be viewed as a spark of Light in the omnipotent Mind of the Lord of All. It thus represents the primeval Buddha residing resplendently in the *dharmakāya* environment prepared to wait the aeons for its projection (*tathāgatagarbha,* the Buddha-Womb) to be freed from defilements.[24] It is the Real, whereas all else is illusional, transient.

Monads can be considered as the cellular constituents, units of Logoic Thought, but they are not rigidly bound to any structure, as are the cells in a human brain. They have their own mode of evolution in cosmos.

---

24 As per the concept of *samalā tathatā* (Suchness covered over or concealed with impurities, *saṃsāric* defilements) and *nirmalā tathatā,* (Suchness apart from pollution), explained in volume 3 of *A Treatise on Mind,* and associated with the Buddhist conception of the *tathāgatagarbha.*

The Sambhogakāya Flower stands in relation to the Monad as the personality does to the Flower. The Monad's nature is thus immeasurably beyond human conception, known only by analogy and then direct perception. It is the source of all Life, Love and Light internal to the human unit, the triple energy that sustains the Egoic manifestation and to which the illumined human prodigal Son must eventually be abstracted.

Even though it is singular, a unity from the point of view of the personality, it can also be seen in terms of a trinity, (the three in One) for its true structure is that of an Eye. This is the organ of Vision of an ineffable cosmic traveller that is evolving to become a Logos. The Eye allows that Entity to look into the corporeal realms associated with human evolution via the Sambhogakāya Flower, as well as outwards into the vastness of cosmic space. The Monad's true home is the evolutionary field of expression that is the 88 constellations, the stars seen in the heavens at night (plus the etheric constellations) which constitute the Body of Manifestation of the ONE about Whom Naught may be Said (THAT Logos). The Monad is a cellular unit within the body of a solar Logos. It evolves within That Being for the cycle of a human becoming a Chohan. It has gained the characteristics of a Lord of Sacrifice from former cycles of expression, producing a type of inclusive expansiveness through repeated Incarnations in cosmic space that are concomitant with the 'footsteps' of the presiding Logos of Whose *maṇḍala* it is a part.

The human form reflects the stages of development of the Monadic sphere on its own level. This relates to the truth of the ancient aphorism 'as above, so below'. The Monad has projected the Thought-Form of That which evolved previously in cosmic space to help construct the Sambhogakāya Flower. In its most reified form we have the Head centre governing the human conditioning, which is so constituted to be able to bear the impact of Monadic Energies and Revelation.

The *Monadic Eye* can be viewed in terms of three tiers of circles that evolve with passing evolutionary time.

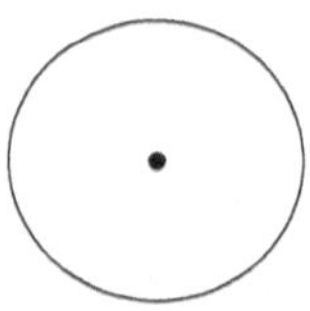

1. The Monad during incarnation into the mineral and plant kingdoms.

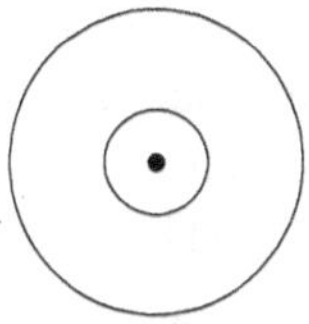

2. At the time of Individualisation and early human evolution.

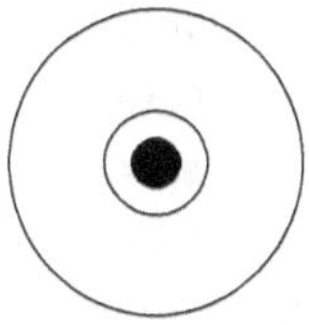

3. The opening of the Eye during various stages of the path of Initiation, from the third to the fifth Initiations. This Eye can be interpreted thus:

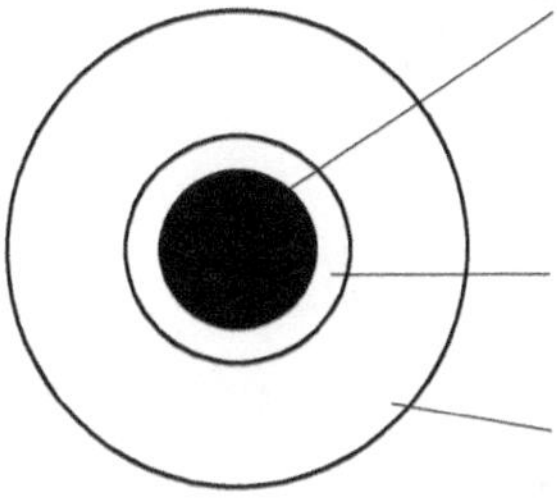

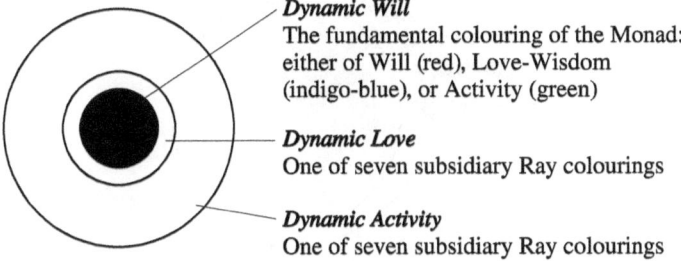

Figure 7: Stages of expression of the Monadic Eye

4. The Monadic Eye is depicted for an Initiate of the fourth degree. Note that the Ray colourings are far more vibrant, and can manifest different hues than are seen for instance in the Sambhogakāya Flower. For instance, the silver-white from the cosmic astral plane can be reflected through its form.

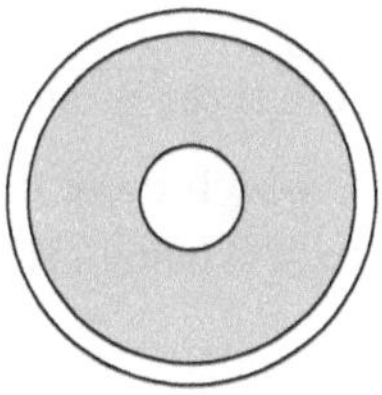

Figure 8: The door to the cosmic Paths

5. Figure 8 depicts the 'door' that opens to the way of the cosmic Paths, which leads the Initiate out of the confines of this solar system and its activity altogether. These Paths gear the Initiate for his/her role as a Causative aspect of the embodying Logos in a future solar system.[25]

   This 'door' can also be viewed as shown below, where we have a depiction of the serpent of time (or rather, a Dragon of Wisdom) biting its own tail. This is symbolic of the nature of spiral-cyclic motion that consumes 'time' in ever-increasing cycles of revelatory awareness. From another perspective it refers to the ability of this cosmic serpent to consume a quanta of dark substance by digesting it in its Fiery stomach, thereby converting it into the variegated arenas of Light.

Figure 9: The serpent of time

The central disc becomes the symbol or name of the avenue of approach (*antaḥkaraṇa*) to any specific star system in the cosmos. It is the Monad's destination in the next great cycle. This destination leads eventually to the Heart or Eye of THAT Logos. It becomes the means whereby the Monadic Eye can gaze into THAT Body, for All are but attributes of That Mind, the in and out-breathing of the One LIFE processes.

Monads are said to exist in three groupings:[26]

a. *The Monads of Will*, of which there are said to exist about five thousand million in the present human Hierarchy in our solar system.

25 The nature of these Paths are explained in *The Rays and the Initiations* and *A Treatise on Cosmic Fire* by A.A. Bailey.

26 See page 579 of *A Treatise on Cosmic Fire* by A.A. Bailey.

b. *The Monads of Love-Wisdom.* These are the most numerous, as this is the keynote of evolutionary expression in this solar system—about 35 thousand million units.

c. *The Monads of Activity*—about 20 thousand million units. They constituted the mainstream of evolution in the past solar Incarnation.

The central Heart of each Monadic sphere is coloured by one of these Rays of aspect. It indicates the nature of the path of ultimate return, or eventual destiny of the Monad. The aeonic process of Monadic evolution concerns changing these colourations from one hue to another, or even the main colouring of the central disc. For instance, green can evolve to blue-green, or it can develop its complimentary colour, red. This solar incarnation facilitates the movement to blue and lays the foundation for all to move towards the red, signifying the demonstration of Logoic Will. Monads with a deep blue-green or deep green-blue are most ancient, and heralded from the past solar system.

The colouring of the outermost, activity sphere, governs the Monad's active involvement in the realms of form. The Sambhogakāya Flower for instance often takes on this hue as its fundamental colouring. By example, if the colouring is violet then the tendency is for the Soul to have the seventh Ray as either its major or subsidiary colouring. There are however no hard and fast rules, for everything depends upon Monadic purpose, the Law of Monadic Return, the need from a Logoic perspective. The colouring of the outer, or activity sphere, however, conditions the Monad's unfoldment for the major portion of its active incarnation in the cosmic dense physical plane. There are many scintillating hues beyond description in terms of physical plane colouration.

The Monads of the human kingdom can collectively be viewed as members of the cosmic plant kingdom—with their 'roots' in the 'soil' of the cosmic lower mental sub-planes. The human Souls can be viewed as the flowers of the Monadic 'man-plants'. The earthly plant kingdom thus manifests as an inverted image of what exists on cosmic levels. Alternatively, the Monads can be viewed as 'seeds' planted in the 'soil' of the cosmic dense physical plane wherein the human Soul (Sambhogakāya Flower) eventually flowers its resplendence as the nutrients *(saṃskāras)* gained by the evolving personality are absorbed into the Flower's form. This cosmic plant kingdom evolves, just as does its terrestrial counterpart.

The evolution of Logoi

The higher cosmic Initiations that govern the process whereby Initiates gain the attributes that qualify them to be Logoi have been explained in *The Constitution of Shambhala.* An analogous process happens with respect to the evolution of the species of Life on earth. Such an evolutionary process can also be used to indicate the nature of evolutionary developments in the universe. The analogues are transposed to Life in a cosmic landscape wherein Initiates (Buddhas) find themselves evolving relative to all the Lives that have evolved therein from prior solar evolutions. They have now manifested as Logoi of various degrees in the vast cosmic landscape within which a Buddha finds himself.

Upon the attainment of the third Initiation by a human unit the Monad then effectively becomes a single-celled unit ('amoeba') of the cosmic animal kingdom. This depiction is obviously viewed relatively, because we are here trying to comprehend the nature of a 'Buddha' that is evolving upon an earth sphere, that has had prior evolutionary development in cosmos, where it had evolved to the stage of being represented as a member of the plant kingdom in the cosmic landscape. As such the Monad incarnated upon the second *dharmakāya* level (*anupādaka*) prior to the Individualisation process of the human Soul-form.

To the human unit the Monad's level of development is that of a Buddha. In comparison, the type of emotional consciousness of the average human unit would maybe relate to the Monad to that of a carnosaur or sauropod of the Jurassic era. This analogy, though crude, nevertheless presents the basic idea, as it is difficult to make proper comparisons. Upon its own level of expression however the Monad has far yet to go. The *gain* of Monadic evolution on earth is accomplished by the evolution of the units of consciousness, the solar Angels and the minds of the reincarnating human personalities as they evolve to become Buddhas. The human personality is in itself the gain from the far distant past in the ocean of time of the encapsulating and conversion of the darkness of primeval cosmic dust into arenas of light as part of a Logoic Body of manifestation.

Upon the attainment of the fourth Initiation by a human unit, the Monad takes on the analogous rank of the equivalent of a member of a species of ant in the cosmic Landscape, a leaf-cutter ant for instance, that utilises bits of the nursery of the plant kingdom for its own purposes.

Upon the attainment of the fifth Initiation by a human unit, the Monad becomes effectively analogous to a worker bee that inhabits a bee hive and helps produce the honey that sustains the entire life of the hive. The hexagonal patterning of the beehive is the symbol of the *maṇḍalic* structure underlying the constitution of Shambhala, the Head centre of the Lord of the world.

A sixth degree Initiate (as was the Buddha after attaining his *parinirvāṇa*) effectively manifests the analogy of a fish swimming in the waters of the cosmic astral ocean.

A seventh degree Initiate takes the analogy of a bird that can fly above that ocean to where it wills in the cosmic Landscape. In the case of the Christ the symbolism was that of a *Dove of Peace*.

An eighth degree Initiate (the present degree of attainment of the Buddha) takes the guise of a small mammal, such as a rabbit, whilst a ninth degree Initiate becomes a larger animal, such as the 'beasts' mentioned in the fourth Chapter of *St. John's Revelation*. Such a One is on the verge of taking cosmic Individualisation.

The tenth degree Initiate is One who has Individualised as a member of cosmic Humanity. By analogy at first the early cosmic Lemurian stage of development is experienced. Such then becomes the Logos of a planet as is the earth. This is then followed by the specifics of all the levels of evolution of the members of a cosmic Human Kingdom, which involves the way of activity of the milliards of stellar spheres in the cosmos. All Logoi are members of a cosmic Humanity undergoing the plenitude of the vastness of their respective cycles of evolutionary journeying.

The student must be careful not to concretise the above information (or anything else concerning the esoteric doctrine) for such concepts cannot be properly understood by one below the rank of a fourth degree Initiate, for only then can the true cosmic Landscape be properly seen and perceived in terms of esoteric relativity. Even then there are limits to such a one's vision. Obviously the higher cosmic Way can only be understood by a fully awakened Buddha who must travel therein, as such a one prepares to break all *saṃskāric* bonds to the earth. He/she has learnt also to view things in terms of transmuted correspondences, not in terms of the substance of the forms of this earth, which is but one of billions in which there is evolving Life. Each sphere has its own intrinsic Life patterning and fundamental characteristics, seen in terms of sound, colour and symbol. Upon an even higher level of transmuted correspondences the great evolved Buddhas from past solar attainments have taken the guise of any of the symbols attributed to the constellations of stars seen in the night sky above us. They were named thus as part of the myths of past aeons as depicted by the enlightened in the early epochs of human civilisation, or else seeded by the enlightened into the minds of the early sailors as they navigated the southern seas.

There are seven levels of transmuted correspondences of cosmic Humanity to our universe, plus a synthesising three. Each level is inclusive of the levels below it, embodying them as the sum of the forms through which they incarnate and experience the phenomenal universe.

1. That which represents *the humanity on this earth*—here viewed as the kingdom of the Sambhogakāya Flower and not in terms of the human personality. We must remember here the multi–dimensionality of our vision. Groupings of human Souls exist upon every planetary sphere in our solar system.

2. *The Logoi of the planetary bodies* to a solar system, non-sacred and sacred.

3. *Logoi of stellar spheres.* They come in all different Ray and subray lines and stages of evolution. Our Monads can be viewed as constituting part of this level of expression.

4. *Logoi of constellations.* They are, as all else, at varying degrees of development. Thus there are those that are essentially members of the animal kingdom, such as Taurus the bull, others as members of a human kingdom, such as Orion the hunter. Members of the plant kingdom at this level are symbolised by the qualities of many of the inanimate objects found in the listing of the constellations.

   The qualities and colourings of the Lives animating these constellations can be clearly seen by those with an awakened Vision, as had the ancient seers who first gave us a pictorial description of them. They knew their true qualities, and thus pictured the various, oft seemingly unrelated stars under the names of the constellations known today. Materialistic scientists, not possessing this ability, and deriding those who have it, have yet to emulate the wisdom of past sages with their cosmological understanding.

5. The Logoi of equivalent Age to *the ONE about Whom Naught may be Said.* This ONE embodies the sum of the constellations known to us, and all the associated stars. They are the 96 constellations, when the etheric ones are also counted. Something can be said about the constitution of His Body of Manifestation, for we are all residing and evolving within it, as Monads, but Naught may be said, presently, about That Purpose or cosmic Direction. Revelation may be forthcoming to the Hierarchy in this present cycle of unfolding Will in systemic space. (Note that systemic space is that 'space' that is incorporated by the Logos of our solar system to form His/Her body of manifestation.) The word 'Naught' must be interpreted esoterically here.

   This Entity, one of many such in our galaxy, can be said to be representative of the true cosmic Humanity therein. Each such grouping of major constellations has theoretically many earth-like planets wherein human units,[27] such as those upon our earth can evolve. Life is not a matter of mere chance, but is purposefully made manifest thus according to the verities of cosmic Law and the consequent application of Intelligent Design. *Maṇḍalic* structures are applicable at every level of cosmos. Level after level of Life can be found bound within the intricacies of the cosmic physical permanent atom.

6. *Groupings of many THAT Logoi* according to Ray and subray considerations.

7. *The Logoi governing galaxies.*

27 The evolution of human-like forms constituting a humanity is more common than the scientific community postulate. For instance, in our solar system there have been over the course of its evolution four such human streams that have evolved, of which our present humanity is the fourth, and one more yet to come, that of our present animal kingdom, who will Individualise in a planetary sphere that is presently forming to accommodate this development.

In the universe such Logoi would be considered vastly abstracted cosmic Men and Women, singular unities that share interrelationships with each other. They are organised into groupings, the major ones being:

8. *The Logoi of galactic clusters.*
9. *The Logoi of major galactic clusters.*
10. *The Grand Logos governing the interrelation of all such clusters.* They are embodied by the Logos of our universe. This LOGOS one can presume to be ONE of a small group thus evolving.

Astrologically it is possible to extrapolate the attributes of the esoteric rendering of the planetary rulers and the synthesising Schemes to the above-mentioned levels of the universe. Consequently Mars, the god of war, conditions the appearance of a humanity because of the testings they must undergo in the material domains. Mercury, the 'messenger of the gods' bearing the caduceus staff, relates to the evolution of the planetary spheres within a solar system. Venus, 'the morning star', who governs the manifestation of the fifth Ray attributes of mind/Mind, would generally rule the evolution of the solar Logoi, the disseminators of Mind to their constituency. Jupiter, the god of wisdom, would then govern the manifestation of the Logoi of constellations, as this represents the gain of stellar evolution. Saturn, the lord of *karma,* would consequently govern the expression of a THAT Logos for the sum of such a One's body of manifestation. A THAT Logos would represent the primal source of *karma* for all lives upon an earth sphere such as ours, including the presiding Logos and for the solar Logos incorporating them. THAT Logoi would logically be the main governing Logoi within a galaxy. The qualities of Neptune, governing the sixth Ray dispensation of the Waters, here considered in terms of the manifestation of the energies integrating all into Unity, would then govern the appearance of groupings of THAT Logoi. Uranus, governing the seventh Ray of ceremonial aptitude, the law of cycles and materialising Power, would then govern the general mode of manifestation of galaxies.

The attributes of Saturn Synthesising Scheme would then govern the karmic appearance of galactic clusters. This Synthesising Scheme is an expression of the Solar Plexus in the Head tier of petals for a Logoic Head centre. This tier organises the sum of the *saṃskāras* gained through mental-emotional activity of a personality. In this case the 'Personality' may be considered the Lord of a Universe. The Logoi of major galactic clusters would be governed by the attributes of the Neptune Synthesising Scheme, who embodies the functions of the Heart in the Head centre for That 'Personality'. The Logos of a universe then would embody the functions of the Uranus Synthesising Scheme, hence the attributes of the Throat in the Head centre,[28] and the entire Head lotus, which collates and organises the sum of what must be, and what is evolved. Obviously what is herein said presents abstracts, an *idea* of cosmic organisation, as viewed relativistically from the earth, by using the law of correspondences.

Also, it should be obvious that the esoteric consideration is concerned with more than just the physical appearance of things, which is but an effect of the subjective inner plane happenings. Hence esoteric astrology, which is the science of energies and their relation to consciousness and the enlightened states of Mind, has its say as to the nature of cosmological conditionings. The true attributes of the Logoi embodying these three synthesising Centres, Causal aspects, or levels, lie beyond our comprehension. They are related to the Abstract Triune Deity and we can really only make basic analogies.

Many of the Logoi on all these levels of transmuted correspondences are out of Incarnation at any particular time. This produces therefore a bewildering and staggering sum of interrelated Lives that are abstracted in increasingly sublimely transcendent supernal Domains.

It can be hypothesised that all cosmic Logoi undergo an evolutionary progression along what represents for Them the Way (to Initiation). Every Being manifests via the symbolic 777 Incarnations that produces the perfection of the Sambhogakāya Flower and its liberation. Thereby we have the redeeming or 'lifting up' of the substance matter, the Elemental lives that constitute the Logoic body of manifestation.

All lives spiral upwards, inwards, and outwards into realms Supernal in accordance with the metered steps of the greatest cosmic Logos. These 'Steps' (or Incarnations) are however conditioned by the sum of

---

28 The meaning of the terms Solar Plexus in the Head, Heart in the Head and Throat in the Head centre were explained in volume 5A of *A Treatise on Mind*. This information can be used to extrapolate the nature of any evolving self-conscious entity in the universe, according to their level of expression, taking transmuted correspondences into account.

the actions of all the groupings of Human units within the macrocosmic Body. There is no ending to the chain of interdependent Life.

As our analysis increasingly incorporates the vaster Logoi, so we progressively move further back in time, from where our present humanity stands. One must think esoterically to properly accommodate this statement. One main reason for this is because of the vastly slower rate of unfoldment (evolution) that such an Entity makes in comparison to that of humanity. I shall try to give an indication of this progression below, comparing their evolution to the Root Races of humanity, however omitting any serious explanation of them. They are explained in H.P. Blavatsky's *The Secret Doctrine* and the works of Alice Bailey, hence the reader needs to refer to these sources for better comprehension. A summary of this evolution has been provided in Appendix 2 for those that are interested to pursue this subject.

I postulate seven levels of cosmic Mind below. By this is meant seven increasingly transmuted cycles of Initiation undertaking where what can be considered as cosmic Mind are attained. These seven levels also fall under three broad categories related to the evolution of the Root Races, the present Aryan dispensation (where the rulership of the mind dominates), the Atlantean cycle, wherein the emotions were predominantly developed, and the Lemurian epoch, which was focussed upon the development of the physical body and its associated sense-perceptions, with a major fixation on sex and sexuality. They can be seen to be governed by the three *guṇas,* or three stages of cosmic evolution, explained below. From this perspective the earliest (Lemurian) stage of cosmic evolution, of vast duration, is governed by *tamas* (relative inertia). The next, Atlantean level of the evolution of cosmos is *rajaistic* (comparatively mobile). The lowest level of expression of the manifestation of cosmos that includes life upon our earth is Aryan in nature and *sattvic* (dynamically active).

Obviously I deal with abstractions here, as it is difficult enough for an average disciple to gain a basic comprehension of the nature of the Mind of a Master of Wisdom, thus it is well nigh impossible to gain more than a fleeting idea as to the nature of a first level Logos (such as Sanat Kumāra) Who is five Initiations ahead of such a Master. What then can really be said or comprehended of a Logos at the second level of cosmic Mind, to Whom the first level Logos may be conceived of as an entity comparatively represented as being a member of the animal kingdom in the greater One's 'Mental environment'?

The first main division of cosmic Mind – Aryan development.

This concerns the humanity of earth-like planets, plus the attributes of the Logoi that overshadow and direct the evolutionary development of the representative humanity. Cosmically they represent the general attainment of the characteristics of the fifth Root Race. Two of the levels of cosmic Mind are represented here.

The *first level* of cosmic Mind concerns the entire evolutionary development known to us, from the mineral kingdom upwards through the various categories of the plant, animal and human stages. It therefore incorporates the process of Individualisation, hence the formation of the kingdom of the Sambhogakāya Flower. We then have the entire gamut of human evolution through the stages of the evolving Root Races via the 777 Incarnations. This is capped by undertaking the Initiation path to the third Initiation, when the Sambhogakāya Flower is thoroughly identified with. Two more Initiations to the fifth (hence embodiment upon the *ātmic* plane) seals the demonstration of Mind at this level of expression.

The stages of evolution of humanity to the attainment of the fifth Initiation can be considered to be the precursor for the development of cosmic Mind, where the fifth degree Initiate is receptive to the lowest sub-plane of the cosmic mental plane. This development can be used as a paradigm to pattern our speculation upon the nature of the evolution of the various levels of Logoi in cosmos.

Here the attributes of the *systemic mental* and *ātmic* planes are explored by the evolving humanity inhabiting any globe capable of bearing human-like Life in a solar system. The level of development relates to the attainment of the characteristics of the fifth Root Race by a humanity. This is esoterically labelled Aryan. The attributes of the sixth and seventh Root Races, though presently attained by many Initiates, is still in the future for humanity. The coming new age will birth the appearance of the sixth Root Race. There will not be many noticeable changes in the physical characteristics, but rather those pertaining to the proper awakening of the attributes of the Heart centre.

The *second level* of cosmic Mind, attained by Logoi of planetary systems, involves a repetition of this pattern. The Sambhogakāya Flowers of humanity are then seen as single celled entities at the microbial level. The higher levels of Initiation attainment in cosmos relate to the evolution of the various levels of the cosmic kingdoms of Nature, as explained above. This process leads to Individualisation of a Logos as a cosmic Human unit at the ninth Initiation, with a completed Sambhogakāya Flower properly appearing upon the higher cosmic mental plane at the tenth Initiation. This level thus incorporates the Initiation process as explained in the section on the cosmic Initiations in *A Treatise on Mind, Vol. 7, Part B,* 155-178.

These two levels represent what is immediately knowable on earth once one undergoes and passes Initiation testings, plus what confronts a Buddha travelling upon his cosmic Path, to undergo the further testings that the Lords of various star systems impose for the Buddha to gain the higher cosmic Initiations.

The second main division of cosmic Mind – Atlantean development.

The third and fourth transcended levels of the expression of cosmic Mind are implicated here.

The *third level* of cosmic Mind relates to Logoi attaining the equivalent of their cosmic third Initiation, consequently the concern here is with Logoi embodying a solar system. This third level brings us back in time to the Logoic correspondence of the Atlantean (fourth Root Race) stage of evolution. The solar Logoi can be considered members of the latter stages of Atlantean development who, though immersed in the general psychicism of the time, developed the ability to think.[29]

The *fourth level* involves a repetition of the process from the third to the ninth Initiations productive of a version of cosmic Individualisation for attained Logoi ('Heavenly Men and Women') manifesting from their unimaginably vast and rarefied strata of 'Beingness', where what in known as 'cosmic Mind' to us becomes a form of empirical dense

---

29 During the Atlantean period in human evolution, as a consequence individuation, self-assertive psychic predation and extreme selfishness, members of the dark brotherhood arose in their ranks, whose activities caused the eventual doom of the fourth Root Race epoch.

substance to a far higher level of evolutionary attainment known only to the Logoi undergoing the process. We are here concerned with the evolution of Logoi embodying *constellations of stars*. At this level the developed cosmic Mind of Masters of Wisdom might comparatively be considered to be similar to amoebic singular celled entities. The Logoi governing constellations relate to the early Atlantean development in the cosmic landscape, when the emotions were developed, without the divisive, separative attitudes of the mind. These Atlanteans possessed a type of child-like lack of the use of the empirical mind, instead there was group-conscious clairvoyance. The constellations can be considered similar to the Ray groupings conditioning the Atlanteans.

The third level of cosmic Mind – Lemurian development.

The fifth, sixth and seventh transcendent levels are represented here, being the major constituency of the stupendous embodying Lives governing galactic evolution.

The *fifth* transcendent level of cosmic Mind would relate to Logoi taking whatever might be considered to constitute their third Initiation. This level of cosmic Mind relates to the latter portion of Lemurian development of humanity upon the earth. They are the THAT Logoi, of which 'The One about Whom Naught may be Said', embodying all the constellations of stars seen in the night sky, as His/Her *chakra* and *nāḍī* system is a representative. The Lemurian evolution spanned a large period of time, as it is said that they first Individualised on earth about 18 million years ago, and overlapped the Atlantean dispensation (which lasted for approximately four and a half million years) right until its end. (Our present Aryan dispensation properly began with the fall of the last portion of the Atlantean continent about 12,500 years ago.) The THAT Logoi are the base level 'Humanity' inhabiting a galaxy. Many are 'Cyclopes' possessing the single Eye (third Eye), betaking them as the Initiates of that Lemurian epoch. The high Initiation in Lemurian times is now equivalent to our first Initiation, whilst the development of enlightened Atlantean characteristics relates to taking the second Initiation. In this first Initiation complete Mastery of the physical domain and its etheric substratum is accomplished. The Lemurian Cyclopes had the ability for instance to levitate huge blocks of stone as their building material. What development of MIND that exists for these Logoi would therefore be focussed upon physical plane governance, hence the evolution (activities) of the local physical universe known to us. They express the reified attributes of the third lowest sub-plane of the mental plane as viewed upon this vast, fifth level of transmuted correspondences of the substance of MIND.

This *Tamasic* level of the development of cosmic MIND completes the pattern of unfolding the stages of Initiation undertaking as established for the first two levels. Some THAT Logoi may have developed to the Atlantean stage, which would be considered the high Initiates amongst them, members of the high Council of what is equivalent to Hierarchy for them. The odd THAT Logos working towards the third Initiation would be equivalent to being a member of their level of Shambhala.

The remaining levels of cosmic Mind relating to Lemurian development go further back in evolutionary time through to the point of cosmic Individualisation at this vastly abstracted level of MIND. One must think of a rarefied level of dark Space, vast expanses of Logoic Dark Brotherhood form of MIND, or so it would seem to our level of perception, but indeed at their level that MIND would be perceptive of it being the Knower and Manipulator of the substance of all the evolving forms existing within its Ken.

The *sixth level* of cosmic Mind would repeat the pattern upon an even more rarefied and base form of Mind, again producing a version of cosmic Individualisation for a *group of THAT Logoi* in a galaxy. This level of cosmic Mind relates to the early to mid third Root Race cycle, that can be considered the average Humanity of this Lemurian epoch. The Lemurian focus concerned establishing themselves in the material environment in which they existed and mastering their physical urges, strong sexuality, and survival in a dangerous environment. The reified attributes of the second lowest of the mental sub-planes is developed by them. The physical procreative forces projected by these Logoi would be responsible for the force of the energies causing the birthing of star systems from out of planetary nebulae.

The *seventh* level of cosmic Mind would relate to the attainment of what would be considered the third Initiation for a Logos governing a galaxy. This relates to the earliest stages of the Lemurian epoch, wherein etheric-physical Incarnation needed to be established. Quite strong and violent activities had to manifest by the earliest Lemurians in order to force physical plane involvement, to incarnate via the *chakras,* to become materially and sensually focussed, rather than being in an etheric form. The reified attributes of the lowest of the mental sub-planes at Their level of expression is developed by them. Such forces relate physically to the intense energies needed to be expressed to establish galaxies in the formative period of cosmic evolution.

Concluding statements concerning Logoic evolution

From this point on it is difficult to speculate just what constitutes MIND for the Logoi of a galactic cluster, for a major galactic cluster, or for a universe. The correspondences would relate to the evolutionary development of the animal, plant and mineral kingdoms respectively. This relates to the forces needed to be brought together at the point of Individualisation to form the kingdom of the Sambhogakāya Flower from out of a cosmic animal kingdom. Individualisation is an Initiation process, and in this case the eventuation was in the form of the 'creation' of the physical universe, which orthodox cosmologists have considered in terms of a Big Bang, but in reality the process is more gradual in terms of the formation of a cosmic Egg.

With respect to the above schema the first two levels presented can be considered as below the threshold of Awareness of a Logos embodying a galaxy, as these levels are but a constituency of a solar Logos. From the vaster perspective we therefore need to start our consideration with the level of the Logos of a solar system, the true base level cosmic Man or Woman evolving in the galaxy. The five remaining levels would then manifest as the characteristics of a pentagram bearing the five Elements, considered at this level, each representing a transmuted correspondence of the qualities related to that Element at the level of correspondence of cosmic perception that the Element presides over. Also one would consider an inversion of the qualities of the pentad, with the Aetheric characteristic at the bottom and the Earthy at the top. This is because of the evolved characteristics of the lower levels would be more akin to this arrangement.

Having rationalised an overall comparative Earthy quality for the Logoi of galaxies then we can also think in terms of the three sub-planes of the dense physical plane. Galaxies would manifest via the Airy-Fiery aspect of the dense physical. Galactic clusters will be conditioned by a Watery-etheric aspect of the dense physical, whilst the super clusters will be governed by the energies conditioning the dense physical level. From this perspective they constitute the nervous system and the brain of the Logos of the universe. This then represents what the Logos of the cosmos incarnates into. By deduction such a Logos is a Soul, a Sambhogakāya Flower unfolding the attributes of the Knowledge petals of its constitution.

At the level of expression concerning the Logos embodying a universe we have retrogressed so far back in time that what is considered here as 'Mental substance' is literally that of a cosmic physical environment. The foundation of the substance of the universe is Mind at its most primitive level of expression. A universe, like all other incarnating entities, exists in order to redeem primal substance, cosmic black dust, though at this stage of evolution that 'dust' and the substance of Mind are virtually identical.

The concept of Individualisation also presupposes that there were other Logoi of universes similarly engaged, and that there is a Soul-form established. The three triads of petals of this Sambhogakāya Flower would barely exist and we would expect only the Knowledge—Knowledge petal to be in any way active, abstracting the *prāṇas* of the mineral kingdom. If the Knowledge—Love-Wisdom petal is active then it would be concerned with the abstraction of the *prāṇas* absorbed from the development of plant-like expression cosmically. If the Knowledge—Sacrifice petal is active, then that petal would absorb the *prāṇas* from animal-like development in the early universe. The concept of Will or Sacrifice would be so embryonic as to be hardly noticeable, except so far as it goes to impel the form of a cosmos into incarnation. However, one is so far back in time that physical Incarnation and the substance of the petal of this Soul form are well nigh identical. We are brought to the time of the attainment of Individualisation of this Sambhogakāya Flower that is the MIND of God. This pre-supposes the existence of the three abstracted levels of Mind at this level of expression, but which would barely contain any discernable qualities, other than rudimentary mineral, plant and animal-like *prāṇas*. This is speculative, as all one can make is informed guesses at most.

The above is implied in the concept that the ten stages of the Initiation process for the seven levels of cosmic Mind that produce the development of a Sambhogakāya Flower for the Logoi concerned is repeated at *three* corresponding transmuted levels. (Not counting the appearance of the Sambhogakāya Flowers for a humanity.) Our vision therefore concerns the development of the three groups of petals of a cosmic Sambhogakāya Flower for the three main levels of cosmic Mind. The ultimate 'God' of this schema consequently is that MIND that embodies such a Sambhogakāya Flower, incorporating a vast and incomprehensible duration of Space.

What therefore Individualises, or rather the teleological question is, for what purpose would have existed at that stage for a MIND to appear? A MIND in this case referring to a member of a HUMAN Kingdom. One can thus postulate that before this MIND there was only *Deva* substance. *Deva* at this stage referring to the innate Intelligence, or rather, the sentience of matter, of the interrelating substance of Space interacting to produce a form of evolutionary time, a sequence of events producing a *descent,* or increasingly materialisation and evolved complexity of experience of that substance. One can conceive of DEVA alone as primal black mineral inchoate elementary empirical Intelligence, which at this stage can be considered as the Mind of a Dark Brotherhood, a Black King, existing at that primeval time upon the *ātmic* realm of cosmic Space. Its primal instinct being simply to Know, hence to assert itself over the vastness of Black Space. This represents the appearance of the precursors of the Agnichaitans, being the elementary mineral-like subordinate entities within that Black MIND.

As it reaches out to Know, its lines of approach *(antaḥkaraṇas)* produce patterns of energy vortices in that substance, swirls of interrelatedness of more Fiery (coloured hues), hence the appearance of *chakras,* the Flowers signifying the manifestation of a 'Vegetable' kingdom. The Flowers represent the mechanism of Thought, of the attributes of Knowingness and differentiations into elementary Thought-Forms, signifying the patterning of what is to be. They represent a step downwards of condensation of Thought to create the conditions of the

Logoic buddhic plane. This causes the appearance of the Agnisuryans, being more fluid (Watery) than the Agnichaitans, hence producing a faster Thought-process.

As the swirls of energy-vortices become faster and more intense they become increasingly Fiery, and so we have the formation of the substance of the cosmic mental plane and the appearance of the Agnishvattas. Their mode of interrelation, and complexity of the differing types of Thought-structures ('animals'), sometimes quite ferocious, sometimes more benign. These animal-like qualities, personified force-vortices of differing types, are then organised in accordance to the attributes of the five basic instincts: of self-preservation, of sex (e.g., chemical attractiveness), of group or herd (the amalgamation together of individual units into larger structures), towards knowledge (further experiential awareness), and of self-assertion of the individual.

We have three levels of cosmic mineral kingdom.

a. *Ātma* at the highest level, the originating Black cosmic Logoic Thought projection into:
b. The mental plane. Here the Sambhogakāya Flower of the three levels of cosmic Mind is formed at the point of Individualisation,[30] which projects Fiery force to cause the appearance of:
c. The universe upon the dense physical plane.

One can also speculate, by taking the *chakra* system as a universal paradigm, that there might be more than one universes in existence, because we would be concerned with the evolution of the main petals of the three major *chakras* existing below the diaphragm, related to purely material and psychic evolution.

One can view *chakras* in terms to being integral to a specific entity, and also Entities, such as for instance planetary Regents, being the component parts *(chakras* or major petals of *chakras)* of a greater presiding Lord, such as the Logos of a solar system. Similarly for a Logos embodying a universe. Galaxies, or clusters of them, would

30 The swirling energies from the three lower sub-planes of the *ātmic,* six *buddic* sub-planes, plus three from the higher mental go to the formation of the three bud petals and nine major petals of the Causal form of this Flower, where the energy of *śūnyatā* becomes its central jewel.

be the externalised effect of the petals of a *chakra* within the etheric Body of the Logos of a universe. (I am looking at petals here because of the vast number of galactic groupings in the universe.) Similarly and logically, the Logos of our universe would likely be a manifestation of a particular *chakra* system of a vaster Entity of which the Logos of our universe would be a part. Because the above analysis has brought the consideration of the state of Awareness of this Logos to be equable with that of a cosmic mineral kingdom, so there would be a major limitation of what would be expressed in such a One's *chakra* system. Most *chakras* would be dormant *laya* centres. The highest of the *chakras* that one could consider consequently is the Solar Plexus centre, with ten major petals, next is the Sacral centre with six major petals, and then the Base of Spine centre with four major petals. The Solar Plexus centre however will deal with the subjective energies preceding objective formation, whereas the Sacral and the Base of Spine centres manifest the objectivity, hence from this objective viewpoint ten manifest universes may suffice to account for the appearance of phenomena.

One would also have to take into account the combined Splenic centres, with the twelve plus eight petals, similarly making twenty, but they would exist more subjectively than the petals of the three main *chakras*. Counting all of the petals of the *chakras* concerned makes forty force vectors. The twenty-two minor centres would at this stage be represented as minor points of energy cross-overs organising the primal black substance into some coherence of form.

The other main centres to consider that would assist in directing the primal *prāṇas* are the Stomach and Liver centres. These centres are needed to convey the dual nature of the energies expressed, hence of the electrical nature of phenomena. They are the adjuncts to the Solar Plexus centre. The Liver centre organises the positive *(piṅgalā)* energies generated by a human kingdom, at whatever level of expression such is found, whilst the Stomach centre organises the negative (*iḍā*) energies associated with the *devas*. Depending upon whether the *iḍā* or *piṅgalā* energy flow that dominates at any time, then there would essentially be 40 + 10 = 50 main force vectors for the dissemination of the energies of cosmic MIND at this most base level of expression. The number 50 presents the base number for the major pentads that appear with respect to the demonstration of mind/Mind. Ten such pentads are thus needed, and one can speculate from this perspective that ten universes are needed to produce the appearance of phenomena, if each universe manifests predominantly as pentads, being the higher correspondence of galactic evolution, where, as previously stated: 'five remaining levels would then manifest as the characteristics of a pentagram bearing the five Elements'. The pentads will incorporate the nature of Deva Intelligence, as the originating organising principle of whatever IS, as all is incorporated as Their Bodies of manifestation. The two outer tiers to the Head lotus, explained in volume 5A of *A Treatise on Mind,* working via the Throat and Heart centres, may play a subsidiary role to assist in the appearance of phenomena, and are needed for interrelation with a Sambhogakāya Flower. Their downward manifesting energies which would then implicate twelve universes (ten major, two subsidiary) manifesting at the beginning of the empirical evolution of whatever could evolve to cognise the existence of the appearing phenomena. This pattern is then reflected in the Head lotus of a human unit, or of any 'Thinker' as its twelve main petals that are organised to process the activities of mind/Mind with respect to the organised *māyā* of forms. Twelve is the basic number governing the *maṇḍalas* of the *chakras.*

Hence, depending upon how one would speculate is the 'advancement' of the Lord or Lords of the universe/s, we can conceive *minimally,* of four universes existing simultaneously, if only a Base of Spine centre existed physically, ten universes if the main centre for energy organisation, the Sacral centre, is included, and twelve, if the consciousness-gain of the evolutionary process is to be processed. A dual Splenic centre should also exist to help in the transformation and transmutation of the evolved characteristics *(prāṇas).* The Solar Plexus centre would deal with only subjective energies, hence its forces would help externalise the Sacral and Base of the Spine effects re the appearance of the physical domain. This 'physicality' may be upon a transcended version of the *ātmic* plane substance, elementary Fire, which because of like nature can be reified downwards to become the substance incorporated as a physical universe.

The originating state of the universe could also be at such a low mineral level that even the *chakras* are in the process of formation.

The innate motion of chaotic atoms in motion would inevitably produce such a formation for a Base of Spine centre, but one would expect prior incarnations of the universe, in which case the scenario earlier given is the most likely. This then brings into consideration an elementary MIND that has evolved to generate the Will to Know, to produce the initiating Thought-Form of what is to BE.

This is all but *speculation* of course, and other scenarios can be conceived, implicating many more primeval universes coexisting with ours. Though the Logos of the universe may exist Mentally at seven Transmuted Levels of Transcendent Perception relative to the enlightened ones upon our earth, nevertheless the potent Mineral-Fires from this Domain will reverberate through the levels of cosmic Mind-substance as the Force that precipitates the appearance of the physical plane universe we observe. Between the Logos of a universe and the consequent appearance of phenomena stand a host of intermediaries relaying the originating Thought or Command to Manifest, thereby awakening the needed *laya* centres as the Energies descend. When activated the *laya* centres are the seeds of the galactic clusters, etc. They are nodes of emanatory expression.

More than this I cannot endeavour to speculate as we are far outside the bounds of what is of relevance to the evolution of humble humanity upon this tiny speck of matter constituting the earth in our galaxy. Speculation aside, what is relevant to the enlightened ones and the highest Dhyān Chohans upon the earth are the integrated orders, magnitude of expression, Ray aspects, and nature of Initiation attainment of the cosmic MEN and WOMEN within the Body of our THAT Logos, and perhaps within That One's compliment (polar opposite). Vast and complex is the matrix of interwoven interdependence between all of the Lives within That Body, and many levels of Initiation testings must be passed as the aeons progress before the Mysteries concerning That Body have been realised. Herein then lies the challenge for all of us to set our sights upon. Let us all dare to strive to attain Logoic Vision and thereby to Know the true ordering and structure of the universe.

6

The Spiral of Consciousness, the Energy View

The general description

All geometrical patterns can become *maṇḍalas,* seed forms, that like all symbols can be evoked to develop direct revelation or insight into aspects of the divinity underlying all manifest Life. The hexagram for instance is a shorthand representation stating the relationship between the principles constituting subjective and objective substance, consciousness and form. Properly elaborated it unveils the symbolism of space as an Entity.

When the process of causation (the evolution of consciousness) is observed the related symbol would be seen as a spiral flow of energy. The energy emanates from a seed point and radiates out in continually spiralling, ever-increasing cycles, until eventually it reaches the periphery of the manifested universe on all possible realms (or dimensions of perception). Such a process of evolutionary continuum is possible where consciousness evolves. An idea of the process can be seen in the symbol of the rock that is thrown into the centre of a still pond, the resultant ripples expand in ever increasing concentric circles until they eventually reach the edge of the pond (or the knowable universe), then keep reverberating almost forever. To the concentric progress we must also add a spiralling one to complete the imagery.

The central or seed point (*bindu*) has no locus at any particular point in space, but rather occurs wherever an observer has focussed an observation. That point is experienced as it flashes in the field of observation. The near instant appearance of a large number of such points then constitutes a picture image held in the mind. This is the basis to individuation in consciousness that allows a reader, for instance, to focus upon the words and meaning of any text. Such becomes what people perceive to be 'real' in the phenomenal world. It only exists so in the eye of the beholder, however it constitutes the basis for one's conscious growth through time and space. (By 'eye' here is meant the object of the particular sense that is actually experiencing the 'point'. This also incorporates the use of the transmuted senses, accessible via the *chakras*.) The near instantaneous correlation of myriads of such points into a picture determines some 'thing' that we are experiencing.

Such a collective energy field, a flux (or tensor field) of such points that are held together in Nature by the Will of a vast Thinker are the atoms known to modern empirical scientists. Between that Thinker and the atoms of substance are a host, a hierarchy of intelligent mediators, the *devas,* existing upon the subjective domains, who embody the substance of manifestation. Together they manifest a corporeality of various natures and shapes that have limited duration, according to the energy put into that field to sustain the illusion. Each individual atom can appear in or out of the time-space zone according to the fluctuations of the grid of the field of which it forms a part. Such fluctuations can move as moving *bindus* of creative expression.

In relation to the *bindu* Govinda states:

> It signifies the concentrative starting-point in the unfoldment of 'inner space' in meditation, as also the last point of ultimate integration. It is the point from which inner and outer space have their origin and in which they come again.[1]

All such points have a relative stability of motion and have a separate existence only in relation to their exterior forms. Everything is transient. This is also illustrative of the process of thought. Each different thought (utilising light energy) sets up an eddy or turbulence which modifies the basic mental matter, forming picture images that govern the resultant physical action. Thoughts come into existence and can rapidly disappear again, to be replaced by another thought or image. The matter of the solar system can also be seen as the result of concretised modifications of mental substance, effected by the meditating creative Logos. Every

---

1 Lama Anagarika Govinda, *Foundations of Tibetan Mysticism* (Samuel Weiser, New York, 1975), 117.

form therein has its cyclic coming and going. Such an appearance and disappearance from inner and outer space is also seen in the world of subatomic particles, of electrons and photons emanating and disappearing in the quantum vacuum, providing the phenomena of the simultaneity of a wave event and a particle of matter.

In terms of the appearance of a human being the *bindu* of its manifesting form is directed by the Śūnyatā Eye of the Sambhogakāya Flower.[2] It is the aperture through which the Flower focusses upon the consciousness of the individual allowing it to retain its control of the manifesting *skandhas* and *saṃskāras.* When related to the manifestation of the *dharmakāya* type of *karma*[3] the *bindus* of all manifesting Life are controlled by the Eyes of the meditating Buddhas, the Lords at Shambhala, and the Shambhallic correspondences in the universe.

The *bindus* activate their component *bījas*, causing these individuating seeds to flower into the full plant of substance-energy denoting manifesting phenomena. The *bindu* is the seed for the entire form to be, whereas *bījas* are the seeds for the various manifold and oft very fleeting constituents of that form. They have been explained in volume 2 of *A Treatise on Mind,* where I quoted the following:

> There are two kinds of *bījas*—(i) natural *bījas* and (ii) *bījas* born of perfuming. The natural *bījas* are the potentialities which have existed innately in the *ālaya* by the natural force of things *(dharmatā).* They produce mental elements, sense-organs and the seeming external objects. The other kind of *bījas* are those which have come into being as a result of the 'perfuming' of actual *dharmas,* the 'perfuming' being repeated again and again from beginningless time. The seeds stored in the *ālayavijñāna,* being perfumed by seven other consciousnesses,[4] are caused to grow, resulting in the appearance of things.[5]

---

2 Explained in volume 3 of *A Treatise on Mind.*

3 The *karma* governing the appearance of the material world.

4 These are the five sense-consciousnesses, the intellect and *kliṣṭamanas,* afflicted mind (the emotional mind). There are said to be nine mental factors to *kliṣṭamanas.* Five of these are expressions of the five sense-consciousnesses, that are the five Watery *skandhas* of the five Elements, of which the five senses come to be expressions. The emotions and the factor of desire are governed by the Watery Element and thus *kliṣṭamanas* is really 'afflictive' whenever the mind is conditioned by this emotional-desire aspect in any way. The remaining four factors are those related to the production of a concept of 'self' or 'I'.

5 Swati Ganguly, *Treatise in Thirty Verses on Mere-Consciousnes,* (Motilal Banarsidass,

The concern here is specifically with 'the appearance of things' in the phenomenal universe, thus with what is here termed "natural *bījas*, that produce 'mental elements, sense-organs and the seeming external objects' by means of the 'natural force of things *(dharmatā)"*. (The 'perfuming of natural *dharmas*' associated with the second type of *bījas* depends upon whether one believes in the existence of *dharmas* or not.[6]) How this *dharmatā* actually functions is not explained in the text. It is however obvious that some agency must exist to call all of the *bījas* of the phenomenal universe to manifest in the ordered evolutionary sequence that they do, and to obey the strict laws of physics.[7] If this was not so then chaos and quick destruction of all sequenced progressive progress would be the result. It should be obvious to all who have meditated upon the miraculous mathematically precise interdependence of all the outer seeming that some sort of Intelligence has made it all to come to be, and to obey the laws and principles of the nature of the meditation-Mind unfolding.

Dharmatā can be considered the means of the application of such a meditation-Mind upon the entire sequence of things that must appear at the right time in the right sequence of space in order to play its proper role in the ever-changing phenomenal scenario. This Mind contains all of the *bījas* that must exist, and intelligently sequences their activities in the phenomenal worlds by the means of lesser agencies of Minds. *Dharmatā* incorporates the *ālayavijñāna,*[8] whose mode of awakening the *bījas* is given below.

Delhi, 1992), 40.

6 *Dharmas,* factors of existence, a doctrine found in the *Abhidharma* of Theravāda Buddhism. The elements of mind. Briefly, each *dharma* is a separate entity or force, there is no substance apart from the qualities of a *dharma*, they have no duration, but flash as new appearances with each moment. The *dharmas* cooperate with each other. (There are said to be 72 of these *saṃskṛta-dharmas.)* Thus they stem from causes and proceed to extinction when influenced by wisdom, but when influenced by ignorance they are continuously generated. The gaining of liberation therefore is what produces their extinction. Though there are problems with this doctrine, nevertheless, the Sarvāstivādins that promoted this doctrine two millennia ago were amazingly prescient with respect to modern concepts of the subatomic world.

7 This statement is not concerned with the manifestation of the individual subatomic particles, for instance, but with the overlying law or principle that sets the pattern for the appearance of the sum of the phenomena.

8 The universal store of consciousness.

One can view the topic of 'perfuming' in terms of the original *bījas* being given added qualities as a result of the activity of the mind or of the evolutionary process. Specifically however, one can look to the functioning of the Sambhogakāya Flower. It 'perfumes' (seeds) into the mind characteristics that need to be developed by its vehicle (the personality) at an opportune time, when the personality is ready to develop the *saṃskāras* which then manifest as subjective urges and images. This path will lead inevitably to the development of the Wisdom attributes of the Dhyāni Buddhas by means of transmuting all of the base qualities of the *saṃskāras*. This is the way of the manifestation of the evolution of prime causative agents upon the path of eventually becoming Logoi.

Massed human consciousness can also be viewed as a prime causative agent, affecting many things in the mental-emotional and physical worlds because of the way that people can alter natural processes. As a consequence of evolutionary progress the 'natural *bījas*' are changed into new forms of *bījas*, which then become the seeds for future potentiality as new cycles, new incarnations of actions, proceed upon higher arcs of progression. All of this can be viewed in terms of phenomena manifesting into and out of the event horizon that is the border between the empirical universe (the mind) and the transcendental, liberated universe of the higher Mind. This has its lower correspondence in the 'event horizon' of the sub-atomic world between etheric and phenomenal space, which produces the paradoxes observed by physicists.[9]

The spirals of consciousness

As the meditation-Mind unfolds it produces eddies of interaction of thought sequences within the confines of its circumscribed thought space. As eddies interact with others they produce increasingly complex and diversified patterns and images. Manifest space thus appears as an immense harmonious conglomeration of rhythmic force vortices ('atoms') of different sizes, patterns, strengths and effects, due to the nature of the originating intellectual thought sequence. (The human atom is but one of these.) They become part of an expanding synergistic

9 This subject shall be explored in a later chapter.

wave of energy flow. This outward flow of energy interaction when highly simplified can be illustrated as shown in the figure below.[10]

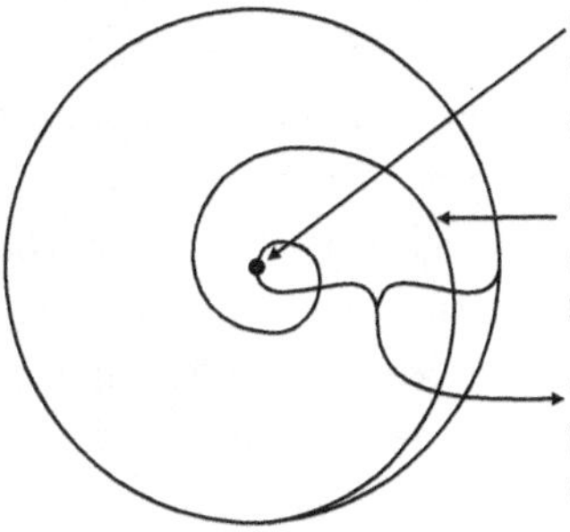

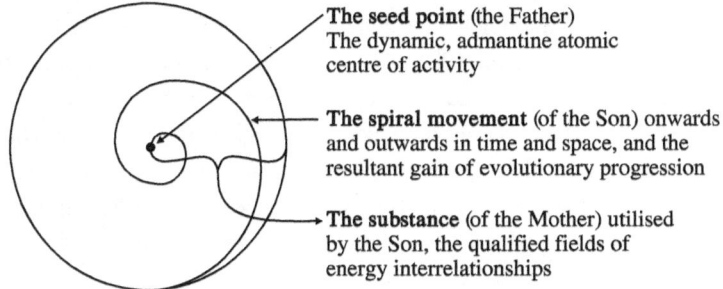

Figure 10. The spiral of consciousness

The ubiquitous spiral symbol should not be quickly dismissed. It has been universally used by the ancients of the world, from the Greeks, Trojans, Egyptians, Celts, Vikings, Druids and Hindus, etc.

The original cause (the Father) is symbolised by the seed point of the gyration. The Son, the accumulator of experience, of conscious awareness, is represented by the swirling line of force, and the Mother symbolises the totality of the manifestation, the substance which the energy of the Son acts upon and moves through.

From the Waters of space, the unknown, comes the circle of time, the activation of *karma* for the manifestation of the universe in right ordered sequence. This originating circle of time is the primeval universal *karma* that caused the manifestation of our solar system. It embodies manifest space yet remains aloof from manifestation. Absolute cosmic time is dethroned, to become limited time in terms of finite solar cycles. Through circumscribing time the Logos manifests His/Her purpose in a solar system. All the forces (the various Gods) of our solar system either stem from, are married to, or else are expressions of this process. This grand Logos is represented in the diagram by the central dot, the seed point from which all proceeds. In Buddhism we have the Ādi Buddha (Vajradhara), the primal Buddha from which all of the other (Dhyāni)

10 See also Diagram 4, Bodo Balsys, *The Revelation*, Vol. 1, (Ibis Press Sydney, 1989), 116.

Buddhas and Bodhisattvas emanate, as well as all of the other forces delineating the phenomena of a solar system. The Buddhist concept however does not generally pertain to a 'God' concept as a Creator, rather, all Buddhist deities are considered personifications of Mind. The origination of primal Mind is only inferred in the concept of an Ādi Buddha, a personified coalescence of *śūnyatā* and its relation to *saṃsāra*. Nevertheless Buddhists have had to account for the originating appearance of the phenomena of *saṃsāra,* hence we have the teachings of Vajrasattva (another name for the Ādi Buddha) as a 'Creator' in Tantric Buddhism, which was explained in the essay 'Does a God exist' in my book *Considerations of Mind – A Buddhist Enquiry.*[11]

The actual process of causation can be further illustrated if we now imagine the centre of the 'ineffable' circle that represents the Father aspect to have a positive polarity, which is the source of the dynamic energy that is the cause of manifestation. The circumference to be must then assume a negative charge. We therefore have the concept of an electrical interrelationship.

Energy flows from the positive centre to the negative circumference, attracting the negative particles to it. However it does not stop there, for it must either be converted to another form of energy or return to its source.[12] The source thus assumes a positive polarity to the returning consciousness, which is negative with respect to the Source's potency. The spark of mind returns to its central luminary (from which it was never in fact separate) as a blazing Light.

This type of motion can be represented as the driving forward of the Father-Mother to fill abstract space and then to return back to the originating centre as the motion of the Son. However, it is unfeasible for the motion of this energy to be arrested there. It must continue its

11 See pages 195-203. Much of that section is a commentary of the work by Eva K. Dargyay, 'The Concept of a 'Creator God' in Tantric Buddhism', *The Journal of the International Association of Buddhist Studies,* Vol. 8, No. 1, (University of Wisconsin, Madison, U.S.A., 1985), 31ff.

12 The alternative concept of a continued outward dissipation would mean annihilation of the complete cohesive unity or cellular structure, which is not a consideration in the field of consciousness. Even in the case of liberation (*śūnyatā*) the integral unity persists, there is no dissipation of the existent Life.

momentum through the original impetus given to it, though on a higher octave of expression.

As the entire action is also propelled forward in space, so it will be seen to be spiral-cyclic in motion, which also indicates the progress of the evolution of mind/Mind. Thus is pictured a wave of energy travelling forward from its source in a cyclic manner, which continually expands to form the space encapsulated by a Logoic Mind. The motion also eventually turns in upon itself towards the higher dimensions of perception. A spatial representation of this idea produces the concept of fourth dimensional motion and the possibility of multidimensional planes of existence that are at 'right angles' to each other. This concept dispenses with the Big Bang theory of modern cosmology, as that theory proceeds in a linear fashion, and is open ended. We are looking instead to a non-localised start of any sphere of activity, such as the appearance of a sun or of a galaxy, though the point of seeding of such is constrained or activated by factors, such as the mass of primal substance to be utilised in the formation of the sphere.

This can be illustrated as shown in the Figure 11 below derived from the book by Perkins.[13]

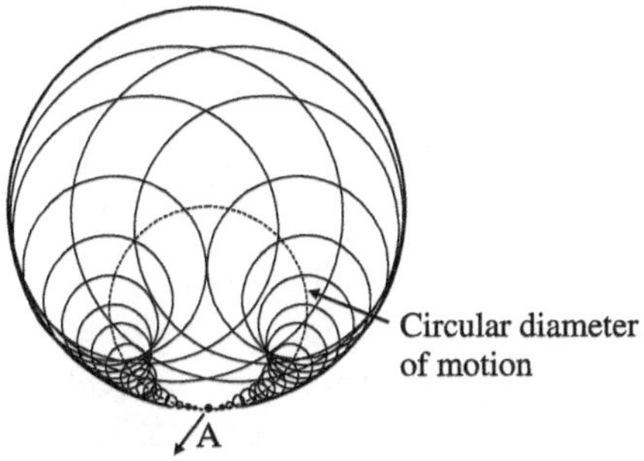

Figure 11. Spiral-cyclic motion

The cause (A) moves outwards to become the total volume of the sphere, then decreases and at A turns at right angles to eventually form another sphere of activity upon a slightly higher turn of the spiral. The

13 James S. Perkins, *A Geometry of Space and Consciousness*, (Theosophical Publishing House, Adyar, third Edition, 2004), 131.

movement from inner to outer space at point A describes a figure 8. The arrow pointing upwards indicates the circular nature of the motion (the dotted sphere). The arrow pointing outwards indicates the forward progressive motion of the cycles of activity.[14]

Perkins endeavours to show motion as it moves from the inner multidimensional space to the outer universe, taking the 'universe' to be an enclosed sphere, literally a bounded Thought-Form of a Logos. As the movement of the spirals cycle from the higher dimensions to the more reified ones, to finally reach the dense sphere, the cycles become increasingly smaller and eventually becoming atomic in scale. Atomic consciousness then spiral-cycles from this point-like space back through the layers of the higher dimensions in ever-increasing spheres of 'awareness'. He states that accordingly the redshift (the Doppler effect) is observed. The redshift is the change in wave length towards the red end of the spectrum that is detected when a wave-motion travelling at high speeds is receding from the viewer. The Doppler effect is used to measure the distances in space, of how far a star or galaxy is away from us. The further back in time the greater the apparent motion of the recession and contraction of space. From this then is deduced the concept of an originating Big Bang, or start of the universe, as postulated by modern physicists. As Perkins is analysing from an esoteric perspective it may be useful to quote him in some detail as a template from which further, hopefully more correct, deductions may be derived.

Perkins states:

> Suppose that light-year space does have the property of acting upon observations of distant objects as, say, an electron microscope upon a tiny speck of matter: present knowledge, as noted above, will no longer apply, and the phenomena of the "red shift" will have to be re-interpreted. Perhaps the galaxies are not moving away from us at all. Perhaps space is not physically infinite. The physical universe may not be as large as it seemed to be[15]...It is possible, then, to reason that the measurements of light-year space that have indicated increasing velocities of supposedly receding galaxies may be revealing instead, an

---

14 This is an adaptation of figure 16 of Perkin's book. (Ibid., 131.) He has A and B at the top and bottom of the dotted circle and omits the forward-progressive motion, as well as the 'figure eight' motion I have described above.

15 Ibid., 115.

increasing nuclear activity inside the galaxy being observed. Galaxies seem to be moving away in increasing velocity because those that are actually further away, are being observed through a space-lens with greater magnification, consequently with greater recorded velocities. The motion indicated through the lens would not be a movement in space, but a velocity of nuclear activity. With such a theory, it follows that except for local orbiting, the galaxies are not moving away at all, relative to our position in space. Where does this view lead us?

As we peer further into the distances of space, the "lens" is becoming more powerful. Observations might be comparable in general with those made through microscopes of increasing magnification. Brought under observation would be activity in the force-fields of the atomic nuclei, where particles obtain velocities that approach the speed of light and beyond, appearing and disappearing in billionths of a second. Where light-year space is acting as a lens far out phenomena will be recorded of bewildering galactic motion and intensities of light that are outside the possibilities of physical explanation.

If such galaxies are not receding, the "red shift" phenomena are providing evidence of some unknown law that is operative in large-scale space, a law that pertains to the transition of forces from material to non-material levels of life. This would explain the fact that with extended vision the galaxies that show increasing velocities are in their local orbits, providing data of another order of life that is beyond the physical, namely, the Astral form of the galaxy.

The most exciting aspect of the foregoing is the bearing that it has on the geometry of space.

For if space is acting as a lens there will be available another approach to the problem of the geometry of the universe. In the space-consciousness conception, as illustrated in Figure 9, matter is continuously appearing in the universe as radiation in nuclear particles, force fields, and finally in physical forms. In the cycles of time all of these forms are born, grow and disappear. The galaxies too, disappear across the rim of space at ultimate velocities. Are these velocities the interlocking bridge within physical mass and subtler material? This point of disappearance—movement at the speed of light—bears immediately upon the geometry of space since theoretically, there can be no movement of objects in physical matter at velocities beyond the speed of light, and observation does reveal that some galaxies are approaching it, the dawning of great revelation seems about to

> take place. For if objects are contacted that are apparently in motion beyond the speed of light, they must be objects that are obeying laws of another order of life beyond the physical plane. If space is spherical, the explanation is impelling that the light impulses are reaching the observer from around the full sphere of physical space, having traversed the great circle of the physical universe. Galaxies at that distance would be seen through the lens of light-year space as having the Astral form of the observer's own galaxy. The object recorded would be seen as obeying laws of a subtler level of life—those of Astral matter. When the super telescopes of tomorrow are produced, and we are able to view a galaxy supposedly moving at the speed of light, it will be our own physical galaxy that has disappeared into its astral form, because we are receiving light impulses that have circumnavigated the great circle of space to the point of observation. Such a conclusion must necessarily be reached if space is spherical.[16] Furthermore, the observer will find that he is viewing the locale as it was at the beginning, and as it will be at the end of time, now![17] He will be seeing the galaxy before and after its appearance in physical form[18]...Let us now assemble the ideas that have been introduced: (1) Any observer is ever at the centre of the universe if his consciousness is seven-dimensional;[19] (2) Large-scale space is acting as a lens;[20] (3) Physical space is spherical; (4) Ultimate space, in terms of velocity, is transformed from material objectivity into non-material subjectivity and mergers with the space-consciousness plenum—the Totality.[21]

The movement from the inner to outer space at point A in figure 11 forms the figure eight (8), where the top sphere of the eight relates to inner space and the bottom sphere to the phenomenal space we observe.

---

16 But space is also spiral-cyclic.

17 Not exactly, because everything will have evolved on to a higher cycle of expression.

18 Ibid., 116-20.

19 By 'seven dimensional' Perkins is referring to the seven planes of perception. Most people's consciousness however do not function seven dimensionally, they view from a three dimensional perspective.

20 It only acts as a lens when viewed multidimensionally. When viewing the physical plane phenomena with modern telescopes then the universe as depicted by scientists applies.

21 Ibid., 121.

The 'lens' relates to the Logoic Eye peering towards the material domains wherein we reside. This Eye effectively possess microscopic Vision to view activities upon our earth. As the Logos thus peers, so intense energy is transmitted from the Logoic Eye to drive the motion of the Thought Construct—the phenomena of world and solar spheres—onwards through space. The motion manifests in terms of progressive evolutionary spiral-cyclic activity. Looking inwards we must develop telescopic vision to view Logoic domains. Scientists similarly look through telescopes to view the universe at large, but make the necessary calculations to comprehend the true sizes of the objects they are viewing.

The statement by Perkins that 'The motion indicated through the lens would not be a movement in space, but a velocity of nuclear activity' needs commentary, as this indicates an accelerated rate of motion of the subatomic particles constituting atoms (and hence the things composed of atoms) moving into fourth dimensional space, where the speed of light is transcended. Viewed from a large-scale universal perspective this view implies a form of 'steady state' universe, because all forms move into and out of this subjective space, rather than from a Big Bang. The motions of galaxies, etc., comes from the initial Thought-Form expansion (a form of non steady state) of the originating Logoic Thought, within which the appearing and disappearing galaxies (etc.) are conveyed. The universe thus consists of both steady state and expansionist elements, but there is a limit to the Thought-Bubble of the Logoic Mind, because all is abstracted again into subjective space, preparatory to the next Out-Breathing. All lesser forms within the universal whole manifest similar cycles of the expression of disincarnation and rebirthing.

Perkins' idea is that the motion emanating from the astral plane is 'faster than the speed of light', as the substance of this dimension of perception is far subtler than that of the elementary particles constituting the sub-atomic universe. This may be so, but as the astral substance approaches the 'event horizon' between the fourth dimensional universe and the three dimensional one, so then condensation and materialisation occurs, which slows down the subjective particles (Anu's) to light speed and forces their combination into atomic unities.

Perkins' statement that 'Galaxies seem to be moving away in increasing velocity because those that are actually further away, are being observed through a space-lens with greater magnification, consequently with greater recorded velocities' applies maybe to a Logoic Eye acting as a 'space-lens', but not to ours. We do have the capacity to look into the atomic world clairvoyantly or with the aid of instruments, and so consequently affect the appearance or 'a velocity of nuclear activity' of that world by directing energy thereto. Similarly a cosmic Logos can effect the substance of the vast distances of space by means of energy moving through His Eye, which sees with 'greater magnification' than ours.

From this perspective however, galaxies can move away from us, relative to our positioning in space, hence the redshift phenomena is applicable for the most part. (Perkins also admits this in his statement 'except for local orbiting'.) There are however other effects influencing the frequency of the light observed, such as the Wolf Effect, the theory that light itself loses some of its energy as it moves through vast distances, accounting for some of the observable redshift, and even that some of the cosmological constants may not be invariant, and so altering what is observed.[22] Hence there is no recession to a primal Big Bang, as well as making the observable universe objectively smaller, more condensed, than conjectured by the Big Bang theory.

Within the ineffable sphere that represents the containment of our universe then, there is a continuous birthing and dying of stars and galaxies. They cyclically come in and out of etheric space and the astral universe. At the appointed cycles an entire universe also can thus emerge. Everything dies and is reborn, but the *creatio ex-nihilo* as conceived by modern physicists and the Roman Catholic theorists does not happen. There is no Big Bang, but rather a gradual emergence from subjective to objective space of energy forms that are pre-existent, but not yet objectivised. The phenomena appearing upon the physical plane is in reality that which has been condensed and reified from the domain of the mind/Mind. All we see is but an extension of Logoic Thought, taking a vast hierarchical succession and increasingly vaster and subtler Logoi into account as we move from a planetary or solar scale to that embodying a galaxy or cosmos.

---

22 See for instance the paper: *Variation of Physical constants, Redshift and the Arrow of time* by Menas Kafatos, Sisir Roy and Malabika Roy.

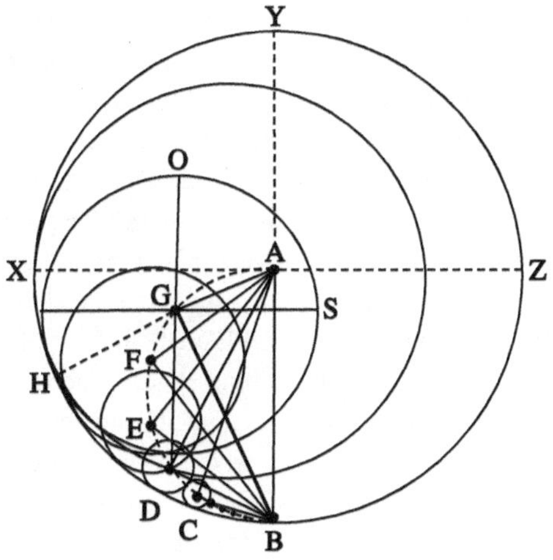

Figure 12: The geometry of spiral-cyclic motion[23]

Perkins then describes geometrically the basis to spiral-cyclic motion via the formation of a number of right angled triangles moving in a curved arc from the centre of the sphere 'A' to the circumference, 'B', producing ever smaller spheres of conscious activity. They are depicted in figure 12 as the loci of the spheres at G, F, E, D, C and B.

> (I)t will be found that *any* sphere that is tangent to a circumscribing sphere is centred upon the apex of a right-triangle whose hypotenuse is a radius of the circumscribing sphere. This fact becomes enormously significant if an infinite *series* of right angles are constructed upon that same hypotenuse with the apex of each triangle being a centre of a sphere that is tangent to the circumscribing sphere. Referring to Fig. 14,[24] such a series of spheres is shown with centres "GFEDC" etc. marked. Each one is the apex of a right-triangle such as "G" in the right-triangle "AGB" whose hypotenuse is the radius "AB" of the sphere "BXYZB". The series of infinite centres will merge into, and become, the line "AGFEDCB". At every point (such as "G") the spheres constructed (such as "HOSH") are at the apexes of right-triangles. The altitude of a right-triangle is perpendicular to the hypotenuse. Such an altitude is included in the line "GS". If "GS" is extended, it becomes

23 This figure is figure 14, of Perkin's book, (Ibid 130).

24 My figure 12.

a horizontal diameter of the sphere "HOSH," thus determining its vertical diameter. Now, the hypotenuse of right-triangle ABG is a radius of the circumscribing sphere, and extended, becomes it vertical diameter "BY", determining thus the horizontal diameter "XZ." The diameters of both spheres are therefore perpendicular to each other by construction. Similarly, every other point that composes the lines "AFEDCB" lies in both diameters of a sphere that is perpendicular to all of the spheres in the series, as well as to the circumscribing sphere itself. Therefore the line "AGFEDCB" traces a direction that is perpendicular to the circumscribing sphere, and is the seventh perpendicular direction from the central point in the sphere to its outer circumference. The series of spheres whose centres move on that line, therefore move from a totality of volume in decreasing spheres to zero volume as a point on the surface of the circumscribing sphere[25]...If the curved movement now continues (see Figure 16[26]) from point "B" on the surface, inwardly, in the completed movement of increasing volumes, their centres will ultimately reach and merge with, the centre of the Great Sphere, "A", and their volumes will become the volume of the circumscribing sphere.

This completed movement generates a four-dimensional spherical form. (Figure 16). The four-dimensional form can be glimpsed only by realizing its motion. This requires imagination. There must be a transition of consciousness from concentration to realization[27]...If we will concentrate upon the drawing in Fig. 16 *thinking* about the logical movement of the point expanding in volumes that become one with the whole, as already explained, and reversing the movement to become the point on the surface, we may realize the logic of the proposition by just thinking about it intently.

Now, if we will bring about a stillness in the mind—stop *thinking* about the proposition—and with the cessation of thinking push consciousness beyond the mind to *become* the whole movement itself, from the whole to the point and return, we may glimpse intuitively the reality of the fourth dimension of any three-dimensional form (the sphere being the ideal, containing all other forms.) Glancing at Fig.

25 Ibid., 129-133.

26 My figure 11.

27 Ibid., 134.

17,[28] you find that there is no central point "A", because the centre is the *whole*—the centre is everywhere. The whole spherical form is glimpsed in four-dimensional consciousness when we see both the inside (dark circle), and the outside (white circle), projecting from within, along the true perpendicular curve, the total volume that becomes the manifested point "B" on the surface...[29] It can now be realized that every point in the manifesting universe is emerging along its true spiral curve. The return of the point from its manifestation on the outer rim inwardly, to the Centre is in terms of the evolution of spheres of consciousness. In expanding spheres of consciousness all living things return to the union of Totality, merging finally in the Central Point—the Logos consciousness of the Whole.[30]

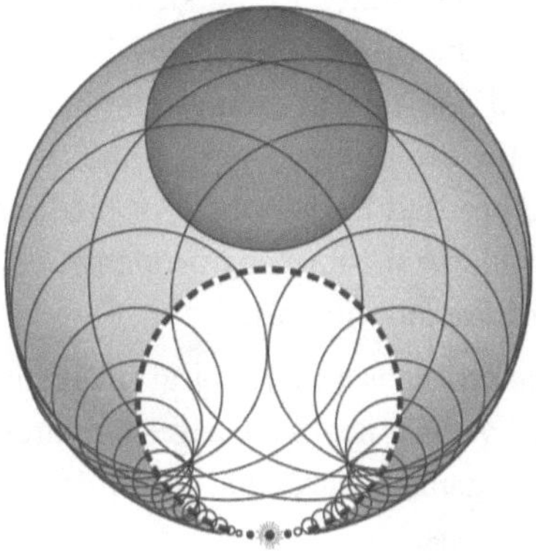

Figure 13. Fourth dimensional motion

Perkins' depiction is somewhat limited, as it omits the true nature of spiral-cyclic motion in relation to the forward progressive motion, but nevertheless provides an idea as to the mode of expression of fourth dimensional motion, and of the movement of consciousness-bits from the inner to the outer universe, and back again. Also he has simplified the dimensional relationship between the seven planes of perception, as he has not included the concept of transmuted correspondences as

28 My figure 13, taken from Perkins book (Ibid., 132). He labels this diagram 'Movement in Fourth Dimension Glimpsed Intuitionally'.

29 Ibid., 134-135.

30 Ibid., 136.

one moves from one dimension to the next. The concept of the universe as a closed sphere is correct from the perspective of the enclosing of a circumscribed sphere of activity by an ineffable Logos, yet is incorrect from the point of view of what is IT that a Logos encloses a Thought-Form within. From this perspective Space is infinite.

Using the simple geometry as outlined by Perkins as a basis his model can be expanded, to hopefully get a truer comprehension as to the nature of cosmos and the appearance of things.

There are three types of motion indicated:

1. *A rotary type of motion* at the circumference.

   With respect to the process of the evolution of consciousness this type of motion relates to the slow repetitious activities of humanity where the mind is sluggish, desire-filled, avaricious, self-centred and resistant to change. It characterises by far the greater part of the evolutionary period and conditions the masses of people who are tied to the wheel of birth and death because of ignorance and *karma*-forming propensities. It conditions the general form of the atomic unit.

   As evolution proceeds a slight spiral fluctuation manifests in the revolution of the great Wheel as it drives forwards (and occasionally backwards), and finally inwards and outwards, upwards and beyond, throughout the aeons.

2. *A spiral-cyclic* motion forming the 8.

   This refers to the mode of the progress of consciousness as it sequentially expands with each turning of the wheel and progresses onwards with the passing of time. The nature of the motion inside the atom is thereby conditioned, indicating therefore the nature of the binding energy that interrelates the components of the nucleus of the atom.

3. *A forward-progressive* motion propelling this 'atom' of energy forward in abstract space.

   This represents the nature of the manifestation of the enlightened as they pierce the barriers of all that has bound them to considerations of form, and then to the limitations of all 'ring-pass-nots' of temporality and illusion. The transmutation of atoms and the radioactivity of certain isotopes is caused by a similar energy within the atomic sphere.

We therefore have:

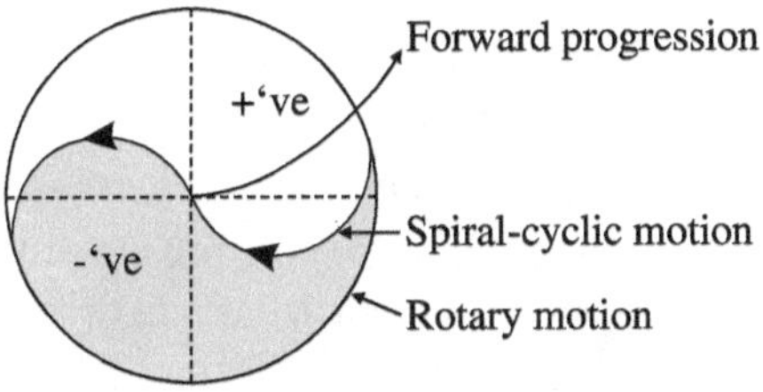

Figure 14. The three types of motion

In this figure the spiral-cyclic motion is depicted for both evolutionary streams on this planet, the masculine human stream, denoted here with a positive polarity (inherent compassion, signifying the utilisation of the human will dominating Nature's forces), and feminine *deva* stream (inherent intelligence, denoted here with a negative polarity because they are the receptive agencies that fecundate and sustain the processes of Nature).[31] The two streams of lives symbolically gyrate around each other, forming a yin-yang. The human stream carries the force of the Will and Love-Wisdom, whilst the *deva* stream represents active Intelligence, the creative Builders in Nature. The Will directs, Love-Wisdom visualises the Plan, and the *devas* build as directed. Whatever humans build, in thought, emotions or in concrete manifestation, there is the *deva* substance utilised to accomplish the act. Whatever appears in Nature there is the effect of *deva* Intelligent Design and karmic weaving according to directives from Logoic Thought. The Logoi are but evolved humans. Thus the play of *saṃsāra* is instigated and enacted.

The imagery accompanying spiral-cyclic motion is that of a tunnel that is created by a manifestation of the movement of consciousness through space. Though the units of consciousness are individualistic, they

31 Upon the physical domain the negative electrons are attracted to the positive polarity of an atomic nucleus, causing the atomic structure to manifest, whereas in the above schema the positive force directs the negative, receptive substance, cohering that substance into form.

attract to them other strands of thought of similar nature, forming energy patterns that spiral ever on in unison and take the form of tunnels in space.

Because of the movements that causal conditions create, consciousness views the manifestation of the positioning of objects in space in differing ways, but works to identify the exact location of each manifested thing, even in motion, at any particular instance of time, such as sun, stars and moon, or moving vehicles. Hence it particularises 'things' into a continuously changing moving image that spirals onwards through time. Such spirals will manifest unless the consciousness is inchoate, in which case there is a moving cloud of ill-digested and vaguely formed images that are caught up within the force vortex of a much larger consciousness-stream that carries in its embrace a vast panorama of collectivised thought-streams. Such is the movement of a Logoic Thought, and of such an Entity's immediate Co-operators. They are the major factors of the evolving *maṇḍala* created by the prime Thinker. Inchoate consciousness can be viewed as the substance of primal cosmic dust.

Each consciousness views an individualised form from a (slightly) different perspective because each views from a different point in space. Each mind/Mind is also capable of modifying what is viewed according to internal forces within their consciousness-space. The sequence of the appearing images, their timing, and of their different qualities, happens because of the causal conditions of consciousness moving through a much larger space (world or cosmic view) than it possesses internally, allows it to continually expand as it moves to assimilate the new.

Whilst consciousness exists it moves, but when it is asleep it is in temporary abeyance *(pralaya),* with a new *manvantara* of activity manifesting upon awakening that builds upon the images from the earlier cycle. Thus we have spiral-cyclic motion, wherein cycles of waking movement are interrupted by interludes of passivity (the moments before differing thoughts, sleep, death) before a new cycle begins. Each new cycle continues on from the previous epoch of thought (seeing anew), even though what has passed may not be remembered, nevertheless the imprints (force vectors) continue, and they drive the consciousness-flow along the same temperament that was sustained previously, though the scenario observed might differ. 'Temperament' here is viewed in terms of appearing *saṃskāras* manifesting from past cycles, energy predispositions that condition the way things are experienced in the present activity. The five sense-consciousness manifest their impressions in consciousness, but the *saṃskāras* move to alter the new raw data in terms of the predispositions emanating from past consciousness-volitions. They form habit-patterns, which can only be changed by the imposition of the will, producing a new forward-progressive motion. Many *saṃskāric* streams manifest to produce the overall predisposition, and which modifies the way the consciousness views the images of the now, whilst all spiral-cycle into an indeterminate future.

The time-space continuum

A complete, existing individuality that was not formally known to consciousness but which finally appears to the consciousness-eye manifests as an image that is integrated via various attributes in its possession, e.g., emotions, *saṃskāras*. Consciousness often alters the image in such a way that will generally befit what is pleasing to its empirical environment. The image is rationalised so that it can be understood. A logical sequence of ideas derived from it can then be formed with respect to what consciousness wishes to obtain, whether that be the truth of what is, or something else, as manipulated and interpreted by the mental-emotions. The image has then been embellished and this pre-empts the next moves of consciousness.

As the eye moves its gaze it sees what was formerly potential, an already existent individuality or scenario having various separated components that are connected by a series of interrelationships making a completed form. The form becomes part of a series within a pattern that consciousness visualises. The mind then formulates a plan of action as to what to do with it, to place it in context with what is appertained to be important, or to disregard it. This guides consciousness as to what is to be viewed next, which may not necessarily appear as desired. What appears is often conditioned by what consciousness desires to see, the image may manifest as an evanescent form, or become more fixed, a focussed exactitude that fits into a pattern of already existing existents. A new *maṇḍala* of desirability can then be formed. All manifests within the *tunnel of the time-space* continuum that is the collective awareness of consciousness as it spreads out to fill more space through time. This tunnel is not linear, but spiral-cyclic, as patterns, images, *saṃskāras*, from the past are innately recalled to condition the new view.

Past events spiral backwards to a singularity and are mostly forgotten, whilst the future expands into ever-greater arenas of luminosity. This creeping luminosity is not normally recognised by consciousness, but it betokens of developing wisdom, of enlightenment vectors developing from the sum-total of the *maṇḍalas* that have developed over the vastness of the consciousness-time. This means that though past events may have been forgotten the accumulated 'weight' of the momentum *(vāsanā)* of their effects are carried through to influence the present as the appearing *saṃskāras*. The images these former '*maṇḍalas* of desirability' contain can be retrieved, but for this the developed luminosity must be bright enough to be directed by the conscious will to move backwards along the paths coming from darkened consciousness-spaces. Memory therefore is obtained.

Ordinary human consciousness is too dull in nature, it has not the luminous penetrative ability to follow the *antaḥkaraṇas* back to past life activity, which represents an earlier *mahāmanvantara* for it.[32] The intervening *pralaya* period has produced a hiatus in the continuum of mind, but not of the causes of the perception *(pratyakṣa)* to be. These seeds *(bījas)* containing the completeness of former *maṇḍalas* of activity can expand (spiral) into the new consciousness-space if activated by a luminous projection. This means that each time-space continuum is but a collectivised unit of many such *bījas* of former cycles of activity that were once the focus of the consciousness-space. All *bījas* are carried forward by the *vāsanā*[33] instigated by their creation and are automatically recalled when the mind generates an energy characteristic synonymous with what they contain. The *bījas* help build the new image that consciousness is creating, being the tools it utilises to create that image. They represent the bricks and mortar that clad the house of the new consciousness view.

32 Here a *mahāmanvantara* is considered as a particular life, whereas a manvantara will then be viewed as a complete thought sequence, or waking consciousness that precedes sleep *(pralaya)*.

33 *Vāsanā*, karmic predisposition driving the *saṃskāras* to come to the surface of consciousness. Potency, a driving force of *karma* and consciousness (*citta*), said to be generated or accumulated from within the *ālayavijñana*. Technically the *saṃskāras* are stored in the *chakras*, and *vāsanā* is the force that drives these karmic aggregates to fruition in *saṃsāra*.

The image seen in the mind therefore utilises the substance of the past, moulded with the new input from the sense-perceptors.

The detail of the images contained in the old substance formerly utilised but now not needed by consciousness is not recalled, but the energy-qualification *(saṃskāra)* from the past is used to build future thought-forms with. The images contained in the substance of the *saṃskāras* can be revealed to consciousness, but it needs the will to project the *antaḥkaraṇas* backwards to awaken the *bījas* that are building the 'now'. If the energy of the *bījas* are strong enough there can be spontaneous memory recall. If the luminous intensity of the will is strong enough retrieval through many *pralayas* and *manvantaras* of past life activity is possible. The Sambhogakāya Flower contains all such images in its grasp and at times deems it necessary for its instrument, the personality, to be aware of past precedents, thus uses its will to awaken a *bīja* in the mind of the instrument, and so a flash of intuition manifests, or revelatory awareness happens, that helps awaken the 'personal-I' to greater understanding than would otherwise be possible. This process is facilitated if that 'I' has a refined consciousness approaching the energy qualification of the Sambhogakāya Flower, or if at any point the mind is generating lofty thoughts, or focussed deeply in meditative analysis. Telepathic impression is of similar nature. Not just awakening *bījas* of the past, but the seeding of that mind with new revelatory awareness from the *ālayavijñāna* is then possible. Thus is genius, or high creative inspiration born.

Logoic Thought manifests in a similar mode, but upon a far vaster scale possible to humans. Here the *saṃskāras* utilised are collective groups of human and *deva* units.

The figure below depicts the time-space continuum. It can also be pictured as a *trumpet* that conveys *sound,* (as also does the shell).[34] This indicates the nature of the conveyance and potency of mantras,

---

34 We thus have the esoteric interpretation to such phrases in the Bible, as presented in *1Cor. 15:51-52:*

> Behold, I shew you a mystery; We shall not all sleep, but we shall all be changed;
> In a moment, in the twinkling of an eye, at the last trump: for the trumpet shall sound, and the dead shall be raised incorruptible, and we shall be changed.

This passage refers to the process of transition of consciousness from normal waking consciousness to that of the inner realms, and then enlightened perception. Note also that the conch shell's sound emanating from its spiral form has always symbolised the *śabdatanmātra*, the subtle field *(ākāśa)* of sound, the Space that liberates.

when each tunnel is viewed as a *nāḍī*. This continuum graphically depicts the development of the meditation-Mind as it builds its path in consciousness. It starts from the infinitesimal point in time and space that represents the self-conscious individual focussed upon the object of perception (the central dot in the diagram, the *bindu,* the focal point of meditation) to the outer limits of possible cognition or revelation, represented by the circumference. The path is spiral, as consciousness travels this way, through repeated cycles of activity, onwards and outwards in space through time, becoming increasingly expansive as it evolves.

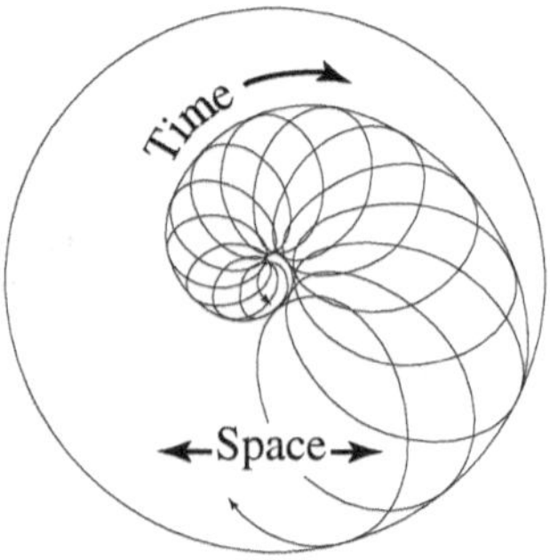

Figure 15. The time-space continuum

This is but an elaboration of the spiral ◎ that Govinda depicts in the centre of the *vajra* (dorje),[35] the 'diamond sceptre'[36] that symbolises immutable power in Buddhism. The *vajra* organises consciousness in terms of the five Rays of Mind when consciousness travels out to encompass space. The outward movement represents the process of expansion of the petals of the *chakras*, and happens in consciousness via the most developed form of meditative absorption *(dhyāna)*. Its power is the basis to the process that gains enlightenment.

The five Rays organise the movement of the petals of the flowers in accord with the expansion, the movements of energy (*prāṇa*) within a person's *nāḍī* system. The expansion accelerates via the meditative

35 Govinda, 63.

36 Ibid., 62.

cleansing process of substance (of the physical body, emotions and mind) and through the disciplines *(tapas)* that allow the meditator to increasingly channel more intense forms of energy within consciousness. Inevitably whatever binds one to the phenomenon of three-dimensional space is eliminated. The meditator is then liberated from it and enters into conscious identification with the multidimensional universe.

The spiral motion depicts the motion of a 'wind', the *prāṇa* moving through the *nāḍīs*, which is the principle of Life *(jīva)*. The inherent vitality of the *prāṇa* sustains the body's life. When viewing this process from above-down, or from within without (from the central point in the diagram to the circumference), we observe the power of divinity descending and manifesting outwards, causing effects in the material realms.

Comprehension of the nature of a spiral is also important cosmologically because most of the galaxies observed in the universe possess this shape. This is important because it indicates that all is an expression of mind/Mind and that the appearance of a cosmos obeys the same rules as an organising Mind that calls forth *saṃskāras* of previous cycles of activity in order to move forward (to exist). Other galaxies, such as a globular cluster, indicate the nature of an Entity that is in the process of abstracting its Thoughts before a new beginning. Other Entities may be in the process of newly establishing themselves in physical space.

As stated, the spiral motion does not depict an inchoate thought, but rather a developed, directed thought process moving through time. Such is the expression of Logoic Thought.

The process of the expansion of consciousness via the time-space continuum with respect to the evocation of *bījas* and the expansion of thought*s* was explained in my book *Maṇḍalas, Their Nature and Development.*[37] To illustrate this expansion I shall present some figures, which are derived from my earlier books. They show the relative nature of progressive existence and thus awareness from microcosm to macrocosm.

---

37 See the section on 'The perception of things', pages, 184-191. Also the evolution of the basic movement of the spiral was treated in pages 429-33.

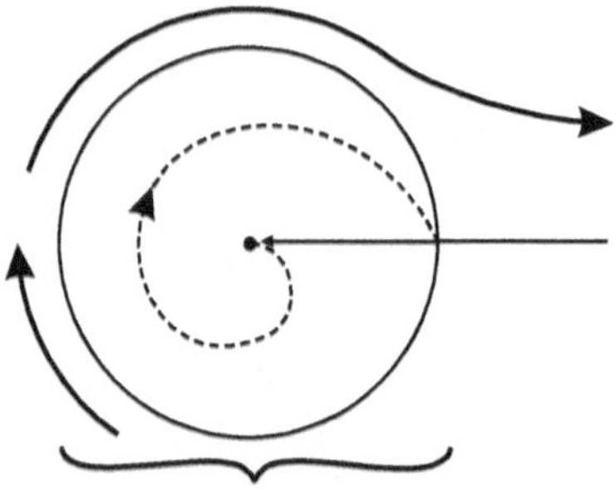

Figure 16: The movement of *bījas*

This figure depicts the basic cellular state of a thought construct within the mind of a human unit. The impact of the present is shown by lines representing the circulating of energies sustaining the integrity of the thought. The past is represented by the central dot from which the *bījas* can be procured to influence the present activity, where a continuous movement allows expansion towards the future. Everything external to the cellular structure equals *māyā* (illusion) to the cell's experience. Thus is depicted the gradual expansion of a cell of consciousness from a central point to the circumference of one unit of thought. As the attributes of the thought diversify, becoming thereby its 'organelles', so then more qualities come into being, allowing categorisation. The central dot (containing the *bījas* of past actions) expands by means of spiral-cyclic motion in the form of the present mental consideration. They manifest as the *saṃskāras* of a thought stream brought to active fulfilment as the future expression unfolds. The *saṃskāras* expand outwards to fulfil the completed sphere or containment of the qualities that the thought represents. The thought spirals through time from the centre to the circumference as it draws like substance to it from the *ālayavijñāna* environment.

Figure 17 posits that the environment external to the cell manifests in the form of an ultimate reality to it. It is that within which consciousness evolves and from which it gathers the information and substance necessary for its future expansion. Such substance is expressed in terms of colouring, tonality of hue, intensity of energy, the qualifying nature of the originating *bīja* and of the reciprocating substance that has been drawn from the general external environment to clothe itself. These seed *bījas* can be derived from any of the seven Ray and subray lines.

Arrows of thought can penetrate the external wall of the present thought construct, influencing its expansive progression as well as its forward motion. Expansive progression, plus forward motion through time produces a curved, spiral motion of the construct, plus for each new thought or idea. Inevitably the outward expansiveness of the present idea becomes the seed *bīja* for a vaster future thought sphere.

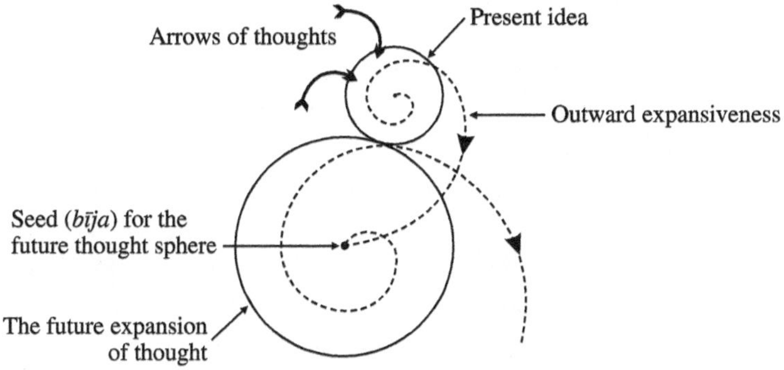

Figure 17: The expansion of thoughts

Depicted here is a well-formed articulated thought form, with a definite boundary delineating an idea. Inchoately constructed thoughts also exist, with hazy indefinite boundaries that are formed through lazy thinking. We can also have sensual brooding thoughts of non-descript desires. These can be likened to planetary nebulae, the cosmic dust from which suns are eventually born.

What exists outside the cell's perimeter represents ultimate reality to it, which terminates any thought sequence (before the start of another differently sequenced idea or thought). Every thought is the indicator of the development of an eventual inclusive realisation of enlightenment.

If the above figure is understood to represent the activity of the evolutionary order of consciousness then the hypothesis can be presented that most states of consciousness progressively expand from a previous one. (Each new truly independent originating thought is but the seed for such a future expansive progression.) There are always links between one thought stream and the next, allowing the retrieval of thoughts and the manifestation of a non-chaotic mind. (Chaos being synonymous with a primitive or deranged mind, or insanity.) This means that the thought streams progress ultimately to the one reality and are expressions of that reality, though they fall under their own limitations. Though some thoughts are negated, proved wrong, by means of the expression of higher revelatory thought, nevertheless all thoughts have gone into producing that enlightenment.

It also proves a *hierarchical structure,* an order to the appearance of consciousness and its relative states of expression. Consciousness evolves from uninformed states, versions of ignorance, to the enlightened stance. It moves from small spheres of perception to macrocosmically large *maṇḍalas* of ideation, from the limited to the expanded, in ever-increasing spirals of Life.

Figure 15 of the time-space continuum endeavours to show the nature of the expansion of consciousness throughout space. The seed *bīja* from the previous idea adds more that just what is held within its 'blueprint' to the new thought-form, because it acts as a magnet for the assimilation of new thought substance from the forthcoming *saṃsāric* contact. This modifies the *bīja* and produces an outward expansion.

Each thought construct can be viewed as a representation of one or other of any of the eight directions of space (explained in *A Treatise on Mind*). For *maṇḍalic* purposes however we can add the conception of the projection of the spiral-cycles from the *bījas* of the past to the arena of future expansiveness for each of the directions.

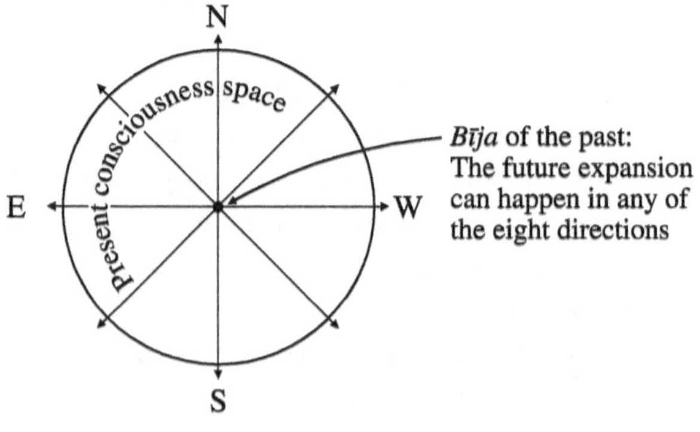

Figure 18: The wheel of the *dharma*

Figure 18 shows that the present is viewed as everything inside the diagram and the future as everything outside of this self-contained sphere. The circumference is the delineating factor, as it alone defines the containment of the integral unity of the cell for any present cycle of activity. (Such a circumference for instance can be considered the outer bounds of a solar system or galaxy.) What it contains defines the consciousness of the personal-I in the now. This consciousness is projected into the future, and depicts how others may perceive it to be.

The time line for each direction, which is normally depicted in a linear fashion should be considered here.

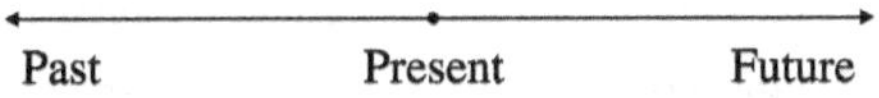

However, with respect to evolving consciousness this time line actually manifests in the form of a spiral:

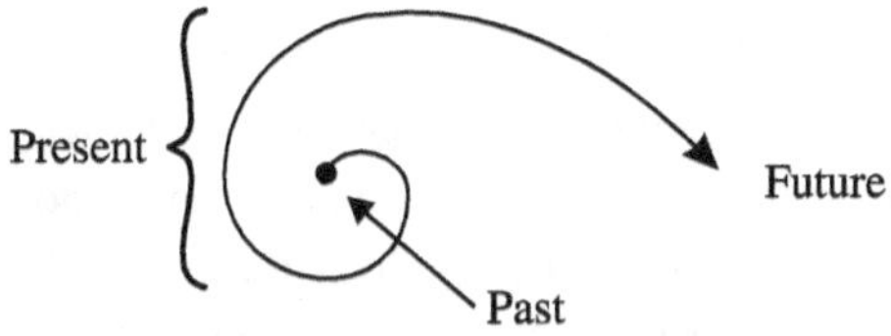

Figure 19: The time line

Each spiral must also be viewed in terms of a trinity, and inevitably in terms of five dimensional space[38] and will expand according to the nature of the thought input, or stunted in its movement if there is a constricting crisis associated with that particular thought stream. There can also be a rapid movement in all eight directions in the case of an enlightened Mind. For the average person however, not all arms are equally progressing, many are stunted, and some may only potentially exist.

The figure below of a swastika expands the basic idealised concepts presented above in terms of the four main directions of energy flow, of moving forces and related aggregates. The circle represents the cellular unit. (The identifier at any level of a unit of consciousness, from what may be considered an amoebic mind to the vast serene Mind-space of a

38 If time is considered a dimension.

Buddha.) It encapsulates the present. The representation here is thus of the ten directions of space. The central point can be seen to go inwards and outwards from the paper, where the inwards direction represents the past and the outwards direction the future. The directions north, east, south and west, represent a cross of energies that are fixed in intent, but which move upwards or downwards, inwards or outwards according to the orientation of the thinker. The intermediate positions represent the moving arms of a swastika, pushing the entire wheel in either of two directions:

- The direction right to left (counter clockwise), producing a backwards motion into the past arena of *saṃsāric* activity. This concerns deepening the path of materialism and self-concern. This path is normally that of ordinary self-focussed consciousness, whilst its complete potency is evoked by the black magician through extreme one-pointed intent.
- The direction left to right (clockwise) toward the expansiveness of future potential. This produces the gain of evolutionary development, adding to the qualities of consciousness. As a consequence the higher stages of the path, leading to elimination of ties to *saṃsāra* will inevitably be produced. The swastika increases its momentum as increasingly energised thoughts are incorporated into it. This is the path followed by the practitioner of the white *dharma*.

Figure 20 summarises all that has been so far presented concerning the nature of the evolution of consciousness. It should be understood to be the basic generative dynamo for all movement of energies (*saṃskāras, nāḍīs*) directed by the *chakras*, and constitutes the internal dynamo for the activity of each *chakra*, without which the *chakra* could not exist.

We can similarly summarise the world of the personality, a self-conscious individual, manifesting a path in life via a conscious field of awareness as shown in figure 21.

The spiral path integrates a set of components coordinated by consciousness, of the way things appear and recede through time. The sequencing and combining of things is determined by *manasic* activity, alchemical Fire, where the philosopher's stone of creative imagination works its magic. *Manas* (the attributes of mind) does not manifest in a

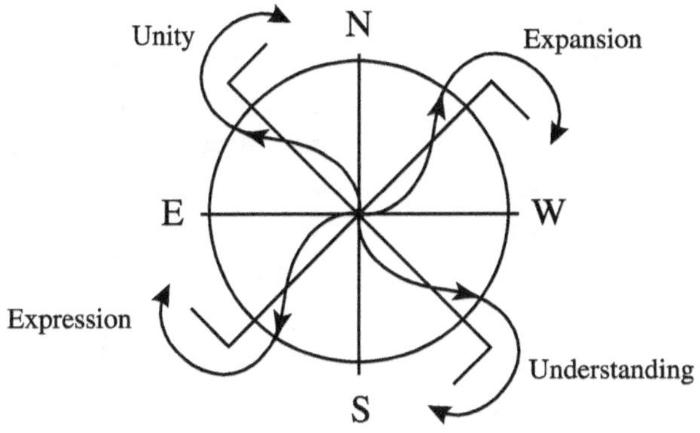

Figure 20: The swastika[39]

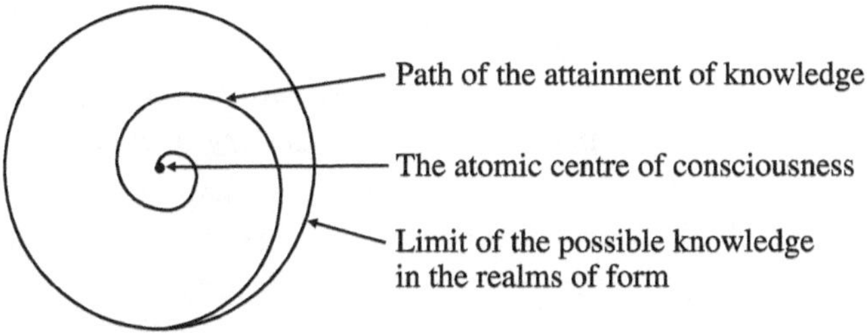

Figure 21: The self-conscious individual

simple cause-effect way like gravity (although it has a form of gravity of its own), it is *alive,* because what it is concerned with is what makes consciousness create, transform or destroy things and the order of magnitude that can thus be wrought. It manifests first via the procreative forces of desire, of the desire to see, to know more, to possess and to manipulate. There is a driving momentum that makes consciousness move, evolve and to create ever more images to festoon its munificence. Desire is later transformed into the will to improve, to love and then to escape the bounds of the limitations that consciousness sets upon any thought structure.

As the images grow in sophistication and complexity, so the 'tunnel' in space developed by consciousness, by means of which we view things,

39 See figure 24 (page 245) in my book *Maṇḍalas, Their Nature and Development.*

moves onwards and upwards into an ever vaster future. Its movement, like a butter press, churns up the undetermined (the *bījas* of all forms of manifestation yet unborn) and gives birth to the new. The potentially manifest (seed idea) is thereby projected into manifestation. Everything appearing in the mind is therefore set into motion, is elaborated by it and grows in space as the thought manifests through time. It then manifests its apogee of usefulness and is finally discarded, or acts as a seed for other evolving thoughts moving through space. Thoughts move at a speed determined by the quality of the substance of the thought and according to the energy it is invested with. This speed is not limited by the 186,000 miles per second limit to light, which only conditions the physical universe, but rather by fifth dimensional conditionings, whereby this limitation is transcended. Highly refined thoughts can nearly instantaneously travel far in space, whilst sluggish thoughts, filled with selfish desire, hover around the object of desire.

The spaces between thoughts contain rarefied substance that can be combined into appropriate forms, allowing interrelating and thus cataloguing of separating parts of the whole in the process of thought. Individualities manifest because separations have appeared in consciousness, like the moon, sun and the stars. Once differing appearing phenomena have been categorised, they can be comprehended by mind. The presumption being that clear thinking has ensued.

Humans catalogue similar types of phenomena differently and this causes the character traits of their personalities. They manifest particularised tunnels or vortices of energy (lines of *manasic* energy) at different rates of motion, colour and sound, because of the differing speeds of the minds utilised. One can also consider the related nature of the *skandhas* and *saṃskāras* called forth via such vortices. The *saṃskāras* are sequenced through time and appear appropriately when the frequency of the consciousness-state impels the qualities they convey, being of the same frequency. Like calls forth like. Thus, whatever is in the process of being developed by consciousness automatically compels the appearance of such *saṃskāras* as vortices ('tunnels', or lines of energies) from past volitions (stored as *bījas* of past actions of like attainments). As consciousness evolves in the manifest personality old *saṃskāras* are left behind in different time-space continuums, because the energy qualifications no longer suffice to attract them. New ones then surface to be integrated into the conscious awareness of the greater 'I'. Thus different things (appearances) are created in consciousness.

The thought sequences of an individual move with those of like nature from other individuals. They attract each other as they move through space and there is theoretically no limit to the size generated by such combined thoughts. All are contained within the Logoic Mind. However that Mind's own Thought sequences obeys similar laws, causing Logoic Thought to be differentiated into Ray and subray categories, which conditions the stars and constellations we see in the night sky. The concept can be extended to the filamentous lines of extension observed between clusters of galaxies and of their superclusters, though what is deduced by astronomers relates only to the physical plane reflection of the causative mental activity.

As the perception of time is created in this process so there are different appearing forms, with differing potentials. Different potentials exist because of the varying intensities of the energy imbued into them. Consequently there are different vibrancies of the imagery in relation to the *maṇḍala* that existed when the forms were materially manifested. The *saṃskāras* appear from a past time, a past life's activity, and bring with them the tendency to keep the personality there (in the past), to repeat such activity and states of mind. (Most humans act thus throughout their life, in their thought, sensual and emotional habit patterns.) The overshadowing Sambhogakāya Flower however has a different agenda, it wishes to propel the past into the future, by setting the paradigms of conscious aspirations and images of what to aspire to. It exists as the store of the future projection of every past *saṃskāra* by properly organising the sum of past conscious volitions in accordance with group purpose. It is assisted in this activity because it also draws its purpose from the *dharmakāya*. If it were not for such a Soul-form overshadowing the personality, then that 'I' would not evolve, it would perpetually tend to utilise the same *saṃskāras* of the past to repeat similar habits over and over. The reality however is that the containment of the future propensity propels the past *saṃskāras* forward, according to an already established *maṇḍalic* pattern, creating the momentum of the eternal Now.

The spiralling tunnel of consciousness moving through space creates a force-field around it, keeping consciousness contained in the form that it possesses so that it does not dissipate into the vacuous all. This force-field has an intensity, which determines the nature of the quality of consciousness, of how fast it can move and the nature of the thought-forms it contains. The thoughts can be dull and sluggish, sensual in nature, or highly vibrant and liberating. The thought field is capable of (instantaneously) attracting to it substance of a like nature from the *ālaya*,[40] and can thereby create new forms (of thought). It also automatically repels thought potencies (energy qualifications) that are dissimilar, incongruent with the energy it contains. (What a thinker is attuned to.) All living things contain similar force fields, the energy body containing the *prāṇa* that sustains their beings. Inanimate matter also obeys similar processes, with vast aggregates of atoms swayed by tensor fields, gravitational and magnetic attraction, and the like.

In this process whereby consciousness moves out to fill space it is human personalities, or Logoi,[41] that manifest the creative process in a moving consciousness through space, for only they can consciously use the factor of time to correlate the different things that have appeared as separate entities, e.g., thoughts, stars, moons. The factor of time is supplanted by the law of cycles in the perspective of Logoic Thinkers.

The force field created through moving consciousness sustains the experience contained inside it, as it determines the length and strength of the image that is held in mind and the patterns interrelating the objects of the image. It sustains the Life of each *maṇḍala* of expression.

In the higher dimensions (the *arūpa* universe of the higher mental plane and beyond) the self-focussed image building faculty of consciousness is not manifest. Here exist universal Idea-forms, force fields of astounding propensity governing group evolution and conditioning whole kingdoms of Nature. The energy forms from the *arūpa* domain become 'flattened', materialised, when manifesting in a lower dimension that can express formed structures. One must consider the transposition from fifth dimensional (enlightened thought) to three dimensional (concretised thinking). Viewing from below up

40 *Ālaya* means storehouse of substance.

41 Logoi also incorporate the *deva* Builders.

we have the appearance of abstracted images. What appears fifth dimensionally to the higher Mind manifests in a vast spacious Vision of simultaneous discernment. Upon the domain of the empirical mind however each 'thing' takes a differing position in space and generally possesses an unknown quality based on the separation of characteristics. Things must be intellectually analysed to be understood. Thus we have the appearance of the forms of individual things, of 'selves' that are combined to make a whole, constituting of the segregated images found in Nature, the stars and the visible universe. The position of the stars manifest through the gravity of space, which denotes Mind moving concretised images. Not just the positioning of stars manifest thus but also the movement of galaxies. All is analysed by individualised human minds (spheres of containment) within stellar spheres. The way space is formed, segregated by mind/Mind, makes the possibility of 'things' having separate existence.

The gravity of consciousness moving through space is established in the form of a time sequence. (By 'gravity' is meant that which interrelates forms into a universal law of attraction, giving them 'weight', the illusion of solidness.) The formless becomes concreted in time, and time moves on to attract the next image or moment of consciousness to form the celluloid of memory of what once was.

When a consciousness moves through a vaster Mental environment then a certain effect is created upon the consciousness that is moving. The vaster Mind affects the smaller consciousness that is part of it. It provides an environment, complete with images of what is to be observed and of vast spaces, of stars and galaxies and their similitude in the higher domains for the small unit to experience as it travels onwards to its perceived future. These 'similitudes' are but the Idea-forms of the vaster Mind. For that Mind they may last but a short duration, but for a relatively infinitesimally small unit of consciousness therein, (whose rate of motion may be for a very brief moment or 'flash' for the greater Life) the Idea-forms can represent vast, very long-lived Entities. Within consciousness concepts of time form, but the speed of time is related to the rate of motion of the Thought structure of the all-inclusive Mind of which it is a part. The consciousness inside has a much faster time expression than the encompassing Mind. Rapid is the small unit of consciousness's rate of changes of ideas, and small is the impress of the effect of its energy impact.[42]

The smaller consciousness or mind is but part of a collectivised Idea within the greater Mind, and forms its own vortices or tunnels of cause and effect. It has its own weight, or force behind it, is appropriately coloured, and resonates with an inherent sound. On its own it would be but a moment of an image in the greater Mind, barely noticeable, unless part of a larger picture, part of an amalgamated *maṇḍala* of such force vortices, all aligned through similarity of characteristics. Together they produce a more permanent structure veiled by the symbolism of imagery. Because of the spiral nature of the movement of consciousness through space, and if all components of the *maṇḍala* have developed similar characteristics, then the force field that is developed spins at a rate according to the overall characteristics of the *maṇḍala*. This can be visualised in terms of *chakras* turning. (They exist as wheels that are the containments of consciousness and different sentient states.) Smaller wheels (constituting of the grouped impact of minute sentient entities) turn quickly and may complete many cycles of revolution before even one petal of a much larger *chakra* has revolved to its next position in the sequence of things. Timed events for those of the small wheel are therefore greatly speeded up, because of the quickness of its motion within the comparative life-span.

We measure our time on the earth because of the revolving and spinning nature of the earth's movement with respect to the sun (giving us our days and nights), and because of the precession of the equinoxes. (That is the earth's overall movement with respect to the stars.) An average human life span can be considered to be approximately 72 years of days and nights measured in this way. Within the cycles of that time frame the human 'I' can learn much. But compared to the motion of a solar system, moving around its galactic centre, such a time frame is miniscule, less than a millisecond of time of the 'clock' that registers its 'day and night'. In this idea we have a consideration of Astrological reckonings of timed events that could be much elaborated.

---

42 Similarly in the world of subatomic particles with respect to their relative durations as they flash into and out of existence relative to the experience of time in the mind of a human observer.

There is another motion not depicted here, which is in a forward-progressive direction, and which carries the *saṃskāras/prāṇas* onwards into space, which manifests in a serpentine form associated with *kuṇḍalinī*. This motion delineates the shape of the petals of each flower and projects the *nāḍīs* from each petal. The points of intersection of the *nāḍīs* produce the locus for the appearing *bindus* of the *chakras*. The spiral-cyclic motion concerns the internal motion of the *chakras* and the *nāḍīs* the mode of liberation of the energies they contain.

An analogous view to the concept of the nature of the functioning of a Logos, is to observe the life and death of a cell in the human body (thus the life-span of the sentience of the cell), which has a much shorter duration than that of the entire body and its overriding consciousness. This overriding consciousness carries the momentum of the smaller sentience of the cellular unit within it, therefore the 'time frame' of the vaster body conditions the time ration of the inner unit. The greater Mind thus propels the smaller units of mind through space within the context of a vaster universe, unbeknownst to the infinitesimal sparks of mind that together constitute the One Mind. That which unifies the intelligences and sentience constituting that greater Mind is termed *bodhicitta*, which can be considered the *jīva* (life force) thereof. Accessing *bodhicitta* thereby becomes the mode of release and ascension for the sparks of mind within the greater Mind. This way is that of the Bodhisattva path.

Chakras

Having elaborated somewhat the concept of the existence of a primal creative Logos and the nature of its expression in relation to a human thought sequence, I can now focus upon what sustains consciousness through its cycles of expression. This involves the spiral nature of the motion of consciousness through space and concerns the mysteries of how *chakras* come to be. *Chakras* are the penultimate mystery because everything that is and is not can be viewed in terms of them and of their motion.

When the motion of the time-space continuum is projected in the four directions of space, according to the way of movement of a swastika, then a 'lotus blossom' is formed, as indicated in the following figure.

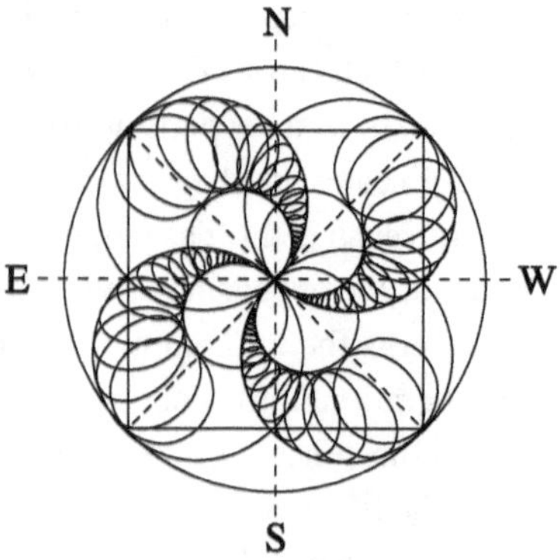

Figure 22: The lotus as an expression of energy and consciousness

There are two orientations to this 'blossom' (or swastika). What is depicted is technically the orientation for the feminine principle, that for the *deva* evolution. The masculine principle in Nature, the human kingdom, is oriented in the cardinal position (north, east, south and west). This makes eight spokes to this wheel, or petals to a lotus, which from another perspective is symbolised by the prongs of a *vajra*. There is a central coordinating prong of integrating substance (expressing the *dharmadhātu* Wisdom of Vairocana) enclosed within the unifying factor of the sphere of expansive inclusiveness (representing the function of the Ādi Buddha). The eight orientations plus the above two factors produce the number ten of perfection. The focus of the figure is in the present, hence two further factors can be added, namely, the moving time line of the past moving to the future.

This diagram depicts in a simplified form how consciousness evolves, and represents the esoteric foundation for the construction of a generalised *maṇḍala*.[43] The lotus fills the boundaries of the circumscribed Thought-Form of the Logos. Each one of the petals conveys *saṃskāras* conditioned by a different Element. North for the Fiery, west for the Watery, south for the Earthy, and east for the Airy Element.[44] What is presented here concerns the source of where the Buddhists and Hindus derive their

43 Detailed teachings of what is presented here can be found in my book *Maṇḍalas, their Nature and Development.*

44 The orientation of these can change according to the focus of consciousness at any time.

'four continents', the central basis of their cosmological system, with Mount Meru in the centre of the *maṇḍala*. Also the four Mahārājas, the guardians of the four corners of the universe in the Hindu pantheon, and their analogies in all the various mythologies, such as the four winds, or powers (forces), or the four seasons.

Each of the quadrants is really dual, because in terms of human evolution there is a left and right hand way to travel. The left refers to that pertaining to materialistic incentive, with a prime focus downwards and outwards attachment to *saṃsāra* and the pleasures of the senses. The right hand way refers to the process of gaining wisdom, where one detaches from *saṃsāra* and aspires upwards towards liberation via the eastern direction of the Heart. As they do not have the type of free will associated with humanity, the feminine *deva* kingdom travel only in one direction, the right hand path of evolutionary progression, but can be manipulated by forceful minds into left hand constructs. (Also certain of their order embody the substance of human thoughts and desires.) In effect, what constitutes the 'left hand' for them is the embodiment of those *deva* forces productive of the process of dying, of disintegration and death of all forms, minute or cosmological. All forms must die to be reborn.

I have explained the eight petals in terms of the attributes of the human and *deva* kingdoms, but this is not quite correct. It is best to think in terms of the evolution of human consciousness, whereby the three *guṇas*: *sattva*, *rajas* and *tamas*, dynamic, mobile and sluggish energies, must be taken into account. When multiplied by the four petals of the Base of Spine centre, or the four directions shown in figure 22, then we have the twelve petals of the Heart centre, and the basic number whereby the entire *maṇḍala* of the *chakra* system is constructed. It presents a basis as to why the Heart is Life, not just for the human form, but for the entire evolutionary progression. From this can be derived why traditionally there are twelve months to a year and twelve signs of the zodiac.

The Sanskrit word *chakra* in its simplest connotation means wheel, the wheel of motion, of the Law. *Chakras* are vortices of energy and depending upon the qualities of the particular *chakras* that are activated they connote the different qualities of a manifest human or Logos. The *chakras* in the bodies of planetary and solar Logoi have astrological and cosmological inferences. Each Logos subjectively manifests in the form of one or other of the *chakras* and consequently embodies the functions ascribed to that particular centre within the Body of a far greater Logos.

The *chakras* can also be perceived as Eyes, allowing the entry of light from one dimension of being into another. They are thus doorways to and from the realms of being (depending upon the Element or aspect thereof that each *chakra* controls) through which the *yogin* can leave and enter at will. This depends upon his/her degree of attainment.

Chakras can also be seen as swirling saucer-like depressions in a person. They stem from points in the spine and are divided by means of spokes of energy into regions that have been likened to the petals of lotus blossoms. The seven major endocrine glands are said to be their physiological externalisations.[45] The *chakras* are in effect eddies of Fiery energy that gradually increase in luminosity from a dull glow to a brilliant incandescence as one is able to increasingly utilise the *(prāṇic or kuṇḍalinī)* energies that are the result of one's spiritual development. This happens very slowly, as the course of normal evolutionary development, or else it can be greatly hastened by means of meditative practices.

The unfoldment of the lotus concerns the summation of the motion associated with the process of evolution. It represents an expansion in the four, six, eight, or twelve directions in space. (If the zenith and nadir and/or the intermediate positions are taken.) The petals of all the other *chakras* are really compounded from the four of the Base of Spine centre, multiplied by three, to take into account the three *guṇas*. Even the Solar Plexus centre with ten major petals, are drawn from the *maṇḍala* of twelve petals with two petals removed, or the Sacral centre, with six main petals, where six is but half of the *maṇḍala* of twelve petals. In the centre of all major *chakras* exist 96 smaller petals. The attributes of the Elements which the *chakras* convey is determined by the number of major petals the *chakra* possesses. Accordingly the seven major *chakras* are also categorised as five, where the Head lotus and the Ājñā centre, being overlapped, count as one. Similarly the Sacral and the Base of Spine centres have one petal from each overlapped, and are accordingly counted as a unit.[46]

---

45 See Alice A. Bailey's *Esoteric Healing* for discussion on this subject.

46 Much added detail concerning the *chakras* was given in *A Treatise on Mind* to

The lotus shown in figure 22 is also in the form of a *swastika*. The swastika and its permutations is found in nearly all the ancient religions.

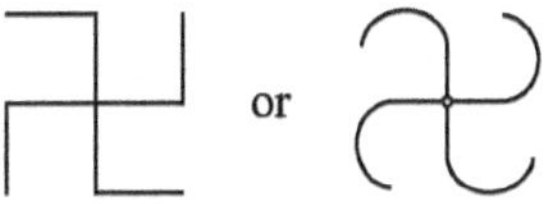

The swastika represents the four winds or forces causing and sustaining the universe, the four Mahārājas, the guardians of the four continents, gates, or directions of the heavens and their analogies in all mythologies. When placed in a circle, the swastika represents the forces pertaining to the formation of the solar or earth system. Outside of a circle it represents the forces pertaining to human activity. It is a very ancient symbol signifying fourfold power, activity, and of material construction found throughout the world's mythologies.

Nearly every ancient theism has a variation of the spiral-cyclic type of motion as an important element in their iconography. For example, it has been depicted as below by the Greeks, Celts, Vikings, Hindus, Buddhists, Aztecs, Polynesians, Egyptians, etc. It shows the three prime aspects or types of energy of Deity (Father, Son, Mother). The circles surrounding each of the symbols indicate that they have cosmological inferences, rather than just relating to humanity. This symbol thus indicates the nature of the laws associated with the projection of thought-forms and of motion.

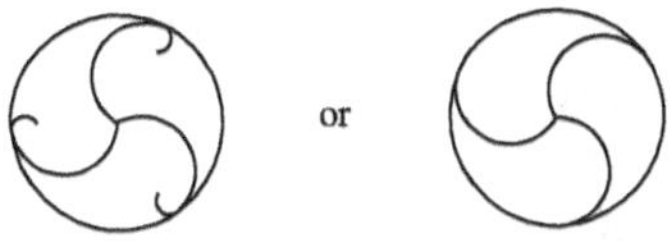

There is also a basic duality, a male-female polarity in the arrangement of the *prāṇas* that constitute the swastika, symbolised by the yin-yang symbol ☯. At humanity's present stage of evolution the Element Earth is negative regarding Fire, whilst Water, though still dominant, is slowly becoming negative regarding Air. The *prāṇas* conveying the Elements Air and Fire assume a masculine polarity, whilst those conveying the Elements Earth and Water assume a feminine

which the reader can refer for further information. See figure 1, 'The seven major *chakras*', in volume 6 for their depiction.

polarity. Upon the attainment of enlightenment all the Elements become subordinate to Air *(buddhi),* and later, with the *dharmakāyic* vision to Aether *(ātma),* signifying the central point of the swastika, from whence all the Elements emanate. In the intellectual person the Element Fire is dominant, and in the predominantly emotional person, the Watery Element dominates.

Overall, the qualities of a person are determined by the relationship of these Elements when viewed in terms of the types of *saṃskāras* that convey them. Logoi similarly deal with *saṃskāras,* but in their case the *saṃskāras* are constituted of the sum of the forces constituting the Lives embodying the kingdoms of Nature, incorporating therefore the human and *deva* streams. In order to determine relative Logoic (or human) age we need to look at which Elements are dominant, to what degree, and to what extent is the expression of the others retarded. All this determines the type of *prāṇas* that flow up the *iḍā* and *piṇgalā nāḍīs* and through the minor ones. The figure of the lotus as an expression of energy and consciousness represents a cross section of a *nāḍī.* The marriage (or blending), refining, and transmutation of these forces occur in the major and minor *chakras*, which are formed where a number of *nāḍīs* intersect. Buddhist and Hindu texts state there are 72,000 *nāḍīs* in the body, but the correct figure is symbolised by the number 96,000, as explained in my book *Maṇḍalas: Their Nature and Development.*

The *chakras* refine and elevate the *prāṇas* in the *nāḍīs* from one level or dimension of perception to the next. They can do so because these vortices of energy reception and transmission are designed to store and transform different *prāṇas,* whilst at the central hub *(bindu)* of each major wheel exists *śūnyatā,* which acts as a transmogrifying agent, and also represents the means of liberation into the next higher dimension when the mode of the spin is fourth dimensional (omitting time here as a consideration of a dimension).[47] They also redirect baser, negative energies to a lower level where the *prāṇic* qualities find an outlet for expression. They are receptors for the energies emanating from the dimension of perception or kingdom of Nature that they are the gateways to, and to which the person is consciously or subconsciously receptive.

47 The view here is only for the major *chakras,* whilst the energy of the central point of the Head lotus is *ātma.* The Base of Spine and Sacral centres are viewed as a unity as also the Ājñā and Head centres.

The energy that flows up the *iḍā nāḍī* is predominantly the Fire-Earth aspect of the various Elements (related to perception in the external universe), and that in the *piṅgalā nāḍī* is the Air-Water aspect (related to perception in the internal universe). The energies in the *suṣumṇā nāḍī* predominantly convey the Fire-Air aspect (mind-intuition), a fusion of the most refined qualities of the other two *nāḍīs*. The *suṣumṇā* path is the strait 'narrow razor-edged path' to enlightenment. It is effectively the middle way that is the epitome of Buddhist philosophy.

The figure of the lotus also shows why the Base of Spine *chakra* is represented with four petals (each of which expresses the energies of one or other of the Elements). The increasing complexity of the number of the petals of the higher *chakras* indicates the extension of this basic pattern in all possible directions in space, allowing them also to fully express all the possible qualities associated with the Element that the *chakra* embodies.

The cyclic forces symbolised by the lotus petals can be seen in all possible directions on all planes of perception. *Chakras* thus take the form of immaculate multidimensional flowers when fully developed. The petals have many different shades and hues and awaken their full floral splendour and perfume during the enlightenment (i.e., Initiation) process. The 1,000 petalled lotus at the top of the head blossoms fully upon enlightenment. When the Fiery Element predominates in the *suṣumṇā* then this indicates the control of the *chakras* by the Sambhogakāya Flower. The dominance of the Element Air in this *nāḍī* allows abstraction into *ṣūnyatā*.

The fifth Element, Aether *(ātma),* integrates all forces in the entire swastika into a unity. It permeates all of space, and together with the other four Elements, explains why the Hindu and Buddhist texts on meditation state that each of the *nāḍīs* convey five different *prāṇas* (*prāṇa, samāna, apāna, udāna,* and *vyāna).* Aether is conveyed in the *suṣumṇā nāḍī* when a person has reached the attainment of the *dharmakāya* at the fifth Initiation. At the sixth Initiation the Initiate rises out of the ranks of the fourth kingdom in Nature (humanity) by having perfectly merged the qualities of both the human and *deva* streams of consciousness, though such a one can still incarnate in human form. The qualities or energies that the *chakras* can convey by themselves within the form are thereby surpassed.

The four petals of the Base of Spine centre represent the foundation stone of manifestation, the square, four directions or pillars supporting the universe. When the central portion of the figure is taken into account and related to the development of enlightened perception, then these five energies also express the attributes of the five Buddhas of Meditation. They manifest the transmuted correspondence of the five instincts and therefore the acquiescence to the energies obtained by the process of meditation. Much has been presented in *A Treatise on Mind* concerning their qualities.

The petals of the lotus blossom are formed as earlier described in terms of serpents (sine waves) of energy *(prāṇa)* that cross each other to encapsulate space for each evolutionary cycle. Here one must consider the nature of fourth dimensional motion manifesting over time, to make a fifth dimension. The internal and external motions are viewed simultaneously in the eternal NOW, in the four (or eight) directions in space. The mind/Mind's energy is the encapsulating mechanism. It also generates the moving images of each thought sequence in any of the directions in space, which become the *prāṇas* in the *chakras.*

Chakras form at major nodes of intersection of differing lines of energy *(nāḍīs)* bearing *prāṇas (saṃskāras),* thereby becoming spheres of containment, whorls of energy drawing many atomic unities *(bījas)* into a coherent form. All is delineated by the integrated force-fields of the *saṃskāras.* The *chakras* are thus the integrating and directing forces controlling the movement of the *saṃskāras,* according to the associated Element, hence the substance of the plane of perception, of which the *chakra* is the custodian. The number of petals to the *chakra* determines its capacity to express the energies of a particular plane of perception.

The human unit can be considered to be constituted of something like a distillation column, with seven major and 22 minor fractional units, which are the *chakras.* The *nāḍīs* from the foundational unit, the Base of Spine centre, convey the most dense *prāṇas* that spiral up the column from major *chakra* to major *chakra,* each dealing with more refined energies as the *prāṇas* move up the column. They move from the Base of Spine to the Sacral, the Solar Plexus, Throat, Ājñā and Head centres. The Heart centre is slightly offset from the main sequence in the human body, as it represents the central point of the coming and going energies. It specifically conveys the energies of *śūnyatā,* the Void, which drives the transmutative process of the entire column. Hence it embodies the integral Life of the unit, which sustains its physical plane livingness.

As well as radiating out to the major and minor centres, the main *prāṇas* can also be seen to turn inwards through the central organising column, where the two main *nāḍīs, iḍā* and *piṇgalā* spiral around the central *suṣumṇā* torrent of energies, which represents their fusion and integration with energies from beyond the system.

From this perspective each *chakra* is also a centre for the distribution of streams of *bījas* that are the seeds for the forthcoming action that consciousness deems necessary. The *bījas* pertain to the Element of which the *chakra* is the custodian. The *prāṇas* of a higher dimension are conveyed by the major *chakra* situated above the lower one. The *chakras* can also be likened to a form of *bīja* because of the lines of intersection of energies, the *nāḍīs* that criss-cross through them. They are the central organising template for entire fields of activity in the subjective domains and in the world of material effects. Cumulatively they are the incarnating mechanism of a human unit, as the physical form is but their outer clothing, a materialised consequence. As for the human unit, so also for Logoi. The *chakras* for a planetary Logos exist upon the four highest systemic planes of perception.

The *chakras* are normally organised to convey the *prāṇas* of the five lower dimensions (from *ātma* to the dense physical), but can accommodate the energies from all seven sub-planes of the cosmic dense physical plane.

The *iḍā nāḍī* effectively conveys the *prāṇas* of the three worlds of human livingness (corporeal, astral and mental), whereas the *piṇgalā nāḍī* conveys the full spectrum of the attributes of the seven Rays.

Kuṇḍalinī and the serpent symbolism

To better understand the type of the motion associated with figure 22 we must look to the nature of *kuṇḍalinī* energy, which is understood to manifest as 'serpent power'.

The symbolism of the serpent is inextricably interwoven with that of the staff, for in hermetic and yoga philosophy it symbolises the force that is released by right meditation techniques in conjunction with the quality of the life lived. When released, *kuṇḍalinī* flows up the major psychic channel (*suṣumṇā nāḍī*) centred in the spinal column. *Kuṇḍalinī* is the energy of matter or the Mother aspect, the primeval causative or formative energy that lies coiled 'in potential' at the hearth that is the central animating dynamo of every form. It is the central reservoir of heat, or internal energy, that sustains the Life of any material form. *Kuṇḍalinī* is effectively the Fiery substance (*cittaprakṛti*) of the universal innate Mind, which is differentially evoked throughout the Womb of the Consort of the Ādi Buddha upholding the vital Life of the Thought Construct according to a preordained pattern. It is instigated in the expression of the union of a Logos with his Consort, and is the 'seed of mind', or at least the latent (potent) Fires that become innate in every atomic or cellular unity that takes form, or will do so. It sustains the potential of all seed *bījas,* integrating them as a unity of the totality of the forthcoming *maṇḍalic* construct. This is the natural result of the 'breathing of the divine Thought' to thrill *mūlaprakṛti* (universal substance matter) with the energy of manifesting Life.

Within that Womb the Consorts of the Dhyāni Buddhas are similarly impregnated with differentiating Fire, and so *mūlaprakṛti* thrills with the sounds of awakening *arūpa* and *rūpa* lives.

- In this way *saṃsāra* can be sustained to follow the paradigm of the purpose of the Ādi and Dhyāni Buddhas.
- In this way intelligence can be wrought as a consequence of the evolutionary progression of all of Nature.
- In this way human consciousness can evolve and *yogins* awaken the inherent *kuṇḍalinī* hidden in the base of their spines, being the basis for attaining the *siddhis* that bespeaks of their quest for liberation and the total control of phenomena.
- From this procedure *saṃsāra* (that within the Womb) can be related to *ṣūnyatā* (the foundation of the Buddha-Mind).
- Only through mastery of this Fire can the potential of the *dharmakāya* be realised by an enlightened Being. This means awakening the qualities of the Dharmakāya Flower[48] through becoming a Buddha.

48 See volume 3, chapter 6 of *A Treatise on Mind* for an explanation of this form upon *dharmakāyic* levels.

Kuṇḍalinī is the inherent liberating Fire that consumes the forms of things with the primacy of the energy of Mind, hence awakens the Eye to see multidimensionally as it passes through the *chakras*, causing them to spin fourth dimensionally, when awakened by the wise. When integrated with *śūnyatā* it liberates the form by burning away the dross *saṃskāras*, leaving only the Void Elements. In Buddhist terminology the substance is made 'empty and void of all self nature'. It has the inherent capacity to bring the meditator to the omnipresent Source of all, once the layered cloaks of differentiated Fires are progressively stripped from the primal Fire. The meditator becomes aware of the reality of universal *cittaprakṛti,* the basis of one's self-focussing capacity, which though identified with the plenitude is empty of substantiality. The meditator evokes a Fiery liberating, compassionate *dhyāna* (*tathatā*[49]) to eliminate the darkness found at the root of all substance matter, which is elementary mind, by meditating upon and demonstrating the compassion of a Buddha. The resultant conflagration consumes the limitations of mind by awakening the complete potency of Mind as the Head lotus explodes with incandescent Light. Those of the left hand path alternatively direct *kuṇḍalinī* downwards to reinforce the *cittaprakṛti* and so become bonded to the vicissitudes of the separated illusional forms of *māyā.*

Kuṇḍalinī is the potency of the Fires found deep within the bowels of the earth. It sustains all inner warmth and can be considered to be not, and yet is. It was seeded in the past, and in the last resort is the most material or concrete expression of the Will energy of the Logos, and so sustains the phenomenal illusory world around us. Thus it 'is' as long as this material world is sustained. It is 'not' upon liberating the forms of things, as it becomes an integrated expression of the moving spirals of awareness of the all-encompassing *chakras. Kuṇḍalinī* can be considered the integrated unity of all the thermal spirals of the Anu's[50]

49 *Tathatā,* suchness, thusness, thatness, characterised as omniscient wisdom. There are two types, *samalā tathatā* (Suchness mingled with pollution), which needs to be converted into *nirmalā tathatā* (Suchness apart from pollution). Absolute Suchness *(tathatā)* is not a static latent and neutral entity of reality, but is a dynamically operating efficient permeation. This concept gains reinforcement through being the 'germ of the Tathāgata' (*tathāgata-gotra*). *Tathatā* becomes the mechanism of containment of the *dharma* in the form of the *tathāgatagarbha* (the Buddha Womb).

50 The 'ultimate', foundational 'atom'. See chapter 8 for explanation.

constituting the embodied form of any evolving 'thing'. A permutation of the energy of *kuṇḍalinī* can be seen in the spiral helix patterning of DNA (deoxyribonucleic acid) found in the nucleus of cells, which is the major constituent of the genes responsible for the genetic code and thus of Nature's immense diversification.

The place of storage of *kuṇḍalinī* in a person is said to be the Base of Spine *chakra*, from where it psychically sustains Life by means of the *nāḍīs*. They roughly correspond to and underlie the nerve cords. The place of awakening of *kuṇḍalinī* is the conjoined petal of the Base of Spine and Sacral centres. In terms of awakening *kuṇḍalinī* from a Buddhist perspective Govinda states:

> The *Kuṇḍalinī*, which is likened to a coiled serpent (the symbol of latent energy) blocks the entrance to the *suṣumṇā*. By awakening the *Kuṇḍalinī's* dormant forces, which otherwise are absorbed in subconscious and purely bodily functions, and by directing them to the higher centres, the energies thus released are transformed and sublimated until their perfect unfoldment and conscious realization is achieved in the highest centre. This is the aim and purpose of the *Kuṇḍalinī Yoga*, of *prāṇayāma*, and of all other exercises through which the *cakras* are activated and made into centres of conscious realization.[51]

Govinda further states:

> In the *'Yoga of the Six Doctrines of Naropa'* the seat of the *Kuṇḍalinī* is excluded from the path of visualization, and the *Sādhaka* is advised: 'Meditate on the four *cakras*, of which each is formed like an umbrella or the wheel of a chariot.' The four *cakras*, however, which form the wheels of the fiery chariot of the spirit (which reminds one of the fiery chariot in which the prophet Elias went to heaven!) are: the Crown and Throat Centres, as the front, the Heart and Naval Centres, as the rear pairs.
>
> In place of the *Kuṇḍalinī Śakti* the opposite principle occupies the centre of the meditation, namely that of the *Ḍākinī*: in this case the *Khadoma Dorje Naljorma (rdo-rje rnal-ḥbyor-ma;* Skt.: *Vajra-Yoginī).* This does not mean that the Buddhist Tantrics denied or

51 Lama Anagarika Govinda, *Foundations of Tibetan Mysticism* (Samuel Weiser, New York, 1975), 156-157.

> underrated the importance or the reality of the forces connected with the *Kuṇḍalinī*, but only that their methods were different, and that the use which they made of these forces was different. They did not use them in their natural state, but through the influence of another medium[52]....the Buddhist *Tantra Yoga* concentration is not directed upon the *Kuṇḍalinī* or the Root Centre, but on the channels, the main power-currents whose tension (or 'gravitational' force) is regulated through a temporal damming-up and modification of the energy-content in the upper Centres.
>
> Instead of the natural power of the *Kuṇḍalinī*, the inspirational impulse of consciousness *(prajñā)* in the form of the *Khadoma* and her mantric equivalents is made the leading principle, which opens the entrance into the *suṣumṇā* by removing the obstructions and by directing the inflowing forces.
>
> *Khadomas*, like all other female embodiments of *'vidyā'*, or knowledge, have the property of intensifying, concentrating, and integrating the forces of which they make use, until they are focused in *one* incandescent point and ignite the holy flame of inspiration, which leads to perfect enlightenment. The *Khadomas*, who appear as visions or as consciously produced inner images in the course of meditation, are therefore represented with an aura of flames and called up with the seed syllable HŪṀ, the mantric symbol of integration. They are the embodiment of the 'Inner Fire', which in *Milarepa's* biography has been called 'the warming breath of the *Khadomas'*, which surrounds and protects the saint like a 'pure, soft mantle'.
>
> Just as knowledge has many degrees and forms, so the *Khadomas* assume many shapes, from those of the human *Jigten Khadomas (ḥjig-rten,* the world of sense-perception) to the female forms of *Dhyāni Buddhas,* who as *'Prajñās'*[53] are united with the latter in the aspect of *'Yab-Yum'.*[54]

Khadomas (ḍākinīs) are of significant importance, as they are but the Buddhist terms for *devas* (of a specific rank), without which the phenomena of *kuṇḍalinī,* or indeed of any manifest form, could not proceed.

52 Ibid., 193.

53 The wisdom aspect of the Buddhas of meditation.

54 Ibid., 193-194. Yab-Yum is seen as the coital embrace of the Buddhas and their Consorts depicted in Tibetan Thangkas.

The *symbol of the serpent* needs appropriate analysis, as it is intricately woven with the way of unfoldment of the creative energy of *kuṇḍalinī*. The symbol is based upon the serpentine motion of energy, and is found in the spiral path that the two outer *nāḍīs* take around the central *suṣumṇā* column. One of these paths, on the left, the *iḍā nāḍī,* is negative in polarisation. It conveys the psychically receptive, feminine, material, Fiery intelligent energy (an attribute of *kuṇḍalinī)* that sustains the corporeal form and its sentience, and gives humanity its animal vitality. The *piṇgalā nāḍī* on the right is positive in relation to *iḍā* (at humanity's present stage of evolution). It conveys the 'Son energy', the consciousness-engendering factor, the result of the experience-gathering activities directed towards the generation of wisdom. This energy can also be thought of in terms of the driving force of *bodhicitta.* This path concerns a holistic comprehension of the synergy of the entirety of Nature. It integrates the vitality *(prāṇa)* absorbed from the air and obtained from 'food' (utilising that term in its broadest possible sense), and manifests as the pure radiance of the sun, as the giver of vitality to all sentient beings.[55]

The symbol of the *iḍā nāḍī* is *the moon,* because like the moon, it conveys reflected light, which is associated with the form nature, the energies of the personality, and that of the psychic world. The symbol of the *piṇgalā nāḍī* is the sun, the greater luminary, which denotes the illuminating light of wisdom *(prajñā)* the radiance of enlightenment. When the sun and moon are therefore spoken of in mystical, esoteric, alchemical, mythological or religious texts, one can always assume that it is a veiled reference to the energies associated with these *nāḍīs.*

The central channel, the *suṣumṇā nāḍī,* conveys the dynamically active Father energy that impels all onwards in time and space. This energy betokens the *dharmakāya* that unites the highest aspect of being/non-being to the lowest, and is the gain of enlightenment. *Dharmakāya* is experienced when the consciousness aspect is fully awakened via deep meditation. The mind has developed the serenity that allows receptivity to intense energies and the vast reaches of all-encompassing space. Mind has become a fully prepared and endowed womb, with the ripened ovum as the seed of its potential sending forth its own note

55 The teaching concerning the *iḍā, piṇgalā* and *suṣumṇā nāḍīs* are provided throughout *A Treatise on Mind,* from which detailed information can be obtained.

or sound. This is embraced by a reciprocal Sound, the Monadic Word consisting of the rarefied Fire that reaches down to the deepest layer of matter at the base or root of manifest being. There it fuses or blends with the Fires of matter (that sustains the Life of every atom) and with the Fires of the evolved consciousness. The triune united Fire, *kuṇḍalinī*, then rises up the central *suṣumṇā* channel, fully vivifying the various psychic centres in the awakening one. The son of Mind is born, fully enlightened and liberated from the throttlehold of the form. The Father has consequently retrieved what was always his own.

This is but an outline of the nature of the raising of *kuṇḍalinī,* but the true picture is of great subtlety and is highly esoteric. The entire story of Creation, of evolution, and the meditative unfoldment of a person or a Logos is therein hidden. The reader should however be warned that the information in the various meditation and occult texts on the subject is generally veiled and misleading, contradictory, or else cursorily treated, because of the potential dangers awaiting the unwise in their premature attempts to awaken this force.

One can speculate that Logoic *kuṇḍalinī* is expressed as the subjective Fire sustaining the furnace of the sun,[56] and will also be found as the internal heat sustaining the Life of each living planet. The *iḍā nāḍī* and the *piṇgalā nāḍīs* therefore can be considered to represent the sum of the forces associated with the human and *deva* evolutions, looking here at multidimensional space and the types of Lives informing the Shambhalic correspondences of all major planets in a solar system.

Kuṇḍalinī is many layered and in its most material aspect it will burn and destroy the form or wreak havoc upon the psychic constitution of the person who has not adequate knowledge, or the moral and psychic purity to rightly direct it. Its tendency will always be to reinforce or distort whatever subtle, uncontrolled desire, or base quality is in the person. (For it manifests through the path of least resistance, as it, being feminine, allegorically seeks to unite with the Father energies.) This is symbolised by the burning poison of the viper.[57]

---

56 The implication therefore is that nuclear energy is but the externalised (exoteric) expression of *kuṇḍalinī.* There the focus would be its *piṇgalā* form, whereas the internal heat of a planet would represent its *iḍā* form.

57 See also Govinda, 139, where he gives the warning: 'Only with perfect self-control and clear knowledge of the nature of these forces, can the Yogi dare to arouse

This introduces the secondary implications of the meaning of the serpent, as that of the *ability to poison;* which concerns engendering the qualities of desire, lust, spite, hatred, enmity and vituperation. It thus also symbolises what causes one to be bound to the life associated with the sensual, illusion-forming material world, which is death esoterically considered. This is the domain of the feminine aspect of *kuṇḍalinī*, and is the most often used symbolic rendering of the term 'serpent'. This term can thus relate to the serpent's form of motion, as well as symbolically to the ability to poison, or else to give wisdom (as the result of gaining enlightenment through the right evocation of *kuṇḍalinī*). As the energy awakens it first becomes Fiery, and then as it intensifies it grows wings of Fire. At its highest level the serpent transmogrifies into a dragon. Thus is the story of *kuṇḍalinī* energy at differing stages of its arousal.

At the beginning of the arousal of *kuṇḍalinī* the term 'serpent' can apply. When it has begun to definitely stimulate the higher psychic centres (*chakras*), then the term 'Fiery serpent' applies. When all major *chakras* have been fully vivified and the person is completely liberated from the material realm and thus can travel in consciousness to all directions in time and space, then we have the 'Fiery flying serpent' stage. It signifies one who has developed the occult power of a *siddha*.

Esoterically, a 'viper' can be considered one whose *kuṇḍalinī* energy has been expressed to the degree that the centres associated with the emotional desire nature (the Waters) have been stimulated, but not controlled. He/she has developed psychic powers that can be used for good and for evil (selfish) purposes. The 'Fiery flying serpent' stage represents an enlightened being, and in esoteric terminology, 'the Fiery Dragon of Wisdom' is considered one who has attained the *dharmakāya* experience.

Another rendering of serpent is that of the *cycle of time,* and in this context the serpent is swallowing time, its own tail signifying the past. The head represents the present moment, of consciousness acting, which

---

them. The directions for their awakening are therefore given in religious literature in such a way, that only those, who have been initiated by a competent Guru, can practice them, in accordance with the rules which have been formulated in the course of millenniums of meditative experience'.

has grown substantially since the beginning of the cycle. It spirals on to another higher cycle of activity, thereby swallowing and sending into the abyss (i.e., sending into the stomach, or rather Solar Plexus centre) the images of the past, thus showing the alpha and omega of all Life. Within that centre emotional-mental *saṃskāras* are stored in the form of *bījas* that can be recalled when need be to colour the new spaces within consciousness with the needed images, which one will come to know and add to. The space enclosed by the serpent can represent either lunar or solar imagery, depending upon whether the present thought construct is enlightening, liberating in nature, or sensual, desire-filled illusion clinging. A similar imagery is conveyed in many creation myths as the cosmic or world Egg from which came the material universe.

All Life is governed by cycles and these cycles spiral within vaster increasingly spiralling cycles. This can be illustrated if we picture the overall course of the moon revolving around the earth, whilst the earth revolves around the sun, and the sun propels the solar system forward in a progressive manner, as shown in the figure below.

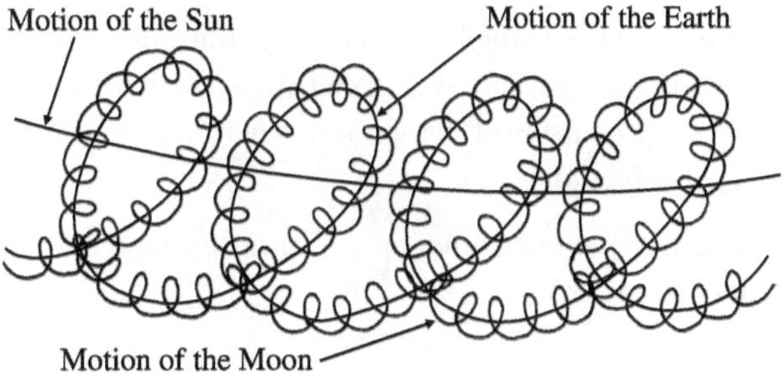

Figure 23: The sun, earth, moon relationship

There are four main levels of spirillae moving around each other and the main spiral to account for the five Elements *(vayus)*. Here we have depicted the orbit of the moon around the earth, which is similarly spiralling around the sun. The sun similarly revolves around the central constellation, and there is also that constellation's revolution around the galactic centre as part of the spiral arm of the galaxy.

This analogy can also be extended to include consideration of the nature of the atom from an esoteric viewpoint and to how humanity gathers experience. We saw also that spirals are incorporated in the way the *chakras* unfold. By this means all entities progress forward in time and space to the Heart of the Infinitude. Love lies at the Heart of all being/non-being, something all the major religious scriptures postulate in one form or another. There is no conceivable beginning, nor a conceivable end, only a timeless duration that is eternally conditioned by the relative beginnings and endings of lesser cycles within the greater whole, and which is itself a lesser cycle within an even greater whole.

By its movements the serpent symbolises all the prime types of energy qualifications underlying the movements of all entities, of the motion that sustains the visible universe and all in it. These motions can be depicted thus:

1. The *moving serpent,* symbolising the basic sine wave motion found in the mode of the manifestation of energy, such as the propagation of light.

It also depicts moving forward through time, the movement that allows consciousness to travel from here to there. It indicates that the forward propulsion of the spiral line may also be in the form of such a wave-like form.

2. The *coiled serpent* symbolising the spiral-cyclic motion (as well as potential energy).

It involves an *outward expansion* from a centre to the circumference of an ineffable circle that connotes the outermost field of contact. There is also an equal and opposite direction that the person can simultaneously travel, and that is *inwards,* back to inner space. Incorporating the two motions implies the figure 8, a symbol of infinity. This is done by means of the Mind that interrelates the subjective, inner universe, with the manifest phenomena of the universe we experience by means of the five senses. The inner and the externalised universe are interrelated by means of the mind/Mind, allowing experience of multidimensional consciousness. The Mind is the doorway from the archetypal, the divine, to the corporeal, phenomenal universe. Consequently it is the means whereby all can be made known.

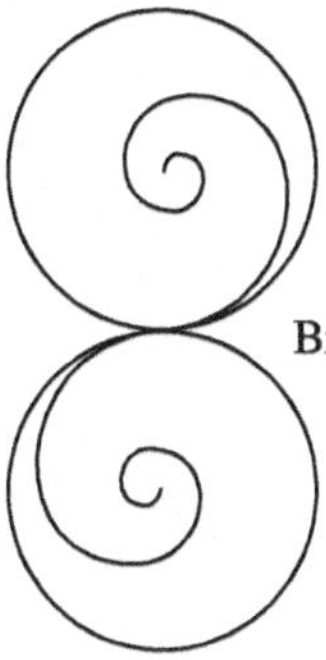

Figure 24: The inwards and outwards motion of consciousness

The focal points of these spheres of awareness are really the one point, as there is an illusional separateness existing between them, having significance only with regards to time, of the ability of consciousness to move in its orientation from one direction to the other. The evolution of consciousness proceeds inwards and outwards from event to event in the universe. The outward motion generates the great illusion associated with *saṃsāric* involvement, the inward motion provides the methodology for eventual release from such involvement. The overall motion in each of these spheres can thus be depicted progressing in both directions. The cycles of endeavour of the outward motion produce evolutionary attainment, whilst those of the inwards motion awaken the inner vision. Upon the enlightenment path motion in both directions can happen simultaneously, thus transcending the concept of time. This results in the projection of fourth dimensional vision.[58]

The circles in the figure can also be positioned on a horizontal plane, an east-west position. The western direction of outwards to humanity is representative of the type of energy associated with the *iḍā nāḍī* (the lunar vessel), whilst the direction *east* (inwards to the

58 The fourth dimension here being considered as right angled spaciality rather than the inclusion of time to the three dimensions of length, breadth and height. The addition of time incorporates a fifth dimension.

divine) represents the type of energy associated with the *piṅgalā nāḍī* (the solar vessel). When more 8's are produced and superimposed upon each other, indicating a continuum of the north-south motion (upwards to the divine and downwards into manifest life), then an idealistic picture of the spinal column is produced, therefore, of the energies associated with the triune *nāḍī* system, as depicted in figure 25, which tries to depict diagrammatically the nature of such energy flow. Not depicted is the way these *nāḍīs* spiral around each other, or branch *nāḍīs*.

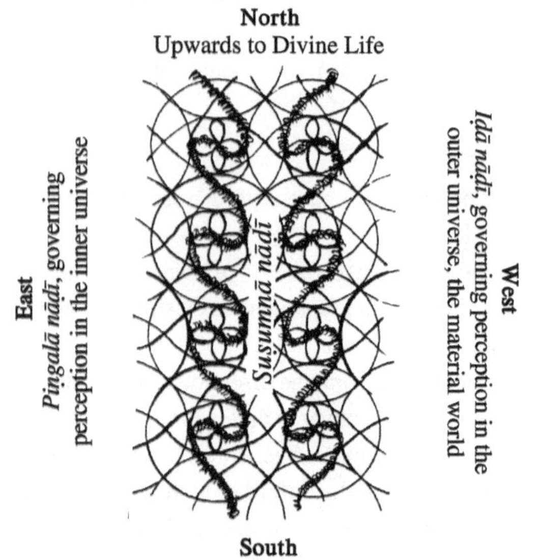

Figure 25: The three principal *nāḍis*

The various segments (or vertebrae) of this column represent steps or gradations of consciousness in the ladder of being/non-being that is evolution. They signify different quantum leaps of energy. After consciousness has completed a complete cycle of endeavour in the material domain, then through meditative focus fourth dimensional motion becomes possible. This motion manifests at right angles to the external motion and at the same time can be inclusive of it. This signifies awakening the *suṣumṇā* path, where consciousness is lifted to the next rung or cycle on the ladder of multidimensionality. Such motion effectively moves tangentially to the originating motion at the centre of the sphere. The radiating tangents form the circumference of the sphere.

The *suṣumṇā nāḍī,* is composed purely of the energy of the Mind, where the Fiery Element has been fanned to its greatest intensity by the Element Air until it becomes *ātmic,* universal Mind.

The two circles in figure 24 can also be considered to overlap, signifying mutually complementary and interrelated activity. The Mind produces the point of fusion. One circle therefore relates to the archetypal, the abstract, and the other to the concrete, the manifest. The interblending produces the sphere of the Sambhogakāya Flower from one perspective, and allows the liberation of *prāṇas* via the *suṣumṇā nāḍī* from another perspective, which incorporates the fusion of the *iḍā* and *piṅgalā nāḍīs.* This point of balance that is both finite and infinite can also be considered the *saṃsāra-śūnyatā* nexus. To experience this nexus the consciousness-engendering aspect must be turned inwards to gain intuitive perception *(sahajajñāna).* Consequently meditation and the development of the intuition has a prominent place in the religious scriptures of the East. Nevertheless, an essential basis for complete enlightenment necessitates development of the powers of the empirical mind *(saṃvṛtti),* the intellectual capabilities that have been the focus of the attainment in the West. We see also why Amitābha plays such a central role in Buddhist philosophy. He is the Buddha of the Discriminating Inner Wisdom, of Boundless Light, embodying the essence of the Mind. He embodies the western direction in the *maṇḍala* of the Jinas. Birth into His paradise-realm, Sukhāvatī, being a major goal of many Buddhists.

As one becomes increasingly enlightened, a process ensues of developing continuous mindfulness *(samādhi smṛti),* so the coarser internal energies become correspondingly refined, distilled, as the energies travel up the retort of the spinal column. The process is one of an increasingly subjective, sublimation and transubstantiation of *saṃskāras,* the changing of the 'bread and wine'[59] of our experiences and form nature into the elixir of immortality, the lighted substance of Life itself.[60] As one

59 This relates to the first of the miracles manifested by Jesus at the marriage supper at Cana. *(John 2:1-11.)*

60 See volumes 5A and B of *A Treatise on Mind* for detail concerning this process.

does so one becomes increasingly aware of the nature of Logoic Mind.

When the top rung of the 'ladder' (or fractional distillation unit) is reached, whereon stands the 'God' of our conscious awareness, then only the most intense type of energy remains. It allows the electric Fire that is an expression of the *dharmakāya*, (the Buddha nature) to fuse with the Fiery distillate from the person. Electric Fire then descends like lightning down the *suṣumṇā nāḍī* (which no longer contains anything that is antithetical to this energy) to the source of the causative Fires that sustain the form *(kuṇḍalinī)*. The interrelation awakens the process that liberates and infuses with Fiery Light the entire constitution. Supramundane *siddhis* then manifest.

Physiologically the analogy is found in the spinal column, where the vertebrae are placed in a ladder-like fashion, whilst the grey and white matter of the spinal cord (the physiological functions of which are analogous to those of the *iḍā* and *piṅgalā nāḍīs*) surround a central hollow tube filled with spinal fluid (which is roughly homologous to the *suṣumṇā nāḍī)*. Moreover, the spinal column is curved, which indicates the wave motion associated with the 'moving serpent' and the *kuṇḍalinī* energy in the central column.

If we view these spirals as spirals within spirals that expand as consciousness progresses onwards in time and space, then each part of figure 25 can be modified, as depicted in figure 15 of the time-space continuum. When clothed in materiality it forms a symbolic *sea shell,* one of the major forms of religious symbols in eastern philosophies, often depicted in the hands of *yogins* and of the major Hindu and Buddhist Deities. It can also be pictured as a *trumpet* that conveys sound, as also does the shell.

When both the inner and the outer directions are depicted, then we see that it can also be taken to symbolise the horns of a ram.[61] (Which therefore symbolises the beginning and the ending, the alpha and omega of the evolutionary process.) The outward motion represents the progress of evolution, the inwards motion signifies a turning about in consciousness at right angles to the outwards motion, producing the

---

61 The ram is here depicted, as astrologically it embodies the initiating will that impels the entire zodiac through space, and also terminates its cycles, whilst the force of Taurus the bull drives the cycles onwards.

process of return to the higher dimensions of perception. The point of turning also signifies the place of union between positive and negative forces in the body, when the two energy directions (*iḍā* and *piṅgalā*) in the field of consciousness are in the process of merging, the point of the awakening of *kuṇḍalinī,* thus becoming fourth dimensional in motion, or the fifth with the inclusion of time.

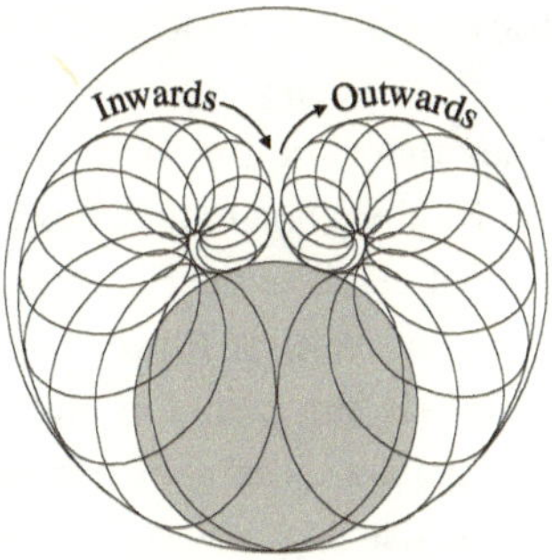

Figure 26: The symbolic horns of a ram

Each 'horn' also signifies one of the arms of the mutable cross (the swastika), be it the right to left or left to right motion. Each arm also can be viewed as principally conveying the qualities of one type of energy or dimension of perception. There are four main Elements associated with manifested space—Earth (the material domain), Water (the subjective, psychic emotional realm), Fire (the mental domain), and Air (the rarefied strata of the intuition, liberated space)—that together constitute the qualities of the manifest universe. Each cycle or sphere of endeavour is normally focussed upon developing or expressing one or other of these qualities, which are unfolded for each cycle of activity, and upon the advanced stages of the cycles (upon the inward way) all qualities can be expressed or developed simultaneously. (Through literally fifth dimensional thinking.) Though appearing separate in time, the originating points of the inwards and outwards motion are in reality the one point, as realised by an awakened one. Their sphere of interrelation (shaded grey) represents the *saṃsāric* domains, and when mastered becomes the eternal Now.

The cycles of birth, death and of perpetual regeneration are also symbolised by the ability of the snake to periodically shed its skin, thus this symbol exists in most creation myths in one form or another.

The activity of the Heart centre

With respect to causation (the appearance of phenomena) three of the Dhyāni Buddhas (Jinas) are directly associated with the formed realms, sustaining the conditionings of *saṃsāra*; Amitābha, Ratnasambhava and Amogasiddhi. They therefore function as the Mother aspect of the trinity, Father, Son, Mother. Their attributes directly condition the activity of each triplicity of petals of the Flowers in the *sambhogakāya* (higher mental) realm. The Will or Power petals are governed by Amogasiddhi, the Love-Wisdom petals by Ratnasambhava, and the Knowledge petals by Amitābha, as explained in volume 3 of *A Treatise on Mind*. From this perspective Akṣobhya is concerned with the process of the transmutation of consciousness and its abstraction into *śūnyatā*. Hence His Mirror-like Wisdom that offers a way of escape from *saṃsāra* via *śūnyatā*, therefore signifies the attributes of the Son. Vairocana embodies the central sphere of the Flower that abstracts all to the *dharmakāya*, which is the Father of all Life.

When related to the way of expression of the *nāḍīs* then we have the imagery depicted in figure 27 of the lotus blossom. The spirals in each quadrant come under the auspices of one or other of these Jinas.

Each Jina can be viewed to be triune, signifying the Dhyāni Buddha concerned, his Consort and the Son of consciousness, the wisdom that is expressed in her Womb. This Son aspect is literally the Dhyāni Bodhisattva associated with the Jina, which human consciousness evolves into. This allows the expansion of the diagram into the twelve main petals of a Heart centre. The way of awakening the Heart centre concerns the mode of vitalising an appearing form, as that centre is its intrinsic Life and governs the unfolding purpose for its existence. The *maṇḍala* that is caused to appear by the projection of thought via the actions of the Jinas can thereby be sustained to fulfil the purpose of the existence of an embodied form, human or Logoic, before dissolution again into its elementary substance.

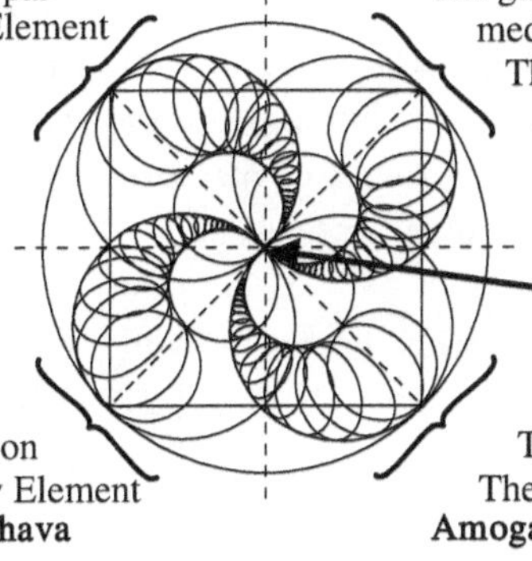

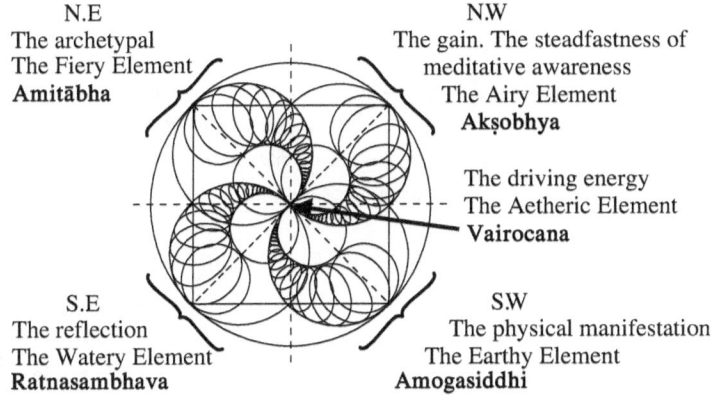

Figure 27: A cross-section of a *nāḍī*

Heart centres can also be seen as Flowers appearing external to the human unit, of which there are two main types, which outline the evolutionary process.

1. The Dharmakāya Flower for all of Nature.[62] This primarily concerns what is governed by the Jinas, producing the involutionary and evolutionary processes. Everything coming into being producing the awakening self-consciousness in a humanity, plus the environment wherein humanity evolves, comes into the sweep of such activity. The formed universe, the sum of *saṃsāra*, is thereby created. Here spheres of activity project their power via triangles that tend to manifest in the form of interrelated squares. The basic *maṇḍala* of the formed universe then manifests in the form of a series of interlaced 'squares' that will evolve into triangles via human self-conscious activity. This process will be explained in chapter 7.
2. The Sambhogakāya Flower that is the *upādhi*[63] of each human unit, from which the path to liberation can be gained. The five Dhyāni Buddhas now manifest as the force assisting the liberation of human consciousness into *śūnyatā* and beyond. The process of enlightenment transforms the squares of *saṃsāra* into the triangles

62 Explained in volume 3, chapter 6 of *A Treatise on Mind.*

63 Vehicle, that which limits something.

of liberated space, and then into the great transcendental spheres within the *dharmakāya*. This is the upward arc of evolutionary ascension. It concerns the transformation and transmogrifying of the factors of *saṃsāric* conditioning via *śūnyatā* into eventual Buddhahood. The dark substance of *saṃsāra* generates light by means of the action of human consciousness.

In Buddhism, such 'darkness' is equated with ignorance, which brings to the fore the twelve *nidānas*[64], the Buddha's twelve-fold formula of Dependent Origination (the philosophy of *pratītyasamutpāda*), which starts with the factor of ignorance.[65] Ignorance is considered the root of all *karma*-formations. The *nidānas* relate to the process of awakening the twelve major petals of the Heart Lotus. The Heart is the seat of Life, and the twelve-spoked wheel of the *nidānas* signify the wheel of Life as far as *saṃsāric* conditionings are concerned. There are however other ways to view these twelve petals, such as relating them to the twelve signs of the zodiac.[66] The formula for the *nidānas* runs as follows. I shall explain them from the broader perspective, omitted by Buddhist philosophers.

1. *Through ignorance is conditioned the saṃskāras, the rebirth-producing volitions, karma-formations (cetanā).*

This phrase refers to the involutionary process, of the formation of the various kingdoms of Nature up to the human. One looks to the *karma*-formations that condition these streams of sentience and of their rebirth as groups and categories of species. It concerns group and herd *karma* and interrelationships leading eventually to the development of human consciousness. Humans awaken individual *karma* as they generate *saṃskāras* through their ignorance, in contradistinction to the group or herd *karma* of animals. One can also posit that for animals

---

64 A *nidāna* is an underlying cause and determining factor, the cause-effect relation. In Buddhism the twelve causes of existence, twelvefold chain of causation (Dependent Origination).

65 The subject of Dependent Origination is explained in depth in *A Treatise on Mind*.

66 See my treatment of *pratītyasamutpāda* in volume 5 of *A Treatise on Mind*, whilst the relation of the zodiac to the petals of the Heart centre is detailed in volume 3.

ignorance as such does not exist, as they do not have the intelligence to even comprehend the meaning of nescience, or any knowledgeable state. They are unaware intellectually, but are aware sentiently and driven by instinct. The associated petal of the Heart lotus is downward focussed below the diaphragm to activate the cycles of material involvement. Consequently the Base of Spine and Sacral centres are awakened.

2. *Through the past life karma-formations is conditioned present life consciousness.*

This is concerned with the eventual development of a human kingdom and its intelligent capacities from out of the animal sentience that generated devotion and elementary mind as a result of evolutionary progression towards the human kingdom. The slow march of evolution found in the progressive group-*karma* formations of the animal species through adaptation of the species to differing environmental conditionings and competition for sex, food, etc, produces this phase of development. Such adaptation is divinely guided by means of the directive agency of the *deva* kingdom, as nothing is left to chance, but rather manifests as ordained by the meditative Will of the originating Logoic Thinker. The *devas* represent the intelligently endowed agents of That Thought process, consequently Intelligent Design is the order in Nature. The various kingdoms of Nature eventually enter a specific form of *śūnyatā* before entering into a new Kingdom or higher level of species, with greater sentient capacity than previously. The petal of the Heart centre that vivifies the dual Splenic centre is now activated.

3. *Through consciousness is conditioned the mental and physical phenomena (nāma-rūpa, the mind that names and the form that receives sensations), causing that which makes up individual existence.*

A human kingdom now appears, with its ability to name things, and to logically deduce the nature of different forms, to command shapes and direct phenomena, all of which distinguish them from the lesser kingdoms in Nature. We have the ability to craft objects, build houses, manipulate all types of forms, and alter the environment as a whole to suit ourselves. Thus evolve tribes and nations of people, and thence civilisations. The animals have only the most rudimentary ability to do this.

The next eight points have a direct relation to the Eightfold Path. They are concerned with the basic *saṃsāric* qualities that are developed by human beings, which keep them actively incarnating until these qualities can be converted into the corresponding virtue. The gain of the originating creative impulse of the meditative Dhyāni Buddha from the *dharmakāya* is accomplished whenever another human 'Son' enters *śūnyatā*. The need for rebirth then ends, apart for compassionate reasons.

The petal of the Heart centre that vivifies the Solar Plexus centre is now activated.

4. *Through the mental and physical phenomena are conditioned the six bases, the five sense organs through which impressions can come, and consciousness which correlates those impressions.*

This refers to the eventual attainment of the first of the eight Paths, *right or perfect understanding*. Empirical mental expression through the five senses produces a tendency to name, and thence to come to know about all phenomenal things. In the early stages of the rebirth process this at first involves learning about sensual pleasures and of the other things desired or craved by the senses. As the first and second parts of the Four Noble Truths[67] are comprehended then the deductive capacity of the intellect is utilised to gain understanding of the nature of the path to enlightenment, and the consequent release from suffering.

The Heart centre now begins to function properly as an organ of consciousness, with a quaternary of petals awakened. They stimulate the four main *chakras* below the diaphragm: the Solar Plexus, Splenic, Sacral and Base of Spine centres.

5. *Through the six bases is conditioned the sensorial mental impressions.*

Mental impressions mean more than just knowing about things, as it also involves developing philosophical analysis whereby wisdom can be acquired. This is the foundation for the second of the Paths, *right or perfect aspiration or attitude of mind*. This aspiration necessitates an

---

67 Everything is transient, impermanent, and as a consequence of attachment to the transience suffering occurs.

enquiry as to which of the many philosophies lead quickest to salvation or liberation. All types of liberating mental impressions can then bear fruit, eventually producing right choice upon the path, similar to the Buddha's attainment 2,500 years ago.

The petal of the Heart centre vivifying the Throat centre is now awakened. The five non-sacred petals of the Heart centre, whereby impressions from the 'bases' of the five sense-consciousnesses are experienced, are consequently fully functioning. The 'sixth sense', the intellect, is here implicated as the sixth 'base'.

6. *Through the impressions is conditioned feeling.*

By 'feeling' is understand the development and utilisation of the emotions, of all types of emotional-mental *saṃskāras*. The more intense the emotions, the more intense the feelings for things or objects of desire. This largely conditions our speech. Most speak in an emotive way. Only a few speak unemotionally, in a carefully controlled meditative manner, but this is what the development of wisdom, through *right or perfect control of the speech* (the third of the Eightfold Path) necessitates. The entire Path here thus concerns rightly transforming emotional verbalisations, and the consequent distortions of the truth that such emotions generally produce, because the emotions prevent thinking to logical conclusion about anything. Feeling also implies basic clairvoyant and psychic abilities, the 'feeling' for instance that many people get of something good or bad about to happen. Once the emotions are properly controlled then it is possible for a *yogin* to awaken *siddhis* without danger of psychic disaster or abuse, which would lead to necromantic practices.

This and the following *nidānas* relate to the energies from the sacred petals of the Heart centre gradually producing their effects via the human psyche. These effects are expressions of the seven Ray energies that lay the foundation for the eventual development of the attributes of love, of the proper functioning of the *piṇgalā nāḍī* attributes. The earlier five *nidānas* predominantly developed the qualities of the *iḍā nāḍī.*

The *nidāna* of 'feeling' relates to a toned down version of the first Ray manifesting via the Solar Plexus centre, the basis for intensifying attachment to the object of desire. This first Ray energy works through all of the remaining *nidānas* to produce the painful results that come as a consequence of attachment to transitory things. The first Ray consequently teaches the necessity of being detached to all attributes of the form, and so to eventually control the Watery output from the Solar Plexus centre. The will is used through 'speech' to drive the mantric or *maṇḍalic* form through to its conclusion.

7. *Through feeling is conditioned craving.*

Craving concerns a strong desire for material things, specifically for all the material comforts associated with a secure home life, and for all forms of associated sensual and sexual pleasures. To acquire these things appropriate actions need to manifest upon the physical plane. Later comes the realisation of the fruitless transitoriness of such cravings, and of the weary continuous cycles of activity to obtain transient things that cause pain when destroyed or dispossessed. Inevitably the actions are then utilised to acquire things of lasting permanence, i.e., of the *dharma*. Consequently the focus of the person is upwards to heights supernal or inwards towards contemplative bliss. This produces fulfilling the fourth of the Eightfold Path, *right or perfect action.*

Here at first the most base aspect of the second Ray of Love-Wisdom, working through the Sacral centre, produces a strong desire, a love or craving for all things that produce pleasure to the evolving personality. The overcoming of such craving necessitates the development of wisdom and so to rightly act to produce the right changes productive of enlightenment in the end.

8. *Through craving is conditioned clinging.*

The resultant clinging to things material and sensual necessitates developing a lifestyle that can sustain such continuous sense gratification. For most this means obtaining a profession or skill allowing them to meet the needs of existence, according to the strength of their desire for the material comforts they possess or wish. The base activity for their livelihood is accordingly selfishly motivated. Later the force of *karma* pummels this sense of 'self' and selfishness with the retributional necessity to pay back whatever was taken from the all to manifest their empires of 'self' and its comfort zones. *Karma* inevitably teaches the individuals that their livelihoods should be selflessly motivated and the results donated for the benefits of all beings. This necessitates treading the path to enlightenment, where every effort is made to assist in the liberation of all people from attachment to material possessions and cravings. *Right or perfect livelihood* is the result.

Clinging concerns a reification of the energies of the mind, the debasement of the mathematical activity of the third Ray via the Base of Spine centre's focus. The spider-like mind at the centre of its web of desire-attachment is intensely attuned to any new prey it might catch. The 'spider' at the centre thus becomes intensely attached to material plane objects and must in time rightly utilise them to manifest the livelihood that appropriately sustains the activity of the personal-I. Inevitably the focus moves away from catching alluring transient objects, and towards building consciousness-links *(antaḥkaraṇas)* to enlightened domains where images and ideas of lasting value are caught, instead of the transience seen.

9 *Through clinging is conditioned the process of becoming, in active and passive activities, thus the rebirth process.*

This statement is concerned with the complete gamut of the Life process in *saṃsāra*, everything concerning one's environmental situation as a viable member of a human society. It involves the needed effort to contribute to that society, to be socially acceptable. It often involves putting oneself in advantageous situations, to adore the glamorous and those in positions of power, or who occupy privileged positions in that society. Later such glamour, and the pride that comes with it, is seen as empty of real meaning, like everything else concerning normal human affairs. The person then strives to achieve something more lasting and valuable in life, eventually leading to the Bodhisattva path and the fulfilment of the sixth of the Eightfold Path of *right or perfect effort.*

Such effort is to then produce harmonious results amidst the strife and conflict (the fourth Ray expression) that abounds all around in the material world. This work manifests via the superimposed Splenic centre's activity to cleanse mental-emotional *saṃskāras* within the *nāḍīs.* This then becomes the main effort by the *yogin* to produce achievements upon the enlightenment path. This subject was explained in detail in volume 5 of *A Treatise on Mind.*

10 *Through the rebirth, karma producing process is conditioned.*

Here we are concerned with all forms of *karma* producing actions. At first humans produce crass, violent, mostly sensual and physically alluring forms of *karma*. Many crimes against all sentient beings and humanity are committed. Later the educational effects of *karma* entices the person to manifest kind and compassionate actions. When one has learnt enough to seriously walk the path to enlightenment then right or perfect *concentration or mindfulness* in thought, word or deed is carefully analysed so that no *karma* is created. That which already exists is annulled, so that eventually *śūnyatā* is achievable.

Here the fifth Ray of Scientific Activity comes into play via the development of the intellect, which at first produces many forms of *karma* producing activities when focussed upon the material world. Later the double-edged discriminative sword of the Mind is developed through right mindfulness in meditation *(vipassanā,* analytical insight meditation[68]) to overcome the karmic proclivity. The most muddied *saṃskāras* associated with the superimposed Splenic centre must now be thoroughly cleansed if the reincarnation process is to be eventually terminated.

11. *A consequent rebirth.*

This leaves us with the eighth of the Eightfold Path, *right or perfect bliss or absorption.* Such rebirth refers to experiencing a higher cycle of attainment, a new state of awareness or developed consciousness and inevitably joyous experience, the bliss of enlightened absorption *(dhyāna)* once all of the processes leading to mastery of the phenomenal world and its *karma* is achieved.

The materialising energy of the sixth Ray of Devotion is here exemplified. When downwards focussed to gaining *saṃsāric* allurements the need for reincarnation continues. When devotion becomes aspiration for liberation, then eventually the activities manifest that will produce the ending of the cycle, hence absorption into the bliss of the Heart centre's awareness, wherein resides *śūnyatā.* The Diaphragm centre, standing between the higher centres productive of liberation and the lower centres of mental-emotional involvement are here implicated.

68 This is preceded by *śamatha,* calm abiding meditation, wherein all distractions in meditation are eliminated.

To be considered are five types of rebirth, each producing its own type of bliss at the beginning of the process, as release from an old *saṃsāric* patterning always means a heightened revelatory experience. The new conditionings may then again occlude proper thought and may mean future unhappiness because of wrong karmic action. The last on this list relates to the attainment of perfected absorption in a bliss state.

i. The conditional rebirth of the sentience embodying the lesser kingdoms of Nature.
ii. Rebirth as an average human.
iii. Rebirth into the Bardo (state of consciousness) experienced after death.
iv. Rebirth as a Bodhisattva.
v. Birth into the realms of enlightenment, being either the Sambhogakāya Flower, *śūnyatā*, or the *dharmakāya* experience.

It should be needless to say that this list has a relationship to the work of the five Dhyāni Buddhas, with Amogasiddhi governing the activities of the first point on the list, and Vairocana the last. Ratnasambhava, Amitābha and Akṣobhya respectfully governing the processes of the middle three points.

12. *Finally the cycle of old age, grief, sorrow, lamentation, despair is repeated.*

What is implied here is not just a repetition of the above, but also to the nature of spiral-cyclic evolution, to the fact that the cycles continue in a gradually ascending order until a way is found out of the realms of *saṃsāra*. This is the evolutionary way, producing the Bodhisattva path, which also manifests via ten levels *(bhūmis)* of spiral activity.

Here the seventh Ray of Ceremonial Ritual, Rhythmic Activity and Power plays its role in producing the cycles as ordained by the play of *saṃsāra* and the nature of the *karma* to be experienced. Such activity inevitably produces *right or perfect understanding* of the nature of the phenomena to be mastered and consequently *right or perfect aspiration* to overcome the material world. This awakens all of the *chakras* in turn, and inevitably the crown of them all, the 1,000 petalled lotus at the top of the Head, which is patterned upon the twelve main petals of the Heart centre. The way of awakening the Heart centre produces absorption into *śūnyatā* and the consequent awakening of the Head centre produces *dharmakāyic* absorption.

With considerable meditation it is possible to analyse the twelve *nidānas* in relation to Logoic evolution when the factors that concern such a One are streams of evolution, sentient, *deva* and human, passing through His/Her Body of manifestation. The *nidānas* are then conditioned by the twelve signs of the zodiac. One must also think in terms of transmuted correspondences for such terms as 'desire, clinging, craving, feeling'. Logoic Desire for instance translates as the intense energy of Love when experienced by enlightened human units, 'clinging' to the mantric Sound that allows the Logos to sustain the form, e.g., of a planet for the needed duration of its existence. 'Craving' concerns wanting the streams of evolution thereon to gain their liberation. 'Feeling' concerns Logoic interrelationships with other Logoi with Whom energies are shared.

The focus of the twelve *nidānas* is upon the evolution of the qualities of a human kingdom. The twelve petals of the Heart centre are arranged in four groups of three. This line of investigation leads to zodiacal considerations. The Heart *chakra* is then a twelve-spoked wheel of Life. At first must be analysed the eight points of the compass (two groupings of four), as explained for instance with respect to the Eightfold Path and the eight Mahābodhisattvas in the book *Maṇḍalas, their Nature and Development.* The are four cardinal positions, which have been explained therein as:

- The *north* concerns the direction upwards to realms sublime.
- The *east* concerns the direction inwards to the Heart of Life.
- The *south* is the direction downwards to the little lives.
- The *west* is the direction outwards to the field of service that is humanity.

The swastika governs the intermediary points of the compass, where the north-east was explained in terms of *unity,* the south-east for *assimilation,* the south-west for *identification,* and *expansion* for the north-west. The swastika's movement here concerns turning from left to right, the normal way of evolutionary development. A swastika turning from right to left relates to the involutionary path, referring to a primary descent of energies and forces through the subtler realms to manifest the physicality of the substance of the various kingdoms of Nature. Such a swastika also manifests in relation to members of a human kingdom who are travelling against the evolutionary impetus in order to sustain the conditions of *saṃsāra* and self-focussed materialism. Such are the constituents of the dark brotherhood, the forces of evil. They are 'evil' because they fight against the spirit of the evolutionary flow. In the general development of human emotional-mental *saṃskāras,* where all objects of perception, of desire-clinging, are projected in relation to concepts of a 'self', then the movement of the swastika is also from right to left. This constitutes the normal mode of travelling until the path of rectification is trod (aspiration for enlightenment), which produces the reversal of that movement to the normal left to right evolutionary momentum of the swastika. They consequently travel from the *rūpa* levels, the formed domains of empirical considerations, to the *arūpa,* formless levels of enlightened consciousness, as depicted in figure 28.[69]

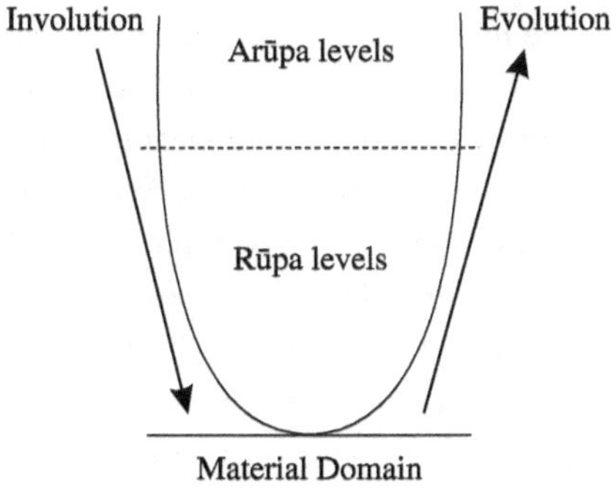

Figure 28: Involution and evolution of consciousness

Eight petals of the Heart centre can therefore be considered to predominantly relate to human evolution, hence the remaining four petals principally relate to the manifestation of the phenomena of *saṃsāra.* They come under the auspices of the activity of the Dharmakāya Flower:

69 Planetary formation is considered similarly, where Logoically our earth represents the lowest point of inversion in the solar system. It is the place of movement from the *ārupa* to *rūpa* aspects of solar evolution. Consequently the earth sustains the longest period of Life-sustaining activity, though all planetary Regents manifest through similar cycles. Similarly for all bodies in the cosmos.

- *North-west,* the first *nidāna*, the formation of the various kingdoms of Nature.
- *South-west,* the second *nidāna*, the development of a human kingdom.
- *South-east,* the third *nidāna*, the development of the ability to name and to command shapes and the appearance of things, i.e., of the rudimentary intellect, and of the early stages of *saṃsāric* evolution. Though earlier explained in terms of human attributes, here one needs also to consider the activities of prime creative agencies in Nature, the *devas,* in the naming of things (mantric sound) to produce the phenomena experienced by means of our senses.
- *North-east,* the twelfth *nidāna*, the spiral-cyclic nature of the general evolutionary patterning.

These four points can also be considered to represent energies that mainly energise the *maṇḍala* of the Heart with the potency of the four Elements. The first *nidāna* concerns the manifestation of the Earthy Element in order to produce the appearance of phenomena. The second *nidāna* projects the energies of the Watery Element, needed by humanity to develop their emotions, aspiration and love. The third *nidāna* incorporates the Fiery Element, the primary force wielded by the *devas* to produce the vicissitudes of manifest life, plus the development of the attributes of mind by humanity. The twelfth *nidāna* relates to the Airy Element that vitalises the sum of the Life process and by means of which all entities evolve. This Element carries the *prāṇas* stemming from cosmic sources.

The remaining eight petals of the Heart centre relate directly to the evocation and evolution of consciousness by humanity in such a way that the energy of *bodhicitta* is evoked, so that eventually Bodhisattvas evolve. The Waters pouring from cosmic sources that evoke this response from humanity make this function of consciousness and its relation to *bodhicitta* the main quality of the energies coming from the Heart centre. The Heart centre is therefore specifically governed by consciousness, Logoically and systemically. Consequently it is the focal point of all Bodhisattvic activity in the realms of form. The Heart is the centre of enlightened perception. It must be emphasised that the

Head lotus *(sahasrāra padma)* also manifests via twelve main petals that denote the extended functioning of a Heart centre.[70]

Much further detail concerning the qualities of the petals and the ways of the expression of the energies of a Heart centre (with respect to human evolution) has been described in *A Treatise on Mind.* What should be noted here is that a *nāḍī* emanates from the tip of each of the petals that interrelate this Flower to other *chakras* in the body. The twelve petals are also custodians of the *saṃskāras* developed as one treads the wheel of rebirth. They are projected in accordance with the way that consciousness must evolve, as directed by the Will of the Sambhogakāya Flower and the effects of the peregrinations of the self-will of the personality. The energies of the twelve astrological signs also work through these petals and their correspondences in the Head lotus to help effect conditioning influences upon the nature of the unfoldment of *saṃskāras*.

As stated, the Heart centre is constituted of five non-sacred and seven sacred petals. Thus five of the petals process specifically the *saṃsāric* activity of consciousness, the five senses and the sense-consciousnesses. The remaining seven convey the luminosity of the seven Rays that help to produce liberation from *saṃsāra*.

The non-sacred petals express the qualities of the four petals that do not embody the Eightfold Path, plus the first of the petals of that Path that produces *right or perfect understanding*, for this petal relates to the process of the perfection of the intellect. Such understanding necessitates a proper intellectual grasp of the nature of things, of what is to be done. The mind (the 'sixth sense') is separative and segregative, which allows it to name, classify and deduce. Its qualities also lay the foundation for the eventual perfection of mind into Mind. Perfection here can also refer to the perfect understanding of a sorcerer as far as his way of darkness and hate is concerned. His is the perfection of the most concrete attributes of mind, which is the way of extreme manipulation of things, to harness power in the material domain.

These non-sacred petals are necessary because they allow the *saṃskāras* of the five Elements to be developed via the sense-

70 See volume 5A of *A Treatise on Mind* for a detailed explanation for the functioning of a major petal of the *sahasrāra padma.*

consciousnesses. Without them life in *saṃsāra* would not be possible. The Airy *prāṇas* of all these qualities, (related to the development of group-consciousness and enlightenment-perception) can also be absorbed and stored in these petals.

The other points of the Eightfold Path admit no such tainting because they are directly associated with pure consciousness unfolding, thus with *bodhicitta*. (Which is non-separative, unifying, bringing all separate parts into an integral whole; even if that 'whole' is termed *śūnyatā* (Void). *Śūnyatā* is devoid of all knowable attributes as far as the mind is concerned. The related base qualities of these petals are thus immediately rejected and directed to the Solar Plexus centre via Splenic centre activity for further processing and refinement. (Thus we have cyclic repetition of base activities until the yogic *tapas*,[71] *dhāraṇīs*[72] and other meditative processes are applied.) This prevents a concept of 'self' from arising, in contradistinction to the petals wherein the intellectual *prāṇas* are conveyed. The intellect generally only knows things in relation to the personal-I, thus thinks in terms of the 'I' and the 'other'.

The petals of the Heart *chakra* have been emphasised to indicate that the *chakras* are more than just flowers of energies in the etheric body. They also relate to consciousness states that have been developed by the individual. They then also can call forth the *saṃskāras* and *skandhas* associated with the Element that the *chakra* is the custodian, the expression for which the geometry of the petals exist. Being the basic paradigm for the *sahasrāra padma* at the crown of the head, the twelve-petalled lotus is derived from the fundamental four of the Base of Spine centre. As stated, the other *chakras* are effectively extensions of this centre, adding two or multiples of two petals below

---

71 *Tapas*, yogic austerities, discipline, warmth, fire, heat; abstraction, meditation. Generally taken to mean the austerities that produce contemplative insights via meditation. However, it literally refers to the practices that kindle the inner creative Fires *(kuṇḍalinī)*, or those that sustain the sum of the contemplative life, rather than mere austerities and asceticism.

72 *Dhāraṇīs*, a means for fixing the mind to an idea, a vision or an experience gained in meditation. They may represent the quintessence of a teaching as well as the experience of a certain state of consciousness, which thereby can be recalled or recreated deliberately at any time. Therefore they are called supporters, receptacles or bearers of wisdom (*vidyādhara*).

the diaphragm, then multiples of four petals and finally multiples of twelve petals above it.

The Logoi embodying planets, stars, constellations, groupings of constellations, galaxies, galactic clusters, etc, are esoterically but different *chakras* manifesting their interrelated purpose within cosmos. They manifest conjoined interrelations according to the *nāḍīs* and direction of cosmic *prāṇas* between them. They can be accordingly catalogued and their attributes determined once one comprehends the nature of the *chakra* the Logos embodies and the stage of development that *chakra* is at with respect to the rest of the system. What is seen physically in the night sky is but the reified effects of the energies manifesting through these cosmic centres. The physical phenomena represents the illusional appearance of the form.

The ten stages of the evolutionary process

It was stated that causation was an expression of the divine Intelligent Mind. The process was likened to an awakening of a Logos from an abstracted *dhyāna* or *pralaya* state to 'Think the divine Thought'. This then resulted in the 'Breathing of the divine Word' as motion. Movement and matter, consciousness and form, action-reaction (the law of *karma),* male and female, positive and negative, are all instantaneous manifestations of that Thought. They are dual results of the same cause, and therefore have an illusive separateness from each other. They are on opposite ends of the same pole of blissful intrinsic awareness.[73]

Through a noetic observation of the known universe, with its psychic, subtle as well as material content, one can deduce that it has a cyclic beginning, and is continuously spiralling onwards to complete a cycle of its containment in an originating Thought. Once that Thought has outlived its usefulness it dies and a new Thought takes its place based upon paradigms established by the previous one. Stars and other unities manifest their appearance in space according to the laws governing the rebirthing process. There are lesser spirals of cause and effect continually manifesting anew, causing the birth and death, sleeping

73 In a similar manner scientists have discovered that to every subatomic particle discovered there is a corresponding antiparticle.

or awakening of every person, atom, or sun. The condensation of matter, the cause of the phenomenal universe, is considered according to the esoteric view to be the result of the formation of a seed, *bīja* (or atom), the golden Egg in multidimensional space. Such is what many mythologies have also indicated that the universe sprang from. For example, *hiraṇyagarbha*[74] (another name for Brahmā), the golden world Egg of the Hindus, and Geb (the 'great cackler') the goose God who laid the world Egg of the Egyptian mythology.

The number *seven* was also seen to symbolise the septenaric constitution of the manifested universe, which when abstracted from the triune Logos, forms the number *ten* with that Logos. This is the number of fulfilment, of the potential and 'final' attainment of all entities. It exemplifies the stages of evolution, of expansions of consciousness:

1. Conception in the Mind of a Logos.
2. Duality, emergence of interrelated opposite forces, the enacting of the *karma* that sets all into motion. *Karma* can only come to be when there are opposing forces to interrelate with each other, something that is acted upon and something that reacts. (The term *karma* in Sanskrit simply means *action.*[75])
3. The balancing forces, producing a coherent form, which is then projected into concretised objectivity by the Will of the Thinker. In doing this the Thinker works with the created patterning of *karma* for the duration of the life of the projected Thought-form.
4. Physical incarnation, the conception is now materialised, e.g., the formation of an earth sphere wherein a humanity can evolve and the related sentient units can manifest their rounds of experiences.
5. The point of greatest divergence. Form has developed its own peculiar traits of great diversity of species and categories of Life. The principle of self identity emerges on its own account,[76] though still focussed within the auspices of group *karma*.

---

74 'He who was born from the golden Egg', the 'God' of all gods and their animating principle.

75 *Karma,* from the verbal root *kṛ,* meaning 'to do, make, act'.

76 Self-centredness (egoism) in people is really the higher correspondence of this stage of evolution, and is an extension (stage 5) of the number 8 in this tabulation.

6. Development of sensory awareness. Nature's experimentation with the expression of the five senses reaches its climax. Many animal forms have appeared wherein one or other of the senses has found favourable expression in heightened development (e.g., the sense of smell in a dog). The process later finds its focus in the development of the faculty of the emotions, the principle of desire and devotion in the higher species of animals. This process leads eventually to the formation of a human kingdom from out of an animal kingdom. This concerns the Individualisation of what was formerly animal group Souls into the Sambhogakāya Flowers of a human kingdom. Each Sambhogakāya Flower is responsible for the *karma* of one individual consciousness stream only, continuously manifesting the 'I's' that create their own streams of *karma* and associated *saṃskāras*. In contradistinction, animals manifest theirs in groups.

   This has happened at a certain time period in earth history on a massive scale with billions of animal forms gaining the rudimentary intelligence of what might be termed animal-man. The Sambhogakāya Flowers project a tenuous stream (*sūtrātmā*) of *manas* into the animal form that has evolved through the normal evolutionary process as outlined by science. A human with the propensity of consecutive thought has now appeared, ready to start the aeonic process of *saṃskāra* formation, attachment to all aspects of *saṃsāra,* then detachment and eventual liberation from *saṃsāric* conditionings as a Buddha.

   Thus a marriage has occurred in this process with the relative masculine (the Sambhogakāya Flower) merging with the relative feminine (the evolved animal form embodied by the *deva* kingdom) to produce the son in manifestation (consciousness). The rest of the philosophy is exemplified in *A Treatise on Mind* and the books by Alice Bailey and H.P. Blavatsky.

7. Development of self-awareness, of an 'I' concept, the intelligence that clearly distinguishes itself from other animal forms around. Humans have the ability to manifest creative bubbles of thought structures around themselves, mimicking the Ideation of the Logos. We therefore have the development of the intellectual faculty that has reached its crescendo in our present materialistic epoch. The turning about in the

seat of consciousness can now happen, changing from an involutionary, self-focussing activity to an evolutionary sequence wherein the sublime nature of the All is consciously identified with and integrated into.

8. First realisation of the other self (or awareness of an unselfish ideal), the loving attitudes of the average person. Gross emotions are transmuted into aspiration and the creative imagination, producing most of the beautiful artefacts and aspects of our civilisations. This necessitates the full flowering of the qualities of a mind.
9. Eventually the above process is sublimated and transmuted in the refined mind of a *yogin,* to produce the full awakening, the enlightened consciousness that a human mind is capable. Such awakening is based upon unselfish idealism and compassionate action. We thus have psycho-spiritual striving to Wisdom through the aspirants, world innovators, helpers of mankind, the emergence of the Bodhisattva path upon a world-wide scale. The gain is the enlightenment of humanity and their consequent travel to cosmos through passing the appropriate Initiation testings.
10. That which produces evolutionary perfection, the process of liberated beings becoming Logoi.

These ten stages were explained in detail in volume 5A of *A Treatise on Mind* in relation to the deities of the *Bardo Thödol* under the heading 'Stages of the development of Nature's *maṇḍala*'.[77] Note also that stages 8, 9 and 10 are the first three steps to a higher cycle of evolution. The process briefly outlined above for those on the earth, has involved many global periods. Such terms as mineral, plant etc., signify states of sentience and awareness, not just to the forms which people are accustomed to viewing. Each stage of evolution is septenary. The septenary is ultimately synthesised into a triad. It recapitulates briefly the preceding stages and then develops that which is to be unfolded for its round of evolution. There is an irregular 'time' interval between the stages.

The *number* 10 can be simplified to:

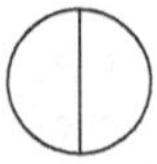

This pictogram shows the first differentiation of the creative process,

77 Pages 51*ff.*

the hatching of the mundane egg (the awakening of the *bīja* of a new Life). This is the stage where the abstract Logos first projected 'His' Thought into the womb of matter. Herein lies the seed form that symbolises the nature of the various androgynous (male-female) deities of antiquity. For instance, the taking of a rib from the breast of Adam by 'God' to create Eve, as depicted in the Book of *Genesis*. There is also the process of cell division (mitosis), and the nuclear fission of atoms.

When this potential expression of duality becomes an actual force in Nature, then it can be symbolised by adding another stroke across thus:

This shows the fertilising of the womb, the union of positive and negative forces: the male (the vertical stroke) and the female energies (the horizontal stroke), producing the birth of consciousness, of a son.

The meaning of these quite well known symbols can be elaborated by utilising the diagram of the lotus blossom (figure 27), as shown in figure 29.

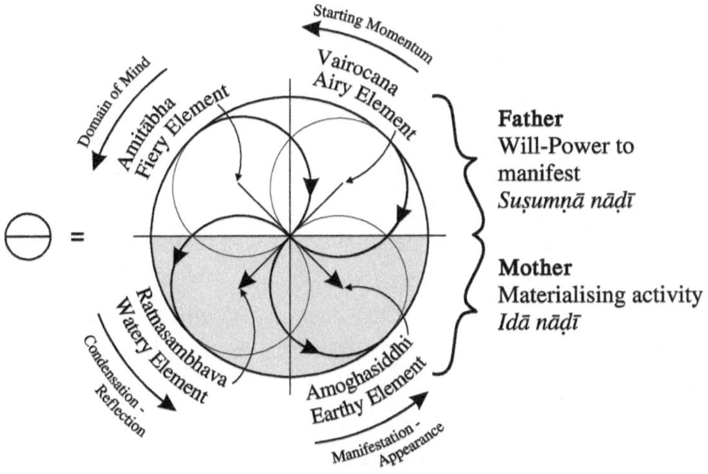

Figure 29: The face of the Waters

The downward stroke of manifestation (The Father energy) is completed when the starting momentum moves from the domain of Mind, governed by Amitābha to that governed by Amoghasiddhi, whereby we have the appearance of phenomena. Figure 29 depicts the basic duality preceding the appearance of phenomena and the motion that produces its manifestation, which takes the form of the yin-yang symbol.

What is depicted as the Watery Element in figure 29 represents the store of past cycles of activity manifesting in 'slumber', a *pralaya* state upon the astral plane prior to projection into manifest incarnation. The Earthy Element represents primal undifferentiated substance *(mūlaprakṛti)* yet to be incorporated by the organising activity of the Logoic Mind. From a Logoic perspective that substance is the concrete portion of the mental plane, hence dense physical to the Logoic Mind, to which energy is added. The energy is then reified in the form that manifests upon or as the dense material domain. In so doing the material domain is moulded into the pattern of the Logoic Mentation, signifying the purpose of the new incarnation. As the reifying energy sweeps through the Waters it incorporates the Lives 'slumbering' therein into the new activity, and the world sphere, or universe, thrills with what becomes an opportunity for further evolutionary growth.

The statement of *Genesis 1:2,* 'the Spirit of God moved upon the face of the waters' represents the state in figure 29 where this 'Spirit' accesses the top half of the diagram, governed by Vairocana and Amitābha, whereas 'the face of the waters' is signified by the bottom shaded feminine portion. At that stage, when only this dual force is flowing, sustaining the *pralaya* (deep sleep, dissolution or disincarnated period), then at first there can be no conscious response from the 'slumbering Lives'. There is just latent duality between the inherent Life and the primal, instinctive, evolutionary urge of matter, there being no field of action relating the two where they can contact or react with each other. At this stage the principle of Life is abstracted from the ethereal and nebulous substance. They have separated into different poles of activity. This signifies stage two of the ten stages of evolution.

When this idea is applied to human evolution (number 7 in the above tabulation) the analogy is found at the era of time when humanity possessed a psychic, unthinking reaction to the Life process. The human psyche during the early evolutionary stages manifested via a process of descent into the animal form via an astral-etheric body that was at first asexual and later became hermaphroditic. Humanity shared a euphoric childlike sentience and had not yet realised the basic duality of their own sexual nature. The act that would give birth to self-consciousness had not yet been committed, they were not 'ashamed' of each other's nakedness.[78] A time came later when humanity walked with the Gods, when 'the sons of God came unto the daughters of men'.[79]

When this early descent of humanity from the subjective domains is relegated to the symbolism of the unfolding petals of the lotus of figure 29, then we have the formation of an inverted Tau cross, comprising of the spheres of activity from Vairocana's domain to Ratnasambhava's, where the Watery (astral) Element dominates. The entire process of the descent from heaven by Adam and Eve signified activating the petal governed by Amoghasiddhi and his All-accomplishing Wisdom. The Garden of Eden represented the domain of Mind whereupon the primal Thought Constructs of Deity existed. Adamic man was one such construct, the great womb of Nature being thereby fecundated by the male seed. A field of action, a plane of perception, the mental, was formed whereon things could be named.[80] At this stage Adam symbolises the formation of the Sambhogakāya Flower upon the higher mental plane. This kingdom then represents the appearance of the masculine principle in Nature. The formation of Eve out of one of the ribs of Adam represented the movement of the creative Impulse into Ratnasambhava's domain.

> And the Lord God caused a deep sleep to fall upon Adam, and he slept; and he took one of his ribs, and closed up the flesh instead thereof; And the rib which the Lord God had taken from man, made he a woman, and brought her unto the man.[81]

---

78 *Genesis 2:25.*

79 *Genesis 6:4.* This was during the Lemurian epoch. The Lemurians are the third (properly human and externalised) human Root Race. They materialised from the etheric domain into the prepared animal forms gained through the evolutionary process, hence can be considered animal-men. The Lemurians at first resided mainly in their etheric body, hence clearly saw the bright luminosity of the high Initiates that administered unto them. Though briefly explained earlier the proper accounting of this early evolutionary period is provided in the Anthropogenesis section of H.P. Blavatsky's *The Secret Doctrine.*

80 *Genesis 2:19-20.*

81 *Genesis 2:21-22.*

The 'rib' can be viewed as one arc of a circle of the diagram moving from one sphere of activity to the next. The 'deep sleep' represents a *pralaya* period from one evolutionary epoch to the next.[82] The human sentience had condensed into a new sphere of sensation that allowed the interplay of sex forces, which in turn could give birth to conscious awareness. Thus we have a triangle of energy, where before just a duality existed (the Spirit of God and the rudimentary mind of Adamic man).

Figure 30 represents the stage where Eve (who embodied the consciousness of the evolving human form) picked the 'forbidden fruit' from the garden of Eden. Here at first the focus of consciousness is upon the inner realms, the building of the Sambhogakāya Flower and the process of the inversion of the consciousness principle so that it can be contained in a vehicle (represented by Eve) to eventually awaken the All-accomplishing Wisdom by mastering physical plane activity (Amoghasiddhi's domain).

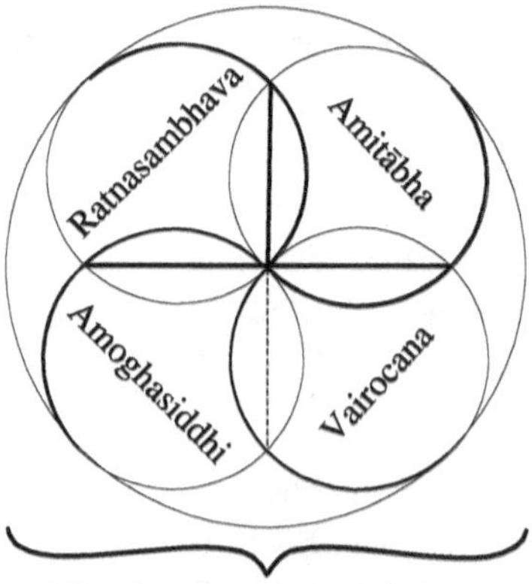

Figure 30: Formation of the Tau cross

The dotted line in the figure represents latent potentiality, the development of self-consciousness via being physically incarnate.

82 Esoterically considered as the movement from the first to the second Root Race of humanity. The Adamic stage, and the hermaphroditic (Adam-Eve before their fall) stages of evolution, on the mental and astral planes represented the first two Root Races. They represented the process of descent of the human Life-force (an 'anima') towards Individualisation, which happened during the Lemurian epoch.

This was still to be attained by the principle represented by Eve. The process moves from the cycles of physical incarnations (the symbolic 777 incarnations, explained in volumes 4 and 5A of *A Treatise on Mind)* towards the liberation signified by Vairocana.

When Adam also ate this fruit, with the consequence that both Adam and Eve became self-consciously aware (knowing of their nakedness[83]) they were expelled from this garden to the material domain to wear 'coats of skins',[84] the various sheaths of substance, including animal skins to keep them warm. Thus started the reincarnation process, with the Sambhogakāya Flower (Adam) overshadowing Eve, the incarnate personality. The process of expulsion inverted the Tau cross. Consciousness individualised and incarnated into an embodied form that cyclically incarnated into the formed realms to gather the necessary experiences as the spirals became increasingly more expansive and turned upwards to the subjective domains.

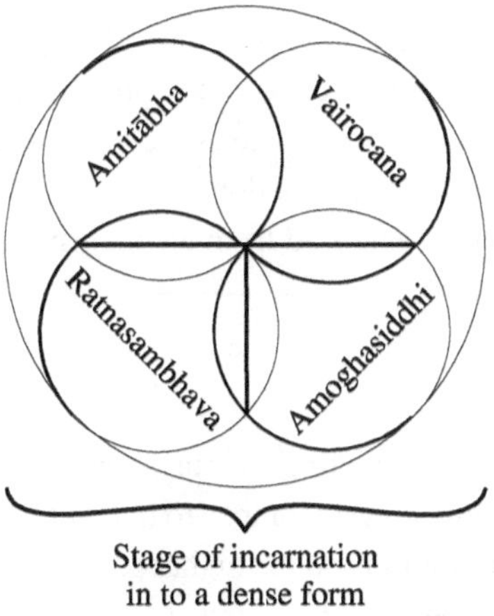

Figure 31: The rectified Tau cross

With respect to humanity the Tau cross symbolises the stage of evolution expressed by the primal (Lemurian) animal-man, who is

83 The symbolism here relates both to the overshadowing Soul-form and the incarnate personality being aware of the 'nakedness' (the paucity) of the developed attributes they possessed, as well as to the sexual awareness between the male and female human forms manifesting upon the inner realms.

84 *Genesis 3:21.*

embedded with the germ of self-consciousness. The *bīja* of the intellect exists, but has not yet been developed, hence consciousness now has the facility of relatively rapid expansion[85] via the awakening of the mind. In the early stages of such development the Lemurian humanity were barely aware of their physical surroundings. Their consciousness was primarily astral-etheric. This stage of development is depicted in figure 31. The focus now is upon evolution as symbolised by Eve, of her ability to bear 'children',[86] which represented the stages of the conscious development of humanity during their formative evolutionary period. The space represented by the Vairocana-Amitābha consequently is empty, and signifies the awakening of the principle of mind that humanity had yet to develop.

As the Lemurian humanity became ever more ensconced in their physical vehicles and became aware of the dense physical plane as their dominant influence, so the forces representing the Tau cross were no longer enclosed within the sphere of collectivism, rather they were self-conscious of an individualised 'I'. Consequently the mind was increasingly aware, conscious of the physical environment and of relationships with others, where the self-centred individual competed with them for sexual and material dominance. This stage is symbolised by the Tau cross, now separated from an enclosing sphere:

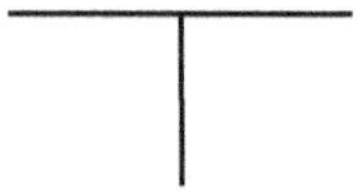

This is the symbol depicting Lemurian evolution. It signifies that physical plane incarnation was completed but that the upward aspiration to the development of mind and the higher principles had not yet begun.

The potency of the mind is at first slowly realised, but eventually, through the meditative process, the Fires of Father-Mother that produce the light of the Son burn with ever-increasing vigour. The person is increasingly able to stand fully in that Light, until eventually the fourth Initiation is undertaken, terminating the Sambhogakāya Flower's existence. All is resolved back from whence it came, though on a higher

85 Compared to the evolutionary development of the animal kingdom.

86 *Genesis 3:16,* 'in sorrow thou shalt bring forth children'.

spiral then before. This is symbolised by the crucifixion of Jesus,[87] and concerns the culmination of the evolutionary process in the field of corporeal manifestation, of the unfolding base lotus (the *mūlādhāra chakra)* awakening the complete potency of the Head lotus. With the understanding of what essentially IS, humanity can return to 'the Father' (Vairocana). This attainment is accomplished by the evocation of the energy latent in the central 'jewel in the heart of the lotus'.

As the earth is the place where humanity has to experience the trials of moving from point to point in space as consciousness develops, so the astrological glyph for the earth as a Logos amongst its peers is the plain cross in the circle signifying the four directions in space.

When the cross is removed from the sphere, making the plain cross, it depicts the 'fall' into generation by humanity. This cross is an extension of the Tau cross, signifying a time when humanity were no longer engrossed with purely physical plane activities, but also began to enquire about their relationship to the external surroundings, incorporating concepts of divinity.

The plain cross becomes the cross of matter upon which people must crucify their lusts and desires, which is what Jesus symbolised by dying upon one. People must relinquish all ties to the manifested planes (a metaphorical death on the cross) before they can ascend to 'heaven', the formless planes of enlightenment.

The circle symbolises the ineffable 'Spirit' (*dharmakāya*) that circumscribes the entire Life process, and as the cross symbolises the material domain, so when the circle surmounts the Tau cross we have the *ankh*, the Egyptian symbol of eternal Life. The ankh can also be considered a version of the plain cross, wherein the northern direction

87 Explained in volume 6 of *A Treatise on Mind.*

has produced conscious awareness of the true nature of divinity, of the way to liberation from formed space. It is the sphere of enlightenment.

The ankh signifies the evolution of mind into Mind, and is an extension of the Tau cross of figure 31 upwards to incorporate the attributes of the sphere of Amitābha and Vairocana. Upon experience of the *dharmakāya* the sphere is then altogether separated from the cross. The ankh shows the triumph of 'Spirit' over matter. The self-conscious one has merged into the All, the bosom of eternal Life. This glyph symbolises the resurrection, the rising of the principle of Life over matter; mastery over *saṃsāra* as the accomplishment of a perfected person.

The ten abovementioned evolutionary stages also indicate the effects of the Logoic Mind governing the interrelationships between all factors in Nature. It manifests at first via a ten petalled Lotus, represented in our bodies as the *maṇipūra chakra* (Solar Plexus centre). In humans this centre is the seat of the personal will controlling the Watery emotional *prāṇas,* but human emotions are not a consideration here, hence in Nature the *maṇipūra chakra* governs the way of evolution of the various streams of sentient Lives.[88] Its attributes are first developed in the animal kingdom. The *maṇipūra chakra* synthesises the *prāṇas* from all of the minor *chakras* in the body. These *chakras* convey the sum of the *prāṇic* qualities specifically developed by the animal kingdom, and are exemplified in the human mental-emotions. These mental-emotions are psychically considered animal-like. They are explained in detail in volumes 5A and B of *A Treatise on Mind,* which interprets the *Bardo Thödol* (the Tibetan 'Book of the Dead'). Therein they are symbolised by animal-headed (theriomorphic) female figures (the Piśāsī, Īśvarī, etc). All of Nature express the interrelated group *karma* governed by the *maṇipūra chakra,* which is but an aspect of the Dharmakāya Flower.

The ten petalled *maṇipūra chakra* can be viewed in terms of two interlaced pentagrams that for the purpose of the causative process are focussed in a north-south orientation. In a later stage, when humanity

88 The interrelation between these streams of sentience signifies a higher interpretation of the Watery Element.

is dominating affairs in the material world, then the orientation is in the east-west direction. Each point of the pentagram represents a petal of this *chakra*.[89] Of the ten petals that fully express the qualities of this interrelatedness, the Dhyāni Buddhas (Jinas) pour their energies through five petals, and the remaining five petals convey the qualities of the response, embodied by the attributes of their *deva* Consorts. The Jina impregnates his Consort with the *bījas* of *saṃsāric* purpose. The combined Wombs of the Consorts convey the evolutionary journeying of the five kingdoms of Nature (viewed esoterically); the mineral, plant, animal, human and divine. This represents the *tathāgatagarbha* (Soul aspect) for Nature's domain.

The first five of the ten evolutionary stages are governed by these Consorts (*prajñās*) of the Dhyāni Buddhas. They are the Divine Mothers, whose concern is with the Lives in their Wombs. At first the involutionary process of the forces that cause the appearance of *saṃsāra* dominates, which incorporates all the forces and entities that share interrelated *karma*. Later the Jinas assist in the evolutionary pull towards liberated space of all the embodied Lives, once human consciousness has been awakened. Thus is the story of causation of sentience in Nature and the evolutionary process expressed esoterically.

The evocation of *suṣumṇā*

Figure 32 endeavours to portray the evocation of the energies of *suṣumṇā,* signifying the gain of the objective of the evolutionary process, wherein the *yogin* can consciously awaken the Fires that hold the body as well as the atomic structure into unity, a composite form. These Fires *(kuṇḍalinī)* are those manifested by the originating causative Impulse of a Logos to form a world sphere. Here the sphere marked +'ve represents the domain signifying the attributes of Vairocana and Amitābha. The sphere marked –'ve represents material plane involvement, wherein all must be accomplished. The eastern direction evokes the Discriminating Inner Wisdom of Amitābha and the Equalising Wisdom of Ratnasambhava which the meditating one

89 The pentagrams manifest as hands that represent pentads of *prāṇic* expression, one downwards focussed and the other upwards. See volume 5A of *A Treatise on Mind* for detail.

must embody through compassionate activity before the inner creative Fire can be awakened. This Fire propels consciousness to the higher dimensions of realisation (the northern direction towards the Ādi Buddha), following along the lines of the initial descent. The initial involutionary process has now been thoroughly reversed, liberating the energies that produced the downward spiral of evolution. These energies, the evoked inherent Fires of matter *(kuṇḍalinī)*, fused with the refined Fires of consciousness produce the All-accomplishing Wisdom of Amoghasiddhi. Mastery of everything concerning the embodied form is thereby produced. At this stage the spiral-cyclic motion has become fourth dimensional, turning in upon itself.

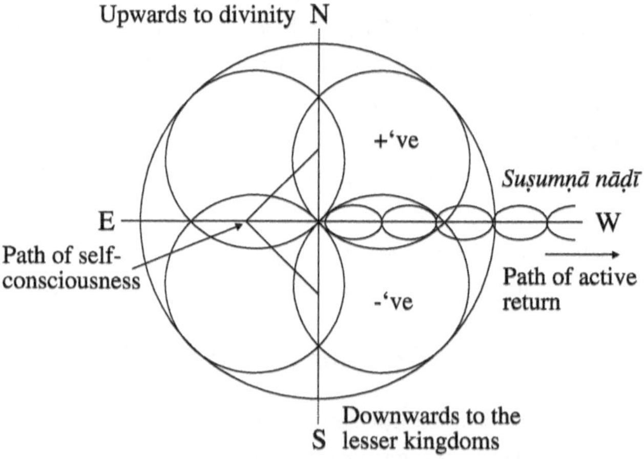

Figure 32: The evocation of *suṣumṇā*

In this figure the *suṣumṇā* is depicted in the western direction, whereas it would normally have a N-S orientation (because it is projected from the Base of Spine centre). The E-W direction emphasises the need for compassionate action, if the *kuṇḍalinī* is to be safely released. It also implicates the mode of awakening an expansive all-inclusive consciousness. The N-S orientation implicates the use of the will for its awakening, which can lead to the mode of activity of the dark brotherhood.

The evocation of *kuṇḍalinī* expresses the gain of three stages of evolution:

- The first is the result of natural evolution, the forces of *the Mother*. This represents rotary or cyclic energy that governs the forces of the formed realms. It incorporates the natural level of liberation of this energy needed to produce the internal heat that sustains the lives of all atomic forms.

- The second is the result of self-conscious growth, the spiral-cyclic dynamics of *the Son*. Here the Fires are integrated with the *prāṇas* evoked by consciousness, to gradually produce a radiatory, expansive multidimensional awakening. Such awakening of the enlightened perception occurs in cycles, wherein periods of revelation appear and then merge into normal consciousness, the cycles become ever-increasingly expansive and more refined until the enlightened stance is the norm. This is the western mode of awakening.

- The third is the result of a definite effort of the will through a purificatory process enacted by a *yogin* over a course of considerable meditative time and yogic austerities *(tapas)*. When the consciousness has sufficiently refined the body then the will evokes a response from the Sambhogakāya Flower, or the Monad at the higher Initiation stages. The *chakra* at the crown of the head *(sahasrāra padma)* is overshadowed and the energies (and corresponding Sound) descend to the Base of the Spine, liberating *kuṇḍalinī*. This represents the forward-progressive energy of *the Father* along the N-S orientation. The Divine Fire of the Father descends to conjoin with the Mother and so en-Flame[90] the atomic lives constituting the embodied form of the individual. The triple Fire (*iḍā, piṇgalā* and *suṣumṇā*) merge into one as they travel geometrically up the spinal column to awaken the latent powers of the *chakras* in turn. They en-Flame the 6, 10, 12, 16, 96 and 1,056 petalled lotuses, as the energies that the various *chakras* convey are experienced by the meditating one. This (the Kingly Way) is a faster path to liberation than the methodology of the Son (the Princely Way), but has more inherent risks, though the will must be evoked at the end of the Princely Way to fully liberate *kuṇḍalinī*.

Figure 32 can also be viewed in terms of the extension of a Tau cross to form the ankh, where the path of active return signifies the

90 This term is better than inflame, as it implicates the generation of a Fiery inferno.

expanding sphere of enlightenment evolving from the Tau. The sphere of attainment here represents the awakening of *dharmakāyic* realisation. What has been ascribed to human consciousness can also be related to the way of evolution of Logoi, but obviously upon a transcendent and vastly expanded scale. The particular Logos of our planet is much hampered by the coarse qualities presently developed by our humanity.

The first seven verses of the book of Genesis

The first seven verses of the book of Genesis state:

1. In the beginning God created the heaven and the earth.
2. And the earth was without form, and void; and darkness *was* upon the face of the deep. And the Spirit of God moved upon the face of the waters.
3. And God said, Let there be light; and there was light.
4. And God saw the light, that *it was* good: and God divided the light from the darkness.
5. And God called the light Day, and the darkness he called Night. And the evening and the morning were the first day.
6. And God said, Let there be a firmament in the midst of the waters, and let it divide the waters from the waters.
7. And God made the firmament, and divided the waters which *were* under the firmament from the waters which *were* above the firmament: and it was so.[91]

In the *first verse* of Genesis what is denoted as 'the heaven' relates to the mental plane, which was formed according to the patterning of the divine Ideation of the *maṇḍala* to be, hence was 'created'. 'The earth' was the sphere of activity, the segregated domain of space, into which that *maṇḍala* was to be projected. We thus have the formation of a world sphere separated into two basic divisions, the *arūpa,* or formless domain ('the heaven') and the *rūpa* or formed domain ('the earth').

The term 'God' can refer to a single entity, as is Sanat Kumāra (explained in volume 7 of *A Treatise on Mind)* or a collective (the Elohim of *Genesis*[92]*)* under a single purpose, such as for instance provided by

91 King James version.

92 For example the statement in *Genesis 4:22,* 'And the Lord God said, Behold, the

the seven Kumāras, of which Sanat Kumāra is one.

Verse two informs us that the yet to be sphere of activity ('the earth') was not yet energised with the factors of embodied Life, hence was 'without form'. It was still an incorporeal Idea in the domain of Mind, hence was Void of the animating Life principle, the Fiery Element that would project the substance of Mind to form manifest Space. Therefore we have the initial inception in the Aetheric Element, associated with Vairocana. The phrase 'the deep' relates to the Watery Element, the domain (cosmic astral and etheric planes) wherein the Life streams were abstracted, either liberated, or of the lesser kingdoms of Nature from the former evolutionary epoch. They are the Creative Hierarchies that will come into objectivity to produce the phenomenal appearance of a world sphere and the Life inhabiting it. 'The deep' is literally the stored qualified substance that a Logos could work with to produce the differentiations of the *maṇḍala* that was to be established. It esoterically manifests in the form of a 'face', which contains the seven facial orifices; a mouth, two nostrils, two ears and two eyes. Each of these orifices relates to a different plane of perception. The mouth to that which speaks the word of power, the *ātmic* plane. The ears hear the corporeal sounds made by dense forms, hence the left ear relates to the dense physical plane, the right ear to the etheric domain. The nostrils convey the breath of life, hence the left nostril breathes in the Watery (astral) Element and the right nostril the Airy buddhic energies. The function of sight relates to the expression of the mind/Mind, hence the left eye sees empirically (the lower mental lane) and the right eye abstractly (the higher mental plane).

These facial orifices can also be seen to convey the attributes of the seven (exoteric) Creative Hierarchies that are explained in Alice Bailey's, *Esoteric Astrology*.[93] It was 'the deep' in relation to the plane of perception whereupon 'the Spirit of God' existed. Darkness was upon this 'face' because as the cycle of manifestation was still to commence, it was not yet in the active process of developing wisdom, hence not illumined by the activating primordial Ray from the Logos. As the Logoic focus was not upon this 'face', so it was not illumined

man is become as one of us, to know good and evil'.

93 Alice Bailey, *Esoteric Astrology*, (Lucis Publishing House, New York, 1982), 32-50.

by Logoic Mind. The Creative Intelligences constituting that 'face' were consequently inactive.

When 'the Spirit of God moved' then the energy of the swastika awakened in the domain of Mind, activating the Fiery Element, as governed by the functions of the Discriminative Inner Wisdom of Amitābha. This 'Spirit' being the impetus of Vairocana and the Dharmadhātu Wisdom[94] reflected by the Mirror-like Wisdom of Akṣobhya via the Airy Element. The attributes of the Logoic Head lotus (the forces of Shambhala) were thus activated. Therein the five higher (abstracted) Creative Hierarchies that bear the *prāṇas* from the 'Mind of God' will activate the seven remaining Creative Hierarchies (predominantly the *deva* Builders) that will cause the appearance of things.

'The waters' implicated here are not those of 'the deep', but rather relate to the cosmic astral Waters. The 'face' thus also refers to the seven Ray energies, the energies from various stars and constellations that would be utilised by the Divine Will to help produce the appearing form. The impact of these energies via the Will activates the seven Creative Hierarchies to produce the appearing phenomena. The energies consequently bear upon the 'slumbering' Lives in 'the deep', illuminating them with the impetus of the Divine Will, thrilling them into the activity of the new *manvantara,* thus producing its dawning early light.

Verse three consequently states 'And God said, Let there be light; and there was light'. This light is that of the seven Ray energies, factors of the Logoic Head centre activating aspects of the Logoic Mind. Therein the *maṇḍalic* construct could be formulated and the Creative Hierarchies (Intelligences) impressed with the appropriate images needed to manifest the new Day. *Antaḥkaraṇas* could then be projected downwards into 'the deep', to thrill the slumbering Lives therein into the activity needed for the appearing *manvantara.*

The light that is seen in *verse four* is the response of the Lives in 'the deep' to the Ray stimulation. The focus is now upon Ratnasambhava's domain, governed by the Equalising Wisdom needed to develop the course of the new *manvantara.* This Wisdom rightly apportions the energies

94 *Dharmadhātu,* meaning 'wisdom of superior qualities'. The 'suchness' or thusness' of being, the universal domain of the *dharma.* It is the pristine cognition of reality's transcendental expanse or sphere of empowerment.

to their respective purposes so that a harmonious whole is eventuated. The response from the Logoic Thought Construct is at first from high level Initiates, the greater and lesser Builders. They received the energies via cosmic astral sources, the Logoic impressions of what to do and the ontology of what is to be. The division of light from the darkness therefore relates to the process of 'condensation', wherein the Builders that were activated could turn their attention to the lesser (unenlightened) Lives, to appropriate forms for them, to en-Soul them. The appearing forms, the appropriation of the substance of the lower planes of perception, hence increasing materialisation, is here called 'darkness'. We therefore have two basic divisions appearing: the en-Souling *(arūpa)* Lives that remain upon the abstracted domains and the manifesting *(rūpa)* forms, who can appear upon the dense physical domain.

The fifth verse mentally consolidates all that has so far occurred. Hence mantric Sounds ('and God called...') is issued from 'God' (the kingdom of Shambhala), clearly establishing the rhythms and cycles of time ('Day' and 'Night', 'the evening and the morning') via which beings must incarnate. Also the *arūpa* (formless) realms via which the energy of Light (the seven Rays) manifested became the Day, whereas the *rūpa* (formed) realms, upon which the Light impacted, became the Night. This clear distinction happened upon the mental plane, which is divided into an abstract and concreted (empirical) portion. Out of the Night appeared the manifest evolving forms, the minds that would reckon with time, and also which will later 'slumber' when *pralaya* appears.

Having clearly consolidated the planes of perception this fifth verse then concentrates upon the process of *manvantara,* 'the first Day'. First we are presented with 'the evening', wherein the attributes of the mind are fully developed and evolve into Mind as a consequence of the evolutionary process. This therefore refers to the prior attainment of the Builders. Next presented is 'the morning', the beginning of the evolutionary process, wherein the approaching Light was to be apprehended by the evolving Lives, and so develop the attributes of consciousness. This 'day' therefore concerns the early period of solar evolution, the foundation for the later Lives to gain wisdom, knowledge of the mysteries of Life as the *manvantara* proceeded. All is governed by the 7 x 7 cycles of expression, the periodical globes, the Root and sub-races, the symbolic 777 incarnations through which evolution proceeds.

The *sixth verse* is concerned with the precipitation of a solid sphere from out of the Waters. This necessitates a further fiat of mantric Sound imbued with the Love-Wisdom that would rightly direct all that was to proceed. The actual manifestation of a 'firmament' however necessitated an intense application of the Divine Will via the outpouring of spiral-cyclic motion in the mode depicted above. This 'firmament' is 'in the midst of the waters', i.e., surrounded by the Watery domain of the (cosmic) astral plane. In terms of our earth we can also view the appearance of land masses and islands in the midst of the oceans. Similarly a human body is surrounded by its Watery astral aura. The phrase 'the waters' also generally relates to the planes of perception surrounding the earthy sphere like the layers of an onion, each layer becoming increasingly dense, material (less fluid), until the most concrete of them all is formed at the centre. Hence we have a *manvantara,* bound by the great year or day of Brahmā.

The next phrase, 'let it divide the waters from the waters' effectively relates to the mental plane, or to the planes of causation *(ātma, buddhi, manas),* that divide the attributes of the cosmic astral plane from the systemic astral plane. (Which is conditioned by our emotions and desires.) A world sphere has therefore appeared within the cosmic astral ocean, that contains within its constitution a reflection of the attributes of the greater ocean external to it. The lesser Waters then can become a reflector into the transient 'firmament' of the energies, the Mantras and Thought Constructs emanating via the greater Waters. All such sounds are carried by the agency of the Creative Intelligence of the Builders of the substance of the forms. Humans evolve later and inevitably learn to emulate the creative activity of form building upon the mental, astral and dense domains.

In *verse seven* the final act of making 'the firmament' was not directly by means of a command, because the previous mantric Sound included the clause related to the appearance of the form, but rather to the act of 'making'. This implies the use of the hands, hence of the *iḍā* and *piṅgalā nāḍīs,* plus the five *prāṇas* manifesting via the fingers of each hand. The tools, or agents of manipulation and creation of the vicissitudes of the dense physical plane are the *devas.* They embody the substance of the mind/Mind, and hence that of the appearance of the form, and of its atomic structure. The lesser *devas,* the 'Blinded Lives', were also the *mūlaprakṛti* that was moulded into the needed forms by the incarnating lives. The act of dividing the Waters that were 'under the firmament' from those that were above it is but a continuation of verse six. It implies the Sacral-Base of Spine centre interrelation. Verse six concerns the process of objectivising the Thought-Form (utilising Sacral centre energies), whereas verse seven relates to the active externalisation of that Thought-construct, hence utilising Base of Spine energies. From this perspective, verse five relates to activating the materialising powers *(siddhis)* of the Solar Plexus centre; verse four to the Heart centre, the Life imbuing organ; verse three to the Throat centre, the organ of the Creative Word and verse two to the Ājñā centre, the directing Eye. Verse one relates to the Head centre, wherein the seed idea originates and the projective impulse emanates.

With the appearance of the firmament this verse concludes the entire process of 'creation'. It signifies the complete descent of the energies of the Logoic Head centre from its place of abstraction, the cosmic astral plane (the Waters above 'the firmament') to the higher sub-planes of the cosmic dense physical plane and thereon establish a Shambhalic correspondence from which all the cycles of Life (the *yugas)* and the Initiation process governing evolution could be directed. The appearing firmament allowed direct physical incarnation of the Lives, as a new globe or sphere of activity was manifest, and 'it was so'.

Oṁ

7

The Spiral of Consciousness, the Geometric View

Analysis of the zero

All *yantras*[1] and *maṇḍalas* of manifestation present a geometric viewpoint of causation, from which the way to liberation can be derived. Liberation progresses along the pathways via which the process of manifestation originated, though now moving upwards through multidimensional space, rather than downwards. Consequently, to properly understand the process producing liberation from form by an individual, even of the consideration by a Logos within His/Her circumscribed causative sphere of limitation, one needs to comprehend how *saṃsāra* originally manifested.

Knowledge of the nature of the method of causation can be gained by analysing the properties of the cipher zero (O). The mathematical concept of the start of any numerological system is the zero, which is really the start of two 'infinities', one proceeding to the macrocosm and the other to the microcosm.

The progression 0...1...∞ (from birth, to present livingness, to the ultimate attainment) must be reversed to complete the cycle of being/non-being. We thus have birth to final attainment (death) and from that attainment to new birth. This movement is spiral-cyclic. We thus have:

1 Visual aids, the systemic arrangement of symbols used to assist the practice of visualisation for meditation, diagrams of divinities, even yogic exercises that facilitate meditation.

microcosm ∞...1...0...1...∞ macrocosm

Here zero signifies 'non-being', the process of birth, the zone or point of origination and symbolises the transitoriness of all manifest phenomena. The number one signifies 'being', the individualised unity, self awareness, that which registers the phenomenon of things in terms of its own self-identity. The infinity signifies the ultimate attainment of complete enlightenment, mergence with the All, which completely transcends the properties of the original birth and its evolutionary progression. (In doing so it incorporates the birthing process or place of origination as part of the process of transmuting all that was, and so continues to a space beyond.) The infinity also signifies death or termination of what once was, and can symbolise the universal storehouse of consciousness, the outermost boundaries of the sphere of cause and effect. The one then concerns the consequent projection of that attainment as a new birth (or *bīja*) on a higher cycle of evolution, e.g., that associated with fourth dimensional perception. The zero then represents the past from which all things have been derived and which now has no substantiality of its own. The infinity can represent the eternal Now, wherein the past, represented by the microcosm, and the future, represented by the macrocosm, merge and produce effects upon consciousness. This consciousness can move backwards or forwards, according as to the tug of the *saṃskāras* that the mind either reinforces or transmutes.

Generally the 0 represents the illusive, ephemeral, ever-changing world we see all around us, which though tangible to our senses, has no significant reality when divorced from those senses. (In this world there is a continuous birthing of forms as they appear into phenomenal activity, and which later recede into the universal storehouse of consciousness from which they emanated.) The infinity represents the real, the permanent, which is said to be unchanging and eternal, *śūnyatā*, the Mind of a Buddha. It lies outside of the cellular ring-pass-not. The one represents atomic perception in the microcosm. Thus the number 1 represents the central point, self-consciousness, the cellular unit of consciousness which relates the zero to infinity (macrocosmic space).

In terms of consciousness growth here is an infinity from:

$$1 \cdots \frac{1}{x+1} \cdots 0 \quad (\text{where } 0 = \infty)$$

which concerns progression towards the microcosm, producing an increasingly narrow viewpoint, a state of ignorance, and from 1...x + 1...∞ (macrocosmic progression, a broadening of viewpoint). Both ends of the summary end in the same thing, 'infinity', symbolised by the ∞, or alternatively by the cypher zero (O), for the zero is a pictograph implying the bounded vastness of the omnipresent universe seen all around us. It is bounded by the limits of our thought or image making capacity. Here x = any particular consciousness state, and x + 1 relates to an addition to that unit of consciousness. When this factor is divided into one it indicates an increasing quotient of ignorance, whilst the 1/∞ implicates the infinitesimally small attribute of consciousness.[2] Such a state happens when consciousness is increasingly impaired, as in developing senility or dementia, but is not normally the human condition. There are however many cases where individuals manifest activities which increase the level of ignorance. There are normally boundaries to such development, and in other arenas knowledge of things may increase.

The statement 1...x + 1...∞ signifies normal evolutionary progression, though progression is slow because there is often regression to ignorant states of perception. 'Progression' here signifies increasing knowledge of the way things actually are, either from a materialistic, religious or esoteric perspective. Such knowledge is therefore relative rather than absolute truth. It is relative because it is progressive. From the Buddhist perspective identifying with transient phenomena as something real is the overall ignorance that most people succumb to. Overcoming ignorance, and so stepping upon the path to wisdom, the gaining of liberation, signifies the move towards infinity. Infinity would normally be considered *śūnyatā,* the Void, however being Void, *śūnyatā* can also be equated with the zero. Moving beyond *śūnyatā* to incorporate what it veils, signifying the attainment of the wisdom of the Dhyāni Buddhas *(dharmakāya)* can be considered the experience of infinity. Here however infinity is conceived of in terms of five modes of expression, and if the Ādi Buddha and Consort are considered, then we have seven. If the Consorts of the Dhyāni Buddhas are added then we have the

---

2 Mathematically it would be considered an impossibility.

attributes of a Heart centre manifesting, wherein *śūnyatā* is experienced. This centre is the place where the transmogrifying energies manifest that convert the transitoriness of things into the Real. Here the zero that is *māyā* (ignorance) becomes the infinity of liberation.

The nature of such perception is transcendent, infinite to normal consciousness. It is spontaneous awareness of all things as they really are. I have also presented in *A Treatise on Mind* that there are levels to *dharmakāya,* hence even infinity is relative, 'bounded'. However, to a normal human consciousness such technicalities are meaningless, as the *dharmakāyic* Mind is beyond conceptualising. It must be experienced in terms of the Wisdoms of the Buddhas.

The sum total of this 'bounded infinity' can be expressed in mathematical notation as the greatest possible number divided by the lowest possible number:

$$\text{Thus}\ \frac{\infty}{0} = \frac{\text{the greatest possible number}}{\text{the lowest possible number}}$$

Now, it can be stated that:

$$\frac{\infty}{0} = \frac{\infty}{\infty} = \frac{\text{the `infinity' of the macrocosm}}{\text{the `infinity' of the microcosm}}$$

so then $\frac{\infty}{0}$ can also equal $\frac{0}{0}$ where the $\frac{0}{0}$ is represented as the start and finish of our concept of infinity, and of primeval causation. The inference here is that the greatest possible number (or consciousness state) could not exist on the finite or three dimensional scale associated with our senses, and thus ceases to be (becoming the zero).

The inference can be extended to include the entire space-time continuum associated with Einstein's theory of Relativity, where mass becomes infinitely great at speeds approaching that of light, and time ceases to be, whilst the path of light itself is curved, to eventually (feasibly) return to its originating source. The beginning is the ending, though the ending is inclusive of the beginning and is embracive of a future expansion that supplants both the beginning and the ending.

The meaning of the fraction $\frac{0}{0}$ (or $\frac{\infty}{\infty}$, or $\frac{\infty}{0}$) is mathematically unresolved and can be taken to have several meanings. For example, it can be understood to represent the number one,

$$\text{e.g.,} \frac{4}{4} = 1, \text{ therefore } \frac{0}{0} = 1,$$

or any feasible number, e.g., 7, or else infinity, for who knows how many times zero can be divided into itself? The expression can also mean zero, or else it can be taken to be an impossible, disallowed operation. Of importance here is that all of these possibilities symbolise the expressed realities of the concept of causality understood metaphysically, as shown below.

The geometry of causation

Hearkening back to the concept of the nature of a cellular consciousness, as symbolised by a circle:

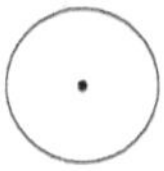

we saw that this indicates the potential of the 'created' universe, as understood by the finite mind. Here is depicted a two-dimensional representation of a sphere whose only boundaries are the limits of the imagination, or the vast expanse of the bounds of the universe. The cyclic nature of manifestation and of the ideal form, a sphere or ovoid, which all Nature tends to acquire (or else has acquired), is also thereby indicated. In this fashion Nature mimics the first act of the creative Logos when that Logos formed a 'circle' circumscribing Thought, as a sphere of self-contained activity, a 'golden Egg', the womb of the cosmos out of which the Elements are fashioned. In a similar manner a person starts as an egg within the womb of his/her mother. The main purpose for each incarnation is the growth of the thinking principle, which is physically 'contained' within the brain and the ovoid sphere of the head.

The symbol of the circle can therefore be used to represent the start and finish of our concept of infinity (the bounds of the creative process). It is the smallest possible thing (or integral form, e.g., the atom) resulting from the greatest possible thing, the universe, and vice versa. Each gives a reciprocal birth to each other.

Putting the information obtained from the mathematical analysis of $\frac{0}{0}$ in diagrammatical form, we get a double triangle within the circle that represents a Logos:

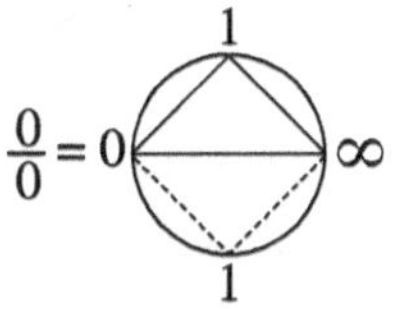

We saw that $\frac{0}{0}$ = 1, infinity, or impossible ($\frac{\infty}{0}$) and that the 1, (e.g., human consciousness) metaphysically represents the midway point between infinity and zero, and from the 'infinity' of the microcosm to the 'infinity' of the macrocosm. Integrating this within the sphere that represents the manifest universe, in the endeavour to comprehend the nature of the *creative expression* of a divine Thinker, a triangle is seen, with another triangle underneath it. The two triangles implicate the positive, and its reverse, the negative aspects, which can also be viewed as the downward projection of the way thought proceeds. This implies that everything in the tangible universe has its opposite. In this theorisation only the positive aspect shall be analysed, indicating that whatever goes for the positive is reversed for the negative.

The positive represents the white *dharma*, the forces of Love and Light, whilst the negative represents the black *dharma*, the forces of darkness, of evil. Darkness is but the shadow of light. It appears whenever something material appears to obstruct the passage of light. Darkness therefore has no substance of its own, for it is of the nature of the great illusion, the state of ignorance, whereas light is of the real, it illumines the way ahead for all consciousnesses aspiring towards future greater luminosity.[3]

The figure thus takes the form of:

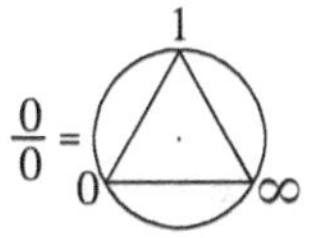

The figure then represents an 'infinite sphere', the manifested form of THAT which is a formless Void. We see therefore that though the absolute Logos is normally undiscernible, the universe is bounded by form that has three primary aspects, here termed potential or unlimited

3 For a fuller analysis on the nature of light see chapter 9 of volume 2 of *A Treatise on Mind*.

driving energy, consciousness (that segregates the unlimited potential into discrete packets of information), and *matter* (or form), the phenomenal appearance of things. Being the universal storehouse of energy, consciousness and of substance is but a version of the statement that a Logos is the emanator, sustainer, and transmuter or terminator of phenomena. This *trimūrti* can also be viewed as Life (energy), quality (consciousness) and appearance (substance/form). Though these aspects are seemingly distinct, they are essentially aspects of the one projection of the original causative Thought.

The figure also primarily indicates that the energy that condensed matter now has the characteristics of form, or rather, of geometric order, for all particles in the tangible universe obey regimented laws of formation, transition, harmony in action, and then a breakdown of that harmony (a disintegration), followed by an eventual reconstruction, hopefully along improved lines where consciousness is concerned. We therefore have:

a. The formation of a sphere (of action) or of the limitation of abstracted Being into a form or patterning of activity (symbolised by the zero).

b. A transitional period of strife or pain, of the interaction between the various positive and negative forces, and thus also of the production of friction or heat and the resultant fire (when the process is relegated to the macroscopic world of forms). Pain is the natural outcome of resistance of a consciousness to evolutionary progress. This is the central pivot of the Buddha's Four Noble Truths, as attachment to the transitory is but a way of not letting go of what keeps one addicted to the past (actions), when the entire momentum of the concourse of things is towards evolutionary progress. The human self-conscious unit allows a fusion between the energies of the Logos and forces of matter. This becomes the way of ascent for the self-conscious unit back to the liberating Source, by transmuting the *saṃskāras* of the entire creative process so that only what is Logoic remains. The conscious unit has consequently identified with the primal causative source.

c. The self-conscious unit consequently undergoes an expansion of consciousness, the intensification of the manifesting light that eventually produces a radiatory activity that at first pierces, and then finally disintegrates the sphere. Identification with the All is then possible, allowing the most intense energies to be expressed. (Symbolised by the ∞.) Eventually a new cycle of activity occurs, but from a higher, liberated, stance.

The above figure therefore provides an explanation of the triune Logos, of the trinity of manifestation, seen for instance in the three bodies (*trikāya*) of a Buddha. The three are a unity, yet each has its own distinct attributes.

- The ∞ represents the Father, the *dharmakāya*, universal Mind.
- The 1 represents the Son, the *sambhogakāya,* the subjective containment of Mind, its organising principle.
- The 0 represents the Mother, the *nirmāṇakāya,* the phenomenal appearance in the world of sense-perception.

The trinity is abstracted on the causal levels, but finds its reflection in *saṃsāra*. This emanation (or reflection) is esoterically described as *the fall of the three into the four.* The 'four' represents the illusional universe, the transitoriness of the appearance of things, which is registered as 'something' by consciousness, and which consequently takes the *nirmāṇakāya* as something tangible. The *nirmāṇakāya* is considered the Mother, because it is the form that can be viewed by a personality, that gives birth to consciousness-attributes and eventually the aspiration to Buddahood. The *sambhogakāya* is the Son because it represents the gain of the consciousness *bījas* stored in the Sambhogakāya Flower. The *dharmakāya* is the Father because it is the originating impulse of everything that is and is not yet in the field of consciousness.

In the world of subatomic particles one must look to the quantum vacuum,[4] the bridge between subjective and objective space. Therein exists an energy dance of subatomic particles moving in and out of subjective space. When the particle is in etheric space then it manifests as the zero. When it individualises as an elementary particle (e.g., a quark) then we can perceive a one, a unitary. When a significant number of such unitaries combine to make the atoms, molecules and

4 This subject shall be further explained in chapter nine.

substance of the physical universe, then we have the appearance of what is signified by the infinity glyph.

The function of the Dhyāni Buddhas

Another way of viewing the process of manifestation is that it expresses the effect of the Mirror-like Wisdom of Akṣobhya reflecting Vairocana's Dharmadhātu Wisdom into manifestation in a triune fashion via the three remaining Dhyāni Buddhas. Ratnasambhava, Amitābha and Amogasiddhi thus represent the reflected trinity of energies and attributes manifesting in the three worlds of human livingness and the *saṃsāric* conditionings therein. They become the guiding lights to the way out from the bounds of the phenomenal realms.

A mirror reflects all things equally to all directions in space. The omnidirectional action of the reflection is symbolised by the Equalising Wisdom of Ratnasambhava and the gesture of giving equally to all concerned. He is represented diagrammatically by the spiral motion, the force productive of manifestation, the eternal cause of the gathering of wisdom. Such wisdom is inherent in the constitution of deity and is expressed as illumination, pure reason onto the plane that represents mind. Ratnasambhava represents the transmuted correspondence of the sexual instinct.

The Dhyani Buddha Amitābha embodies the mental plane and the transmutation of the intellect into illumination by means of the Discriminating Inner Wisdom. Amitābha represents the lotus (of each *chakra)* in the process of flowering to full bloom, allowing the receptacles of knowledge, wisdom and power represented by each *chakra* to be expressed. The body of expression of mind is symbolised by a square on the lower mental plane, as there are four sub-planes to the empirical mind. (The foundation of which is the Base of Spine *chakra.)* When reflected onto the physical plane this square of mind becomes an octagon, the eight spokes of the wheel of the law, the Buddha's Eightfold Path. The perfection of mind is the means to liberation from the trammels of *saṃsāra*. This octagon can also be seen in terms of the symbolism of the eight points of the compass, giving us an idea of direction in space. Amitābha expresses the higher correspondence of the instinct towards knowledge.

When the Discriminating Wisdom is applied as a direct path of action and all the forces inherent in Nature are constructively utilised, then the work of the fifth Dhyāni Buddha Amoghasiddhi, who embodies the All-accomplishing Wisdom, is effected. He expresses the transmuted correspondence of the instinct of self-preservation.[5] Therefore, he has in his palm a double *vajra*, the Buddhist 'thunderbolt', the symbol of indestructible power, indicating his rulership over all things.

Viewing the process geometrically we see that as soon as matter is formed and regulated the triangle becomes 'the four' of a square. When another factor is added to the triangle it must geometrically become a square to keep the resemblance of form. The progeny, the resultant form, is thus objectively differentiated from the cause. The process of manifestation therefore proceeds from the third principle, the Mother (the lowest expression of the abstract triune Logos) and she expresses herself as the fourth principle, the manifest universe, symbolised by a square.

The three aspects of the Logos can be viewed in terms of the meditation process thus:

- The *Seed Thought* (or Will-to-Be), the Father, Vairocana.
- The *Word* (or the Sound which amalgamates all into a coherent shape, the appearing form or object of meditation), the Son, Akṣobhya.
- The *Breath*, movement, the various activities in the totality of the body of being (and which is inherently triune, the mode of activity or fulfilment of that object of meditation), the Mother, embodying the attributes of Ratnasambhava, Amitābha and Amogasiddhi.

This process first involves inbreathing, the ingathering of the necessary energies and the *bījas* of the concepts to be utilised for the new form (Amitābha). Next the breath must be held still in meditative concentration, whilst the *bījas* are appropriately activated with the necessary substance (Ratnasambhava). Finally there is an outbreathing of the activated forms into corporeal activity where the progress of their complete evolutionary purpose can be acceded to (Amogasiddhi).

5 From this perspective Akṣobhya governs the group or herd instinct and Vairocana that of self-assertion.

The attributes of Akṣobhya will manifest via the path of return of the outbreathed form, with the attributes of *Vairocana* governing its final abstraction.

These aspects are archetypally unified, and together they project a triune Thought-Form that has a five-fold interpretation that imbues substance, the omnipresent Waters of space, with the qualities of the creative Logos. This signifies the third stage of the creative process, the formation of 'the square' and its appearance upon the etheric levels. The fourth stage concerns its proper clothing with the necessary *saṃskāras* and *skandhas*, whilst the fifth stage involves its final abstraction back to its creative source. The Thought and the Word are spontaneously co-engendered with the Movement that produces the material universe.

The foundational geometry

As the creative process unfolds to form the material world, so that world tends to reproduce itself as countless different forms, all reflecting (in differing degrees) the triune qualities of the creative Logos. The 'square' is therefore illusory, having only a momentary reality, but no permanency in the scheme of things. It symbolises the transitoriness of the material world, and as a medium is a continuously changing process happening between the archetypal and what is projected into objectivity to manifest as form.

The square, the foundation of transitoriness, thereby becomes the symbol of the Throne or Seat of Power that is the base of all manifestation. This base has its foundation in illusion but nevertheless supports the activities of all Buddhas and Bodhisattvas, being what they sit upon. This indicates the qualities they have mastered and consequently comprehended in totality. Material manifestation represents the arena of service work that will be their accomplishment (encompassed within the *chakras* that are their arenas of responsibility in the world sphere). They have risen out from the transitory phenomenal world beneath them, but must teach others to master the domain of the square and so live in the realms of triangular relationships. Ultimately the triangles merge into great arcs and spheres of attainment that become their arenas of responsibility.

Every form has three basic characteristics: energy-consciousness-form; birth-expansion-death, *rajas-sattva-tamas*, etc., mirroring the qualities of the creative Logos. It can therefore be said that the true characteristics of the material world tend to be reproduced in terms of the smallest geometrical shape, a triangle (or rather, in triangles).

Looking at the overall picture, we find that there are three negative, receptive, limited, reflected qualities compared to the three positive, real (to humanity, permanent), universally effective characteristics of the Logos. There are three attributes that are archetypal, the subjective Plan (that represent the realms of Light), and three qualities that are manifested or formed, the expression of the Plan. (They represent the realms of darkness, into which the light must be reflected or expressed, so that 'darkness' is eventually converted to light.) This makes six aspects in all, plus one that represents the plane that is in equilibrium, the transitional plane. Being neither positive nor negative, it is the mirror through which energy can flow freely from one plane to the other.

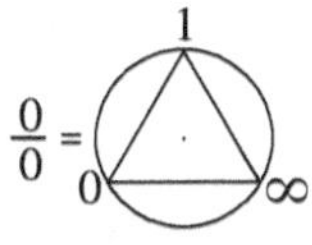

becomes:

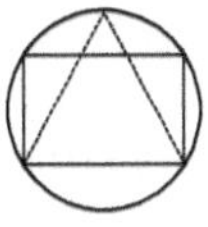

The number seven is thereby generated. This number underlies the construction of the physical universe and the eternal interrelationship between matter and consciousness. The number seven therefore symbolises all the factors that constitute the formed universe, the cause and effect of manifestation and that which interrelates them. These seven principles represent what is sensually perceptible or mentally (psychically) known. We thus have the seven rays of light of the visible spectrum, the esoteric Rays and the seven planes of perception.

It should be noted here that the process of the square tending to form a triangle (the dotted line in the bottom figure) produces the characteristics of a pentagram, the principle of the mind, the development of consciousness, which is abstracted into the triadic form through the evolution of Mind.

The Father aspect, being unknowable to those on the manifested planes, can be taken as a unity (concealing the abstracted triune Logos) and can be seen circumscribing the Son-Mother septenary (the archetypal and the manifest). From this angle we see that the circle enclosing the septenary metaphysically makes the number ten (10), the number of perfection, of completion.

There are thus:

- Three aspects of the unknown, but conceived Logos. The Monadic or primal source of all. The *dharmakāya* (the Father aspect). This has its seed expression in the human intuition.
- Three archetypal aspects. The luminous subjective basis for the appearance of things. The *sambhogakāya* (the Son or Soul aspect). We find its reflected expression in the mental illumination of one aspiring to deep thought.
- Three aspects that are reflections into the phenomenal appearance. (The *nirmāṇakāya,* the Mother aspect.) This corresponds to the normal brain consciousness of the average thinking human. Thinking is relegated in terms of considerations of the form via what is gained through sense perceptions.
- All are contained within the ONE, the Absolute Logos, the Ādi Buddha.

The 'creation' of matter can thus be geometrically represented as a triangle forming or tending to form a square. This tendency is represented as a hexagon with the triangles interlaced:

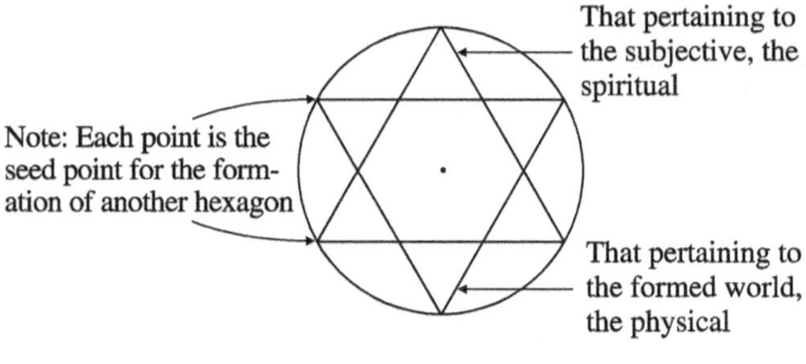

Figure 33: The hexagon

The hexagon is a universal symbol depicting the formation and construction of the matter in the universe. It shows that only six principles out of the seven continually manifest the septenary. In Buddhism the term utilised is *dharmodaya*, the arising of *dharmas*, the source of phenomena. This is depicted by a pair of interlocking triangles, symbolising the non-dual unity of male-female union, as the state of primordial enlightenment and is seen as the *yantra* of Vajrayoginī-Vajaravārahī[6] as the source of all Buddhahood.

Here also lies the foundation of the reasoning as to why the Buddha chose to teach of Six Realms for the expression of the manifest world, being the realms wherein rebirth is possible. Later Buddhists embellished the primary teachings into concrete assertions, forgetting the symbolic import that all is mind-constructed. The relation of the Six Realms to the qualities of certain minor *chakras* was explained in volume 5A of *A Treatise on Mind*.

The hexagon also depicts the triune representation of Deity. The triangle pointing upwards represents the Son aspect and the inverted triangle shows the downward projection, the reflection of divinity into matter. The triangles have been separated and interwoven, showing that matter seems to be separated from its foundation in the *dharmakāya* and yet interwoven with it. The central dot shows the link between the Logos (Who is cyclically manifest, yet aloof) and the rest of the formed universe. It also indicates the link with the unmanifested Absolute.

Upon the domains of human evolution, the upward triangle shows our spiritual nature and aspirations to the Infinite. The inverted triangle depicts our 'fall' into generation, into *saṃsāric* illusionality, the allurements and defilements that must be overcome before enlightenment is possible.

The septenary is an expression of the Absolute, represented by the circle. When the creator, preserver and regenerator aspects of the abstract Logos are included, then is depicted the perfect number ten. This manifestation is essentially an instantaneous expression of the creative Impulse.

---

6 Vajrayoginī, (T. rdo rje rnal'byorma). 'Adamantine *yoginī'*, and epithet of Vajravārahī. Vajravārahī, (T. rdo rje chag mo) the 'adamantine indestructible sow', the Consort of Chakrasamvara. She manifests as the supreme *ḍākinī* of all Buddhas, and represents the purifying inner Fire of the *maṇḍala*, cleansing the subtle body of all hindrances and defilements. She is red in colour and carries a sow's head above her own. The sow eats with its snout in the dirt, signifying the attributes of the Base of Spine *chakra*, that supports the activities of all the other *chakras*.

The ten principles are attributes of the One, yet each has an illusive, separative identity. For instance, there is a difference between a ray of the sun and the sun itself, though both are really the same thing. One defines the existence of the other. The sun would not be a sun without its rays, by means of which we perceive it as an illuminating object in space.[7] Each ray is therefore a embodied ambassador of the Sun via which we can obtain understanding of its expression. Similarly with the 'creation' of the physical universe. The Ray of materiality, of substance, sentience and of consciousness, that was originally seeded in *bīja* form by the embodiment of the universal (cosmic) Mind must eventually return to the central spiritual Sun (the Logos), though it never existed independent from it, because it is imbued with the same integral essence.

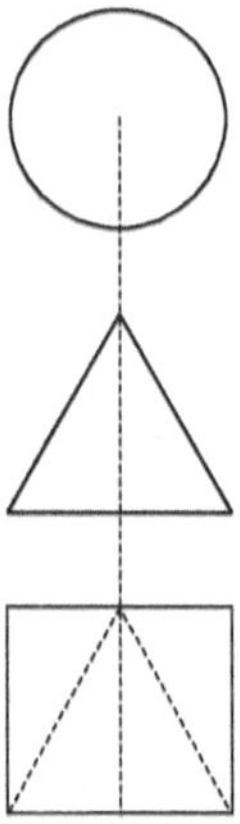

The Absolute manifests the Three
and the Three (the third Logos, the Mother) manifests the Four (the fourth principle)
which manifests another Triad, thus making seven
Thus the fourth principle manifests via the seven Rays, and the correspondences of the archetypes of the septenary upon the corporeal planes

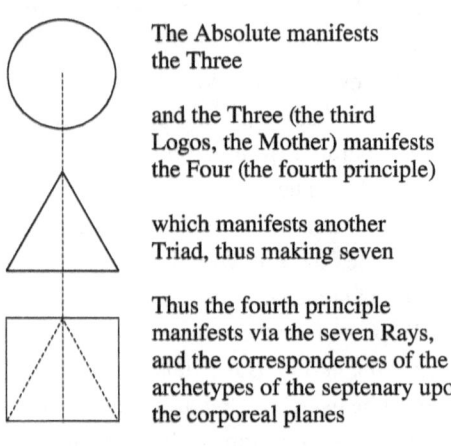

Figure 34: The ten principles

All is but an emanation of Mind and conditioned by its laws, but Logoically that Mind is Love expressed as Wisdom and directed by the inherent Will of the One. The Mind that is Love manifests its seven Ray potencies in the form depicted in the list below.

- First there is the potent Will-of-Love directed to the spheres of the needy ones. This is the One source that builds its circle of abstraction, which becomes the domain of the Absolute manifesting

7 Without the rays a different entity would exist, such as a brown dwarf.

as one of similar interlinked domains in the vastness of space. The 'needy ones' are viewed as an integration of many such spheres of action incorporated within One universal *maṇḍala* of expression. Alternatively, 'the needy ones' relate to the Elemental lives constituting the substance of dark space. The origination of things concerns the movement of that substance into the activity that originates mind.

- Secondly that which interrelates All is Love-Dynamic, the expansive magnanimity of Love's embrace. It may be perceived as the universal flux of the Mind that is Love encompassing all substance as one embracive whole. Within that embrace the Will-of-Love incorporates what is needed into the Thought of what is to be.

- Thirdly this Mind manifests as the Mathematically Active reverberating Sound of Love that incorporates the sphere of the 'needy ones' with evolutionary Law where the past is rightly organised and directed towards the future. This relates to the second verse of Genesis, where it states 'the Spirit of God moved upon the face of the waters'.

- Fourth comes the Serene, dispassionate Beautifying Adornment of Love-Wisdom. This fourth principle empowers the lesser Wills incorporated within the sphere with the vision of what is ordained to be. This level represents the mirror that reflects the energies of the higher three into the lower ones in order to harmonise their forms of activity via wise and loving Purpose. The ordered sequence of the entire *maṇḍala* of being is thus vitalised.

- Fifth comes the Light of Love, electrifying the separating Wills. Here the active attributes of Mind, the Greater and Lesser Builders interrelate with the dark substance that is to be built into the evolving forms to sustain the Days and Nights of the eternal Now. The archetypes of the lower corporeal planes come into existence. The interrelation between the positive Wills and the negative, receptive substance that is acted upon manifests as an electrical interplay that generates light as a consequence of overcoming the resistance of the substance in accordance with the evolutionary pull of the Wills.

- Sixth comes the Magnetically Empowering generative Spread of Love. This relates to the crystallising or condensing effect of the energies upon the dark substance to produce the (illusional) appearance of the forms, collectivised as the 'firmament' of verse six of Genesis. The 'Spread of Love' integrates the sum of the 'firmament' into unity and wisely directs its Purpose.

- Seventh comes the ritually manifesting externalising Power of Love. This relates to the manifesting cycles of expression that organise the constituents of the 'firmament' with a seven-fold mode of activity, continually reincarnating as the activity of the 'fourth principle' of figure 34. This principle generates the four main *yugas* or cycles governing the attributes of the four lower planes of perception that are incorporated as an expression of the trinity of Being. The *yugas* manifest as the forces of the four arms of the swastika. Each *yuga* however spirals or cycles seven times to produce evolutionary perfection.

Finally, with regard to the previous mathematical analysis of the cipher zero, one possible explanation of the expression:

$$\frac{0}{0}$$

has so far been omitted, that it could be *impossible*. This possibility is however included in the rationale of the Buddhist concept of the ultimate attainment of human evolution as being both being and non-being, thought and non-thought, neither of these combinations, or one or the other. Or else it would transcend all such dualities (or their fusion), so as to the intellect being devoid of all classifiable or discernible characteristics (i.e., be 'impossible'). This is the interpretation that has been adopted by Nāgārjuna and his followers in his concept of the *catuṣkoṭikā*, the 'four cornered' reasoning.

Of the *catuṣkoṭikā*, and Nāgārjunian philosophy in general, the limitations of reasoning as normally interpreted is that this philosophy does not allow for the existence of anything but *śūnyatā*, and will not even accede that such a thing as a 'creation of anything', is possible, that anything exists, or that there is even a thinker perceiving such. Note that Nāgārjuna's teaching in the *Mūlamadhyamakakārikā* may be philosophically sound from the point of view of absolutes, if properly interpreted by commentators.[8] The relatively trite interpretations of the

8 See my treatment for instance of chapter one of this text, concerning the appearance

orthodoxy, can however, apart from the acceptance of its basic tenet of what *śūnyatā* is/is not, be further embellished in an investigation of what actually is observed all around. The fact is that our concern is with non-absolutes, with finite infinities, and the process whereby consciousness evolves through transitory quantified units of progressively expanding accumulators of experience. The nature of *śūnyatā* has been thoroughly investigated in my writings, and need not be further elaborated here. The rival Mind-only, Vijñanavādin, school of Asaṇga and Vasubandhu with their teaching of the *ālayavijñāna,* allows many possibilities for the explanation of what is in the world around, regarding the evolutionary path of the all. In the integration of the doctrines of *śūnyatā* and the *ālayavijñāna* comes the answers to many of the enigmas perceived by physicists. This integration concerns the nature of the śūnyatā-saṃsāra nexus and its reflection in the world of subatomic particles.

Einstein's theory of general relativity

Perhaps something should be said concerning Einstein's theory of general Relativity. Briefly, the theory states that gravity is said to move matter along curved pathways within space-time. These pathways are determined by the amount of mass and energy an accelerating object moves. This is the normative behaviour of all interstellar objects. Gravity and acceleration are said to be equivalent. A body's mass, the inertial resistance to changes in motion, manifests as weight, the response to gravity of that mass. The basis to Einstein's theory is the principle that the velocity of light is constant and that inertial and gravitational mass are equivalent. Space does not conform to the lines and angles of the Euclidian geometry found in school textbooks, based upon a flat surface, because gravity distorts the coordinates of space-time, similar to a grid of straight lines on a rubber sheet would curve if you placed a heavy object upon it.

The quantity of mass determines the curvature of space-time and the nature of space-time determines how that mass moves. If gravity curves space a light beam passing near a massive object would be deflected from its course. That deflection would shift the apparent position of

---

of existents, for an example of correct interpretation in chapter 10, volume 1 of *A Treatise on Mind.* Also 'The examination of time in the *Mūlamadhyamakakārikā',* chapter 11 of volume 2 can also be consulted.

the source of the light (e.g., a distant star or galaxy). By bending light large masses act like a lens, the gravitational lensing alters the apparent position of a distant object, creating multiple images of it, or their brightening if the images overlap, thereby unseen masses can be probed. The mathematician Minkowski, working with Einstein, showed that when space and time were combined it yields a mathematical description of happenings where the location of any event can be specified by a set of space and time coordinates. The maths also provided a prediction for the expansion of the universe. There is also an effect of time dilation, which is based upon the effects of two accelerating objects away from each other, whereas the speed of light by which this movement is measured remains constant, making the objects to appear to be moving slower relative to each other.

The curvature of space-time has an analogue to the teachings concerning cyclic and spiral-cyclic motion, however, such motion is the effect of consciousness, the movement of images that have 'weight', a certain force, equivalent to the quality and intensity of the energy imparted, plus the nature of the substance of the thought that is moved. The substance can be highly refined or very sluggish, such as that of self-focussed base desire. Gravity is said to be the effect of acceleration of mass, whereas in the field of consciousness the force or 'weight' of the substance of the thought-form being moved serves the same function. The 'gravity' of the thought—the nature of the intensity of its 'weight'—attracts to it thoughts of similar nature. Upon the mental domain like attracts like.

An expanded universe is consistent with the nature of an expanding Thought-form, which moves from a seed point *(bīja)* to the bounds of a pre-arranged circumference, according to the energy of the 'Idea-form' projected by the Thinker.

As the theory of general relativity is based upon the speed of light being a constant and that nothing in the material universe can travel faster than light, some of Einstein's deductions do not apply on the inner realms because the speed of Thought is faster than that of light. Consequently the geometry of space, said to be causative of gravity, does not apply, neither does time dilation. There is an alternate theory for gravity in the world of subatomic particles, which shall be investigated in chapters eight and nine. All *maṇḍalas* however are embodied Logoic Thought, which are obviously geometric in nature, and the principle of divine mathematics, e.g., the third Ray of Mathematically Exact Activity, rules all of space. Empirical scientists are obviously discovering some of the mathematical formulae and rules that have been set in motion by the Lipika Lords of *karma* in their formulation and expression of what is, according to the originating seed Thought. The mode of manifestation of *karma* underlies the physical appearance of every form and of the way that consciousness interacts with forms and cause objects to move. The appearing thought-forms manifest their own forms of geometry.

There is also a gridwork geometry underlying the *nāḍī* system, from which the *chakras* stem, to be taken into account. This may have a bearing upon the subject of the curvature of space-time, as it is theoretically possible for large masses to affect its observed structure. However, one must clearly define exactly what constitutes 'mass' and the energy fields underlying such an entity, and this is not easily done in modern physics because effectively we have the appearance of 'something' forming out of space that is inherently empty, apart from various states of energy.

Esoterically, energy is all there is, and that energy is bound by consciousness, whilst 'mass' is but the effect of conscious projection downwards via the reification of the substance of consciousness, in ever more congealing cycles of activity. The appearance of such condensed substance in three-dimensional space is then what is viewed as 'matter'. This then has a gravitational attraction to other units of 'condensation', as all Thoughts are unified into one panoptic *maṇḍalic* construct. The force of 'gravity' is the energy the Logoic Thinker must use to keep all aspects of the *maṇḍala* into unity, and consequently 'mass' represents the force vectors, the tensor fields, that are the nodes of activity of that *maṇḍala*.

8

The Nature of Bījas (atomic unities) and Causation

On *bījas*

It has been stated earlier that *chakras* are the places where the *bījas* (seeds, here of consciousness attributes) are stored within an incarnate individual. The *chakras* however draw the *bījas* from the *ālayavijñāna* environment (the Sambhogakāya Flower),[1] whilst the philosophy that all is an aspect of mind/Mind must be taken into account. Within the *manasic* environment exists the consciousness-stream that is the process of the evolution of each personality. Each *chakra* can be considered a lotus blossom that is a vortex of energy that conveys the *bījas* called forth into active manifestation by the mental-emotional activities of the person. They then direct the *bījas* to their target and can transform or transmute the qualities they embody. *Bījas* are not only called forth by humans in the development of their conscious awareness of 'things' and happenings, but also by Logoi in Their considerations of the manifestation of a world sphere. *Bījas* convey the causative forces of what is to be.

Whether we look to the evocation of *bījas* or the awakening of *chakras,* this abstruse subject involves the three basic types of motion explained earlier.

1 From the Sambhogakāya Flower itself for the higher attributes and from the three permanent atoms stored within its constitution for the lower attributes of the human persona. A detailed explanation of the *bījas* can be found in volume 2, chapter 4 of *A Treatise on Mind.*

First there is *rotary motion*. This is the simplest form of motion. It revolves around a central nucleus or axis, the hub of a being's sphere of action. This type of motion therefore segregates and produces an individual, an ego-centred being, or atom. On the macrocosmic scale it produces the separation of aspects of primal undifferentiated substance into myriads of at first homogeneous unities that later agglomerate in various ways to produce heterogeneity. The basic separations of substance are thereby formed. Inevitably all aspects of the diversity of Life around us will consequently manifest. In humanity this motion produces self-centredness, selfishness, desire for sensation, where all things are appropriated for a self, the personal-I. Rotary motion manifests as an expression of the Mother in the various stages of world formation, the nurturing of separate species of things and unities. The *tamasic* (inertial) qualities of this motion produces slow evolutionary change, friction and heat.

Cosmologically this type of motion is seen in the internal movement of the great gas clouds (nebulae), the orbits of celestial spheres around each other, in the formation of clusters of stars and the internal movement of some of the galaxies. It emanates from the 'fire mist' stage of occult philosophy.

The action of the impetus of the exertion of the Will of the Father, depicted previously as a *forward progressive* type of motion, is *sattvic* (rhythmic intensity) in nature. The impetus of this energy can be seen in the movement of the universe as a whole; in the projection onwards and outwards of all the clusters, galaxies and stars from a primeval cosmic centre. Exoterically physicists have likened such motion to the Big Bang. It represents the movement of substance-matter as a unified whole, be that of star systems, constellations, or of the galaxies of which they are a part. All apparently separated parts are propelled forward in space in a definite direction and move to a distant centre to fulfil a purpose. The quality of this motion can be described as spontaneous power, radiatory penetrative ability, and directed Will.

The integration of these two types of motion (rotary and forward progressive) produce *spiral-cyclic* motion. This motion is denoted by the term *rajas* (kingly, passionate action). This is personified by the action of the Son, who not only revolves (metaphorically) around himself, but also around other beings, stimulating all kindred entities (atomic unities) to manifest a forward, positive, more inclusive expansion of consciousness. It is that cohesive quality that attracts individual units to each other, forming coherent groups or organisms; the consciousness of the many becoming fused into an all-embracing One. It therefore helps to modify the various shapes, forms and molecules found in the natural world with each change in consciousness. New *saṃskāras* are thereby evoked. In humanity this motion results in the evolution of group awareness; the awareness of other beings with similar motives and activities as oneself. On the solar level it is the response to and identification with the many different energies that the solar Logos contacts upon His path within a constellation through the galaxy.

Spiral-cyclic motion produces the complete diversity of the qualities of the five Dhyāni Buddhas throughout space, as Life can be then imbued with the characteristics of great wisdom. It is the driving momentum of the energy of *bodhicitta* that produces compassionate understanding and the integration of unitary consciousness into that of the group of which it is esoterically a part. The sum of their evolutionary attainment then spiral onward into ever-expanding inclusive arcs of awareness.

Spiral-cyclic motion can be depicted as motion starting from a point and gradually cycling out in an expansive forward-progressive direction with an ever-increasing momentum via the energy contained in its torque. Its quality can be described as fluidity, light, mobility and rhythm. The energy inevitably becomes liberated and the motion turns in upon itself. This period can be described as a concentric inward focus (*dhyāna*, or absorption) as part of the meditation process, to produce absolute quiescence *(pralaya),* when relating to the condition preceding the creation of our solar system.

When Deity awakes from a 'deep sleep' state (a re-emerging from *pralaya)* and impels the new cycle with a new mantric Command, the spiral-cyclic motion that ensues happens on a higher octave of expression, and is generally coloured by another quality. Deity (or the Sambhogakāya Flower) has resumed another cycle of outward focussed activity and expansiveness for further gathering of experience. The creative process is again thereby demonstrated.

It was previously stated that the ideal shape of any form is the sphere or ovoid, and that the planes of perception (or 'days of creation' in the

Christian Bible), the number of *chakras*, and rays of light are governed by the number seven. One can therefore assume seven principal spirals or cycles of evolvement through the totality of time that the creative process manifests by.

The vast number of spiral galaxies indicate the prevalence of spiral-cyclic motion in the cosmos.

The nature of *bījas*

The process of the evolution of consciousness, leading to meditation and consequent enlightenment, be it on an atomic, human, or else cosmic, hence Logoic scale, parallel each other and can be depicted by the same set of symbols. The depiction of spiral-cyclic motion, when described pictorially in terms of seven spirals of evolution, presents a basic diagram of the internal motion of a *bīja,* an atomic unit of storage. An 'atom' can also be considered in terms of the physical atom, or a human, planetary or solar atom. The atom is a unitary entity conveying a principle or set of attributes and organised according to set laws that integrate other such entities into a unified field of expression. The atom can incorporate other entities within the bounds of its singularity.

Bījas are described as elemental seeds, each one being a store for the containment of a specific *saṃskāra*. *Bījas* are therefore the repositories of *karma* which are brought to fruition whenever a person manifests a particular action in thought, word, or deed. From each minute *bīja* (which can be conceived as infinitesimal points of energy qualification) can spring a full *maṇḍala* or world of action-reaction, of cause and effect. Such *maṇḍalas* can even grow to encompass the size of a galaxy or universe. Much has been said concerning *bījas* in Buddhist philosophy from certain perspectives, but actually very little is known in that philosophy as to how they are constructed, and hence can accomplish their feat of storage and conveyance of *saṃskāras*. Without such knowledge the entire philosophy is obviously tepid, left incomplete. This causes the reader to faithfully acknowledge the necessity of their existence, but to profess near absolute ignorance as to *how* they work.

Much in Buddhist philosophy depends upon the fact of their existence: the workings of *karma*, the nature of the activity of *saṃskāras,* the nature of the reckoning of time; in fact, the very foundation of the *śūnyatā-saṃsāra* nexus depends upon them. Thus the ignorance as to the way they function and how they store *saṃskāras* needs rectification. I shall endeavour to do this here and in the process of the unravelling of this abstruse mystery to the extent that it is possible, will consequently lead Buddhist (and indeed most) readers into very unfamiliar philosophical territory.

The *bījas* manifest in the form of elementary 'atoms' within the *chakras* that call them forth and convey the *saṃskāras* that help qualify the *chakras*. The *bīja* is the store of all the qualities associated with the *saṃskāra,* its background and subsidiary qualities. It is a conveyor of the potency of the *chakra* from which it is derived. In corroboration with other *bījas* it is an animating dynamo from which stem the lines of qualified energy that vivify the petals of the *chakra* in turn, according to the qualities generated by an incarnate 'I'. Their energy fields become the basis of the *prāṇas* that course through the *nāḍīs* stemming from the petals of the *chakras.*

In the economy of its constitution the *bīja* thus contains the capacity to store and transmit the types of *skandhas* and *saṃskāras* associated with the *chakra* and its activities. *Bījas* can also be called forth into activity by the driving impetus of the will of a thinker building a thought construct. *Skandhas* represent the elementary substance and the *saṃskāras* the attributed consciousness quality of the action utilising that substance. There are five types of *prāṇa* which represent the moving streams, 'winds', of *saṃskāras* coursing through a body of manifestation, and effecting its attributes. For the major portion of human evolution the *skandhas* convey the substance of the three worlds of human livingness:

a) That which is purely *manasic.*

b) That of the desire-mind *(kāma-manas).*

c) That which is fundamentally an expression of the five sense-perceptions.

These qualities are represented as three spirals of activity within the atomic structure of the *bīja,* though the effect at first involves rotary motion. There is virtually an infinite amount of cross-exchange of qualities associated with these three fundamental streams. The concern here is with the substance of the formed side of things (*skandhas*), and which convey the attributes of consciousness *saṃskāras*. Pathways exist for the conversion of the *skandhas* into more refined attributes.

The *skandhas* convey the internal heat, the basic energy qualification, of the atomic structure, and the attributes that exist prior to the development of self-consciousness. As stated in volume 1 of *A Treatise on Mind,*[2] the *skandhas* are bundles or groups of attributes that together constitute the human personality and are responsible for the evolution of consciousness. Exoterically, there are five *skandhas:* 1. form, or body, the sense organs, sense objects and interrelationships *(rūpa),* 2. perception or sensation, feelings and emotions *(vedanā),* 3. aggregates of action, or the motives to thus act *(saṃskāras),* 4. the faculty of discrimination *(samjñā),* 5. revelatory knowledge *(vijñāna).* Effectively, all of these forms of activity are attributes of the *saṃskāras* that are carried through from life to life collectivised in their various groupings. Of the *skandhas, rūpa* represents the sense-consciousnesses, whilst *vedanā* and *samjñā* are together the *kāma-manasic* (desire or emotional mind) aspects of consciousnesses. The *saṃskāras* are expressed in the form of the five different types of *prāṇas* conveyed throughout the *naḍī* system. They are one's karmic accumulations that must be worked with in that life, and are eventually transmuted into the seeds of enlightenment *(vijñāna).*

Esoterically I relegate the term *skandha* more purely to consideration of form *(rūpa)* and of the mental substance *(manas)* that incorporates the body of expression of the material world. Indeed most people view things with a materially focussed consciousness. However *yogins* (practitioners) manifest a multidimensional view, where experiences via the outer sense-perceptors are not as important as those derived via the subjective perceptions of the inner organs *(chakras)* by means of which they are experienced. The focus therefore is upon the inner world of experiences of *yogins,* consequently the other attributes of the *skandhas* have been treated ontologically as categories in their own right. Hence *saṃskāras* are differentiated from (yet incorporate) the *skandhas,* because all interrelations with phenomena by means of the faculty of discrimination is carried by them.

---

2 See pages 55-56.

When viewing the *skandhas* directly in relation to human evolution we can look to the aspects coursing through three spirals of activity in terms of *rūpa,* the sense-consciousnesses, *vedanā,* the feelings and emotions, and *samjña,* the faculty of discrimination. These three are responsible for the normal expression of empirical consciousness gained by humans, and collectivised under the term *kāma-manas.* These *skandhas* also have their correspondences for the lower kingdoms of Nature (mineral, plant and animal[3]). These spirals are depicted as the outermost, or courser spirals of the esoteric depiction of an atom. The three spirals for the lower kingdoms eventually manifest at a higher octave of expression conveying the refined attributes of what was originally generated. These higher octaves develop as the *skandhas* of the three worlds of human livingness, hence the generation of *saṃskāras.*

Saṃskāras convey the *bījas* of all that is gained by human consciousness in the form of the five *prāṇas,* whilst *vijñāna* expresses the seeds of enlightenment, wherein is found the attributes of the higher Mind, the higher dimensions of attainment beyond empirical considerations. *Vijñāna* is defined as the faculty of distinguishing, discerning or judging. It represents consciousness in all its attributes, from which is appropriated the mind (*manas*), discriminative knowledge. When the 'five-ness' of the *prāṇas* governing the expression of mind are coupled to the attributes of Mind then the 'seven-ness' governing the sum of manifest Life comes into play. This manifests in the form of seven spirals of expression, whose energies run counter to the three mentioned above, because focussed upon enlightenment, rather than the attributes that bind one to *saṃsāra,* the prison house of the senses.

To effect this conversion consciousness must work to transmute basic *saṃskāras,* to therefore act opposite to the flow of the main stream of *skandhas.* This topic shall be later elucidated, as it is concerned specifically with the nature of consciousness and its evolution. This

3 Note that esoterically the kingdoms are viewed differently than in normal biological texts. We think in terms of the mineral, plant, animal, human and the divine (enlightened beings, including the *devas)* rather than Monera, Protista, Fungi, Plantae and Animalia. Monera, Protista and Fungi can be viewed as a transition between the mineral and plant kingdom, whereas there are other transitions between the plant and animal and animal and humans. Domesticated animals for instance can be considered a transition stage between the animal and human stages.

involves spiral-cyclic motion governing the quality and colouring of the *saṃskāras* involved. We also saw in my earlier writings that consciousness is conditioned in terms of the seven Ray qualifications, which manifest in the form of seven spirals of energy. The base characteristics of the seven Rays would likewise undergo cycles of refinement as the personalities concerned go through the process of becoming enlightened. There are three main cycles to consider.

a) The cycle of being an average human unit possessing mundane consciousness, where many *saṃskāras* are produced due to attachment to the various lures of *saṃsāra*. The attributes of the *skandhas* then dominate. This stage can be considered to represent two phases. The first wherein the human is mainly Earthy, focussed totally upon the material world, with a barely perceptible intellectual development. The second being where the individual is much more Watery (emotional and psychically orientated), but again with little mind developed.

b) The process where the mind is developed. At first the focus is the self-focussed desire-mind or mental-emotions *(kāma-manas),* which is where average humanity presently resonate. The next stage concerns engendering the intellectual curiosity to Know above all things, the activity on the path to generating light. Inevitably it works to overcome or to transmute former lethargic and erroneous mental-emotional thought processes.

c) The stage where consciousness engenders only that associated with *bodhicitta,* hence the drive to enlightenment that produces the compassionate Mind of the pure white *dharma.* Effectively consciousness turns in upon itself to produce heightened meditative vistas of expansive illumined awareness, a new vaster form of direct perception of the nature of cosmos.

These cycles are incorporated into the seven streams of levels of multidimensional expression, governed by the seven Ray attributes.

There are therefore two main streams of energy to consider in the panoply of incarnate life, one is consciousness and the other constitutes the substance (the unconscious Elemental lives) that consciousness utilises. The process of Life concerns the conversion of that basic substance into conscious awareness and then to expand the reach of its revelatory expression. One must also remember that all phenomena is imbued with Life. Diagrammatically these two streams are depicted as two different groups of spirals. The first is a slow and relatively sluggish movement in one direction, producing basic heat because of the relative inertia of its motion and because of the effect of the friction of resistance to change of that through which it moves. This movement relates to the energy of matter, of the Mother. In many ways this energy of the *skandhas* is antagonistic to the evolution of consciousness, for consciousness evolves by fighting and overcoming the lethargic force or pull of matter. At first however, people are attached to the allurements of that matter, hence to the various trappings of *māyā*. Many *saṃskāras* are thereby created.

The other stream is fast, electrical in nature (the energy of vibrant consciousness states), producing light as it overcomes the sluggishness of the darkness (ignorance). This stream therefore moves in the opposite direction to that of the *skandhas* and manifests in seven different octaves of expression, or states of consciousness that every person cycles through during their evolutionary unfoldment. This represents the Son energy manifesting through seven cycles of activity, generating a magnetic, attractive, field associated with each cycle, indicating the extent of the reach of consciousness into space. The spiral-cyclic activity represents seven consciousness states coloured by the seven Rays and their subray conditionings. The consciousness states represent the attributes of the converted *skandhas* into *manas* and the development of the seeds of enlightenment *(vijñāna)*. In the field of consciousness what is attracted is of like nature, rather than an exponent of opposite polarity. However, when consciousness manifests an effort to work with substance, in the building of thought-forms, then that substance manifests as a negative or receptive polarity to the positive force engendered by consciousness. Esoterically this is viewed as a human-*deva* interrelationship.

The interrelation between the two sets of spirals, of the *skandhas* and *saṃskāras,* produces the phenomenon of the form into which a person incarnates, being bundles of aggregates that are collated and distributed by the *chakras* and *nāḍī* system.

With respect to the expression of human consciousness *per se,* the substance of the Mother consists mainly of *manas* in its five subdivisions, as worked upon by the five Dhyāni Buddhas to produce the perfection of wisdom. Reified *manas* is the elementary substance of the Mother's Womb, whilst the *skandhas,* the substance of the three worlds of human livingness, are its basic expression. Therein manifests the *māyā* of people's dreams, aspirations and actions. They are the expression of desire-mind *(kāma-manas)* and purely sense based forms of generated *saṃskāras*. Overcoming these *saṃskāras* of attachment constitutes the nature of the path of release. The *saṃskāras* are conditioned by the five aspects of mind, but there are also two esoteric aspects indicated in the seven spirals:

a. That which awakens the Son of Life within that Womb, the development and refinement of consciousness that eventually liberates the substance of the Womb of time and space through the evocation of *bodhicitta*.

b. The awakened, enlightened stage of realisation, the liberated *kuṇḍalinī* Fire merged with the energies in the *suṣumṇā* that abstract all life back to its emanating source. This abstraction process represents the energy of the Father. This energy inevitably drives inwards the cycles of expression of the spirals, producing a vortex of energies moving at right angles to the main spirals. The Father energy in the central stream, represents a dynamo of generated energies.

The basic trinity of Father, Son, Mother (manifesting also as the three *guṇas,* and *trikāya,* three bodies of a Buddha) is found throughout Nature and must be factored in all computations of energy flow.

The *nāḍīs* are streams of energies in the body of a human or Logos, and though not normally considered in the spiral form of the elementary atoms depicted below, nevertheless the energies within them *(saṃskāras, prāṇas)* cycle, spiral through the seven stages of expression. Each *prāṇa* therefore can develop the seven stages and can accordingly be coloured by any of the seven Ray modes, bearing one particular major colouring at any time. They manifest different intensities of hue, depending upon the stage of expression of the cycle concerned.

Bījas as 'atoms'

The entire course of the flow of energies in the *nāḍīs* within one unit, considered as a complete system, can be considered a *bīja*. There are *bījas* within *bījas* within *bījas*. We can view such from a Logoic scale and the levels of the various Creative Intelligences (of which humanity is one) right through to the flash of an elementary atomic unit appearing and disappearing in the quantum vacuum, from whence the substance of our forms emanates. The *bījas* are atoms with their own intrinsic spirals, which turn inwards at the central column, where the *prāṇas* manifest either in the *iḍā* or the *piṅgalā* form.

Every integral unity in Nature can be viewed in terms of an atomic unit of force conveyance. Because the subhuman kingdoms have not yet developed the capacity of mind, so the complete seven spirals of conscious evolution are not applicable. The number of spirals of awareness here being four (for the four kingdoms in Nature, including humanity[4]), with a fifth spiral being fully attained as a consequence of the development of the abstract Mind by a humanity. Two other spirals can be considered to complete the septenary of manifest being.

As stated, the energy of the Mother manifests as a dual group of spirals that assume a lower and higher force relationship to each other. These spirals develop the *bījas* of the forces of Nature. The lowest concerns the energy of the lower three planes of perception, and are resistant to the general evolutionary pull. Resistance causes friction and heat, hence is the basis to material plane Fire, which produces expansion of the spirals. Three levels of expression of such basic Fires have been considered, relating to the triune domain of *māyā*.

The second grouping, evolving from the first, is more *manasic*, the developed Fiery substance (sentience) of the three kingdoms of Nature (the mineral, plant and animal) below the human, and which is also incorporated as the substance of the human mind. The attributes of the empirical mind are slowly developed as a consequence of such development. The developing consciousness is dull, resistant to change,

---

4 The four relates to the four Elements that govern material manifestation, also to the esoteric fact that we are in the fourth Round of the evolutionary impulse. A Round relates to the specific major spiral (cycle) of energy that is being vivified within the world sphere. The fourth Round governs the evolution of the fourth kingdom in Nature.

and rotary in nature. This means that the basic thought-forms developed occur again and again, as they relate mainly to basic survival instincts, self-focussed selfishness and the sex impulse.

In this way the internal motion of the atom facilitates the development of the base nature of the *skandhas* of the three lesser kingdoms in Nature, as well as of the general psychic constitution of a human unit. It represents the substance of the periodic vehicles of humanity, the personality equipment, that can also be viewed in terms of the development of the three *guṇas; tamas, rajas,* and *sattva.*

One reason why humanity can be considered Lords of Sacrifice and Compassion is because by incarnating into a form, (utilising the substance of *māyā)* they automatically work for aeons at the conversion of these elementary streams of sentient lives into higher states of *manasic* viability. At first this activity is unconscious, but later, upon the Bodhisattva path, the process becomes entirely conscious and is greatly hastened for all concerned.

One must remember here that all that concerns us is differing energy states of embodied Lives and *manasic* variability. The impact of one upon the other constitutes Life and the inevitable liberating evolutionary process.

The two types of spirals implicate the inherent duality associated with manifest space, signified by the dual nature of the activity of the *deva* hierarchy, who are directly concerned with the evolution of the formed universe. The duality also indicates the nature of chemical affinity in substance and the sexual instinct in all kingdoms that impels the evolutionary urge towards the perfection of the forms which will act as containers for the developing mind. At first we are confronted with the early stages of evolutionary attainment when only duality exists, the positive and negative of electrical interrelations, spirit-matter, Father-Mother, Buddha as a Logos and Consort (the *deva* kingdom). Later the child, the consciousness principle carried by humanity, evolves from the substance of the Consort's Womb. Consciousness awakens in the form of a septenate evolutionary stream. Consciousness manifests in the form of the variegations of light, hence humanity is collectively Lucifer,[5] the light bringer, yet fallen (solar) Angel into the materiality of the relatively sluggish *saṃsāric* form.

5 *Isaiah 14:12 ff.*

One can picture the concept of the spiral in terms of a sun travelling in a curved path around a central luminary. As it moves in time through space, so the planets revolving around it move in a spiral pattern, with their satellites moving around them in a similar spiral pattern.

Figure 35: The spirillae of an atom[6]

The clairvoyant investigation concerning the spirals given in the book *Occult Chemistry* has shown that each of the major spirals has six lesser spirals (spirillae) progressively getting finer in nature revolving at right angles to each other in a tiered fashion. The subtlest of these spirals is said to consist of minute dots, 'specks of nothingness' floating in the ether of space (which the authors of this book call koilon) and further state that 'the interior of these space-bubbles is an absolute void to the highest power of vision that we can turn upon them'.[7]

The following figure of the 'ultimate physical atom', which the authors call an Anu[8], is taken from *Occult Chemistry.*[9] The positive Anu conveys energy from fourth dimensional astral space to the physical

---

6 This image is taken from: Annie Besant & C.W. Leadbeater, *Occult Chemistry,* third edition, editors, C. Jinarajadasa and E.W. Preston, (Theosophical Publishing House, Adyar, Madras, India, 1951), 21. Their caption reads 'Formation of bubbles into 1st spirilla and 2nd and 3rd spirillae'.

7 Ibid., 17.

8 The Sanskrit term is properly spelt *aṇu,* meaning minute, the infinitesimally small, not further divisible. As a noun it signifies an atom. I shall continue using the anglicised term Anu to be consistent with the published texts.

9 Ibid., 13.

domain and the negative projection directs energies from the physical domain inwards. There are two types of spirals depicted, two bands of three thicker ones and three bands of seven thinner spirals.

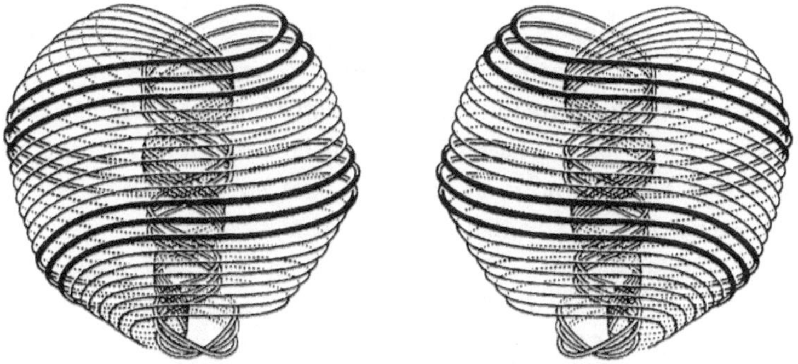

Figure 36: The ultimate physical atom

In his books D.K. calls the Anu of the highest plane of perception *(ādi)*, and its correspondences on the remaining planes 'the atomic level', but does not elaborate what is meant by 'atomic'. These Anu's consist of the 'specks of nothingness' that the authors of *Occult Chemistry* designate to the highest level of spiral. We can presume that at this level the atomic form constituting what is known as 'substance' first differentiates from the ethers.

There are four etheric sub-planes to the cosmic physical plane as well as to the dense physical plane of our world sphere. Concerning the dense physical plane the first level of the spirals conveys the energies of the most refined of the ethers, those directed by the Head lotus and the Ājñā centre.[10] This represents the formative and directive forces for the spirals that follow. The second spiral conveys the *prāṇas* from the second ether as directed from the Heart centre. The third spiral conveys the *prāṇas* from the Throat centre via the third ether and the fourth spiral conveys those from the fourth ether as directed by the centres below the diaphragm. The lower triad of the thicker spirals convey the Fires of the three planes of human livingness. These are the concreting, condensing, organising forces manifesting via the three major *chakras*

10 In speaking of the *chakras* here one must visualise both planetary centres, as well as those incorporated in a human form.

below the diaphragm and the twenty-two minor centres. Consequently they reify what manifests via the fourth ether. The boundary between the fourth ether and the reified formative space ('gaseous substance') is what is known to empirical scientists as the quantum vacuum.

Viewing the appearance of the 'atom' in terms of the esoteric four plus three[11] arrangement and taking the two larger 'extra-spirals' into account,[12] then there are three levels of four spirals to consider, and five levels of three spirals. Each of the levels manifest at a higher octave of expression to the previous one. Each main spiral also manifests as a grouping of seven spirillae as explained above.

Taking the human *nāḍī* system as such an 'atom', the grouping of 3 x 4 = 12 spirals manifest as consciousness attributes in accordance to the qualities of the three *guṇas,* and the three types of crosses (the cardinal, fixed and mutable) via the experience of subjective space. This concerns the integration of consciousness with the universal whole when freed from the limitation of form. These *prāṇas* are conveyed by the *piṇgalā nāḍī* and the Heart centre.

The 3 x 5 = 15 spirals manifest in terms of the attributes that govern the development of mind/Mind in evolutionary space (Nature) as conveyed by the Throat centre and the *iḍā nāḍī.* This produces the expression of the five sense-consciousnesses and the *saṃskāras* they generate. These attributes are effectively reified expressions of the energies from the five Dhyāni Buddhas, hence their Wisdoms will eventually evolve from these expressions of mind. Each of the triads manifest the attributes of the three *guṇas* for the domain of expression for one or other of the Dhyāni Buddhas. The highest of the five triads relates to the conveyance of the *prāṇas* associated with Vairocana and his Dharmadhātu Wisdom. This manifests via the Head lotus and is directed by means of the Ājñā centre. This level therefore relates to the integrating organisational activity of the Mind (the Sambhogakāya Flower) bringing about the manifest appearance of the entire 'atom' from out of pure energy (the 'specks of nothingness') and the functioning of the mind. From here also the *saṃskāras* of liberating will developed upon the yogic path and that of sacrifice are abstracted into the Sacrifice-Will petals of the Sambhogakāya

11 Effectively the fall of the 'four into the three' explained in chapter seven.

12 See figure 37, for a description of these.

Flower. From these spirals *saṃskāras* are abstracted to the four etheric level spirals of the three octaves of seven spirals.

This triad relates to the evolution of the attributes *(saṃskāras)* of the smell sense-consciousness, the next triad to the taste sense-consciousness, the third triad to the sight sense-consciousness, the second lowest triad to the touch sense-consciousness and the last to the hearing sense-consciousness.

The second triad of spirals comes under the auspices of Akṣobhya's Mirror-like Wisdom, which reflects the formative energies from the highest triad into the remaining spirals. However they are mainly agents that reflect the attributes of Love-Wisdom *(bodhicitta)* from the Sambhogakāya Flower via the etheric spirals of the three octaves of seven into the aspiring personality still evolving via the five sense-consciousnesses and the grouping of 3 x 5 spirals. They therefore help engender the qualities of devotion and aspiration of the empirically minded ones. The attributes of the Heart centre are consequently developed.

The third triad of spirals engenders the Fiery attributes of Amitābha's Discriminating Inner Wisdom. They synthesise the Watery and Earthy *manasic* attributes developed in the last two triads, and reap the development of the *saṃskāras* of the intellectual pursuits of a person. They also project the more refined loving energies of Mind developed to the two higher groupings of three spirals. The *saṃskāras* developed are abstracted into the Knowledge petals of the Sambhogakāya Flower. The attributes of the Throat centre are thus developed.

The fourth triad of petals, as far as a human is concerned, relates to the development of Ratnasambhava's Equalising Wisdom, hence engendering plant-like thought-forms, those that come as a consequence of social and group interrelationships, plus the loving exchanges between couples in their sex relationships. The *saṃskāras* of manifold emotional colourings of the human aura are thereby produced. The attributes of the Solar Plexus and the minor centres (the Inner Round series of *chakras)* are consequently developed. This relates to forming the attachments to pleasurable things, selfishness, and all mental-emotional exchanges people have. Such exchanges teach them eventually, through karmic consequences, to be loving and kind to each other.

The fifth group of spirals, in terms of human development, incorporate *saṃskāras* mainly of human sensual experiences, those that come as a consequence of physical plane attachments to material things and sexual relationships. Things which people desire to possess are amassed and hoarded. The *saṃskāras* of violence, greed, theft, acts of war, death and material struggle are thus accumulated here, hence needing mastering. The quality to be developed is Amoghasiddhi's All-accomplishing Wisdom, once mastery of and detachment to all of *saṃsāra* is accomplished. The attributes of the Sacral and Base of Spine centres dominate here.

There are 15 energies (3 x 5) related to the evolution of *manas,* the principle of mind in Nature, as conveyed by the *iḍā nāḍī,* consequently the attributes of the Throat centre are developed.

Twelve energies (3 x 4) are related to the experience of subjective space, the integration of consciousness with the universal whole freed from the limitations of form. The attributes of the *piṇgalā nāḍī* and the Heart centre are thus developed. Its potencies manifest via the three crosses, the cardinal, fixed and mutable and the twelve astrological signs.[13]

There are six finer spirillae nestled around each other and each of the seven main spirals, thus making 7 x 7 = 49 vectors for the energies of the Rays and subrays. Now, taking the 3 x 5 grouping of spirals associated with *manasic* development, we therefore have 7 x 15 = 105 spirals altogether governing the evolution of mind/Mind for a human unit and within planetary evolution. This number relates to the appearance of the 105 Lords of Flame at the time of the Individualisation of a human kingdom from out of the animal kingdom, as was explained in volume 7A *(The Constitution of Shambhala)* of *A Treatise on Mind.* These Lords of Flame came at that time with Sanat Kumāra to form the planetary Head centre (Shambhala). The number 105 was also explained in relation to the arrangement of the petals of a Head lotus in volume 5A. When the triune central column of energies is added, then we get the sacred number 108.

When considering the 3 x 4 x 7 spirals and spirillae of the remaining spirals then we get the number 84 (12 x 7). This concerns the mode whereby the energies of the Heart centre can influence manifest space

13 See volume 3 of *A Treatise on Mind* for a detailed explanation of this subject.

via the seven planes of perception, the manifestation of consciousness attributes from subjective into objective space. When the number 84 is added to 108 then we get the important number 192, twice the number 96, the number of petals to the Ājñā centre. The number 192 is also the number of petals of the Heart in the Head centre. When the number 96 is multiplied by 11, then we get the number 1056 that governs the number of petals to the Head lotus. The significance of these numbers were fully explored in volume 5A of *A Treatise on Mind,* hence need not be elaborated here. What is important to note however is the correlation of the petals of the main *chakras* in the body and the spirals and spirillae of the 'atom' (Anu). It indicates the way that the *saṃskāras* conveyed by the 'atoms'[14] can vivify the petals of the lotus blossoms with the attributes that they are the repositories for.

As each spirillae conveys a different combination of Elements utilised by the *saṃskāras* they are all of different density, intensity and colouring. In some there is a barely noticeable flow of *prāṇa,* in others the torrent is strong and forceful, because that is the *saṃskāra* that is being presently experienced. One can therefore only depict an idealised picture.

The main spiral conveys the major *saṃskāric* quality expressed, e.g., the Watery Element. The subsidiary spirals would then convey refinements of this quality, introducing the gradations of the other *skandhas* or *saṃskāras* that might be present, Fire, Earth, Air and Aether. We can see therefore that the Aetheric quality would be conveyed by the finest spiral 'tube', if indeed any such quality is generated by the person. (Torrents of energy are a more correct description than 'tube'.) We see therefore that the Element conveyed in the second largest spiral would represent the subsidiary major quality of the *saṃskāra,* which for the average person would be either Fire, Water or Earth. (Making the *saṃskāra* either *kāma-manasic*, or simply *kāmic.)* The motion within all spirals travel in unity, where a major *saṃskāra* encompasses subsidiary ones. We can be viewing the present, the past, or that which will flow into the future.

The *saṃskāras* do not stay the same all the while but are acted upon by various energies as they spiral onwards. They are called to the

---

14 The 'atoms' *(bījas)* being considered here are the permanent atoms for each human unit, which will be described later.

surface of experiential awareness in the present moment via the *chakra* concerned at the specific stage that the energy they express must be experienced by the consciousness concerned. What is experienced at any consciousness-moment is what technically is called 'a *bīja*', but we see from the above that it is but a small part of a far larger 'atom' of expression. In the minuscule is also seen the vaster perspective.

Esoteric students can also later meditate upon the correlation between the corresponding spirals governing planetary manifestation and the way that the Lords of Shambhala govern the evolution of all lives upon earth by means of them. The sentient and conscious attributes developed by the five kingdoms of Nature being the *saṃskāras* reaped by these Lords of Life.

The clairvoyant investigation of the Anu

As a consequence of their decades long clairvoyant investigation of the structure of the Anu,[15] Besant and Leadbeater state:

> It will be seen that the Anu is a sphere, slightly flattened, and there is a depression at the point where the force flows in, causing a heart-like form. Each is surrounded by a field.
>
> The Anu can scarcely be said to be a "thing" though it is the material out of which all things physical are composed. It is formed by the flow of the life-force and vanishes with its ebb. The life-force is known to Theosophists as Fohat, the force of which all the physical plane forces are differentiations. When this force arises in "space," that is when Fohat "digs holes in space"—the apparent void which must be filled with substance of some kind, of inconceivable

15 The term is derived from H.P. Blavatsky, *The Secret Doctrine,* (Theosophical Publishing House, London, 1888, 2005), Vol. 1, 148, where she states: 'The *Primordial Atom (anu)* cannot be multiplied either in its pregenetic state, or its primogeneity; therefore it is called "SUM TOTAL," figuratively, of course, as that "SUM TOTAL" is boundless. (See Addendum to this Book.) That which is the abyss of nothingness to the physicist, who knows only the world of visible causes and effects, is the boundless Space of the Divine *Plenum* to the Occultist. Among many other objections to the doctrine of an endless evolution and re-involution (or re-absorption) of the Kosmos, a process which, according to the Brahminical and Esoteric Doctrine, is without a beginning or an end.' Blavatsky is concerned with the 'atom' embodying the cosmic dense physical plane, whereas the clairvoyant investigation is focussed upon its minute counterpart on our terrestrial sphere.

tenuity—Anu appear; if this be artificially stopped for a single Anu the Anu disappears: there is nothing left[16]...In order to examine the construction of the Anu, a space is artificially made[17]...the surrounding force flows in, and three whorls immediately appear surrounding the "hole" with their triple spiral of two and a half coils, and returning to their origin by a spiral within the Anu; these are at once followed by seven finer whorls, which, following the spiral of the first three on the outer surface, and returning to their origin by a spiral within that, flowing in the opposite direction—form a caduceus with the first three. Each of the three coarser whorls, flattened out, makes a closed circle; each of the seven finer ones, similarly flattened out, makes a closed circle. The forces which flow in them again come from "outside," from a fourth-dimensional space. Each of the finer whorls is formed of seven yet finer ones, set successively at right angles to each other, each finer than its predecessor: these we call spirillae. (Each spirilla is animated by the life-force of a plane, and four are at present normally active, one for each Round. Their activity in an individual may be prematurely forced by yoga practice.)

In the three whorls flow currents of different electricities; the seven whorls vibrate in response to etheric waves of all kinds—to sound, light, heat, etc.; they show the seven colours of the spectrum; give out seven sounds of the natural scale; respond in a variety of ways to physical vibration—flashing, singing, pulsing bodies, they move incessantly, inconceivably beautiful and brilliant[18]...Force pours into the heart-shaped depression at the top of the Anu, and issues from the point, and is changed in character by its passage; further force rushes through every spiral and every spirilla, and the changing shades of colour that flash out from the rapidly revolving and vibrating Anu depend on the several activities of the spirals; sometimes one, sometimes another, is thrown into more energetic action, and with the change of activity from one spiral to another the colour changes.

The Anu has—as observed so far—three proper motions, *i.e.*, motions of its own, independent of any imposed upon it from the outside. It turns incessantly upon its own axis, spinning like a top; it describes a small circle with its axis, as though the axis of the spinning top moved in a small circle; it has a regular pulsation, a contraction and

16 Occult Chemistry, 13.

17 Ibid., 14.

18 Ibid.

> expansion, like the pulsation of the heart. When a force is brought to bear upon it, it dances up and down, flings itself wildly from side to side, performs the most astonishing and rapid gyrations, but the three fundamental motions incessantly persist. If it be made to vibrate, as a whole, at the rate which gives any one of the seven colours, the whorl belonging to that colour glows out brilliantly.[19]

This clairvoyant description of a normal ultimate physical atom will be helpful to understand the nature of the containment of *saṃskāras*. It should however be noted that the term *Fohat* defined above as 'life-force' is only a correct term to use if considered cosmically, hence needs explication.

Fohat is the coalition of Intelligent forces through which cosmic Ideation, or cosmic Intelligence, impresses upon substance, thus forming the various worlds of manifestation. It is the Divine Thought. H.P. Blavatsky states: 'When the "Divine Son" breaks forth, then Fohat becomes the propelling force, the active Power which causes the ONE to become TWO and THREE — on the Cosmic plane of manifestation. The triple One differentiates into the many, and then Fohat is transformed into that force which brings together the elemental atoms and makes them aggregate and combine.'[20] Blavatsky further states that Fohat 'is One and Seven, and on the Cosmic plane is behind all such manifestations as light, heat, sound, adhesion, etc., etc., and is the "spirit" of ELECTRICITY, which is the LIFE of the Universe. As an abstraction, we call it the ONE LIFE; as an objective and evident Reality, we speak of a septenary scale of manifestation, which begins at the upper rung with the One Unknowable CAUSALITY, and ends as Omnipresent Mind and Life immanent in every atom of Matter'.[21]

Consequently Fohat is the driving energy in cosmos producing the appearance of phenomena. It is a Fiery force (*śakti*) emanating from the cosmic mental through the cosmic astral plane via a triune energy (cosmic electricity) that when impacts via the three higher systemic planes precipitates the latent substance (*mūlaprakṛti*) into activity as

19 *Occult Chemistry*, 13-14.

20 *The Secret Doctrine, Vol. 1*, 109.

21 Ibid., 139.

the demonstration of material power. It is the energising expression of the Divine Thought of the third Logos (Brahmā) that produces the activity of the seven systemic planes of perception. It is the cause for their manifestation. It can be considered the expressed power of Lord Agni (the Deva Lord governing the Element Fire in cosmos).

Fohat is cosmic electricity in its primordial essentiality. Literally it is the Fiery electrical energy that is the basis to the creative potency from the Mind of the embodying Logos, and which emanates as the breath of Light. It is the expression of that dynamic Will that translates as electrical phenomena in all its diversifications when conveyed through the medium of the *nāḍī* system of etheric space. Fohat is the driving Fiery force of the enlightened Mind in action, and *ākāśa* is the substance (cosmic vitality) that is moved by this electrical Wind. Fohat can be viewed as Monadic Ideation, the energy of Logoic Thought, in its creative, sustaining or destructive aspects.

Fohat is ever-present and active from the primordial beginning of a *manvantara* to *pralaya,* and becomes quiescent, or latent, sleeping in *pralaya* until the next *manvantara.*

We see therefore that all manifesting phenomena normally observed can be viewed to be electrical in nature. All chemical compounds for instance, are the result of the exchange of forces between the positive nucleus of atoms and the negative electrons. Atoms interrelate when there is a surfeit or excess of electrons in order to balance the electrical charge. Light is an electromagnetic wave of energy, whilst thoughts are also composed of light energy, though existing upon the mental plane.

With respect to the manifestation of a human unit however, the energy used emanates from the Sambhogakāya Flower as the (positive) Ideating Thought, the Will-to-Be, which manifests as the *jīva,* the Life-force that sustains the Life and manifestation of the (negative) evolving personality. That will seeds the *saṃskāra* bearing *bījas* that propel the course of events of that personality's emotional-thought life because they produce a strong tendency to think or act in a certain way. Thus is the course of the *karma* to be set. The personal free will can however counter this tendency, and so *bījas* of new *karma* can be created.

Concerning the Anu, or 'ultimate physical atom' ('UPAs'), physicist Stephen Phillips states:

> As will be seen shortly, the UPAs making up a proton or neutron are, as subquarks, bound by strings both inside a constituent quark and externally into subquarks in the two other quarks. This means that they are monopole sources and sinks of two different types of hypercolour gauge fields, one responsible for confinement of subquarks in quarks, the other responsible for confinement of quarks in baryons like protons and neutrons. The reported 'change in character' of the force binding UPAs together as it enters and leaves a UPA may be due to these two different hypercolour gauge fields. It may also be due to the differences in the hypercolour flux densities inside two or more flux tubes terminating on the monopole, these tubes carrying different quanta of hypercolour flux because the monopole acts as a source or sink of flux. (see ESPQ, p. 74).
>
> Besant and Leadbeater said that each type of UPA was 'surrounded by a field.' According to the string model, a monopole is the end-point of a vertical excitation of the Higgs field, the vortex consisting of circulating currents of Higgs bosons. This observation therefore probably refers to the Higgs field. According to quantum theory, subatomic particles are surrounded by a field of virtual particles, which spontaneously materialize from the vacuum, briefly borrowing their energy from the vacuum and then annihilating one another and returning their energy to the vacuum. Although all these fields except the field of virtual electron-positron pairs around an electrically charged particle are a short-range field of virtual particles, they are not the likely explanation of the field observed around a UPA because virtual particles exist only momentarily and cannot be observed directly, according to the uncertainty principle.[22]

I quote Stephen Phillip's work here to point out the relation between the ultimate physical atom and the quark of modern QED theory. Those who wish to gain an insight into the relationship between the clairvoyant investigation of the various atoms of the periodic table and modern physics need to study his works, the detail of which lies outside my book.[23]

---

22 Stephen M. Phillips, *ESP of Quarks and Superstrings*, (New Age International, New Delhi, 2005), 62. Note that some of the technical terminology utilised here will be explained in chapter nine.

23 See also his first book: *Extra-Sensory Perception of Quarks,* (The Theosophical Publishing House, Madras, London and Wheaton, Ill, 1980). The writings of physicist Chris Illert can also be consulted, plus the book by H.J. Arnikar, *Essentials of Occult Chemistry and Modern Science*, (The Theosophical Publishing House, Adyar,

Concerning the actual spiral whorls of the Anu, as shown in figure 36, we see (utilising Philip's deductions upon Besant and Leadbeater's *Occult Chemistry)* that the first level spiral (whorl or helix) contains 1680 circular turns (336 per 360 degree revolution, where each whorl makes five revolutions). This number is important because 1680 is 7 x 240, where the number seven relates to the septenaries of Life, and the number 240 = 12 x 20, to the expression of the energies of the twelve signs of the zodiac (or Creative Hierarchies) manifesting via two cycles of perfection (2 x 10)[24]. It is that which generates the perfection of the second Ray of Love-Wisdom (20). The number 1680 is also 105 x 16, where the number 105 is the number of the Lords of Flame that embody the attributes of a planetary Head lotus multiplied by the number signifying the Christ principle, the embodiment of the second Ray of Love-Wisdom ($16 = 2^4$). More specifically the number 16 refers to the major petals of a Throat centre that emanates the creative Fires. The number 336 = 7 x 48, where the number 48, from one perspective, relates to the number of petals to one lobe of the Ājñā centre, the organ of inner vision, focussing and directing power. The number 48 is also significant, as it is the sum of the major petals of the *chakras* below the Ājñā centre, the *prāṇas* of which are absorbed into either lobe of the Ājñā centre, depending upon whether they bear *iḍā* or *piṇgalā* attributes. The number 48 is also half of 96, the base number governing the expression of each of the major *chakras*, and is also the foundation number for the constitution of the Head lotus (to which there are 11 x 96 petals). The number 7 x 48 therefore relates to the integration of the *prāṇas* of the seven major *chakras*. The number 336 x 5 = 1680, further subdividing these *prāṇas* in terms of the attributes of the five sense-consciousnesses, and the entire schema governed by the Dhyāni Buddhas. (The qualities of which are explained throughout *A Treatise on Mind.*)

There are 25 circular turns to a major ('extra-spiral') at the first order or level, producing 176 turns for the subsequent spirillae. The number 25 = 5 x 5 relates to the expression of the mental principle (the fifth plane of perception) and its extension in all fields of mind. The number 176 = 88 x 2, where the number 88 symbolises the expression

---

Chennai, India and Wheaton, Ill, 2000).

24 The number 10 here can also relate to the grouping of 7 plus 3 spirals to the Anu. There are also right and left turning spirals to consider.

of spiral-cyclic energies that produce the evolution of consciousness via the evolutionary process in Nature. Concerning the whorls of the seven electro-spirals there are 25 circular turns to each first order spiral and 175 circular turns to the subsequent spirillae. Here we have the continuation of the principle of mind, cycled through seven levels of expression (175 = 7 x 25).[25]

Rolf Jackson states:

> The anu is observed to consists of 10 spirals, lying next to each other in a highly dynamic flow of energy. Each of the spirals in Figure 10,[26] is itself a spiral. And according to observations this structure of nested spirals continues in seven layers, giving rise to a structure resembling Figure 11.[27] Note however that the actual structure is not spirals wound *around* other spirals, but rather spirals that are themselves wound into spirals. These nested spirals are referred to as spirilla of order n, where the first order is what is seen in Figure 10, and the 7th order is what can be seen in Figure 11 as a line of dots (the innermost spirilla).
>
> This means that the whole spiral can be unwound into an enormous circle of the tiniest imaginable dots. The dots appear to the clairvoyant as the absolute basic building block of all substance (physical, astral, mental etc.). And when analyzed these dots actually appears to be bubbles of absolutely nothing. And the bubbles should not be thought of as a membrane but rather like bubbles in water, where the bubble is defined by the *absence* of water.
>
> When examining the vacuum, or the background in which the bubbles exist clairvoyant observations indicate that the vacuum itself is of a density of an altogether different scale than the substances explored. It thus seems that matter is not "something" floating in "nothing", but rather the opposite. This forces us to alter our notion of space and matter almost to the point of reversing them. Thus matter is the *absence* of the dense substance of the vacuum, and what we consider to be vacuum, may not be empty, but full.[28]

---

25 Most of these numbers have been explained in volumes 4, 5 and 7 of *A Treatise on Mind,* and will be further elaborated in my forthcoming series *The Astrological and Numerological Keys to the Secret Doctrine.*

26 My figure 37, of Babbitt's version of the atom.

27 My figure 35, of the spirillae of the atom.

28 Rolf Jackson, M.Sc.E., *The Conscious Universe,* (Gaia Consciousness Institute, 2002), 8-9. The information is extracted from *Occult Chemistry* by Besant and Leadbeater.

One can surmise that ideally the main spiral of energy travels the way of The Golden Spiral, which is a product of the Fibonacci series: 1, 2, 3, 5, 8, 13, 21, etc. The spirillae will revolve around the spiral and each other according to energy content vitalising them. The law of Economy determines the size and closeness to each other.

Explaining the structure of the entire 'atom' obviously incorporates concepts that are alien to the normal scientific community and to traditional Buddhism. A major source for revelation as to the construction of the 'atom', and from which much of what is here derived, was first presented by Edwin D. Babitt in his book, *The Principles of Light and Color as Revealed By the Material and Spiritual Universe*, first published in 1875. The main diagram from that book is presented as figure 37 below. Babbitt's book explains the nature of the 'atoms' in detail, how they combine with each other, transmit colours and how the various laws of nature can be explained through analysis of their qualities.

In this figure by Babbitt we see that because the extra-spirals conduct a more lethargic type of energy (frictional energy, producing heat that is expansive in its action) than the intra-spirals, they are not so tightly packed and easily come into contact with other atoms. They are thus called 'extra-spirals' or *thermo-spirals* because they convey the heat that sustains the atom, causing it to keep its integral shape.

What Babbitt considered to be the 'atom' the authors of *Occult Chemistry* labelled the 'ultimate physical atom' (Anu). They gave examples of the elements of the periodic table wherein many Anu's are combined in various arrangements to make the different atoms observed in Nature, of which there are *seven* main types: spikes, dumb-bell, tetrahedron, cube, octahedron, bars and stars. The simplest atom (hydrogen) is composed of 18 Anu's. I shall not deal with these atoms at all, as they are adequately explained in *Occult Chemistry* and the works of Stephen Phillips.

The various spirals can be thought of as tubes of etheric matter that convey *prāṇic* energy within them. The most economic number of spirals that can effectively express all the energies of the Mother (the reflected universe) is, as seen from the preceding pages, governed by the number *three*. Various triadic relationships have been presented earlier. This number can also represent the three concrete planes associated with the material world; the solid, liquid and gaseous. We also have the physical, astral and mental planes in the body of the solar

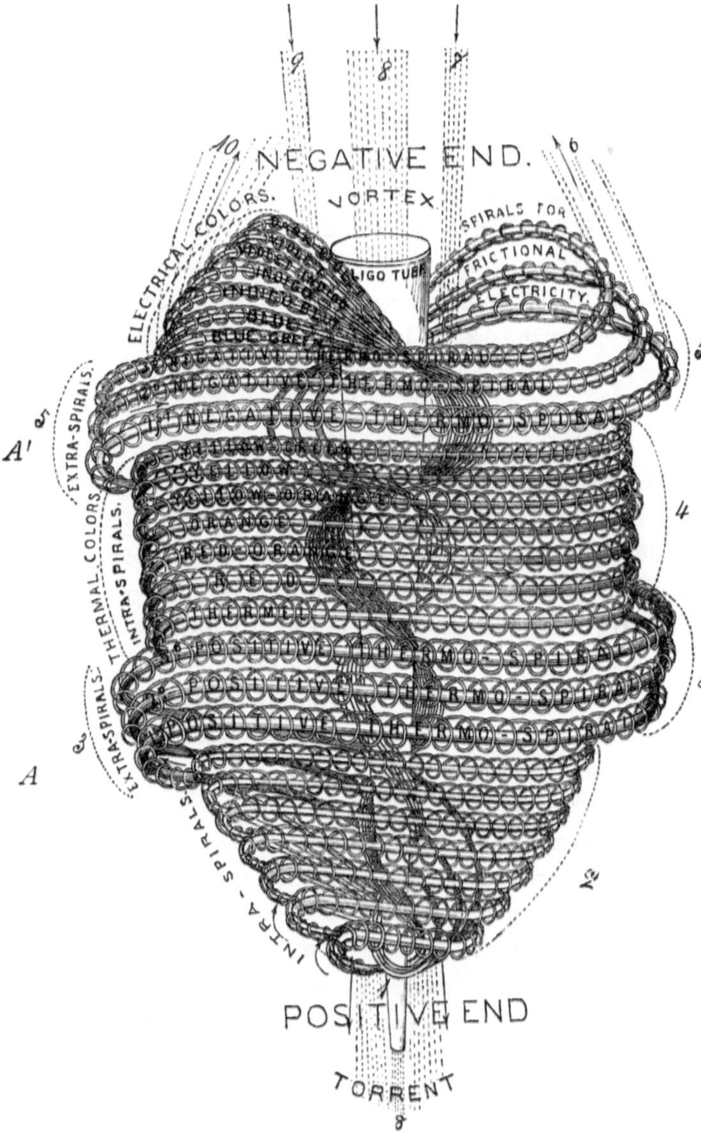

Figure 37: Babbitt's view of the atom[29]

29 Edwin D. Babbitt, *Principles of Light and Color,* (The College of Fine Forces, New Jersey, 2nd Ed., 1896), fig. 135, 102. In relation to this figure Babbitt states that this is the 'general Form of an Atom, including the spirals together with influx and

Deity and their correspondences in all the realms of being/non-being, therefore also in the atomic world. There are thus protons, neutrons and electrons as the major components of that world, as well as gamma rays, alpha and beta particles, as the major components in the disintegration of these atoms.

It was also shown above how the trinities are incorporated as part of a group of five, which accounts for the expression of the energies of the mind, which is intrinsic to the Mother's domain. The number five is the number governing the attributes of mind, which eventually evolve into the wisdoms of the five Dhyāni Buddhas. The mental plane is the fifth counting from above down and also from below up, when the dual aspect of mind is counted, as well as the separation of the physical plane into an etheric and a dense material portion. The mind is symbolised by the pentagram, where its two hands and feet represent the four empirical, concrete sub-planes of the mental, and the apex represents the head, the domain of the abstract Mind.

Spiralling round at right angles to each of the major channels of forces are the smaller channels called spirillae by Babbitt. If the 'atom' is viewed esoterically as embodying the physical form of a solar system, then the spirals and spirillae manifest as the seven planes of perception and their sub-planes. Through these flow the energies to and from the sub-planes of whatever plane of perception that the 'atom' is receptive to, and from cosmic sources, such as the energies of the twelve signs of the zodiac. The spirillae can also be considered to convey the four sub-Elements of what main Element the atom conveys. (Looking again to the significance of the number five above.)

The correlation of this solar 'atom' to the Anu observed by clairvoyants is that its spirillae can channel any of the energies from the four etheric sub-planes, consequently from the Throat, Heart or Solar Plexus centre levels of expression. Such energies then also help determine the properties of the Anu concerned, especially when arranged in the specific combinations determining the different chemical elements.

---

efflux ethers, represented by dots, which pass through these spirals. The spirillae with their still finer ethers are not shown. The spirals must pass around atoms in the appropriate direction in different substances, which fact will account for what chemists call dextro-rotary, rotating to the right, or laevo-rotary, rotating to the left'. In the first edition of his book Babbitt has here what he terms a 'Ligo tube', but is rightfully omitted in the second edition.

In terms of human evolution:

- The triune *Mother* stage (the extra-spirals) governs evolution and related conditionings until Individualisation occurs. These spirals relate to subhuman consciousness, e.g., mediumistic abilities in humanity.
- The *Son* stage (intra-spirals) governs the unfoldment of human evolution from the time of Individualisation to the full development of intellectual faculties and related qualities in any incarnate person.
- The *Father* stage (the central vortex) governs the path of Initiation, relating to the unfoldment of supra-human consciousness.

The two groupings of extra-spirals (or thermo-spirals) of this 'atom' are concerned with the stages of the progress of evolution up to the birth of self-conscious humans. They thus relate to the evolution of the three lesser kingdoms of Nature. From a *deva* perspective, the first series concerns the progress of the Elemental lives animating the bodies of manifestation of the three lesser kingdoms of Nature. It concerns the progress of the forms, the basic substance utilised by the animating Lives. The second series concerns the evolution of the Elemental lives constituting the human periodic sheaths, the dense, astral and mental bodies. It implicates the final preconditioning when the animal kingdom can finally gain their liberation and become human.

From the perspective of a planetary 'atom', the intra-spirals are associated with the general evolution of consciousness upon the seven planes of perception. The first and lowest septernary are principally concerned with the evolution of the intelligence incarnating into the forms provided by the feminine *deva* kingdom. We thus have the progress of the animating Lives, the seven Creative Hierarchies as they manifest via incarnate forms. The middle tier (labelled 'thermal colours') concern the evolution of these Creative Hierarchies via the evolution of the attributes of the fourth Creative Hierarchy (humanity) in the field of consciousness. Consequently the seven Ray attributes are developed and refined. The topmost tier concerns the process of liberating the seven Creative Hierarchies from incarnate expression.

Babbitt viewed combinations of these Anu's as spirals nesting within other spirals, according to certain principles governed by the nature of the associated spirals. These combinations are shown in figure 38.[30]

---

30 The figures are taken from pages 103-119 of Babbitt's book.

Taking an *a priori* view of these arrangements of the Anu's in subatomic space then one can deduce that when combined in vast fields in interstellar space they represent the electromagnetic field extending from each sun, or light bearer. The Anu field is created as the energy of light is projected by a sun into the vast distance of cosmic space. The Anu then takes the *attributes of a photon,* the elementary particle that is the quantum of such a field, bearing the radiation of light and its accompanying heat. The spiral nature of the Anu's constitution when interrelated in the sheets of 'atoms' depicted accounts for the wave motion of light, as energy passes through one spiral to the next from 'atom' to 'atom', and yet can also simultaneously manifest as a particle when an Anu (or rather, a pair of them) is taken as a separate entity. This accounts for the wave-particle duality observed for photons and which has mystified physicists ever since the phenomena was observed by them. As the Wikipedia article states:

> A single photon may be refracted by a lens and exhibit wave interference with itself, and it can behave as a particle with definite and finite measurable position or momentum, though not both at the same time. The photon's wave and quanta qualities are two observable aspects of a single phenomenon, and cannot be described by any mechanical model; a representation of this dual property of light, which assumes certain points on the wavefront to be the seat of the energy, is not possible. The quanta in a light wave cannot be spatially localized.

A photon is said to be stable, have zero rest mass and charge, with a spin of one and moves at the speed of light in a vacuum. The wavelength of the photon is determined by the amount of energy imparted to the Anu's as the wavefront moves, the stronger the energy imparted the smaller the size of the Anu carrying the energy, the weaker the energy the larger the size of the Anu, the Anu's appearing along the direction of propagation of the field of photons. The stronger energy produces a more compact organisation of the spirillae, hence producing a shorter wavelength. Conversely a weaker energy produces a larger or longer wavelength. There is zero mass because Anu's are bubbles of 'nothing', and they have zero charge because positive and negative forces are evenly balanced within them. They have a spin of one because the energies of the Anu are circulating around its spirals, whilst the Anu has been observed to spin in the abovementioned clairvoyant investigations.

If we look at the energies moving transversely from Anu to Anu, from a spirillae of a particular colour (frequency) to the corresponding spirillae, then we have the propagation of a electromagnetic wave, as the motion moves from Anu to Anu in wave formation. (Moving from one complete cycle of a spirillae and then jumping to the next Anu.) The energies moving at right angles to this motion via the central torrent of each Anu produces the lines of magnetisation of a electromagnetic force. As light is constituted of seven subrays or hues, so there are seven spirillae to account for the progression of the full spectrum, whilst the three groups of such spirals to each Anu allows for the demonstration of the three *guṇas* of every energy expression. These *guṇas* represent the infrared, normal and ultraviolet spectrum.

This theory thus implies that the energies from the etheric domain generates light in the form of Anu's, producing electromagnetic waves that move through space to illuminate darkness. The Anu's (photons) are therefore the carriers of the energy as it makes it appearance in the phenomenal universe. When the light is extinguished the Anu's revert back into the etheric space from which they were formed. The atoms of all manifest forms, such as those of rocks, leaves, coloured clothing, have an inbuilt receptivity to the wavelengths of light of which the Anu's of their constitution are attuned to. Hence when the photons arrive the specific spirillae corresponding to a particular hue that is seen are excited, stimulated by the energies passing through. The spirillae corresponding to other hues are not thus stimulated, or else that part of the Anu lies dormant, resistant to the energy input. The fact that we see the reflected light of the hues not accepted by the atomic structure of the material concerned is accommodated by this view, where the spirillae absorb certain aspects of the energy passing through and reject the remainder, being the hue we see with our mind-eye interrelationship. When the Anu's appear the natural motion of the spirals and spirillae produce the vortex of energies through the centre, which draws upon added energies from the encompassing ether.

The sheets of Anu's are also nestled into each other, as depicted in Babbitt's book, with a positive end inserted into the opening of the negative end, therefore leaving the grouping of thermal colours exposed to convey the subrays of light and also the extra-spirals that convey the heat and warmth of the light.

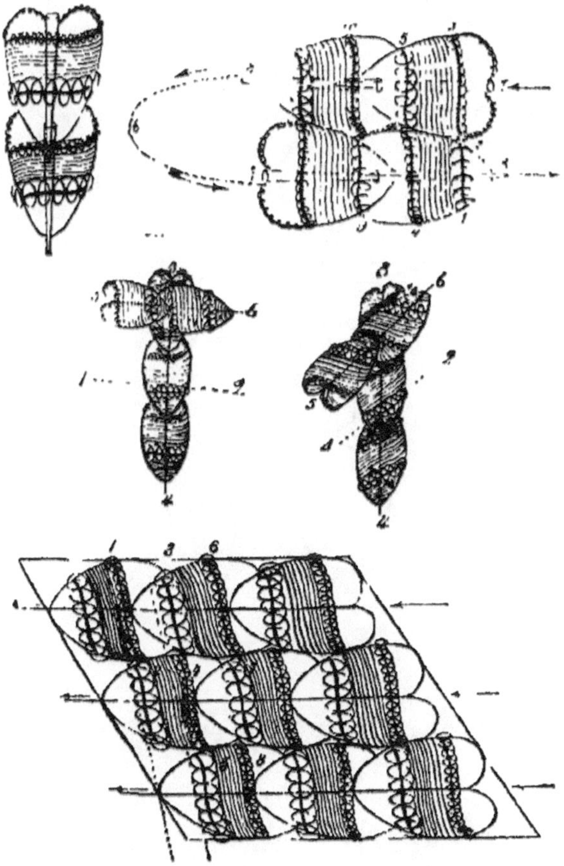

Figure 38. Various views of combinations of Anu's

When single Anu's are separated from the wavefront and interrelated in the manner described in *Occult Chemistry* we get the 'atoms' described by Besant and Leadbeater, which Stephen Phillips calls micro-psi atoms or MPA's.[31] Phillips then further elaborates them in terms of his quark theory, plus his hypothesis that there are three sub-quarks (which he calls omegons) to every quark.[32]

31 *ESP of Quarks and Superstrings,* 9-13.

32 Ibid., preface, x.

The Anu's considered here are the sub-quarks, Phillips' omegons, which if they dissociate from the transverse sheeting of Anu's and join in a triangular fashion, then they form the quarks mentioned in the texts. The nucleus of a Hydrogen atom is clairvoyantly observed with 18 such Anu's to form its structure.

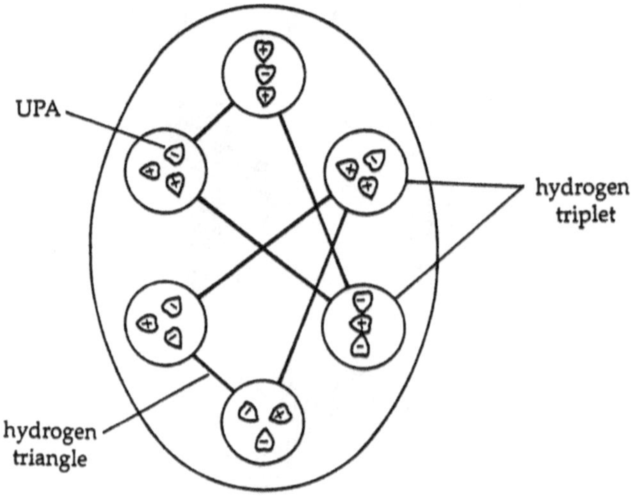

Figure 39: The hydrogen micro-psi atom

Figure 39 is derived from *Occult Chemistry* and taken from Phillips' book,[33] where he argues that this arrangement appears to be that of a deuteron, a neutron-proton integration.

If my hypothesis above is correct then it can be deduced that all observable phenomena is but composed of the energy of light, hence an expression of the attributes of mind, as the nature of thoughts is defined in terms of the quality of manifesting light.[34] What is illumined in the mind is what is perceived and consequently comprehended. What is not thus illumined relates to being in a state of ignorance. The process of evolution is to convert dark space (synonymous with ignorance), 'dark elementary matter' into spheres of light. An illumined Mind that stretches to encompass the vastness of space is considered enlightened.

33 *ESP of Quarks & Superstrings,* 44, figure 3.1.

34 See the various volumes of *A Treatise on Mind* for further elaboration on this subject.

Mind directing the attributes of light through the dimensions of being, projected via the force of mantra then accounts for the marvellous *siddhis* (psychic powers) possessed by the enlightened. They can also control the formation of the quarks constituting the atoms of chemistry, hence can control the appearance of 'things'. Consequently they have mastered all of the attributes of the phenomena of light right through to the physical domain and the atomic structure that composes it. A Master of Wisdom wisely directs *kuṇḍalinī* (internal heat, the energy of the thermo-spirals of the Anu) and the potency of light, via the force of mantra (utilising the powers of Mind) to compassionately effect changes in the subjective and material domains. Such a one has mastered completely all the attributes of mind and resides in the domains of Mind *(dharmakāya)*. Initiates of lesser degree however can become *siddhas* possessing psychic powers without full mastery of the attributes of mind/Mind.

The permanent atoms

The main concern in this chapter is not directly with the atoms of chemistry, but with the mechanics of *the physical and astral permanent atoms* of humans, as well as with the mental unit, and of the way they store *saṃskāras*. Each spiral and spirillae is attuned to a different frequency of energy. The concept is similar to, though upon a higher plane of perception than what was earlier described from *Occult Chemistry* for the individual Anu's: 'the seven colours of the spectrum; give out seven sounds of the natural scale; respond in a variety of ways to physical vibration—flashing, singing, pulsing bodies, they move incessantly, inconceivably beautiful and brilliant'. Each energy frequency is an aspect of a consciousness-attribute. They manifest as *bījas,* attracting to them atoms of similar coloured nature when activated by a conscious-volition. This happens in ways as depicted in Babbitt's book, where the spirals and spirillae absorb energies from the ethers. Hence any particular mental-emotion arises. Everything is after all but qualified energy.

Some of the information relating to the permanent atoms has already been presented in the section where the human *nāḍī* system was taken into account in relation to the spirals of the 'atom'.

Babbitt speaks of four grades of ethers, which he terms:

> Thermo ethers, which are said to flow in the thermo spirals and 'constitute the coarser grades of heat'.
> Electro ethers, the 'element of frictional energy' being connected to the axial portion of the thermo spirals.
> Thermo-Lumino Ether, 'used in the intra spirals which form the thermal colors'.
> Electro Lumino Ether, 'connected with the spirals of electric colors, and may be called the blue-green ether, the blue ether, the indigo-blue ether, and so on with the other colors'.[35]

He then states that these 'color ethers (or in other words light), move 186,000 miles a second, or not widely different from the swiftest electricities.'[36]

This information from Babbitt's book is presented so that the concept of these 'ethers' are not confused with that which is described in our esoteric texts as the ethers of the four etheric sub-planes of the physical plane, though the four above-mentioned ethers would represent the reflexes of the subjective ones. Manifesting via the four ethers of the etheric sub-planes are the mental-emotional *saṃskāras,* which also have their effects upon the substance of the dense physical form via the structure of the Anu's that substance is constituted of.

As the name suggests, the *permanent atoms* are permanently retained by the Sambhogakāya Flower (the Soul or Ego) for the duration of the symbolic 777 incarnations that it projects into manifest expression after the death of each personality.[37] They are the seed *bījas* via which a new incarnated individuality can be formed. This is due to the constitution of the spirals and spirillae, where the seeds of all *saṃskāras* of past karmic volition can be stored via their energy qualifications. This includes all forms of action and elementary substance utilised by the incarnating *jīva*[38] even before the

35 *Principles of Light and Color,* 109-110.

36 Ibid., 110.

37 See volumes 4 and 5A of *A Treatise on Mind* for an explanation of the number 777 in this respect.

38 *Jīva,* embodied Life or 'self', inherent Life force. An integrated conscious entity embodied by a Life force. It is the *prāṇa* that has become assimilated by an entity

Individualisation process of a human unit occurred. As outlined above all stages and levels of Life of the evolving consciousness can be represented when the 'atom's' structure is related to the reincarnating human *jīva,* and also to that of a solar or planetary Logos. The human Monad anchors its expression upon the second sub-plane of the cosmic dense physical plane and so expresses its evolutionary purpose therein via the *ātmic, buddhic* and mental permanent atoms that condition the activities of the three main tiers of the Sambhogakāya Flower. Herein lies the basis for much future study for esotericists, the Bodhisattvas that struggle to reveal formerly veiled truths to humanity.

The permanent atom upon the lower mental plane is called the mental unit as it only contains four main spirals and deals with the attributes of the empirical mind,[39] whereas the permanent atom upon the higher mental plane contains seven main spirals. The attributes of the four spirals of the mental unit were first developed by means of the aspiring development of an animal kingdom towards gaining the empirical mind prior to Individualisation of that kingdom into the human by means of the formation of the Sambhogakāya Flower. Concerning the mental unit D.K. states:

> The fundamental difference between the mental unit and the other two atoms consists in the fact that it contains only four spirillae instead of seven. This is brought about by the very facts of evolution itself, for the mental unit is the first aspect of the personality triad, or of man functioning in the human kingdom on the three lower planes. At his transference into the spiritual kingdom, these three aspects—the mental body, the astral body, and the physical body—are synthesised into the higher by a dual process:
>
> 1. His polarisation shifts from out of the lower three atoms into the Triadal atoms.

---

and converted into his own Life force, (or *prāṇic* fluid) by the action of the Heart. It is that which distinguishes one's *prāṇic* emanations from all others. In terms of consciousness it can be seen as one whose consciousness is separate, individual.

39 These four spirals respond to the energies of the concrete sub-planes of the mental plane, to the sum of the energies of the personality, symbolised by the quaternary. The mental unit is transient, an effect of a past evolution and allows *saṃsāra* to be sustained in consciousness.

2. The force which these atoms generate and embody is merged and blended into the higher force points.

A permanent atom is the positive nucleus or germ substance to the sheath wherein it is found. It is that which is the basis of form-building, and it is literally a vibrant point of force, emanating from the second aspect of the Monad, which aggregates to itself, and subsequently builds into form, the negative or third aspect. But it must here be remembered that this second aspect is itself dual, and that in considering the permanent atoms we are dealing with the feminine aspect of the second Person. The spirillae therefore are but streams of force, or second aspect vitality which circulates geometrically within the circumscribing wall of substance, composed of third aspect force or substance. What has been said of objectivity, or of the cosmic atom can be equally well predicated of the permanent atom of man the microcosm:

"The primordial ray is the vehicle of the divine Ray."[40] *Negative force forms a receptacle for positive force. Atoms are but force centres, and the centres as we know of them are but aggregates of force points which have reached a specific point in evolution, and are responding to the first great aspect in some degree, or to electric fire.*[41]

From the above we see that there is a distinction between 'the circumscribing wall of substance, composed of third aspect force' and the vitality pouring through that wall. This wall can be considered the substance of the fourth ether that contains the energies pouring through the Anu. Hence it should not be the focus of meditative attention, but rather the manifesting *prāṇas*.

The *chakras* are known because of the way that they appear in our consciousness when vivified by the Fires of mind. Unawakened *chakras* are *laya* centres. The *laya* centre is the point where primordial substance differentiates and gives birth to formed space. It is the place of emergence of manifestation, the *maṇḍala* of things. A *laya* centre is also a point where substance is absorbed back to its ineffable source, etheric space. Hence it can be also seen as the dividing point between one dimension of perception and the next. All *chakras* contain at their

40 The footnote given here is: *Secret Doctrine,* 1, 108.

41 *A Treatise on Cosmic Fire,* 526-27.

hearts a *laya* centre, a concept that can be extended to include being part of the planetary, systemic or cosmic *naḍī* system. This incorporates any sacred spot or site on the earth that has not yet reached its full potential, or any place that is to become a place of importance in the future evolution of humanity or of any kingdom of Nature. A *laya* centre can also represent the ultimate state of quiescence, *śūnyatā*.

We can therefore glean that the centres, once awakened, pour forth electric Fire via the etheric domain to form the central torrent of the Anu's. This torrent gives them their overall intensity or expansiveness and rate of motion, the vibrancy of the colouration of the spirilla and spirillae that are vivified. Thus are produced the overall characteristics of the 'atom(s)' concerned. The 'force points' are the minor and major petals of the *chakras*. The 'aggregates of force points' are literally units of the expression of Mind, *maṇḍalas* via which the energies of mind can effect changes in the realms of form. The 'force points' can thereby sweep (galvanise) the 'force centres', Anu's, into activity by incorporating them into aggregate forms, and condensing them into the appearances that we perceive via sense contact. The 'specific point in evolution' happens at the appropriate cycle, in accordance with the manifesting *karma*. This is directed by the Lipikas (agents of *karma*)[42] via the representative *devas* that appropriate the Anu's upon a mass scale to produce observable phenomena. The Sambhogakāya Flower acts as a Lipika Lord for the human unit that it 'flowers' into manifestation via the *chakras* that are its mechanism of control of the human form.

The four spirilla of the mental unit are directly related to the four lower sub-planes of the mental, and convey the *saṃskāras* of the empirical mind. They account for the reifying nature of that mind, and thus for the inability for most people to manifest the clear logical abstract thought associated with the enlightened Mind. A bridge in consciousness must first be built between the spirals of the mental unit and the mental permanent atom to do so. The bridge building process is accomplished mostly through logically achieving focussed thought upon abstract issues, such as the cause of the universe, and is accelerated through yoga-meditation techniques *(dhāraṇīs)*. This process

42 They shall be explained later.

is provided in Hindu and Buddhist texts on meditation, and formalised in Bailey's books under the rubric of 'building the *antaḥkaraṇa',* the *antaḥkaraṇa* being defined as the consciousness-link. Until such bridge building has been accomplished by average humanity via the gap in consciousness one cannot expect of them to manifest the clear compassionate considerations of a Bodhisattva.

The energies manifesting through the permanent atoms, and consequently the substance they attract, allows building the sheaths through which the entity incarnates, be that entity a Logos or a human unit.

The mode of retrieval of *saṃskāras* in the spirillae from the permanent atoms is via the directed will of the Sambhogakāya Flower. They are also pulled into activity by means of the mental-emotional activity of the personality, who is manifesting activities bearing the same, or similar, energy qualification as that possessed by the *bīja* of the *saṃskāra* found in a particular spirilla of the permanent atom. These *bījas* convey the impression or expression of the mental-emotional and physical activities manifested in previous lifetimes.

In a person the physical permanent atom is found in the region of the pineal gland, the physical externalisation of the third Eye, the organ of clairvoyance.

Each spiral channels a different octave of expression of energies to its companions, coloured by a characteristic hue. The magnetic impulses of the energy following through these spirals (controlled by the *deva* Lords of *karma*) attract other normal Anu's to them, causing a symphony of mutual vivification, producing a manifest form. Each spiral responds to group energy flowing from the various kingdoms in Nature, depending upon the quality of the spiral and of the energy that it can receive.

The hues of energies then denote the qualities of the *saṃskāras* conveyed, which, once drawn upon (energised) at any particular moment, magnetically attracts to them the Anu's of similar frequency. The massed field of expression of these minute 'atoms' produce the flavour of the emotions, mental imagery or physical urge noted by the individual at any particular moment in time. The massed field of Anu's produce the auras of individuals that can be noted by any clairvoyant. In the consideration of this subject matter it is important to note a warning to all esoteric students. It is important to get an empirical understanding of the nature of how permanent atoms function, but not to manifest an undue emphasis upon them. The permanent atoms are concerned with that which is material in nature and undue focus can materialise forces relating to the physical plane or the mental-emotions which are antithetical to the proper spiritual advancement of the individual. Undue obsession with a certain *saṃskāric* quality can result if the specific atom is unduly stimulated, and this is in the nature of the mode of activity of the forces of materialism, the dark brotherhood, who deal exclusively with the formed side of things. The forces pertaining to the higher domains should instead be invoked, to produce a positive down flow of force.

D.K. states:

> A close concentration upon and study of the spirillae and atoms would be scientifically and technically interesting and possible, but would not lead to increased spiritual development but to personality emphasis and, therefore, to increased difficulty in the treading of the Path. The more advanced a disciple, the more dangerous such emphasis and preoccupation would be, whereas the scientist or the aspirant upon the Probationary Path could study such matters with relative impunity because he would not bring in the energy which could galvanise these "points of force" into dangerous activity.[43]

Much will be confusing if one does not remember that both a person and the atom can be considered to be an aspect or unit of energy. As such, one continually receives energy of different intensities and hues, and then retransmits that energy after colouring it with one's own quality. People must learn over time to refine and intensify their energy fields, the vibrancy or hue of the aura of their own minds and personality equipment. In this way they evolve. The vibrancy of the auras of most people are quite dull because they spend very little time in refining their mental-emotional equipment or to achieve high aspiration of thoughts. It takes many millennia of evolution to transmute the vibration of thermal reds and associated greyish hues into the electrical hues of the first group of spirals.

43 A. Bailey, *Discipleship in the New Age, Vol. 1* (Lucis Publishing Co., New York), 766.

When a person or Logos incarnates they do so by means of a physical permanent atom. A being's qualities and possible conscious development depends upon the number and quality of spirillae of the spirals that were energised in former lives and which are now re-energised, with a view of what is to unfold in the future ones. According to the type of energy that such an atom can channel, so it attracts the appropriate physical matter to it (by empathy of vibration). This 'matter' consists of atomic substance that is non-periodical and which is stored in the universal store of all such substance found in Nature. A physical body is thus formed, be it subtle or gross, that can fully express the qualities desired by the incarnating entity according to the empathy of what manifested in the past.

The Sambhogakāya Flower utilises three permanent atoms, one each for the physical, astral and lower mental planes respectively, to store the detailed information needed. Such information concerns the qualities gained in past lives, upon which the future life is based.

Each of seven spirals can draw energies from the particular plane of perception out of the seven that it is attuned to. They are thus associated with different stages of evolutionary development, and empathise with the expression of the associated Ray energy. The first stage of evolution refers to when the Mother aspect is dominant, the second stage to the Son aspect, and the third stage to the Father aspect.

A prototype generalised *bīja* can be considered for the human kingdom that depicts the nature of the evolutionary journeying of humanity moving against the direction of the base substance of *saṃsāra.* Overcoming the push of the oncoming stream of substance is what constitutes the nature of the path to enlightenment. This stream of substance, the *skandhas,* constitutes the composition of the forms that we incarnate into. When aspects of consciousness condition this substance, making it a unique part of the experiences of an individual then the associated *saṃskāras* are formed. The generated *saṃskāras* recede into the past when new qualities are built from the base *manasic* substance by consciousness. These qualities are stored in the spirals and spirillae of the primary *permanent atom* that represents the point of power of the major *chakra* that is drawn upon to express the major characteristics of the *saṃskāra.* The past volitions are thus stored therein, and which can automatically be drawn upon as a person spirals on by producing new mental-emotional characteristics.

One moves forward upon the spirals of evolution as old *saṃskāras* are transmuted into new, more vibrant characteristics. Until then one reincarnates by drawing upon the same spirillaic conditions, consisting of substance already characterised by *kāma-manasic* qualities generated in past cycles. To build new, more vibrant substance is therefore the onus of the path to light. The Lord of *karma* (the Sambhogakāya Flower) chooses from all available qualities stored in the spirillae that it wishes to try to convert into increasingly vibrant *saṃskāras*. Depending upon the success achieved in that life so it projects its next step on the path of eventual liberation by utilising the gain from that past life, in conjunction with a thought projection to the outstanding qualities still needed. The spirals of the relevant spirillae are thereby stimulated. Old *saṃskāras* are transformed, transcended or transmuted into the next desirable characteristic according to the dynamics of colour interrelationships. Thus consciousness flows from one vibrant state to the next, but the process in terms of human mental-emotional development is long and arduous, involving many cycles of pleasurable experiences mixed with suffering, and thence a conscious striving to overcome all difficulties. Such is the nature of human evolution.

In such energy flow much of the future has already been established. The fact is that generally the *saṃskāras* created are of a higher energy content than what was previously lived through. That which is the norm moves on with added qualifications to vivify spirillae and spirals that are capable of conveying the higher energy qualification. These spirillae are activated as energy moves in to vivify them. They call consciousness to them, being the future expression, and presage the time when the consciousness can reside in this arena more prominently.

Until then the consciousness-stream generally regurgitates and recycles states of awareness of a sluggish nature to it (people's inebriation with the past). As more lines of energy are thrown forward, due to inspirational, creative and aspirational endeavour, so eventually sufficient energy has been built in these forward spirillae to pull the consciousness to them. The personal-I then breaks with past lines of activity by letting go of attachments to them. The old types of *saṃskāras* are then transmuted, allowing residence in the new. Movement has therefore occurred in the consciousness-flow to another sector of the *bīja*/atom.

This analysis concerns the nature of the flow of *skandhas* and *saṃskāras* from life to life. The changes in costumes for each historical period one wears, as well as body types one incarnates into are of relative unimportance. What is of importance are the life situations and experiences wherein opportunity manifests to change old habit-patterns, attachments, desires and thought life.

In the new cycle of higher energy states, sluggish *saṃskāras* from the past have to be recalled (recycled) so as to be re-experienced in such a way that (hopefully) the related energies become more vibrant. The *saṃskāras* are thereby transmuted into new qualities. If this is not possible then oft painful repetitive action is the norm for some lives. Life aspects, such as addictive drug experiences or forms of emotionality, are then lived and relived until the personal-I concerned utilises personal will to change for the better. Change is always possible, because there is always the inner impetus from the Sambhogakāya Flower to be taken into account. The new intensified qualities to which one aspires may also have previously been generated as the high point of activity in past lives and will surface at the appropriate time in the new cycle. Such high points need to be stabilised in a new habit-patterning that becomes the norm. If such activity is not possible because of too much attachment to the ancient pull of sluggish *prāṇas* then the Sambhogakāya Flower normally shortens that life. The person is recalled to experience the astral heaven or hells, preparatory to rebirth from an higher energy perspective.

The nature of the flow of a human consciousness-stream has characterised the polemic of Buddhist syllogisms for ages, but how consciousness is stored and comes to manifest in each succeeding personality has never been properly conceptualised. The reason is obvious, because of the complexities shown here. Armed with this new key of the esoteric nature of the 'atom', plus many other new insights found in my writings, such as the nature of the Sambhogakāya Flower and the concept of Logoi, the Buddhist philosophy can finally be developed further than was ever possible before. The formerly veiled 'ear whispered truths' are consequently being revealed in a logical and transparent manner.

The blood's circulation, the nervous and *nāḍī* systems, reflect the spirals and spirillae of the physical permanent atom in the human body. (The heart's shape is similar to that of the Anu, and the circulatory system has a similar shape to the buddhic permanent atom, which is in the form of a figure eight, according to D.K.[44])

When the spirillae are taken into account, then it can be seen how each atom is 'tuned' in such a way as to predispose one to respond naturally to certain energies, the frequency of which helps generate personality traits, mental-emotional attributes. The attributes of the main spirals are attuned to the seven planes of perception and the corresponding spirillae to the substance of the seven sub-planes of the respective plane. The energies that the permanent atom of a human unit can accommodate and which flow in the spirals and spirillae thus relate to the level of perception that has been attained previously. For example, some people will have intellectual dispositions because all the spirals up to and including the mental spirals are vivified, the principal *prāṇas* flowing in them being Fiery rather than Watery. When the atom is vivified mainly to the (Watery) emotional spirals, then the person will naturally respond to astral influences and conditionings without true mental abilities. There will however be somewhat a susceptibility to buddhic impression, because of empathy of vibration between the respective buddhic and astral spirals and spirillae.

The various forms of purification, meditation, breathing exercises, and contact with the impulses of the Sambhogakāya Flower can greatly hasten natural evolution because they increase the vibratory rate of the energies coursing through the permanent atom, and that vivify the higher spirals in particular. This enables that atom to draw down and anchor intensified energies from the higher planes.

Inevitably the energy which flows through the spirals becomes greater than the atom can bear, which then becomes radioactive and finally disintegrates. Such a being is then liberated from *saṃsāra,* as was the Buddha.

The placing of the permanent atoms within the Sambhogakāya Flower, which D.K. terms 'the Ego' or 'the causal body', is described by him thusly in relation to the process of the gaining of enlightenment:

44 *A Treatise on Cosmic Fire*, 531.

> The causal sheath is to the clairvoyant therefore a sphere of vibrant living substance; within it can be seen three fiery points. At the heart of the sphere is a central blaze of light, emitting rays; these rays are given as seven in number, and play upon these points or circles (analogous to the electrons in the atoms of science[45]) and *at this stage* produce most effect upon the astral permanent atom. The physical permanent atom has a position relatively close to the positive centre, and the force plays through it, and passes on to the astral permanent atom in the form of five rays of parti-coloured light which blend with the intensely vivid hue of the astral permanent atom, and increase its intensity until the blaze is so excessive that it appears to the onlooker as if the two points blended, or the two electrons merged, and (in merging) produce such an intensity of light that they are seen as dissolving. The mental unit, having a position within the causal body analogous to the planet furthermost from the sun, becomes vibrant likewise, and the two other points (considered now as one) begin to interact with the mental unit, and a similar process is set up and is pursued until these two points—circulating around their positive centre—also approach each other, blend, merge, and dissolve. The centre of positive life gathers or synthesises the three points, and thus *the three fires of the personality* repeat on their tiny scale the microcosmic procedure as seen in the synthesis of electric fire, solar fire, and fire by friction, and only a blazing unit is left. This blazing unit, through the combined heat of its being, burns up the causal body, and escapes back on to the planes of abstraction. Thus man is the Path itself, and also the pilgrim upon the Path; thus does he burn, but is also the burning-ground.[46]

D.K. elaborates the nature of the permanent atoms in *A Treatise on Cosmic Fire,* pages 507-536, and states in part that:

> The permanent atoms on each plane serve a fourfold purpose as regards the central or egoic life:
>
> They are the distributors of a certain type of force.
>
> They are the conservers of faculty or ability to respond to a particular vibration.

45 In the early 1920's when D.K. was writing this book the electrons were the smallest atomic particles then discovered, and bear an electrical force. Nowadays many more sub-atomic particles are known, hence as I stated the better analogy is to photons.

46 Alice Bailey, *A Treatise on Cosmic Fire,* 513-14.

> They are the assimilators of experience and the transmuters of that experience into quality. This is the direct result of the work of the egoic Ray as it plays upon the atom.
>
> They hold hid the memory of the unit of consciousness. When fully vibrant they are the *raison d'être* for the continuity of the consciousness of the man functioning in the causal body. This distinction must be carefully made.
>
> We must always remember in studying these difficult matters that we are dealing with the logoic dense physical body and that:
>
> The mental unit is found in logoic gaseous matter.
> The astral permanent atom in logoic liquid matter.
> The physical permanent atom in dense physical substance.
>
> And they therefore have their place in matter of the three lowest subplanes of the physical body of the Logos. Consequently when in the process of evolution, and through initiation, man achieves the consciousness of the Spiritual Triad, and transfers his polarisation into the three triadal permanent atoms, he is simply able to function *consciously* in the etheric body of the particular planetary Logos.[47]

The Ray energies manifested by the human personality vivifies the physical permanent atom, those from the Sambhogakāya Flower act upon the astral permanent atom, whilst the Ray streaming from the Monad have a connection with the mental unit (via the Soul form). D.K. states that:

> The effect which they have is threefold, but is not simultaneous; they work ever, as do all things in Nature, in ordered cycles. The stimulation, for instance, that is the result of the action of the monadic Ray upon the mental unit is only felt when the aspirant treads the Path, or after he has taken the first Initiation. The action of the egoic Ray upon the astral permanent atom is felt as soon as the Ego can make good connection with the physical brain; when this is so the egoic ray is beginning to affect the atom powerfully and continuously; this occurs when a man is highly evolved and is nearing the Path. This threefold force is felt in the following way:
>
> *First.* It plays upon the wall of the atom as an external force and affects its rotary and vibratory action.

47 Ibid., 508-09.

Second. It stimulates the inner fire of the atom and causes its light to shine with increasing brilliancy.

Third. It works upon the spirillae, and brings them all gradually into play.

II. THE PERSONALITY RAY AND THE PERMANENT ATOM

The *Personality Ray* deals with the first four spirillae, and is the source of their stimulation. Note here the correspondence to the lower quaternary and its stimulation by the ego. The *Egoic Ray* concerns itself with the fifth spirilla and with the sixth, and is the cause of their emerging from latency and potentiality into power and activity. The *Monadic Ray* is the source of the stimulation of the seventh spirilla.[48]

The *devas* and the permanent atoms

Because the 'atoms' are concerned with the manifestation of physical phenomena, so they can be considered the corporeal forms of certain classes of *devas*. The law of *karma* therefore manifests via the expression of these forms with the certainty that it does. The 'atoms' consequently come under the auspices of the Lipikas, under whom work the Rāja Lords governing the manifestation of the planes and sub-planes of perception. Ultimately the source of this law and its effects stems from the third sub-plane of the cosmic mental, counting from below upwards. From here emanates the cosmic law of Karma. To gain further insight into this most esoteric of subjects it is necessary to quote at length from *A Treatise on Cosmic Fire*. D.K. states:

The permanent atom of the astral and physical planes is a sphere of physical or astral substance, composed of atomic matter, and characterised by the following qualities:

Responsiveness. This is its inherent power to respond to the vibration of some one of the Heavenly Men,[49] as it is transmitted via the deva, or Brahma aspect, of His threefold nature. The permanent atom finds its place within the sphere of influence of one or other of the great devas who are the Raja-Lords of a plane.

Form Building Power. These devas sound forth two syllables of the threefold microcosmic word and are each (on their own plane) the

48 Ibid., 71-72.

49 Planetary Logoi.

> coherent agency which gathers substance into form, and attracts matter for purposes of objectivity. The *astral* sound produces the microcosmic 'Son of Necessity,' and when it reverberates on the *physical* plane produces physical incarnation, and the sudden appearance on etheric levels of the seven centres. The building of the dense physical is the result of consequent automatic action in deva essence, for it must ever be borne in mind that man is essentially (as regards the physical plane) an etheric being, and his dense physical body is esoterically regarded as 'below the threshold' and is not considered a principle.
>
> *Relative Permanency.* In the seventh principle of all manifesting entities is stored up and developed capacity, acquired ability, and the atomic memory, or in other words the *heredity* of the Thinker, viewing him from the physical standpoint or from the emotional. There is no permanence whatever in the sheaths; they are built into temporary forms, and dissolved when the Thinker has exhausted their possibilities, but the seventh principle of each sheath gathers to itself the achieved qualities and stores them up—under the Law of Karma—to work out again and to demonstrate as the *plane impulse* at each fresh cycle of manifestation. This permanency is itself likewise only a relative one, and as the inner fire within the atom burns more brightly, as the external fires of the ego or solar fire beat upon it with ever increasing intensity, the atom in due time is consumed, and the inner blaze becomes so great that it destroys its encompassing wall.
>
> *Heat.* Herein lies the distinction between the permanent atoms on all planes, and the atomic matter of which they form a part. It is not easy to make this distinction clear, nor is it desirable at this time; the true facts of the case are one of the guarded secrets of initiation.[50]

'The threefold microcosmic word' is the Aūṁ and its permutations that effect the conditionings upon the three worlds of human livingness. For the *word* to have effect in form building necessitates the existence of the *deva* legions that *hear* and respond accordingly to the energies of the sound waves organising their building capacity. Of the senses hearing is the most limited and manifests in response to physical plane substance.

The 'heat' of which D.K. speaks is the internal heat of the incarnating *jīva* (solar Fire) coupled with that heat *(kuṇḍalinī)* generated through the appearance of the integrated form. Little more can be said because

50 Alice Bailey, *A Treatise on Cosmic Fire,* 515-16.

of the possibility of the misuse of such information by those of wilful and separative intent to utilise esoteric knowledge in order to control the manifestation of substance, and for personal gain, to the detriment of the evolutionary progression of all manifest Life. The categories given: responsiveness, form building power and relative permanency can be considered an integral triad embodied by the *devas*. That which gives Life to that form, its sustainability in terms of evolutionary progression, is the incarnate *jīva*. It generates the heat of incarnate activity and the eventual emanation of light.

D.K. then states that the permanent atoms constitute the substance of the *chakras* of a Logos, whilst the atomic matter forms 'other parts of His great body of light'.[51] The permanent atom also comes 'under the attractive power of the second aspect, whilst atomic matter itself is vitalised by the Life of the third aspect'. There is a trinity of expression implied here, where the permanent atoms respond to electric Fire, the will of the incarnating entity, which affects the 'other parts', the darkened substance of the mental plane. This substance has its reflex upon the physical domain, and from it emanates a 'great body of light' (a sun). Such a body can be seen and hence comprehended by the consciousness principle that evolves from the darkness to perceive the light. The light emanates from the darkness (elementary substance) and both types of matter are controlled by the energies passing through the *chakras*. These energies represent 'the attractive power of the second aspect', the Love-Wisdom principle, which binds all atomic unities into an integral form, producing an integrated oneness from the separated atomic lives (the activity of the 'third aspect'). Manifesting via the *chakras* the permanent atoms therefore come directly under the control of the consciousness principle of a human unit (Logoic or otherwise) whose purpose is to convert the darkness of the primal matter into emanations of light.

D.K. then states that:

> A permanent atom comes under the direct control of the lower of the three groups of Lipika Lords, and is the agency through which They work in the imposition of karma upon the particular entity who may be utilising it. They work directly with the permanent atoms of men, and produce results through the agency of form until they have

---

51 Ibid., 517.

> exhausted the vibratory capacity of any particular atom; when this is the case the atom passes into the stage of obscuration, as does the seventh principle of any sheath. It comes under the influence of the first aspect, manifesting as the Destroyer.[52]

We see here that the *karma* that a unit of consciousness experiences is directed by the Lipikas who activate the *bījas* of the substance that was formerly generated. These *bījas* are of two types, a) of that energy that was directed to others, b) one's mental or emotional perturbations. The energy directed to others must now be experienced in the way it was directed, e.g., in the form of anger or loving-kindness, hence must manifest in a reciprocal expression. One's own mental-emotional perturbations become the basis of the present personality equipment, which must then be amended or adapted to the presented life situation. A higher level of Lipika will work to rectify the karmic effects of group interrelations, e.g., the effects of the selfish extraction of wealth from others, through rectifying the perturbations of the permanent atoms of the Logoi within whom all lives reside. The next level of Lipika will deal with planetary *karma* as a whole.

Exhausting this 'vibratory capacity' is but another way of stating the evolutionary and meditative process whereby human units evolve the capacity to reflect the contact with the higher planes of perception into the higher order spirillae of their permanent atoms. As more intense energies are directed from the higher planes, so inevitably it strains the vibratory capacity of the spirillae of the atom to handle the intensity and so the atomic structure begins to disintegrate. Its Life force merges back into the pool of elementary substance, but now upon a higher, more refined level than before.

Form building

In the quotes below from *A Treatise on Cosmic Fire* the subject of the Pitris is broached. The term Pitris *(pitṛ)* means 'fathers, ancestors'. They are progenitors, the ancestors or creators of the human forms, said to be born from the side of Brahmā in the Viṣṇu Purāṇa. There

52 Ibid.

are seven classes of these *devas,* where three are incorporeal or *arūpa*, (thus building our *ātmic*, *buddhic* and *manasic* sheaths) and four are corporeal *(rūpa)*, who are the builders of our lower mental, astral, etheric and dense physical sheaths. The four lower classes are called the *lunar ancestors (barhiśad pitṛ).* They are the 'Fathers' who build the types of the forms that we incarnate into during each of the great epochs, Races and Rounds of evolution. The term *lunar pitris* refers to the elementals (Elemental lives) constituting the embodied form. The *solar pitris* are collectively the *devas* embodying the substance of the Causal form of the Sambhogakāya Flower.

Their work concerns the substance of the forms from the higher mental plane to the etheric and dense physical appearance. Very little can be added to what D.K. has presented in his monumental work hence I shall quote at length from this book.

D.K. states:

> We can now trace the progress of egoic energy as it passes down from the abstract levels to the permanent atoms. On each plane the work is threefold, and might be tabulated as follows:
>
> 1. The response within the permanent atom to the vibration set up by the solar Pitris; to word it otherwise: the response of the highest group of lunar Pitris to the chord of the Ego. This definitely affects the spirillae of the atom, according to the stage of evolution of the Ego concerned.
> 2. The response of the substance to the atomic vibration upon the particular plane involved. This concerns the second group of Pitris, whose function it is to gather together the substance attuned to any particular key, and to aggregate it around the permanent atom. They work under the Law of Magnetic Attraction, and are the attractive energy of the permanent atom. On a tiny scale each permanent atom has (to the substance of a man's sheaths) a position relative to that which the physical sun holds to the substance of the system. It is the nucleus of attractive force.
> 3. The response of the negative substance concerned and its moulding into the desired form through the dual energy of the two higher groups of Pitris. Some thought of the unity of this threefold work has been given in the differentiation of the substance of any plane into:

a. Atomic substance.

b. Molecular substance.

c. Elemental essence.

This differentiation is not entirely accurate, and a truer idea of the underlying concept might be conveyed if the word "energy" took the place of "substance and essence." This third group of Pitris is really not correctly termed Pitris at all. The true lunar Pitris are those of the first and highest group, for they embody one aspect of the intelligent will of Brahma, or of God-in-substance. The third group are literally the lesser Builders, and are blind incoherent forces, subject to the energy emanating from the two higher groups. Occultly these three groups are divided into the following:

a. The Pitris who see, but touch nor handle not.

b. The Pitris who touch but see not.

c. The Pitris who hear but neither see nor touch.

As they all have the gift of occult hearing, they are characterised as the "Pitris with the open ear"; they work entirely under the influence of the egoic mantram. If these differentiations are studied a great deal may become apparent anent a very important group of deva workers. They are a group who only come into manifestation as a *co-ordinated triplicity* in the fourth round in order to provide vehicles for man; the reason for this lies hid in the karma of the seven Logoi, as They energise the fourth, fifth and sixth Hierarchies. In the earlier round in each scheme these three groups attain a certain stage of necessitated growth, and embody the highest evolution of the substance aspect. Only the highest and most perfected of the atoms of substance find their way into the vehicles of man,—those which have been the integral parts of the higher evolutionary forms.[53]

Cumulatively the 'Pitris with the open ear' manifest as the symbolic 33 *crore*[54] Creative Intelligences as far as the building of a form is concerned. This phrase therefore has a far wider connotation than just building the forms of human units because this gift of occult hearing

53 Alice Bailey, *A Treatise on Cosmic Fire,* 781-83.

54 A crore is 10,000,000

also includes listening to the mantras emanating from cosmic astral sources so that a planetary or solar form can be built.

These three groups of *pitris* hence can be viewed to manifest as groupings of the Agnishvattas (mental plane *devas),* Agnisuryans (astral plane *devas)* and Agnichaitans (physical plane *devas).* These *devas* manifest as the three thermo-spirals (in their two groupings) when viewed in terms of the Logoic permanent atom.

The 'Pitris who see, but touch nor handle not' convey the energies from the higher mental plane in relation to form building. They see the pattern of the *maṇḍala* to be and set the forces in motion whereby this *maṇḍala* can objectivise. Their work collectivises *manasic deva* substance once the spirals of the mental unit are activated and sound out their note. The energies from the mental sub-planes moves via them to activate the *bījas* of manifestation for the second group, 'the Pitris who touch but see not'. The astral permanent atom responds with its own vibratory note. The sense of touch is an expression of the Watery (astral) domain and conveys the concretising, form building emanations as a vibratory response to what is touched upon the higher domains.

The concept of 'touch' involves the ability to mould and to manipulate, to build with substance of various types. A condensation or concretisation of mental plane substance is now possible where what is built can similarly be 'touched', hence made known, by means of the 'five fingers' conveying the *vayus,* the *prāṇas* that vitalise the formed life. Forms appear astrally that can fluidly change or be sustained in accordance with the impetus of a thought. Such reified forms have more 'weight', thus are more empirically tangible. They can be sculptured and patterned as the images or prototypes of what can appear physically. This level of expression for our planet is symbolised by the 'garden of Eden' in Genesis, which speaks of the building of such a Watery, astral world, prior to Adam and Eve being expelled from that garden to live in 'coats of skins'.[55] These 'coats' refer to the periodical sheaths of a human unit, signifying life in the dense physical plane. This happened as a consequence of having eaten 'the tree of knowledge of good and evil'.[56]

---

55 *Genesis 3:21.*

56 *Genesis 2:6-9* states: 'But there went up a mist from the earth, and watered the whole face of the ground. And the LORD God formed man of the dust of the ground, and breathed into his nostrils the breath of life; and man became a living soul. And the LORD God planted a garden eastward in Eden; and there he put the man whom

The concretising energy manifests as mantric sound to the 'Pitris who hear but neither see nor touch', the lesser Builders that produce the emanation of the atomic structures found upon the physical domain. They work via the spirillae of the physical permanent atom upon the highest (atomic) sub-plane of the physical.

The 'dual energy of the two higher groups of Pitris' manifests upon the four etheric domains to externalise the *chakras* (that are consequently emanations of the energies from the four main spirals of the mental unit). All three groups of Pitris are the builders of the *chakras,* being effected by the energies manifesting from the mental plane. The *chakras* are however principally the work of another set of *devas,* the *solar devas,* who set the originating energy of the forms (the petals) into motion. The work of these *devas* is needed because the *chakras* are the consciousness-bearing principle in a human unit. The four petals of the Base of Spine centre consequently reflect the energies of these four groups of *devas* in that each of the four petals channel the attributes of one or other of the four main Elements.

The 'Pitris who hear but neither see nor touch' are manipulated by the sum of the above energies manifesting via the petals of the *chakras,* the substance of which they embody. They cause the appearance of the 'atomic' forms as an automatic reflex. Being thus energised they become the Elemental lives that also manifest as the spirals and spirillae of the Anu's. (Consequently these Pitris are dual in nature, as are all *devas.)* Hence all that we see is the product of the emanations of sound, vitalised by the *prāṇas* generated by mind/Mind, and coloured by the originating thought expression from the mental plane. This produces the appearance of the substance that can be touched, hence sounds that can be heard and the radiant energy that can be seen by the manifest human unit.

All of this work happens under the auspices of the Lipikas, the agents of *karma,* who are explained in volume 7A of *A Treatise on Mind.*

Utilising the terminology from Bailey's *Esoteric Astrology* it can be said that in relation to the seven manifest Creative Hierarchies[57] that

---

he had formed. And out of the ground made the LORD God to grow every tree that is pleasant to the sight, and good for food; the tree of life also in the midst of the garden, and then the tree of knowledge of good and evil.' The phrase 'the ground' here refers to the lower mental plane, which is 'dense physical' from a cosmic perspective.

57 See Alice Bailey, *Esoteric Astrology,* 35.

(counting from above-down) the fourth Creative Hierarchy is humanity, the fifth Creative Hierarchy is that designated 'Makara, the mystery', who are 'The Pitris who see, but touch nor handle not'. The sixth Creative Hierarchy are the Lunar Lords (or 'Sacrificial Fires'), the 'Pitris who touch but see not'. They work with and upon the seventh Creative Hierarchy, 'the Elemental lives' (the 'Baskets of Nourishment', 'Blinded Lives'), who are the 'Pitris who hear but neither see nor touch'.

Under the heading of '*The Work of Form-building*' D.K. states:

> This work of form-building proceeds under definite laws, which are the laws of substance itself; the effect is the same for human, planetary and solar vehicles. The different stages might be enumerated as follows:
>
> 1. *The Nebulous.* The stage wherein the matter of the coming sheath begins to separate itself gradually from the aggregate of plane substance, and to assume a nebulous or milky aspect. This corresponds to the "fire-mist" stage in the formation of a solar system and of a planet. The *Pitris of the Mist* are then active as one of the many subsidiary groups of the three major groups.
> 2. *The Inchoate.* Condensation has set in but all is as yet inchoate, and the condition is chaotic; there is no definite form. *"The Pitris of the Chaos"* hold sway, and are characterised by excessive energy, and violent activity, for the greater the condensation prior to co-ordination the more terrific are the effects of activity. This is true of Gods, of men, and of atoms.
> 3. *The Fiery.* The internal energy of the rapidly congregating atoms and their effect upon each other produces an increase of heat, and a consequent demonstration of the spheroidal form, so that the vehicle of all entities is seen to be fundamentally a sphere, rolling upon itself and attracting and repulsing other spheres. *"Pitris of the Fiery Spheres"* add their labours to those of the earlier two and a very definite stage is reached. The lunar Pitris on every scheme, and throughout the system, are literally the active agents in the building of the dense physical body of the Logos; they energise the substance of the three planes in the three worlds, the mental, the astral and the dense physical planes of the system. This needs much pondering upon.
> 4. *The Watery.* The ball or sphere of gaseous fiery essence becomes still more condensed and liquefied; it begins to solidify on its outer surface and the ring-pass-not of each sheath is more clearly defined. The heat of the sphere becomes increased and is centralised at

the core or heart of the sphere where it produces that pulsation at the centre which characterises the sun, the planet, and the various vehicles of all incarnating entities. It is an analogous stage to that of the awakening of life in the foetus during the prenatal stage, and this analogy can be seen working out in the form-building which proceeds on every plane. This stage marks the co-ordination of the work of the two higher groups of lunar Pitris, and the *"Pitris of the Dual Heat"* are now intelligently co-operating. The heart and brain of the substance of the slowly evolving form are linked. The student will find it interesting to trace the analogy of this, the watery stage, to the place the astral plane holds in the planetary and systemic body, and the alliance between mind and heart which is hidden in the term "kama-manas." One of the profoundest occult mysteries will be revealed to the consciousness of man when he has solved the secret of the building of his astral vehicle, and the forming of the link which exists between that sheath and the astral light in its totality on the astral plane.

5. *The Etheric.* The stage is not to be confined to the building of the physical body in its etheric division, for its counterpart is found on all the planes with which man is concerned in the three worlds. The condensation and the solidification of the material has proceeded till now the three groups of Pitris form a unity in work. The rhythm set up has been established and the work synchronised. The lesser builders work systematically and the law of Karma is demonstrating actively, for it should be remembered that it is the inherent karma, colouring, or vibratory response of the substance itself which is the selective reaction to the egoic note. Only that substance which has (through past utilisation) been keyed to a certain note and vibration will respond to the mantram and to the subsequent vibrations issuing from the permanent atom. This stage is one of great importance, for it marks the vital circulation throughout the entire vehicle of a particular type of force. This can be clearly seen in relation to the etheric body which circulates the vital force or prana of the sun. A similar linking up with the force concerned is to be seen on the astral and the mental planes. *"The Pitris of the Triple Heat"* are now working synthetically, and the brain, the heart and the lower centres are co-ordinated. The lower and the higher are linked, and the channels are unimpeded so that the circulation of the triple energy is possible. This is true of the form building of all entities,

macrocosmic and microcosmic. It is marked by the active co-operation of another group of Pitris, termed *"The Pitris of Vitality"* in connection with the others. Group after group co-operate, for the three main bodies are distributed among many lesser.

6. *The Solid.* This marks the final stage in actual form building, and signifies the moment wherein the work is done as regards the aggregating and shaping of substance. The greater part of the work of the lunar Pitris stands now accomplished. The word "solid" refers not solely to the lowest objective manifestation, for a solid form may be ethereal, and only the stage of evolution of the entity involved will reveal its relative significance.

All that has been here laid down as to the progressive stages of form construction on every plane is true of all forms in all systems and schemes, and is true of all thought-form building. Man is constructing thoughtforms all the time, and is following unconsciously the same method as his Ego pursues in building his bodies, as the Logos follows in building His system, and as a planetary Logos uses in constructing His scheme.

A man speaks, and a very diversified mantram is the result. The energy thus generated swings into activity a multitude of little lives which proceed to build a form for his thought; they pursue analogous stages to those just outlined. At this time, man sets up these mantric vibrations unconsciously, and in ignorance of the laws of sound and of their effect. The occult work that he is carrying on is thus unknown to him. Later he will speak less, know more, and construct more accurate forms, which will produce powerful effects on physical levels. Thus eventually in distant cycles will the world be "saved," and not just a unit here and there.[58]

There are five levels of descent of Fire from the domain of Mind indicated here.

First the 'Fire-mist' stage, which concerns the organisation of the elementary substance by means of the reflex action of the five Dhyāni Buddhas, with the instigating energy emanating from Vairocana. This stage occurs upon the plane *ādi*, and the curdling process begins upon the plane *ātma,* from whence originates the primordial *karma* that organises what is to be.[59] The Lipikas play a role here to integrate the newly forming solar or planetary sphere with the geometry of the cosmic whole.

58 Alice Bailey, *A Treatise on Cosmic Fire,* 783-86.

59 This entire process is explained in detail in chapter 6 of my book *The Astrological and Numerological Keys to the Secret Doctrine, Volume 1.*

The milky appearance manifests as the undifferentiated substance that issues upon the higher mental plane. It continues to curd, to aggregate into forms as the energies from the Mind and of the cosmic Laws are brought to bear via the buddhic plane whereby chaotic elementary forces are organised in order to impose a *maṇḍala* of what is to be. The focus is inevitably upon the third mental sub-plane, which acts as a cosmic 'event-horizon'. The action of controlling the chaos and the condensation of the substance comes under the auspices of Akṣobhya's Mirror-like Wisdom.

The work of the next three levels of *pitris* concern the evocation of the Fires of Mind in order to clothe the 'Fiery Spheres' *(maṇḍalas* upon the mental plane) created. The ring-pass-not of the form is constructed, clearly defining the bounds of what is to be. We are considering spheres because the work ensues via a group or mass scale upon the third mental sub-plane under the auspices of Amitābha's Discriminating Inner Wisdom, producing the generation of light. (Amitābha is also known as the Buddha of Infinite Light.) This Wisdom is that of the mind/Mind *per se*, which generates the spheres of containment of thought-forms. As the Fiery attributes of mind are developed so light is generated and the seven Rays manifest their full potential.

Kuṇḍalinī is generated as a consequence of the Fiery form-building force of the Logoic Mind organising the primal atomic substance into coherent form. The internal heat of the atomic structures manifests and is sustained for the duration of the Logoic Will, giving life to the *maṇḍalas* generated. (The form building attributes of the Taurean impulse is now at full swing.) *Kuṇḍalinī* manifests in triune fashion, inevitably giving birth to the three thermo-spirals of the appearing Anu's. *Kuṇḍalinī* here is also inherently a septenary: that of 'the Fiery Spheres', 'the Dual Heat', 'the Triple Heat', plus that which integrates the sum of the form from out of the 'mist'. Hence, everything is born in Fire, is clothed with Fire and projected upon their errand by means of Fire. Fire is the energy of the mind/Mind.

D.K. then states that the 'ball or sphere of gaseous fiery essence becomes still more condensed and liquefied'. By 'liquefication' is meant that the Fiery energy becomes more fluid, less rarefied. This then is the work of the 'Pitris of the Dual Heat'. The astral plane that is now our focus consequently can be considered more a Fiery fluid rather than 'Watery', as the 'wetness' that is known to humans regarding this plane is a consequence of their generated desire-emotions, but does not exist outside of human activity.

Ratnasmabhava's Equalising Wisdom governs this stage and works via the sixth mental sub-plane, which condenses the Watery energies as a flux that tends to harmonise the spheres into a unified energetic field. The heat is dual because it incorporates the Fires of the fifth mental sub-plane plus the more subdued Watery aspect of the sixth sub-plane, which dampens the Fires, but retains the heat. The heat is the gain of the spheres as they rotate and generate friction through incorporating the elementary substance into their forms. The psychic content of the forms is built preparatory to being projected into dense objectivity. As they condense the spheres become smaller, more compact, preparatory to precipitating into manifestation.

'The Triple Heat' refers to the manifestation of the energies from the three planes of human livingness that will vitalise the three thermo-spirals and spirillae of the Anu. At this stage *kuṇḍalinī* is ensconced as a principally triune Fire sustaining the Life of the form as its internal heat. Utilising this Fire these *pitris* produce the appearance of the form from the ethers. They are the lower reflex of the 'pitris who hear but neither see nor touch'. The All-accomplishing Wisdom of Amoghasiddi governs this final accomplishment of form building.

The 'Pitris of Vitality' principally vitalise the groupings of seven spirals of the Anu, with the five *prāṇas,* but also of the seven Ray attributes. Consequently they can also be conceived of manifesting the radiance of the sun, and embody the vital energy captured by the plant kingdom. They represent the sum of the five groups of *pitris* so far explained, plus the first Ray organising energy of the Creative Will of the Thinker. The second Ray demonstrates the Love-Wisdom that integrates the entire process of form building into unity. The vitality is essentially an effect of this first and second Ray combination. The 'Pitris of Vitality' are governed by the fourth Ray, those of 'the Mist' are governed by the second Ray, those of 'the Chaos' by the third Ray. The fifth Ray governs 'the Fiery Spheres', the sixth Ray 'the Dual Heat' and the seventh Ray the nature of the manifestation of 'the Triple Heat'.

D.K. then continues with his thesis on form building as it emanates from the domain of the Sambhogakāya Flower and concludes with the following.

> From the standpoint of the lower self, the two most vital moments in the work of the reincarnating Ego, are those in which the mental

> unit is re-energised into cyclic activity, and in which the etheric body is vitalised. It concerns that which links the centre at the base of the spine with a certain point within the physical brain via the spleen. This is dealing purely with the physiological key...[60] This appropriation of the lowest body is distinguished in several ways from the approach to the other sheaths. For one thing, there is no permanent atom to be vitalised. The physical plane is a complete reflection of the mental; the lowest three subplanes reflect the abstract subplanes and the four etheric subplanes reflect the four mental concrete planes. The manifestation of the Ego on the mental plane (or the causal body) is not the result of energy emanating from the permanent atoms as a nucleus of force but is the result of different forces, and primarily of group force. It is predominantly marked by an act of an exterior force, and is lost in the mysteries of planetary karma. This is equally true of man's lowest manifestations. It is the result of reflex action, and is based on the force of the group of etheric centres through which man (as an aggregate of lives) is functioning. The activity of these centres sets up an answering vibration in the three lowest subplanes of the physical plane, and the interaction between the two causes an adherence to, or aggregation around, the etheric body of particles of what we erroneously term "dense substance." This type of energised substance is swept up in the vortex of force currents issuing from the centres and cannot escape. These units of force, therefore, pile up according to the energy direction around and within the etheric sheath till it is hidden and concealed, yet interpenetrating. An inexorable law, the law of matter itself, brings this about, and only those can escape the effect of the vitality of their own centres who are definitely "Lords of Yoga" and can—through the conscious will of their own being—escape the compelling force of the Law of Attraction working on the lowest cosmic physical subplane.[61]

Later on in his Treatise, with respect to the work of the building *devas,* starting with those of the permanent atoms, D.K., states:

> This particular group of devas are the aggregate of the lives who form the mental unit and the two permanent atoms. They, as we know, have their place within the causal periphery, and are focal points of egoic

60 Ibid., 788.

61 Ibid., 789.

energy. They are the very highest type of building devas, and form a group of lives which are closely allied to the solar Angels.[62] *They exist in seven groups connected with three of the spirillae of the logoic physical permanent atom. These three spirillae are to these seven groups of lives what the three major rays are to the seven groups of rays on the egoic subplanes*[63] *of the mental plane.* This phrase will bear meditating upon, and may convey much information to the intuitional thinker. There is a correspondence between the three permanent atomic triads, and the appearance of man in the third root race. A curiously interesting sequence of the three lines of force can be seen in:

a. The triads of the involutionary group soul.
b. The appearance of triple natured man in the third root race.
c. The triads in the causal bodies of any self-conscious unit.

These building devas are the ones who take up the sound as the Ego sends it forth through certain of the transmitting deva agencies, and by the vibration which this sets up they drive into activity the surrounding deva essence in their two groups:

a. Those who build the form.
b. Those who are built into the form.

They only affect those of analogous vibration. The stages of the building of any of the four forms through which lower man (the Quaternary) functions, follow exactly analogous stages to the building of the dense physical body, for instance, of a planet, or of a solar system. This can be traced all the way from the nebulous and chaotic stages through the fiery to the solid, or to the *relatively* solid where a subtle body is concerned.[64]

The Logoic permanent atom

In the solar system the three groups of *devas,* denoted Agnichaitans, Agnisuryans and Agnishvattas, which are explained in *A Treatise on Cosmic Fire* and volume 7B of *A Treatise on Mind,* embody the substance

---

62 The Sambhogakāya Flower, plus the *devas* that are incorporated into them, having built the forms.

63 The higher mental plane. These three sub-planes constitute the dense physical substance for an incarnating Logos.

64 Alice Bailey, *A Treatise on Cosmic Fire,* 939-40.

(energies) of the Logoic thermo-spirals. The *deva* kingdom manifests in a dual fashion, where there are those that are the liberated lives, the controlling, co-ordinating factors, the Greater and Lesser Builders, and those that are the 'Blinded Lives', the 'army of the Voice', etc., that are built into the substance of things and which are automatically galvanised into activity by the mantric directives of the Builders. This accounts for the two levels of thermo-spirals when it concerns the development of the sub-human kingdoms.[65] When human consciousness appears then a triune expression comes into manifestation,[66] accounting for the three extra associated triads of spirals explained above. The intra-spirals of the 'atom' represent the domain of the streams of consciousness lives to the development of superconsciousness by humanity.

Solar and planetary Logoi also possess permanent atoms existing within their version of the Sambhogakāya Flower, where the seed of the physical permanent atom manifests upon the first (atomic) sub-plane of the cosmic dense physical. All of the energy ('the attractive quality of cosmic love'[67]) impacting upon our solar system manifests via this 'atom'. Most of what is possible to say has been presented in Alice Bailey's *A Treatise on Cosmic Fire,* from which I shall quote at length, as what is expressed is what comes from the vision of a Master of Wisdom, hence needs careful meditation and logical analysis if one is to gain a proper comprehension of this technically difficult subject.

D.K. states concerning the Logoic physical permanent atom:

> This physical permanent atom (as in the case with the corresponding atom of the incarnating jiva), has its place within the causal body of the Logos on His own plane; it is, therefore, *impressed* by the totality of the force of the egoic cosmic lotus, or the attractive quality of cosmic love. This force is transmitted to the solar system in two ways: Through the medium of the Sun, which is in an occult sense the physical permanent atom; it, therefore, attracts, and holds attracted, all within its sphere of influence, thus producing the logoic physical body: through the

65 There is of course a gradual evolution of one group to the next, as implied in the movement upwards of the spirals behind the 'atom' depicted in figure 37.

66 Monad, Soul, personality.

67 Alice Bailey, *A Treatise on Cosmic Fire,* 1180.

medium of the planes which are the correspondences to the seven spirillae of the physical permanent atom of a human being. Thus a dual type of attractive force is found: one, basic and fundamental; the other more differentiated and secondary. These streams of energy, judged by their effects, are called in human terminology *laws*, because their results are ever immutable and irresistible, and their effects remain unchangeably the same, varying only according to the form which is the subject of the energy impulse.

Thirdly, the student must bear in mind that the seven planes, or the seven spirillae of the logoic permanent atom, are not all equally vitalised by the attractive pull emanating from the logoic lotus via the heart of the Sun. Five of them are more "alive" than the other two; these five do not include the highest and the lowest. The words "the heart of the Sun" must be understood to mean more than a locality situated in the interior recesses of the solar body, and have reference to the nature of the solar sphere. This solar sphere is closely similar to the atom pictured in the book by Babbitt and later in *Occult Chemistry* by Mrs. Besant. The Sun is heart-shaped, and (seen from cosmic angles) has a depression at what we might call its north pole. This is formed by the impact of logoic energy upon solar substance.

This energy which impinges upon the solar sphere, and is thence distributed to all parts of the entire system, emanates from three cosmic centres and, therefore, is triple during this particular cycle.

a. From the sevenfold great Bear.

b. From the Sun Sirius.

c. From the Pleiades.

It must be remembered that the possible cosmic streams of energy available for use in our solar system are seven in number, of which three are major. These three vary during vast and incalculable cycles.

Students may find it of use to remember that,

a. The Law of Economy demonstrates as an *urge,*

b. The Law of Attraction as a *pull,*

c. The Law of Synthesis as a tendency to concentrate at a centre, or to merge.

The streams of energy which pour forth through the medium of the Sun from the egoic lotus and which are in reality "logoic Soul energy" attract to them that which is akin to them in vibration. This may sound rather like the statement of a platitude, but is susceptible of really deep significance to the student, being accountable for all systemic phenomena. These streams pass in different directions, and in the knowledge of occult direction comes knowledge of the various hierarchies of being, and the secret of the esoteric symbols.

The main stream of energy enters at the top depression in the solar sphere and passes through the entire ring-pass-not, bisecting it into two halves.

With this stream enters that group of active lives whom we call the "Lords of Karma." They preside over the attractive forces, and distribute them justly. They enter, pass to the centre of the sphere and there (if I may so express it) locate, and set up the "Holy Temple of Divine Justice," sending out to the four quarters of the circle the four Maharajahs, their representatives. So is the equal armed Cross formed—and all the wheels of energy set in motion. This is conditioned by the karmic seeds of an earlier system, and only that substance is utilised by the Logos, and only those lives come into manifestation who have set up a mutual attraction.

These five streams of living energy (the one and the four) are the basis of the onward march of all things; these are sometimes esoterically called "the forward moving Lives." They embody the *Will* of the Logos. It is the note they sound and the attractive pull which they initiate which bring into contact with the solar sphere a group of existences whose mode of activity is spiral and not forward.

These groups are seven in number and pass into manifestation through what is for them a great door of Initiation. In some of the occult books, these seven groups are spoken of as the "seven cosmic Initiates Who have passed within the Heart, and there remain until the test is passed." These are the seven Hierarchies of Beings, the seven Dhyan Chohans. They spiral into manifestation, cutting across the fourfold cross, and touching the cruciform stream of energy in certain places. The places where the streams of love energy cross the streams of will and karmic energy are mystically called the "Caves of dual light" and when a reincarnating or liberated jiva enters one of these Caves in the course of his pilgrimage, he takes an initiation, and passes on to a higher turn of the spiral.

Another stream of energy follows a different route, which is a little difficult to make clear. This particular set of active lives enter the heart shaped depression, pass around the *edge* of the ring-pass-not to the lowest part of the solar sphere and then mount upwards, coming into opposition therefore with the stream of downpouring energy.

> This stream of force is called "lunar" force for lack of a better term. They form the body of the raja Lord of each of the planes, and are governed by the Law of Economy.
>
> All these streams of energy form geometrical designs of great beauty to the eye of the initiated seer. We have the transverse and bisecting lines, the seven lines of force which form the planes, and the seven spiralling lines, thus forming lines of systemic latitude and longitude, and their interplay and interaction produce a whole of wondrous beauty and design. When these are visualised in colour, and seen in their true radiance, it will be realised that the point of attainment of our solar Logos is very high, for the beauty of the logoic Soul is expressed by that which is seen.[68]

The energies from the great Bear manifest as the interior vortex of the solar atom, plus the force which provides the groupings of seven within the 'atom', the sevenfold manifestation of the Rays. The law of Synthesis governs this activity. When the main stream 'enters at the top depression in the solar sphere and passes through the entire ring-pass-not, bisecting it into two halves' it causes the originating spiral-cyclic motion, the appearance of the forces signifying the yin-yang. The energy of the Great Bear manifests via the fourth cosmic mental sub-plane.

The energies from Sirius manifest as the three types of intra-spirals, viewed primarily in the 3 x 4 arrangement earlier explained. They are governed by the law of Attraction, which applies to the energies moving through all seven spirals and spirillae of the intra-spirals. Here the energies from the sixth cosmic mental sub-plane are invoked.

The Pleiades govern the two groups of three extra-spirals plus the three inner groups of three, making the fifteen groups explained above. Here the law of Economy applies and the energies from the seventh cosmic mental sub-plane come to manifest the appearance of form upon the cosmic dense physical plane. The mode of manifestation of this form is what has so far been described above. The spirals shared by both the Sirian energies and those from the Pleiades represent the places of the conversion of the substance of the planes into the fields of consciousness. They represent the planes of struggle and eventual triumph for humanity over their materialism. Enlightened perception is consequently engendered.

---

68 Ibid., 1180-1184.

The law of Karma manifests via the fifth cosmic mental sub-plane, hence constitutes the expression of the Thought-Form making tendency of Logoic Manasic Fire (Mahat). It consequently concerns the action-reaction mode of the creation of Thoughts and their consequent dissolution at the end of the cycle of their usefulness. *Karma* is said to be a Sirian law by D.K., but here Sirius represents a mechanism of the distribution of this law to our solar system. This needs further explication. To do so one needs to refer to the information I gave concerning the laws of the cosmic mental plane in volume 6 of *A Treatise on Mind.*[69] Looking at the swastika of figure 5 of that text we see that the law of Karma is situated as the dynamic centre of the swastika, causing its movement, and is governed by the signs Pisces and Gemini.

To the abovementioned laws one more was added, which I termed 'the law of Identity', which governs the energies of the Logoic Sambhogakāya Flower manifesting via the third cosmic mental sub-plane. These energies were said to be governed by the signs Leo and Taurus. It represents the downward arm of a cosmic cardinal cross that has its externalisation upon the sixth cosmic mental sub-plane and the law of Attraction, governed by the signs Taurus and Aquarius. The cardinal cross at this level of expression manifests from the fourth cosmic mental sub-plane as the law of Synthesis, governed by Cancer and Aries. This energy is grounded by the law of Economy, governed by Aries and Capricorn, and works to produce the appearing Logoic Bodies of manifestation.

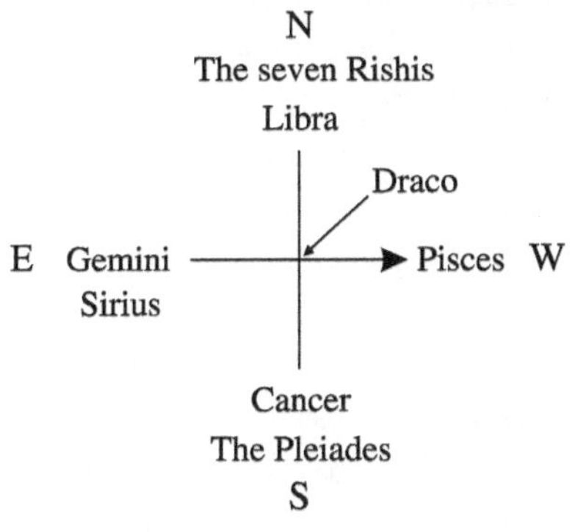

Figure 40: The cosmic law of Karma

69 See *Meditation and the Initiation Process*, 367-97.

In looking at the signs governing these laws we see that two signs are omitted, Libra and Cancer. Together with the Fiery Breath from Draco they form a cardinal cross aspect with the signs Gemini and Pisces, which gives us a better picture as to the nature of the emanation of the law of Karma. Gemini and Pisces represent the mode of the manifestation of the Sirian law of Karma, and manifest along the east-west arm of this cross, as part of the law of Attraction (hence *karma* manifests as the law of Love in systemic space). The energies of the seven Rishis of the great Bear manifest via Libra as the northern arm of this cross, which demonstrates as the Law of Synthesis. Cosmic Karma consequently comes as an expression of the in and out-breathing of the Logoic meditation process. From Logoic *dhyāna* comes the energies that cause the manifestation *(manvantara)* of the new Logoic cycle of expression, and its eventual dissolution *(pralaya)*. The story of *karma* lies in the terms *manvantara* and *pralaya*. This meditative expression impacts upon the sublime 'wives' of the Rishis, the Pleiades, manifesting via the attributes of the sign Cancer the crab, the sign of incarnation and mass movement of consciousness-attributes, of the Lives informing the substrate of what has been 'created'.

The Fiery Logoic *kuṇḍalinī* energies from the constellation Draco the dragon is drawn upon to bring the entire cycle of what is to be into manifestation.

The Sirian law of Karma governs Monadic evolution, hence manifests via the three primordial Rays. The energies enter via Gemini, the sign of the cosmic Christ, preparatory to entering into incarnation via the next sign, Cancer the crab. Gemini governs the energies manifesting in the etheric body, hence the *nāḍī* and *chakra* system. Consequently the karmic impact is via the planetary Logoi that form the body of manifestation of a solar Logos. The seven intra-spirals of the Logoic permanent atom are thereby vivified. Pisces, governing the western direction, is the last sign of the zodiac, hence is the sign of death, of liberation. Pisces thus concerns the reaping of the *karma* of the entire evolutionary process and abstracting it upwards to the higher dimensions, the Waters of cosmic astral space.

In this cardinal cross we see two Watery signs (Cancer and Pisces) and two Airy signs (Libra and Gemini), with Draco incorporating the Fiery Element. All energies are brought into manifestation via the

Earthy Element (Virgo, the polar opposite of Pisces, to which the arrow in the western direction is directed). Gemini and Pisces then act as the two arms of the mutable cross that forms to precipitate the cosmic energies via the open gate of Cancer in such a way that the Logoic physical permanent atom is activated. Via Libra is drawn the karmic forces that effect the conditionings of the permanent atom so activated.

Libra the balances draws energies from the fifth cosmic mental sub-plane (the cosmic law of Karma), hence the complete impact of the Energies of the Logoic Mind, the number five signifying the mental plane and its forces *per se.* From this perspective we see that *karma* is instigated via the cyclic Impulses from the Logoic Mind when the Divine Thoughts are projected into the *maṇḍala* of Being. Libra's polar opposite is Aries the ram, which starts new cycles of activity via the directive Will.

Gemini the twins draws energies from the fourth cosmic mental sub-plane (the law of Synthesis) and hence is receptive to cosmic *buddhi. Buddhi* here acts as a mirror reflecting the higher cosmic energies into the lower domains and vice versa. Gemini's polar opposite is Sagittarius the archer, who fires arrows to direct the energy flow. Gemini is constituted of a mortal and an immortal brother. The latter represents the cosmic Christ (the cosmic Hierarchy of liberated Beings), and the former represents the corruptible substance to be redeemed. (The cosmic dark brotherhood can also incarnate therein and rule.) This is the basis of the cosmic *kāma-manasic* attribute wielded by the Sirian system. The field of Desire *(kāma)* is corruptible, but can be converted to Love. *Manas* (here Mahat) is that which corrupts, but is also the energy that redeems, once the *kāma* is overcome, then is expressed the Love-Wisdom, the energy that is the hallmark of the Sirian system. The purpose of *karma* concerns the process of the conversion of *kāma-manas* to Love-Wisdom.

Pisces the fishes draws energies from the sixth cosmic mental sub-plane (the cosmic law of Attraction), hence is the sign associated with the liberating Avatar, 'the world Saviour'.[70] Its polar opposite is Virgo the virgin, who governs the *deva* kingdom that embody the substance of the

70 Alice Bailey, *Esoteric Astrology,* (Lucis Publishing Co., New York, 1982), 262. Note that the foundation of all of the astrological information found in my works is derived from this book, and those of Bailey in general. I normally embellish and add to D.K.'s seminal information.

appearing forms. An objective of the Piscean energy is to extract (liberate) the streams of Lives ('swimming fishes' in the ocean of substance) from the manifest domains and so offer the seven cosmic Paths for them to follow. The paths first go through Sirius and then to various destinations, but eventually find a resolve in the constellation Orion the hunter (a cosmic Heart centre) for the birthing of a new solar system.

Cancer the crab draws energies from the seventh cosmic mental sub-plane and hence wields the law of Economy via the activity of the seven Pleiades, whereby cosmic astral substance is condensed into the ethers of cosmic space. From these ethers forces can come to bear upon the cosmic 'black dust' that is to be transformed into lighted substance, and so the new form to be is produced. The polar opposite is Capricorn the goat, which embodies the substance of the planes. The path of the transformation of that substance is assisted by the Potency of the forces of the abovementioned cosmic mental sub-planes, which flow through Capricorn. The transformation process attained by the Lives that evolve from the substance causes their Initiation process as they progress step upon step along the spirals of evolution towards the cosmic mental plane.

In all of the signs given above we see that they relate to the juxtaposition of the cardinal cross over the activity of the mutable cross. Omitted above are the signs of the fixed cross: Aquarius, Taurus, Leo and Scorpio. They are concerned with the process of the engendering of Love-Wisdom by the Lives travelling up the planes of ascent. (The seven spirals of the Logoic permanent atom.) In doing so they must learn to overcome attachment to the transience of the material domains by manifesting compassionate intent as they learn to obey the law of Sacrifice along the Bodhisattva way. The law of *karma* then works to free them from their aeonic bondage. Most are liberated as Bodhisattvas, and so eventually find their placing in the domain of the immortal Brother (Gemini). Some continue on under the Piscean impetus as Avatars to try to salvage the incarnate *jīvas* and so bring them eventually, under the law of Attraction, to the sixth cosmic mental sub-plane. The work here primarily concerns the conversion of the most recalcitrant *jīvas,* those that have chosen the left hand path, hence are firmly entrenched in materialistic bondage.

The law of *karma* works upon the mortal Brother, via an Avatar, primarily to convert Watery attributes into Airy ones. (The attributes will thus travel the direction east.) Such cosmic Avatars work under the impetus of the Logoic Sambhogakāya Flower, whose energies manifest from Libra to Pisces via the 'mirror' (Gemini) and Sirius. Hence the energies flow via the fifth and sixth cosmic mental sub-planes and the corresponding cosmic astral sub-planes to impact upon the systemic Watery (astral) domain. Massed desire-attachment to the material domain via the impetus of the separative mind must be converted to desirelessness and love. Such is the path, and an Avatar is assisted in the task by working steadfastly with the law of Love to overcome *karmic* bondage to form.

The Lipikas

Karma manifests via the Lipikas, who are effectively arranged in a 1 + 3 format. (Esoterically manifesting the arrow of the cardinal cross.) D.K. states that 'Three of Them are closely connected with Karma as it concerns one or other of the three great Rays, or the three FIRES, while the fourth Lipika Lord synthesizes the work of his three Brothers and attends to the uniform blending and merging of the three fires'.[71] In figure 40 Libra stands for the synthesising Lipika, in which case the arrow is fired downwards into material plane expression. The arrow pointing in the western direction in the figure concerns the projection of the liberating *karma* (Will and Love-Wisdom) for a human kingdom as it aspires to free itself from the trammels of *saṃsāra*. (The experience in the substance of the Womb, governed by Pisces' polar opposite Virgo, and the general movement of the mutable cross therein.) Such energy will effect humanity during the sixth Root Race.

The energies awakening the Logoic thermo-spirals manifest via the Libra-Cancer interrelation (the descending vortex of the 'atom') that manifests a central point of energisation with the application of the Fiery Breath of Draco, which causes the establishment of the ring-pass-not of the 'atom', to attract the needed substance from the Taurean store (the cosmic astral plane). The energy input also causes the spirals to revolve, producing the upward way of the entire evolutionary path.

The Lipika were explained in volume 5A of *A Treatise on Mind* in terms of being the Guardians of the *Bardo Thödol*. They are the four

71 Alice Bailey, *A Treatise on Cosmic Fire*, 74.

Mahārājas, the four great Deva Lords that embody the substance of the Throne of Deity. These are the four great Rāja (kingly) Lords that embody the four directions in space, the four continents or Elements in Buddhist and Hindu philosophy. They are sometimes equated with the four Kumāras, but more correctly are their Deva compliments. Their technical names in Hinduism are Vessāvana (or Kuvera, north), Virūdhaka (south), Dhataraṭṭha (east), and Virupākśa (west). They are the regents and protectors of the four directions of space; north, south, east, and west, hence are Lipikas that circumscribe the activities of that space and so are agents of the related *karma*. Further detail concerning the Lipikas is provided in volume 7A of *A Treatise on Mind,* from which the following extract is derived:

> The Lipikas represent the binding forces that weld the activities of the evolving Thought into unity, integrating it as part of a greater expansive Inclusiveness. They move from the great Thought to the lesser ones it incorporates with mathematical precision, projecting the exact coordinates of what is to be, according to a prearranged formula. Hence they are the active Intelligent units of expression Logoically considered. They are the forces of Mahat, the embodiers of the third Ray methodology throughout the aeons of time. They are the Aūṁ of evolutionary being, from whence the septenary of life's expression rebounds in cyclic reverberation as it converts into the Oṁ of consciousness cosmic awareness. Here there is a cross integration with the activity of the three Buddhas of Activity. The Lipikas are the recording agents and the Buddhas the directive Wills. Together they manifest the functions of the 4 + 3, where the Lipikas are focussed via the four cosmic ethers and the building forces of Nature, whilst the Buddhas of Activity control the Lives evolving through the three planes of *saṃsāra*.[72]

What has so far been explained is the nature of the 'five streams of living energy' that are 'the forward moving Lives'. They establish the geometric parameters of the incarnate sphere of activity, and incorporate the karmic propensity to move all that will incarnate into that sphere forwards to a higher Initiation standing. They bring into the atomic form the zodiacal and planetary energies (cosmic forces in the case of a solar

72 *A Treatise on Mind,* volume 7A, 485.

Logos) and that, as was stated above, according to D.K., they 'embody the Will of the Logos. It is the note they sound and the attractive pull which they initiate which bring into contact with the solar sphere a group of existences whose mode of activity is spiral and not forward'. This group establishes and vitalises the Heart of whatever is to be. The 'Caves of dual light' are the *chakras,* which interrelate the *prāṇas* from the lower domain with those from the higher ones.

When the members of a Creative Hierarchy can master the qualities associated with a particular *chakra* that condition their lives, the Element that the *chakra* consequently conveys, they can then take Initiation and move to a higher dimension of perception than they were formerly experiencing. The *chakras* convey the animating forces from 'the streams of love energy' that 'cross the streams of will and karmic energy' into the spirals and spirillae of the Anu's that are attracted to them, or which they cause to come into objectivity. Energies consequently move from the larger encompassing Logoic permanent atom to smaller atomic unities that are called into activity because of its existence and the magnetic potency of the associated *chakras.* From the subjective (the *chakras)* appears the objective universe, the lines of force and currents of electromagnetic energies passing through the spirals and spirillae of the atoms of Nature.

We therefore see that each plane of perception is an expression of the Logoic physical permanent atom at the level of the manifestation of the thermal-colours of Babbitt's 'atom'. Each plane in its turn is a similar embodiment to its subsidiary sub-planes. It is through the complete atomic *bīja* that the incarnating principle, Soul or Logos, finds it possible to contact and interrelate with substance and consequently experience the lower manifested domains. Each 'atom' is a vortex of energy that by its constitution allows the projection and formation of formed forces, allowing the illusionality of 'things' to exist in the physical plane. We thus have the appearance of a time sequence in relation to the manifestation of experiential 'things'. (Via any spiral interrelationship, or via groups of such vortices.) The manifestation of any particular phenomena is then a product of the consciousness-space evolving as part of such spirals and spirillae. This is governed by the will of the Thinker, who manifests the activity of his/her thought or creation of ideas and images in the mind by means of them. A consciousness-space is but a way of defining a 'cave' wherein exists the dual light of that which consciousness reaches out towards, and that which is built into its constitution. The spirillae are qualified by desire for the purpose of sensation or contact with a particular plane of perception or realm of existence and the related qualities.

This particular incarnation of the solar Logos is said to be one that is presently endeavouring to express the quality of vibration of the spirals emanating the intra-spiral thermal colours, associated with the Love-Wisdom principle. (The Son in incarnation.) The last incarnation developed the higher correspondence to the infra-red colours (the Mother or Brahmā incarnation) of the extra-spirals, and the future incarnation will produce those associated with the electrical colourings. The fundamental colour of the solar Logos is said to be *blue*, the colouring of the second of the spirals associated with the electrical colours. From this colouration we see that our solar Logos is relatively highly advanced cosmically. In the last incarnation the solar Logos was said to be generally coloured green. The future Incarnation can be postulated to be red plus a correspondence of indigo in this system, making thus a violet. Colourings however can change from the point of view or angle of vision.[73]

All atoms or units of energy (e.g., the Sambhogakāya Flowers), find a place in one of the spirals or spirillae of the Logoic permanent atom. Because our material world is incarnate so the lower spirals of this atom are fully vivified. From one perspective our present earth evolution represents the high point of solar evolution, though quite a number of planetary Logoi have surpassed its attainment and have produced a host of *nirvāṇees*. The qualitative gains of the past have however been incorporated into the earth humanity's evolutionary

73 The colourings of the seven Rays esoterically are for instance red for the first Ray, indigo blue for the second Ray, then green, yellow, orange, pink or sky blue and violet for the remaining Rays. (These Rays are said to be subrays of the second Ray.) In figure 37 of Babbitt's view of the 'atom' we see that his schema for the colourings of the spirals follow that of the spectrum of natural light, which he has divided into fourteen hues, seven for the electrical colours, from blue green blue, indigo blue, indigo, violet indigo, violet, with dark violet being the most intense. For the thermal colours he has yellow green as the most intense, followed by, in descending order, yellow, yellow orange, orange, red orange, red and thermal.

journey. This may not be apparent presently because of the evil forces ascension over the governments of the Western world specifically, but will be demonstrated in the future after that entrenched evil has been mostly eliminated.

The mode of manifestation and the quality of the categories of Life manifesting upon any planetary system is determined by the expression of the coursing of the Lives through the spirals and spirillae. The higher the spiral vivified the more advanced the planetary Life. All entities are therefore defined esoterically according to the type of energy they express. First there is the establishment of the ring-pass-not of the solar sphere, causing the plain dark disc consisting of primal undifferentiated substance. As the Logoic 'Soul energy' passes through it in order to organise this substance it brings with it the organising and informing twelve zodiacal and ten planetary energies needed to cause the appearance of things. They are the energies 'the Spirit of God' needs to integrate the 'face of the waters' (the primal substance) into a coherent form through which Logoic Purpose can be expressed. This causes a 'top depression in the solar sphere' and causes the sphere to rotate. The appearance of this form happens upon the fourth cosmic mental sub-plane for a Logoic sphere, and the corresponding fourth mental sub-plane for our terrestrial sphere. This activity is governed by the law of Synthesis, hence the application of the Logoic Will, and comes under the influence of the energies of the sign Libra the balances. Libra breathes out the meditative impulse, the mantric Sounds for the new Day, and calls forth the *karma* of what is to be. Later it breathes in the results of the Day's experience and abstracts the gain into a meditative interlude.

The Lords of *karma* (the mental forces of the Logos), the Lipika, consequently pass through the prepared sphere and utilise the available (cosmic) forces with which to build the new structure. Their work manifests via the third cosmic mental sub-plane (counting from below up), where, as D.K. stated they 'preside over the attractive forces, and distribute them justly'. The attractive forces are those that come into play via the second cosmic mental sub-plane under the auspices of the law of Attraction. Here the primal substance and the zodiacal and planetary energies interrelate, causing the 'curding' of the substance. (Incorporating therefore new, smaller atomic spheres within the greater one.) To do so the Lipika must incorporate the effects *(karma)* from past actions, hence the appropriate *saṃskāras* that need to be expressed for the new world sphere are called into play. These *saṃskāric* forces then condition the newly forming spheres with their inherent energies. This work is accomplished under the auspices of Gemini, which interrelates the forces of the 'immortal brother' (the energies utilised by the Lipika) and the 'mortal brother', the substrate with which to build the new. (Cosmically, the primal 'dust' is elementary substance upon the *ātmic* plane that has never been previously incorporated with Logoic Thought.)

The Lipika, as D.K. stated in the earlier quote, 'enter, pass to the centre of the sphere and there (if I may so express it) locate, and set up the "Holy Temple of Divine Justice," sending out to the four quarters of the circle the four Maharajahs, their representatives. So is the equal armed Cross formed—and all the wheels of energy set in motion.' Hence is established the Throne or Seat of Power of the Logos at the centre of the sphere, the awakening of the *chakra* system, as indicated by figure 27, 'The lotus blossom and the Jinas'. The Lipikas set the stage for the manifestation of the *karma* of the sphere of activity according to the attributes of the four directions in space.

From the sixth cosmic mental sub-plane Pisces incorporates the energies of the cosmic Waters, the cosmic astral plane, whilst the succeeding sign, Aries the ram, governs the energy of the driving will of initial beginnings that sets the great Wheel and lesser wheels in motion. Upon the cosmic astral plane and the four cosmic ethers exist the *nirvāṇees* from the former cycle of activity. They are attracted to the newly manifesting Logoic form according to the key Note and colourings that it emanates. That of like nature heeds the call.

The 'five streams of living energy (the one and the four)' that 'are the basis of the onward march of all things' can also be considered to represent the activity of the five liberated Hierarchies.[74] They embody the major petals of the spheres of activity *(chakras)* set up on the cosmic astral plane and reflect their Purpose into the cosmic ethers. They are 'the forward moving Lives' viewed as five. Here the Ājñā centre and Head lotus are a superimposed unity, then the Throat, Heart and Solar Plexus centres, whilst the Sacral and Base of Spine centres are also a superimposed unity.

74 See *Esoteric Astrology,* 34, 36-37.

Having established the inner, subjective zone of energisation, the stage is set for the appearance of the embodied form, the spirals and spirillae of the Logoic permanent atom. Seven notes are sounded that attract to the solar sphere 'a group of existences whose mode of activity is spiral and not forward'. They are the seven Creative Hierarchies that as stated 'spiral into manifestation, cutting across the fourfold cross, and touching the cruciform stream of energy in certain places. The places where the streams of love energy cross the streams of will and karmic energy are mystically called the "Caves of dual light"'.

This level of activity is governed by the law of Economy manifesting via the seventh cosmic mental sub-plane, as ruled by the attributes of the sign Cancer the crab.

The transposition of these two crosses (the mutable replacing the cardinal) produces the appearance of a fixed cross, thus the Heart centre is born and consequently the projection of *jīva,* the Life force, into manifestation. The Creative Hierarchies therefore manifest as an expression of the energies of cosmic Love, whereby the septenary spirals constituting the 'atom' can be vitalised with living energies.

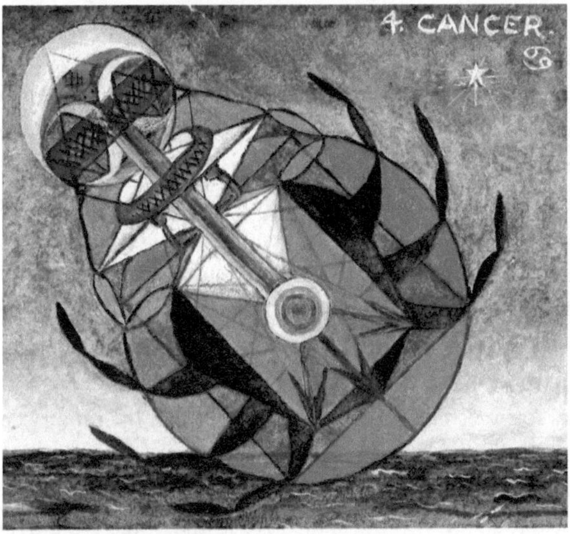

Figure 41: The symbolism of Cancer the crab[75]

75 This is one of the *Tablets of Revelation* painted by the author depicting the esotericism of the twelve signs of the zodiac.

The *jīva* is directed via the *prāṇic* triangle, the *chakras* concerned are the centre between the Shoulder Blades, the two Breast centres, the Diaphragm centre and the dual Splenic centre. They can be counted as five or seven, if the Splenic centre is counted as two and the Heart centre is included. The spiral activity of the seven Creative Hierarchies manifests via their expression. With them the remaining sixteen minor *chakras* (plus the rest of the Inner Round series of small *chakras)* constituting the subjective body can be vitalised. They are the Alta Major centre at the back of the head, the two Eye centres, the two Ear centres, the Mouth centre, the two Hand centres, the Stomach and Liver centres, the two Gonad centres, the two Knee centres and the two Feet centres. Sixteen in all (4 x 4), not counting the above-mentioned centres. With them the complete manifestation of the form is possible, as the organs are the externalisations of the energies manifesting through them. The Alta Major centre also manifests as a triad in the head with two other important minor centres governing the pineal and pituitary glands.

The symbolism of the crab in figure 41 relates to the evocation of the energies of the Base of Spine, Splenic and the two Gonad centres interrelated with the Head lotus (the crab). The energies are in the process of descent into manifestation via the systemic Waters. The crab holds the wish-fulfilling gem, which is also the organ of vision (the Ājñā centre). The rainbow hued *suṣumṇā nāḍī* becomes Neptune's trident needed to control the vicissitudes of the Waters of *saṃsāra.* The crab's legs spin the spirillae of the extra-spirals of the Logoic atom.

Taking the Splenic centre as a unity, and discounting the Pineal and Pituitary gland centres in the head (which come under a different dispensation), then there are 21 minor centres, which have a relation to the 7 x 3 intra-spirals of the permanent atom. As the 'seven streams of energy' spiral round via their three octaves *(guṇas)* of energisation, so they meet the descending 'streams of will and karmic energy' and nodes *(bījas)* are formed. They cause the appearance of the minor centres, via which the physical form can appear. The minor centres attract to them the appropriate *(deva)* substance that enables the appearance of physical plane phenomena. From the subjective to the objective universe does the form-building process manifest, be that a human form, a planetary or solar system, or any other cosmological body.

The minor centres form a network of *nāḍīs* that link petal to petal of the flowers that manifest.[76] They are linked to the petals of the major *chakras,* which govern the forces which embody the planes of perception. They manifest the 'seven lines of force which form the planes', around which move 'the seven spiralling lines'. The traverse lines manifesting from the Lipikas form the nodes from which the *chakras* appear. Upon entering a minor centre ('cave') and mastering its qualities the *jīva* (human unit) then takes Initiation, which leads that pilgrim to a major centre and thence onto a higher level of experience ('a higher turn of the spiral') than hitherto possible.[77]

The nature of the substance that informs the form-building process, and which is transformed in the transmogrifying and transmutation process related to the Initiation undertaking of a human *jīva* was described by D.K. above as:

> Another stream of energy follows a different route, which is a little difficult to make clear. This particular set of active lives enter the heart shaped depression, pass around the *edge* of the ring-pass-not to the lowest part of the solar sphere and then mount upwards, coming into opposition therefore with the stream of downpouring energy. This stream of force is called "lunar" force for lack of a better term.

They embody the extra-spirals of the 'atom' and manifest in the five levels explained above when the associated substance becomes increasingly refined by being incorporated into the evolving Lives, and then the human thought streams whereby they become transformed through the Initiation process. D.K. states that this substance is governed by the law of Economy, and is so until humanity converts it through desire and personal will into the forms they so desire, the building of the edifices of civilisation, etc., then it comes under the law of Attraction. Upon the Initiation path the law of Synthesis holds sway. All the while the law of *karma* governs the interrelation of all the Lives and forms of action until liberation from *saṃsāra* ensues.

---

76 See figure 22 (p. 409), volume 5A of *A Treatise on Mind* for the detail concerning this *nāḍī* system related to the part that mainly concerns the Bardo Thödol.

77 Obviously the technicalities of this abstruse esoteric subject needs the inner vision awakened to properly comprehend, as then one can 'see'.

In conclusion I shall add some further quotes from *A Treatise on Cosmic Fire:*

> *The Logoic Plane.* The first, the second, and the third subplanes of the first cosmic ether respond specifically to the vibration of one of the three aspects, or to those cosmic Entities Whose influence reaches the matter of the planes from without the system altogether. On the fourth subplane comes a primary blending of the three fiery Lives, producing archetypally that force manifestation of electricity which eventually causes the blazing forth of the Sons of Light on the next plane. In this electrical connotation we have the three higher planes ever embodying the threefold Spirit aspect, the lower three embodying the threefold substance aspect, and then a plane of at-one-ment whereon an approximation is made which, on the path of return, marks the moment of achievement, and the point of triumph. This is succeeded by obscuration. Hence on every plane in the solar system we have a fourth plane whereon the struggle for perfect illumination, and subsequent liberation takes place, the battle ground, the Kurukshetra. Though for man the fourth plane, the buddhic, is the place of triumph, and the goal of his endeavour, for the Heavenly Man it is the battle-ground, while for the solar Logos it is the burning-ground.
>
> This differentiation of the subplanes of the systemic planes *into a higher three, a lower three, and a central plane of harmony is only so from the standpoint of electrical phenomena, and not from the standpoint of either pure Spirit, or pure substance, viewed apart from each other. It concerns the mystery of electricity, and the production of light.* The three higher planes concern the central Forces or Lives, the three lower concern the lesser Forces or Lives. We must bear this carefully in mind, remembering that to the occultist there is no such thing as substance, but only Force in varying degrees, only Energy of differentiated quality, only Lives emanating from different sources, each distinctive and apart, and only Consciousness producing intelligent effect through the medium of space.[78]

D.K. further states:

> It should be remembered that all the planes of our system, viewing them as deva substance, form the spirillae in the physical permanent atom of the solar Logos. This has earlier been pointed out, but needs

78 Alice Bailey, *A Treatise on Cosmic Fire*, 520-21.

> re-emphasising here. All consciousness, all memory, all faculty is stored up in the permanent atoms, and we are consequently dealing here with that consciousness; the student should nevertheless bear in mind that it is on the atomic subplanes that the logoic consciousness (remote as even that may be from the Reality) centres itself. This permanent atom of the solar system, which holds the same relation to the logoic physical body as the human permanent atom does to that of a man, is a recipient of force, and is, therefore, receptive to force emanations from another extra-systemic source. Some idea of the illusory character of manifestation, both human and logoic, may be gathered from the relation of the permanent atoms to the rest of the structure. Apart from the permanent atom, the human physical body does not exist.[79]

D.K. also states:

> The units, therefore, in the three lower kingdoms possess no permanent atoms but contribute to the formation of those atoms in the higher kingdoms. Certain wide generalisations might here be made, though too literal or too identified an interpretation should not be put upon them.
>
> First, it might be said that the lowest or *mineral kingdom* provides that vital something which is the essence of the physical permanent atom of the human being. It provides that energy which is the negative basis for the positive inflow which can be seen pouring in through the upper depression of the physical permanent atom.
>
> Secondly, the *vegetable kingdom* similarly provides the negative energy for the astral permanent atom of a man, and thirdly, the *animal kingdom* provides the negative force which when energised by the positive is seen as the mental unit. This energy which is contributed by the three lower kingdoms is formed of the very highest vibration of which that kingdom is capable, and serves as a link between man and his various sheaths, all of which are allied to one or other of the lower kingdoms.[80]

Concerning the statement above that 'Though for man the fourth plane, the buddhic, is the place of triumph, and the goal of his endeavour, for the Heavenly Man it is the battle-ground, while for the solar Logos it is the burning-ground', it can be said that 'the place of triumph', refers to the complete transmutation of the originating elementary substance, into

---

79 Ibid., 693-94.

80 Ibid., 1133-34.

which a Monad incarnated. At first a field of consciousness is produced, and thence its transmutation into the Void Elements, signifying the liberation of the human unit from the thrall of *saṃsāra* and the death of the Sambhogakāya Flower. This happens within the precincts of the cosmic dense physical plane. For a Heavenly Man (planetary Logos) it is a battle ground because the focus here is with the substance of the cosmic astral plane, and concerns the major testings for such a Logos to convert the *saṃskāras* that are the streams of conscious and sentient Lives within such a One's Body of manifestation (the cosmic dense physical) into liberated Souls *(Jīvanmuktas, Buddhas).* The liberated Ones can then travel the far reaches of the cosmic astral domain. For a solar Logos it is 'the Burning Ground' because the focus is now upon the cosmic mental plane, hence we are concerned with the transmogrifying effects within the Minds of the great Lives (constituting the Mind of a solar Logos) passing through Initiations that will burn up the last vestiges of attachment to cosmic dense physical plane substance, so that cosmic Individualisation can occur. Such attachment concerns Thought Impulses related to what has transpired in the past and which no longer serve a purpose.

9

The Atomic Universe, Esoterically Understood

The concept of ether

In this chapter I shall endeavour to incorporate the information so far presented in terms of some of the subatomic particles discovered by the scientific community, and thereby provide a proper framework for thinking along esoteric lines by future scientists. I will only incorporate the main particles associated with quantum elecrodynamics (QED), hence the formation of atomic particles and the energy fields they reside in. The veritable zoo of other particles will not be included, as they are mainly the result of various combinations of Anu's. This investigation concerns the nature of the manifestation of the appearances we consider as matter from out of subjective space. Consequently I do not propose to present an exhaustive compendium of esoteric physics, but hope to present sufficient of an esoteric view for future physicists to use as background theory for the development of further research and theorisation as to the true nature of manifesting phenomena.

Much of this discourse will concern the boundary between the fourth ether, hence the subjective domains, and the dense physical universe. The fourth ether acts like a mirror, reflecting one to the other, and is the transition between energy and matter, the effect of the expression of this transition is termed *māyā,* illusion. I shall call this boundary 'the event horizon'. The concept here is similar to what physicists have described for a black hole, where 'the event horizon' is conceived as a radius around it between normal space and its internal space, wherein gravity is so dense that even light cannot escape, preventing any information manifesting as to what is inside the black hole.

A horizon represents the boundary between two different zones of experiences which appear to exclude each other. In this case we are concerned with the horizon between normal empirical space that determines the world of physical plane phenomena and subjective space that lies outside of normal observation, either by means of our senses, or by means of physical plane instrumentation (except those that are most sensitive to subatomic particles).[1] With the fourth ether however, information can be gleaned, as it simply represents the boundary between the inner multidimensional space and the three dimensions experienced by means of the five senses and cognised by the intellect.

For esotericists, however, the four etheric sub-planes come within the bounds of 'physical' space. They are energy fields containing matter far subtler than what is normally experienced, as the atoms that are the building blocks of three-dimensional perception no longer exist. The 'event horizon' therefore represents the boundary that exists between the substance of subjective energy fields and the appearance of atoms and the many subatomic entities that scientists postulate as existing. From this boundary the known subatomic particles and forces come into and out of existence from various energy states. This horizon explains the wave-particle duality, where a photon for instance can exhibit the properties of both waves and a particle. It can produce interference patterns when passed through a slit, as does wave motion, and yet manifest as particles with a definite finite measurable position or else momentum. The photon becomes a wave upon entering etheric space, and a particle when in empirical space, and moves from one to the other with great ease.

Though Einstein's theory of Relativity appears to have eliminated the need for the existence of an ether to explain certain anomalous phenomena, nevertheless such an ether does exist, and is perceptible to all clairvoyants and *yogins* with developed *siddhis*. The *chakras* that are found in the ethers are the basis to all meditative awareness. This subject, as well as that of multidimensional perception, has been dealt with earlier and detailed throughout my books, hence I will not repeat

---

1 Physicists have discovered aspects of the phenomena associated with etheric space, which has produced their perplexing quandaries.

the information here, but rather continue upon the line of reasoning already established. Consensus regarding the absence of, or no need for, an existence of an ether has lasted for many decades in the scientific community, however revisionist viewpoints have now surfaced in that community. In a paper, entitled *'Quantum Vacuum and a Matter – Antimatter Cosmology',* Rothwarf and Roy[2] stated in their introduction:

> For a long time the concept of an aether as a medium for the propagation of electromagnetic waves has been discredited, even though Maxwell's equations were originally derived based upon the assumption of an aether. Today the need for something like aether is acknowledged in physics by invoking terms such as "quantum vacuum," "vacuum fluctuations," or "zero-point fluctuations." In fact Grossing has recently shown how the Schrodinger equation can be derived by invoking such zero-point fluctuations.[3]
>
> Allen Rothwarf[4] has reviewed the objections to an aether and concluded that it was indeed needed to explain many problems in physics such as wave-particle duality; the nature of spin; the derivation of Hubble's law; electric fields; Zitterbewegung; inflation in cosmology; the arrow of time; the Pauli exclusion principle; the nature of the photon; neutrinos; redshifts; and several other ideas. In that paper he referred to previous aether models but finally chose to explore a model based upon a degenerate Fermion fluid composed of polarizable particle-antiparticle pairs, e.g., electron-positron pairs. This leads to a big bang model of the universe, where the velocity of light varies inversely with the square root of cosmological time, t.
>
> He was motivated in part by Dirac's 1951 letter to Nature titled "Is There an Aether?"[5] in which Dirac showed that the objections to an

---

2 The credentials given for Frederick Rothwarf are: Department of Physics, George Mason University, Fairfax, VA 22030 USA, also Magnetics Consultants, 11722 Indian Ridge Road, Reston, VA 20191., USA. For Sisir Roy they are: Centre for Earth Observing and Space Research, College of Science, George Mason Univ., Fairfax, VA 22030, USA, also: Physics and Applied Mathematics Unit, Indian Statistical Institute, Calcutta, INDIA.

3 Grossing, G., http://arxiv.org/abs/quant-ph/050879.

4 Rothwarf, A., Physics Essays 11, 444 (1998).

5 Dirac, P.A.M. Nature **168**, 906 (1951).

aether posed by relativity were removed by quantum mechanics, and that in his reformulation of electrodynamics the vector potential was a velocity.[6] Dirac concludes his letter with "We have now the velocity at all points of space-time, playing a fundamental role in electrodynamics. It is natural to regard it as the velocity of some real physical thing. Thus with the new theory of electrodynamics we are rather forced to have an aether."

They also state:

In the Introduction we claimed that with the aether model *the expansion of the aether replaces space-time and one should obtain results equivalent to those derived by relativity models.* In support of this contention we showed that the time dependence for the expansion of the universe *R(t)*, given by Eqn (4), is nearly identical to that found from relativity for an Einstein-De Sitter universe dominated by radiation. Moreover, another result that arises naturally from the aether fluid is Hubble's law. Furthermore, the theoretical deceleration of the ether expansion, $a_{Ro,}$ corresponds very closely to the centripetal acceleration a_{Co} given by Carmeli's relativity-based MOND[7] model. Also we have been able to derive from the aether model the Tully-Fisher relationship and to show that it can be related to the redshift of a given galaxy. Thus, we argue that the aether deceleration supplies the local centripetal acceleration needed to account for the flat rotation curves observed for spiral galaxies.[8]....We wish to point out that in a sense the aether is equivalent to "dark matter" that pervades the universe.

At the heart of the discussion in this volume lies the concept of the Anu, the properties of which were explained in chapter 8. The Anu was shown to consist of five main groups of spirals, two of which are larger, and called extra or thermo-spirals. The Anu's were considered the basic building blocks of all manifesting phenomena, whether existing within subjective or objective space. The thing to note when observing phenomena is that everything is deduced by means of the mind and the Fires governing that mind are electrical in nature. It is the electrical

6 Dirac, P.A.M., Proc. Roy, Soc., A209, 291 (1951).

7 Modified Newtonian Dynamics.

8 Saunders, R.H., and McGaugh, S.S. arxiv.astro-ph/0204521.

energy of this Fire that pours through the spirals and spirillae of the Anu. If the idea of the Anu is extended to the size of a universe then that universe can be considered to be Mind conceived, with attributes that are extensions of the properties of the Anu. The Anu's are also collectivised by mind/Mind in various ways to produce the thoughts of a thinker.

The problem of gravity

The subject of the nature of gravity is somewhat of a contentious issue. It is said to come into manifestation with the appearance of mass, and it is difficult to pinpoint exactly what is meant by mass in the scientific literature. When mass does appear then normally also charged particles are associated with this phenomena.

As gravity is the fundamental force that holds a form into a coherent shape and which integrates forms, because of the attraction between their masses, into an integral system or form, so it stands to reason that gravity is but an extension of the property of the Anu. From this concept some postulates shall be made below to try to explain the nature of the subatomic particles discovered by physicists from the esoteric viewpoint. Hopefully, future experimenters should be able to validate whether these postulates are correct or not.

1. A single Anu is a graviton.[9] Consequently they are built into the sum of what is conceived of as matter. Because as postulated below, three Anu's constitute a single quark, therefore it stands to reason why gravitons have not been discovered. It is the attractive forces between the corresponding spirals of two or more Anu's that can be presumed to account for the existence of the law of gravity. Note also the two different spins of the Anu can either project energies downwards into manifestation or upwards towards subjective space.

---

9 Gravitons, the carriers of the force of gravity, have been postulated to exist by Physicists, but have not yet been discovered, their existence is problematic in quantum field theory, and still needs solving. Whether individual Anu's are indeed gravitons yet remains to be seen, but I shall equate the two as a hypothesis related to the appearance of phenomena. In Einstein's theory of relativity however, gravity is the result of space-time being curved, which affects all particles, whether having mass or being massless (as is a photon). Mass bends or warps space-time, thereby effecting the force of gravity.

2. A double Anu is a photon.
3. Sheets of Anu's produce the phenomena of light waves.
4. When the Anu's are combined, producing an even number of them then no mass is discerned in the subatomic particle. They manifest therefore as bosons, force carriers that can occupy the same quantum state, hence are not governed by the Pauli exclusion principle.
5. When the Anu's manifest in the form of triads or other odd numbers then mass comes into existence. An Anu triad is conceived where a pair of Anu's are interrelated through insertion of a narrow end into a wide end of another, then another Anu is incorporated at right angles to the pair, in the manner found in figure 38. The mass is formed through the interrelation of the energies, producing a vortex of reifying forces. With the appearance of mass then such phenomena, such as the action of gravity and the appearance of solid objects can be perceived. These Anu's manifest as fermions, which obey the Pauli exclusion principle, where a fermion can occupy a particular quantum state at any time.
6. When three Anu's are thus interrelated they can manifest either as a quark or an electron. The difference being that an electron derives its energies from the seven intra-spirals, and the quark from the thermo-spirals. When electrons manifests as an electromagnetic wave then I would expect an instantaneous dissociation of this triadic arrangement. Cumulatively the quarks manifest the positive attractive charge of a proton. The electron the corresponding negative electrical charge. Together with protons they constitute the sum of the elements and compounds of manifest space, in the manner described in *Occult Chemistry* and further elaborated by Phillips in his books. The positive proton represents the attractive mass of matter to be experienced. The moving valency electron represents the mechanism of experience and the photon represents the carrier of what has been experienced.

   Their minute size, coupled with the integration of the three Anu's produces the point-like characteristic attributed to quarks and electrons.
7. Two groupings of 3 x 3 Anu's make a proton (a hydrogen nucleus).

8. A neutron decays into an electron plus energy, which manifests as an electron antineutrino, the carrier of the former binding energy. We can consequently consider a neutron to possesses 21 Anu's. The number 21 = 7 x 3 is the appropriate number that would seal the attributes of the physical plane, as there are seven sub-planes to any plane of perception, and we are concerned with the three lowest planes (mental, astral and dense) as the planes of *saṃsāra*. Also there are three sub-planes to the dense physical plane (gaseous, liquid and dense) related to the exoteric appearance of phenomena. There are only 18 Anu's to the protons, hence technically there is a shortage of three to make the base number 21. The protons consequently have a positive charge to attract to them the missing three via the negatively charged electrons that couple with the atom in which the proton exists to produce a balanced zero charge. The electron antineutrino that is emitted would probably not contribute Anu's to the constitution of the neutron whilst the neutron exists in the atom, as it represents the binding energy, but instantly takes form when the neutron disintegrates. If the four Anu's associated with neutrinos are added, then we have the number 25 = 5 x 5, which refers to the primacy of the energies of mind, with which this thesis is primarily concerned.
9. One can speculate a neutrino to consist of four Anu's, two pairs superimposed upon each other. This would account for a neutrino having no observable mass. (Though physicists have recently discovered that they do have a minute mass.)
10. All other subatomic particles can be considered to be the result of various combinations of Anu's.

Having stated the above a more detailed explanation can now be provided.

As the manifestation of triads is central to the esoteric philosophy: Father—Son—Mother, Will—Love-Wisdom—Activity, *sattva—rajas—tamas*, so also there should be three subatomic particles (carriers of energy) that play this role, hence I would postulate that, in terms of their energetic relationships:

- Gravitons represent the Father—Will—*sattvic* role. They manifest the integrating power of gravity that draws all into a unity. Without the functioning of this very weak force (10^{-38} times the strength of the strong nuclear force) the universe could not cohere as a unit, nor would anything that it contains, such as galaxies.
- Photons represent the Son—Love-Wisdom—*rajasic* role. They manifest the electromagnetic wave fields of light. Light carries the consciousness principle, the ability to see, and hence to know.
- Electrons and quarks represent the Mother—Activity—*tamasic* role. (They are viewed together here because their structure is the same.) Electrons represents the negative valence of an atom, allowing the construction of compounds, whereby the world of phenomena is expressed. They also help cause the charged particles of plasma (though most plasma is electrically neutral) and of the electromagnetic phenomena in the universe. Quarks are bonded as part of the process that cause baryons (large particles) to exist.

As the Father aspect is normally abstracted, whose Will coheres all into unity during *manvantara,* and then acts to effect the dissolution process *(pralaya)* when the cycle has run its course, so then the effect of gravitons is similarly abstracted, playing a subjective, hidden pull upon the appearance of the phenomena of the other two. The direct effect of this force however is upon the electrons and quarks, producing the Father-Mother interrelation that produces the appearance of the form. Their 'children', the gain of their interrelation, is the atomic structure with which the material universe is built (the feminine 'child'). We also have the electromagnetic waves of photons, producing the light that allows the appearing phenomena to be appropriately perceived by mind (the male 'child').

From the above perspective, gravitons represent the energy sustaining the equivalent to the *suṣumṇā nāḍī* in Nature. Photons represent the attributes to the expression of the *piṅgalā nāḍī* in Nature, and electrons and quarks represent the effects of the *iḍā nāḍī* in Nature.

Without the development of mind, its deductive, image-building facility and the capacity to memorise, the appearance of a universe would be meaningless, a non-event. No matter how many universes appeared and died, if there were no witnesses to that phenomena that could retain the experience as memory and consequently grow a vaster Mind-scape then what is the point of the appearance of a universe? Whether it existed or not could not even be a conjecture. From this we can postulate that the objective for the manifestation of a universe is to produce bearers of mind/ Mind. Once such bearers appear and evolve through billions of years of evolutionary attainment, then it stands to reason that the development of such Minds would encompass a universe and beyond—because of the ability of Minds to project what has been observed towards future expansion, and to produce creative activity, the building of Thought-forms of varying degrees and sizes. So then it follows, and by taking the previous information in this volume into account, that the super-Minds that have developed will have learned to create such a universe from which they evolved through pure analytical deduction and projection of the creative forces of the mind/Mind. The Mind simply needs the raw materials with which to build. Projecting such a postulate backwards, then the conclusion is that the present universe is the product of the work of the cumulative effect of the Creative work of a grouping of Super-MINDS. Consequently the universe is Mind-Created.

The four ethers

Before a more detailed explanation of the fundamental sub-atomic particles and forces can be given we must first comprehend the nature of the four etheric sub-planes of the dense physical plane, from which the fundamental particles manifest as aspects of their energies. The Ray governing each of the etheric sub-planes shall also be given.

First sub-plane of the etheric.

First Ray of Will or Power. **Ether 1** is the subtlest ether and reflects the energies originating from the plane *ādi* into manifestation. From here emanate the energies that will inevitably produce the force of *gravity,* working via each individual Anu. The entire formed universe manifests from the extension of the 'atomic' structure of the Anu. Each Logos incarnates via their version of an Anu, where the planes of perception that are embodied by the Logoic Mind manifest in the form of its seven main spirals. What is enacted upon a vast scale is mirrored into the microscopic world of appearances. From the highest level to the lowest, energies can then flow in an increasingly reified manner, to eventually produce the concretisation of what is known to incarnate humans.

At the *ādi* level of manifestation the energies of the combined Logoic *Head-Ājñā* centres are conveyed to cohere the universal All contained in a Logoic Mind into an integral form and to direct the observable phenomena towards a unified purpose. These centres organise the *bījas* and *bindus* of former cycles of expression into the forms that are to be. Consequently the potencies that the *bījas* and *bindus* hold in situ are expanded into the *maṇḍalas* at all levels of expression of the All. The resultant activity in terms of reified energy projection manifests through the first ether in order to help effect the phenomena we know so well via the appearing Anu's. Combinations of Anu's produce the atoms of space. In this way gravity attracts physical entities to each other. All is unified thusly as a consequence of the expression of the Will-to-Be of a Mind of a primordial Logos.

Second sub-plane of the etheric.

Second Ray of Love-Wisdom. **Ether 2** reflects the energies of *anupādaka* into manifestation. This is the major energy governing our solar evolution and concerns the experience of the unity of all that is, equated with the force of *bodhicitta* that drives all that has been incorporated in manifest space towards liberation. The energies within the appearing forms now shape the *maṇḍalas* of all that is to be via spiral-cyclic motion. This produces the appearance of *electromagnetic radiation,* the emanation of light rays (hence photons), and the downward motion of an impulse that eventually manifests in a forward-progressive expansion of conscious awareness in humans. Our thoughts are electromagnetic in nature, we think by utilising light rays. Logoic Thoughts are vast and universal in scope.

The qualities of this second etheric sub-plane, which relates to the expression of the *Heart centre* at this level of manifestation, explains the phenomena of nonlocality in the quantum world, where elementary particles can act instantaneously at long distances upon each other. There can be instantaneous effects of an experiment in one laboratory upon a similar experiment in another laboratory in another part of the world. The Heart centre produces the experience of the universality of the Oneness of the All, and expresses the Emptiness that is *śūnyatā,* which is reflected in the phenomena of nonlocality manifested by the subatomic particles.

Third sub-plane of the etheric.

Third Ray of Mathematically Exact Activity. **Ether 3** reflects the energies of *ātma* into manifestation, hence it is the lowest extent of cosmic Fire, which manifests as an aspect of the formative attributes of Mind in manifestation. The energies stemming from this sub-plane, being reified Fire, would manifest both the *strong and weak nuclear forces* that form the protons and neutrons of the nucleus, and integrate the atomic nucleus into a unity. Quarks and hadrons come into existence via this ether to produce the observed mass of each atom. The *weak force* changes *quarks* from one form to another and is responsible for radioactivity. Quarks are organised by means of the mode of the manifestation of mind/Mind, expressing the trinities of organised expression, such as past, present and future of any thought-form, or of origination, sustainability and eventual destruction. (Brahmā, Viṣṇu and Śiva, and their feminine counterparts.) These masculine/feminine pairs respond to the energisation from Ethers 1, 2 and 3 respectively. Being expressions of the mind/Mind we would also expect an appearance of fiveness in their intrinsic properties, such as electric charge, mass, colour charge, spin and particle decay. Via this level of expression the *electron* and the associated electric currents of space would also manifest, as well as all forms of the activity of plasma. Hence the energy streaming from the third ether is the Mother of all the phenomena that we experience with our senses.

The associated *chakra* at this level of manifestation is the Throat centre, the organ of mental creativity.

Fourth sub-plane of the etheric.

The fourth Ray of beautifying Harmony overcoming Conflict – which acts as a mirror for the attributes of the above Rays, hence reflecting the attributes of these three ethers into manifestation. **Ether 4** contains all of the centres below the diaphragm and the Inner Round. The focus of activity manifests via the Solar Plexus, Sacral, Base of Spine, Liver, Stomach, and the two Splenic centres. This makes seven centres, capable of absorbing the energies from the septenaries of manifestation.[10] The dense physical domain then represents the Eighth Sphere. The

10 See figure 7, volume 5A of *A Treatise on Mind*, 207, for a diagram of these *chakras*.

Diaphragm centre acts as an intermediary between the three higher Ethers and these eight centres (the seven, plus the Eighth Sphere).[11]

Via ether four the formative forces appearing in the higher three ethers are crystallised, producing the appearance of phenomena, the complete structure of the Anu, and the combination of the Anu's, producing the objectivisation of atoms, the appearance of the forms existing in our tangible universe. This ether sustains the quantum vacuum, the zone of transference of energies and forces from the subjective to the objective universes. Consequently the forces manifesting through this 'vacuum' are potent and strong.

The violet Ray colours the four ethers, manifesting from a light lavender shade to deep violet.

The Higgs field

Once the quarks come into existence producing the nuclei of atoms (the fermions) then the force field generated through the formation of positively charged protons attract negatively charged electrons. Electromagnetic radiation can then proceed and also the force of gravity, all manifesting via a mirror-like action from the subjective domain via the quantum vacuum that is the boundary between etheric and gaseous space. The (baryonic) matter that appears manifests as the universe we can see and touch, whilst the electrons and photons can move into and out of etheric space because they are intrinsically the substance of the subtler, more energetic ethers. Here the *Higgs field* comes into play, from which all elementary particles are said to gain their mass via the appearance of the Higgs boson.[12] From this perspective this field is but a term describing the properties of the fourth ether.

---

11 The *prāṇas* of a system that are completely rejected, being no longer useful to the incarnate individual, enter a zone of containment, called the eighth sphere. It is styled the 'eighth' because it contains that which cannot be sustained by the qualifications of the seven major *chakras*. It is a hell zone *(naraka)* when experienced because it contains the *prāṇas* that consciousness has eliminated and which would sicken it or stifle its growth if perpetuated in. The physical domain represents this attribute from the perspective of the embodiment of Mind. It is not a principle for the incarnating *jīva*, the human unit, representing that which must be transformed and eventually transmuted upon the enlightenment path. See also volumes 5A and 7A of *A Treatise on Mind,* where it is considered in terms of a consciousness state.

12 A boson is a subatomic particle with zero integral spin, further explained below.

As the existence of the Higgs field has been confirmed in 2012-13, and for which a Noble prize was awarded, so it is important to gain some idea as to its importance, which lies in problems concerning the concept of mass of the subatomic particles. Einstein's formula $E = mc^2$ equates energy with mass, however it is difficult to properly define what exactly constitutes mass in physics, as all exists in different energy states. The subject of where mass comes from, which forms the basis of everything we see and touch, hence needed proper answering. To explain the manifestation of the rest mass[13] of elementary particles physicists have come up with the concept of the Higgs field and the Higgs boson, from which the elementary particles draw their masses. Quarks gain their masses because they interact with the Higgs field. The concept is that they travel through space which is filled with something to slow them down. This something is the Higgs field. Their motion is impeded in this field, which provides them with mass. The Higgs mechanism only explains the amount of mass that a particle has at rest, not the mass it appears to gain at high velocity, due to relativity. The relativistic mass, due to the high binding energy within the nucleus of the atom accounts for most of the mass. Hence most of the rest mass of protons and neutrons is not due to the rest masses of their constituent quarks, rather, about 99% of their mass is due to the binding energy of the strong force (the gluons) between the quarks, that keep them integrated as triads.

In the standard model of particle physics the Higgs field is defined as a quantum field that interacts with the other fields in this model that produce the observed elementary particles so that some of the particles manifest a non-zero rest-mass via the Higgs mechanism. This mechanism generates the known rest masses of the elementary particles.[14] A quantum field is something defined at each point of space-time, where the excited states of a field, the states of higher energy than the vacuum energy, produce the elementary particles. Thus, for every elementary particle, there is a corresponding field. Consequently this makes sixteen such fields to account for the quarks and leptons and an extra one for the Higgs boson.

---

13 The rest mass is the measured mass of these particles when they are not moving.

14 Before the discovery of the Higgs field the problem physicists had was how to appropriately explain the correct rest masses to the elementary particles, without violating the fact that right-handed fermions are not observed in Nature.

The Standard Model is a gauge theory (or local theory, meaning that independent symmetry transformation at different points in space can be made without affecting the physics, the physics being independent of one's coordinate choice). The Higgs field does not provide energy to the subatomic particles, it simply converts the energy of those that travel at light speed into rest mass. It does this by periodically interacting with their spin. The mass (or energy) is conserved because the *Higgs boson* is created from energy that is transiently borrowed from the Higgs field. The energy is returned to the field after its interaction with other particles. The Higgs field imparts masses only to fermions, such as electrons or the top quark. The Higgs field provides a mechanism for electroweak symmetry breaking, which separates electromagnetism from the weak interaction. Consequently it also endows the Z and W bosons[15] with their considerable rest masses, and consumes the Z boson's longitudinal degree of freedom, which would otherwise wreak havoc with renormalisation (elimination of infinities in the theory). Also by interacting with the charged fermions (electrons, muons, taus and quarks) it gives them rest mass.

The value of the Higgs field is constant in space, consequently it possesses no kinetic energy, thus sits at the bottom of the scalar potential. For this reason the Higgs field is described as a complex scalar field, which is conceived of as a two-dimensional space of complex numbers, however when considering quantum fluctuations there is a contribution to the cosmological constant. This applies to all fields, not just to the Higgs field.[16]

The Higgs field does not slow particles, rather it makes particles resist acceleration, not velocity. Hence the Higgs field is compatible with special relativity. The Higgs field gives particles energy, hence they gain mass. The different particles have different masses because of their different strengths of interaction with the Higgs field. Photons have no interaction with the Higgs field. As there is a neutral electric charge for the field, which is what a photon would interact with, there's no interaction, hence photons remain massless.

---

15 The carriers of the weak force.

16 Physicists have a problem with the models of the calculations of the cosmological constant that come from all the fields, which is much greater then the actual cosmological constant measured. This arena thus still needs rectifying.

The Wikipedia article states:

> In the Standard Model, the Higgs particle is a boson with spin zero, no electric and no colour charge. It is also very unstable, decaying into other particles almost immediately. It is a quantum excitation of one of the four components of the Higgs field. The latter constitutes a scalar field, with two neutral and two electrically charged components that form a complex doublet of the weak isospin SU(2) symmetry. The Higgs field has a "Mexican hat-shaped" potential. In its ground state, this causes the field to have a nonzero value everywhere (including otherwise empty space), and as a result, below a very high energy it breaks the weak isospin symmetry of the electroweak interaction. (Technically the non-zero expectation value converts the Lagrangian's Yukawa coupling terms into mass terms). When this happens, three components of the Higgs field are "absorbed" by the SU(2) and U(1) gauge bosons (the "Higgs mechanism") to become the longitudinal components of the now-massive W and Z bosons of the weak force. The remaining electrically neutral component either manifests as a Higgs particle, or may couple separately to other particles known as fermions (via Yukawa couplings), causing these to acquire mass as well.

The above should suffice to provide *an idea* as to the nature of this field. As I am not a physicist, interested readers should consult the appropriate texts for detailed information and explanation of the terms used, especially regarding the mathematical modelling. The information I have provided here is necessary because it is important to see whatever correlation there might be between the scientific concepts that have been proven by experimentation and the esoteric understanding of the etheric body. From my presented theory we see that only those energies manifesting from the third ether (e.g., quarks and electrons) interrelate via the Higgs mechanism upon the fourth ether to produce the appearance of mass. Photons, coming from a higher energy source (the second ether), and whose Anu's are bound as sheets, and not as triads, do not interact with this mechanism, hence remain massless. It may indeed be that the Higgs mechanism partly described above is what causes three Anu's to combine, and hence 'flash' into empirical space as a point-like entity with the aid of the newly discovered Higgs boson. When talking about this field all that is considered is the junction between the substance of the fourth ether, and the gaseous matter of the physical plane enacted upon by the formative forces of elementary mind passing from the third ether. The fourth ether simply represents the sub-plane where crystallisation of energies is enacted.

The fourth ether represents the lowest reflux of the plane *buddhi* (hence *śūnyatā)* into manifestation. Here therefore exists the physical plane Void, hence the appearance of the quantum vacuum and its natural fluctuations of energies as phenomena appears and disappears into and out of manifest space.

The Liver, Stomach and Solar Plexus centres can here be seen as a unity dealing with Watery phenomena, viewed in terms of energy vectors. From the planetary Solar Plexus centre will come the energy, the strong force, that will organise the up and down quarks to form the constituency of the protons and neutrons of atoms. Via the planetary Liver centre will come the energy that generates electromagnetic radiation, the light bearing photons. From the planetary Stomach centre would come the energies that denote the weak force within the nucleus of an atom, plus the force that causes the appearance of electrons. The effect of the planetary Base of Spine centre would then cause the manifestation of the force of gravity. The potency of the planetary Sacral centre helps to bring all forces and particles into active manifestation via the event horizon.

The fourth ether acts as a mirror, reflecting the attributes of the higher Rays into the three lower ones, which condition material manifestation. Gravity manifests as the force of Mind working via the Base of Spine centre to organise the elements of thoughts (Anu's), collectivising them into a picture (idea) of the form of what is to be. This conveys information, setting the stage for the appearance of things that have mass, and consequently weight in the material domain.[17] This is a fifth Ray function. Its mode of conveyance (driving impetus) produces light and sound (esoterically, electrical activity) for the expression of its information. We thus have the photon (*piṅgalā* form of expression,[18] conveying light) and electron (*iḍā* form of expression conveying sound,[19] esoterically understood). Photons and electrons are energised via the seven intra-spirals

---

17 Weight being the force exerted by a mass in a particular gravitational field. Hence a given mass may have different weights, on say the earth or the moon, because of the differing gravitational fields of the two planetary bodies.

18 The *piṅgalā* line being an expression of the energy of the planetary Liver centre.

19 The *iḍā* line being an expression of the energy of the planetary Stomach centre.

of the atoms, where the photon is arranged via sheets of two or more Anu's and the electron as particles consisting of three conjoined Anu's.

The sixth Ray conditions the appearance of the host of subatomic particles that have energetic interrelationships with each other. One can however hypothesise that the neutrinos, that interact with the subatomic world with great difficulty, would be governed by the fourth Ray. The seventh Ray would then produce the appearance of the atomic structure via the interrelation of quarks and electrons, thus the manifestation of the various compounds of the elements.

One can also note that there are three broad categories of the electromagnetic spectrum. First those with a low frequency, low wavelength, low quantum energy, comprising of the radio and television waves, from approximately 105 to 109 Hz. Next there is an intermediate grouping; of microwaves, radar, millimeter waves, telemetry and the infrared. Finally there is a high frequency grouping, starting with visible light, ultraviolet light, X and gamma rays. From them then we can consider a septenary manifesting. The first Ray line consisting of the X and gamma rays. The second Ray line being represented by the ultraviolet rays. The third Ray then conditioning the appearance of the visible spectrum. The fourth Ray would then condition the manifestation of the infrared rays. The fifth Ray would help produce the phenomena of mm waves and telemetry, the sixth Ray would condition the frequencies of microwaves and radar and the seventh Ray the radio and television, with their long wavelengths.

The third to fifth Rays are the Rays of Mind, and in the schema of this electromagnetic spectrum we see that the wavelengths associated with the first two Rays are those that are effectively destructive to phenomenal life. This means that they are liberating to the Life principle enclosed within those forms. The aspects of the Rays of Mind of this listing are those that are the mainstay of modern civilisation enhancing electronics; the computer and entertainment industry, microwave ovens and the like.

Quarks and leptons

Let us now look to the three generations of quarks and leptons. First an explanation of basic terminology is needed. Physicists classify particles according to their mass. Light particles are termed leptons, particles with medium mass are called mesons and the comparatively heavy particles are baryons. (The term hadron is also used to denote both mesons and baryons.) Leptons are also called fermions, which are either charged or uncharged particles. Fermions are elementary particles following the Pauli exclusion principle that have half-integer spin, hence relate to the *iḍā nāḍī* attribute within a human form. This *nāḍī* relates to the expression of the intellect, which dissects, discriminates, segregates things into component parts, in order to comprehend. Finally there are bosons, which have a spin of zero or of an integer value. They therefore relate to the expression of the *piṇgalā nāḍī* in the human form, which conveys the principle of consciousness. Consciousness expresses itself in terms of unities, wholes, universals, integrations of unitaries into the greater landscape of thought, hence the fluidity of abstract and enlightened thought, here symbolised by the whole numbers of bosons.

The leptons are said to be point particles with no internal structure. In my thesis however we see that they are probably superimposed pairs (neutrinos)[20] or triads (electron, muon and tau particles). Each are also said to have their anti-particles, which can be considered to consist of Anu's with their spirals spinning in the opposite direction,[21] hence bringing energies from the lower domains to the more subjective levels, meaning therefore sublimation, liberation of the respective forces. There is obviously a similar schema for anti-particles as for the particles concerned with the externalisation of energies. Hence I shall only deal with the particles, with a proviso that a similar consideration can be made for the anti-particles.

Let us recall the five groups of triads of spirals to the Anu and to consider the fact that the strongest (coarsest) of the energies conveyed manifest via the bottommost spirals. The more refined energies manifest as part of the intra-spirals, thus higher up in the sequence. The energies manifest similar to the functioning of a fractioning column. Also to be considered is that these five spiral groups convey the *iḍā prāṇas,* thus related to the evolution of substance into sentience and the development of consciousness. There is a correlation to the subatomic world, where

---

20 If a neutrino has mass, such as the Tau neutrino, which is said to manifest up to 164 MeV, then an Anu may have been added in some arrangement with the pairs to cause the appearance of mass.

21 Consequently if the opposing spirals of the particles and anti-particles come into contact with each other there would be annihilation of their structures.

we are concerned with purely physical forces and energies. Here we must look to the quarks and leptons (light particles) that are point particles conveying the mass that constitutes the formation of the phenomena of the material universe. They are the building blocks of matter. In terms physicists use, mass here is conveyed in terms of electron volts (eV), the stronger the electron volt, the greater the mass. The particles also come in pairs.

Concerning *quarks,* there are said to be six types of quarks (up, down, charm, strange, top and bottom[22]) that account for the formation of all known intermediate heavy particles (mesons) and the baryons, the heavy particles, such as protons and neutrons. Quarks have never been observed singularly, hence always occur either in pairs to form the mesons, or in triplets, to form the baryons. All quarks have a spin (angular momentum of 1/2), an electrical charge of either +2/3 or -1/3 and highly varying masses. A proton is constituted of two up quarks, each with a charge of 2/3, plus a down quark, with a charge of -1/3, hence producing a net positive charge of +1. A neutron consists of one up quark and two down quarks, whereby the charges balance each other out, hence producing a neutral charge. The combination of neutrons and protons in an atom produce its nucleus, with the sum of the positive charges carried by the protons attracting to them an equal number of electrons, each of which carry a charge of -1, thus making the constitution of an atom.

My main concern here is with the masses of the observed particles. These were first worked out mathematically and later confirmed experimentally, though there is a problem with the discernment of the exact masses of quarks. This is because of their containment, preventing them being isolated or measured directly, hence the masses are implied through scattering experiments. Consequently there are different tables for their masses. They are also bound with a binding energy (gluons), which complicates matters. Wikipedia gives their mass in terms of MeV (million electric volts):

up = $2.3 \pm 0.7 \pm 0.5$
down = $4.8 \pm 0.5 \pm 0.3$
charm = 1275 ± 25

22 The reason for the attribution of these names can be found in any pertinent book on physics. Similarly for the detailed background to the information here presented.

strange = 95 ± 5
top = 173210 ± 510 ± 710
bottom = 4180 ± 30

The masses given for the leptons in MeV are more precise:

electron = 0.510 998
Muon = 105.658 366
Tau = 1776.84
Electron neutrino = < 0.00000022
Muon neutrino = < 0.17
Tau neutrino = < 15.5

In figure 42 I shall represent the five triads of spirals governing the *iḍā* functioning of an Anu as straight lines, with an descending order of density of the energies conveyed, hence placing the Earthy triad at the bottom, and the most refined, Aetheric triad at the top. The Watery and Earthy triads are the extra-spirals of Babbitt's 'atom', and the remaining three the intra-spirals. To them I shall add the corresponding quarks and leptons in terms of their corresponding masses. It should be noted that there is a resonance between the spirals of each triad. Hence the bottom line, of for instance the Watery triad, which resonates with the bottom line of the Earthy triad, and the top line of the Watery triad with the top line of the Earthy triad, and so forth. The quark pairs hence shall be assigned according to this resonance. They occupy the extra-spirals because they are concerned with the appearance of matter, the substance of our forms. The leptons and their corresponding neutrinos shall therefore be assigned to the intra-spirals, effectively therefore being carriers of the 'consciousness-principle' upon an atomic scale.

Such resonance is but a reflection of the way energies manifest through the planes of perception esoterically considered. The line of energisation moves from the odd numbered planes and sub-planes, 1-3-5-7, (the *iḍā* line) and also the even planes and sub-planes, 2-4-6 (the *piṇgalā* line). Similarly for the seven Rays.

The significance of triads is of similar importance in both the inner realms and the exoteric domain of subatomic particles. In the listing of the quarks and leptons for instance, we have three pairs of quarks, whilst the combination of three quarks make either a proton or neutron. There are three main particles to the leptons (the electron, muon and tau) and there are three related neutrinos, and also for their antiparticles. Similarly we speak of three planes of perception concerned with human livingness (the physical, astral and mental) and three abstract planes *(ādi, anupādaka* and *ātma)*, with a central plane *(buddhi)* reflecting the energies from the abstracted three into the empirical three, and vice versa. There is also the consideration of the three major petal groupings of the Sambhogakāya Flower, abstracted by three bud petals, the three main *nāḍīs,* the three *guṇas,* the Logoic trinity of Father-Son-Mother, Brahmā, Viṣṇu and Śiva, etc. The fact is that the groups of three that regulate the ordering of subatomic particles simply reflect the paradigms set by the forces manifesting from the inner universe.

The elementary particles are but conveyers of the types of energy expressed through the various spirals of the Anu. In this schema the topmost set of spirals have no elementary particles assigned to them as yet discovered by physicists, because being too refined. This means that their energies can descend no further than the three lowest etheric sub-planes. They can be considered to signify three stages of the manifestation of the graviton (explained later). Each stage simply represents the conveyance of coarser energies, that sympathetically convey energies from the *ātmic* plane and thence to the domain of mind, from whence comes the emanatory impulse that causes the formation of things.

Consequently the three pairs of quarks are organised similar to the functioning of the mode of the manifestation of mind. They express the trinities of organised expression, such as past, present and future of any thought-form. Like all forms there is an origination, sustainability and eventual destruction. One would also expect an appearance of the fiveness that governs the domain of the mind, in their intrinsic properties, such as electric charge, mass, colour charge, spin and particle decay.

Here the base numbers for the quarks are used and the mass of the leptons is rounded out to three decimal places.

The five levels of these triads are labelled Aetheric to Earthy, relating to the type of energy that empowers the respective quarks and leptons. They represent the forces of the five alchemical Elements. The Earthy relates to that which is most reified, the Watery to the forces of condensation (amalgamation). The Fiery incorporates the electrical force from the mental plane, thus the formative forces that cause the amalgamation of forms. Of these *the Fiery level is paramount* because it expresses the domain of the mind that instigates the formation of what is to be. Here the emanation of the Fires of mind is the electron. It becomes the main carrier of the originating impetus of the energies from the Logoic Mind, which sets the conditioning for the electrification of the universe. The electron's electrically negative charge is then perfectly balanced by the positive charge formed through the appearance of the proton. This produces the conservation of energy needed for the closed system of the universe (and esoterically the balancing of *karma).*

Aetheric		
Airy	electron neutrino	0 MeV
	muon neutrino	<0.17 MeV
	tau neutrino	<15.5 MeV
Fiery	electron	0.511 MeV
	muon	105.658 MeV
	tau	1776.84 MeV
Watery	up	2.3 MeV
	strange	95 MeV
	bottom	4180 MeV
Earthy	down	4.8 MeV
	charm	1275 MeV
	top	173210 MeV

Figure 42: Fermions and quarks

The Airy triad are energy carriers from the subjective domains, literally from the three lower etheric sub-planes. The Aetheric triad represents the point forces that the lower triads draw energies from. Here they are left vacant, but later when the function of the Anu as a graviton is explained, then it can be placed at this level. (Logically three types of Anu's can be conceived, each being receptive to a different type of energy.)

In relation to the levels of resonance therefore, we see that the first generation of subatomic particles is constituted of the electron neutrino upon the highest of the three Airy spirals, the electron for the corresponding Fiery spiral, the up quark for the corresponding Watery spiral and the down quark for the corresponding Earthy spiral.[23] As this first resonance level is what directly pertains to the energies of the mind (the highest of the three planes of human livingness), so the particles of this level accounts for the phenomena we experience. (The atomic structure, and electromagnetic currents.) The second resonant level generation of subatomic particles (muon neutron, muon, the strange and charm quarks) relate to the Watery domain, hence the fluidity of the quark transformations and colour change seen at this level. Muons quickly decay to form an electron or positron. The third resonant level (tau neutron, tau, bottom and top quarks) are heavy particles that relate to the Earthy domain. This Element here simply relates to the transient appearance of mass.

This concept of resonance from one spiral to its corresponding one *demystifies the problem physicists have regarding explaining the existence of these three generations of particles*, i.e., why this is so.

The manifestation of colour charge through strong interactions in the theory of quantum chromodynamics, where there are three colour and their 'anticolours' can also be considered to relate to the three different energy levels of each triad of spirals. The exchange of forces (gluons) as energy moves from one level to the next producing transformations in the associated particles is conditioned by these three energy levels. That there is said to be eight different gluon types to act as force carriers can esoterically relate to the energies manifesting from the four different resonant levels in either an upwards or downwards fashion. The forces

23 Lists of the three generations of subatomic particles are given in such texts as *Elementary Particle Physics* [Universities Press (India), 2000], 24. (Bruce Winstein, Chair of the committee.)

(gluons) moving downwards would tend to reify, produce bondage and the appearance of mass. The forces moving upwards would tend to release from bondage, allowing a rearrangement of mass, consequently the changing of one particle to the next. Such a rearrangement from an esoteric perspective also concerns an inwards (eastern) motion of the energies, whilst the appearance of a new particle constitutes an outwards (western) motion.

Physicists may be able to determine whether the intermediate positions of these directions in space also apply. These eight directions in terms of pathways of consciousness expression have been explained throughout *A Treatise on Mind.*[24] The reified concepts concerning these directions theoretically can be applied to the field of subatomic particles. Here the northeast direction, labelled as 'unity', relates to a common source for the energies. The southeast is 'expression', meaning the actual manifestation of the subatomic particle. The southwest direction is given as 'understanding', meaning here the gain of the energy exchanges as a consequence of the formation of the particles, whilst the northwest direction of 'goodwill' translates in the case of subatomic particles to their liberation from bondage to form.

Quarks have five properties, they are said to have mass, are strongly interacting, they have an electric charge, are weakly interacting and have a spin. These properties can be related to the five alchemical Elements thus:

1. *Mass.* Conventionally, mass of any manifest form is considered as the quantity of matter that form contains and is determined by the force that gravity exerts upon it, or else by the acceleration exerted upon it by a given force. People normally consider a mass as the existence of the quantity of a particular thing, such as a brick wall, that possesses a weight, the weight being the product of the pull of gravity upon it. Here the associated alchemical Element is Earth, which is concerned with the interrelation between forms.

   In the world of subatomic particles, and of quarks in particular, there is the current quark mass of the quark itself, which is small when compared to the constituent mass of its mass when bound to

24 These directions have been utilised to explain various aspects of the *buddhadharma* and esoteric philosophy. See specifically volume 1, 110-120 of *A Treatise on Mind*, where they are introduced.

another quark by means of the gluon stemming from the strong binding energy. (Most of the mass is derived from the binding energy.) The size of the particle is determined by the amount of energy that is incorporated by the mass. (For quarks this is called the quantum chromodynamics binding energy.)

2. *Strong interactions and colour change* relates to the expression of the Watery Element, which determines the fluid mutability, a state of flux or change of a manifesting form within the environment that it resides in. There are three 'colours' attributed to quarks, labelled red, blue and green. (There are also antired, antiblue and antigreen for the anti quarks.) These colours concern a system of attraction and repulsion between quarks which binds to any specific form, such as that of a proton. The laws governing the colour interactions, that are described in terms of quantum chromodynamics and gauge symmetries, will be found in any appropriate textbook. The significance of trinities, and the nature of their manifestation in terms of the three *guṇas* for instance, has been explained earlier, whilst an interested physicist can determine if colour interrelationships are indeed governed by the same general principle as the mode of the manifestation of trinities.
3. The *electric charge* attributed to quarks, which are fractional (-1/3 or 2/3 times the elementary charge) relates to the expression of the Fiery Element and hence to the electrical nature of the action of the mind. The mind organises the electrically charged substance upon the mental plane to produce the images and ideas in the mind, completed thoughts. The atoms, molecules and the myriad compounds observed in Nature are but the correspondences to thought-forms that exist upon the mental plane. In the subatomic world they come as a consequence of the formation of protons in the nucleus with a positive charge, allowing them to attract negatively charged electrons, bound within valency shells (because of the Pauli exclusion principle). The valency laws then produce the interrelation between atoms.
4. Next we have the *weak interaction* (or force) which allows the quarks of the different flavours to change from one flavour to

the next. This is the cause of the radioactive process (beta decay), consequently allowing one atomic element to convert to another one. The associated Alchemical Element here is Air, which is the carrier of all the *prāṇas* governing the evolution of consciousness and of all the kingdoms of Nature. The *prāṇas* convey the *saṃskāras* that are the agents of evolutionary change, as explained in *A Treatise on Mind*. The entire Life process is concerned firstly with the creation of *saṃskāras* and later with their refinement and transmutation into enlightenment vectors.

5. Finally there is *spin,* which corresponds to the most refined of the Elements, Aether, sometimes denoted as space, being that within which everything moves. It is the Element associated with the *ātmic* plane. Upon the physical domain Aether is the most limiting of the Elements, hence governs the sense of smell, for instance, and often is omitted in the description of things. Spin is normally visualised as rotation around an axis, though this is problematic in particle physics because physicists consider the quarks as point particles. Spin is characterised by its degree of freedom, its location in three dimensional space. For quarks the spin is a vector measured in terms of the reduced plank constant, which produces a spin of + or – ½.

   As earlier stated, within the *ātmic* plane world spheres and their associated *karma* come into active manifestation, producing a rotary motion, which can also be characterised as 'spin'.

The Higgs field, dark matter and subjective space

In terms of the three fundamental laws governing evolution in cosmos, *the law of Economy* governs the activity of the entire dense physical plane and manifests as the determining factor for 'the principle of least action' for the subatomic world. This law governs the interrelation between manifesting forms and refers to the quickest, easiest path of interaction, where least energy is expended to produce a given result.

The *law of Attraction* governs the electromagnetic radiation that spreads out from one light source to impact upon another. The energies are absorbed by one or other body of manifestation, with which they interact and accordingly modify. The interchange esoterically produces an exchange of information or of knowledge in the realm of consciousness.

Though gravity manifests as an attractive force it really comes under the auspices of the *law of Synthesis,* which holds everything together in terms of an integral form. Gravity is counteracted by expansion, the apparent movement away of galaxies from each other as determined by the recession of their redshifts. This can be likened to thought-forms projected outwards to fulfil their purpose. The thoughts eventually produce that which they were designed for and consequently must die, be disintegrated into the elements constituting them. Inevitably all must be brought back to the originating source, the mind of the thinker, as conditioned by the law of *karma.*

The *Higgs field,* by means of which elementary particles are said to gain their mass, can be considered to represent the energy of the fourth ether, into and out of which the particles move. As they do so Anu's are caused to combine in various ways by means of the energies imparted from the ethers, and so obtain mass. The Higgs boson then represents the agglomeration of Anu's in the process of their combination in the appropriate forms that allow them to gain the transience of mass in the way that physicists theorise. The Higgs field is said to be everywhere and surrounds and interpenetrates all objects, everything that has mass—our bodies, physical objects, planets, solar systems and galaxies. Similarly for the etheric body (or 'etheric double' as it is sometimes called).

Astral matter can be considered an extension of the four ethers and the two domains are practically indivisible, except through the mental-emotional activities of humanity, who have created the heaven and hell states into which people enter after death. The astral body can extend from the earth to the moon. The mental body can interrelate the planetary bodies in a solar system. People can travel thus in a body of thought-substance. Residence in a body constituting of *śūnyatā* or *buddhi* is needed to travel through cosmic space within the Body of manifestation of a 'That Logos' to suns that have a direct relation to our solar Logos. Logically then a body consisting of *ātmic* substance would allow such a one to travel to many destinations in a 'That Logos'. One could then speculate that travelling to any place within a galaxy would then necessitate being abstracted in the substance of the higher sub-planes in a Monadic form. However, all that is deduced here concerns movement within what is considered the cosmic dense physical plane, hence the equivalence to the mode of locomotion of pedestrians upon the earth. This concerns travel within an embodied form. Initiates of the highest degrees on earth and Logoi can also 'astral travel' within the cosmic astral ocean. Presumably a Logos abstracted upon the cosmic mental plane can travel the spaces between galaxies.

Interstellar travellers have designed 'ships' (UFO's) of buddhic substance, containing an inner mental-astral environment that allows the conveyance of Souls that have not yet been liberated. Such 'ships' simply obey the same laws, though upon a transmuted correspondence, that allow subatomic particles to appear in and out of the quantum vacuum. The UFO's have left their signature in the innumerable very intricate crop circles that have appeared all over the world. The scientific community has put their blinkers on concerning this phenomena, unexplainable in terms of being hoax's (other than obviously crudely made ones), or any other convoluted empirical theory they have come up with. There is no yardstick that materialistic thinkers can use to comprehend how these large intricate designs suddenly appear overnight. So they enter into a process of denial, as if the phenomena simply does not exist, and go on with their CETI projects and other speculations as to what intelligence or life form might be 'out there', but refusing to acknowledge that their query has already been answered. It is time for so called scientists to remove their blinkers and biases and honestly analyse all of the evidence that is plainly staring them in the face. Similarly they should demand *all*, non-doctored, original photos and data sets of what have been obtained from NASA's exploration of Mars for careful scrutiny. Complete transparency should be provided in this field and given to humanity for all investigators requesting the information. The common heritage of humanity should not be in the hands of certain biased individuals or governments with an agenda that is antithetical to humanity's well-being.

Dark matter and *dark energy* from one perspective can be considered to be the sum of the ionic matter, cosmic dust, dying and dead stars, *plus* the substance and energies associated with the planes of perception constituting cosmic dense physical space, specifically of the fourth ether. The phenomena from these planes have been observed, experienced and explained by clairvoyants and the enlightened for untold millennia. They are needed for all reincarnating entities, and are the source of all psychic phenomena and the sum of the expression of mind/Mind, but are not taken into account in the analysis or computations of the biased materialistic scientific community.

There are planets, stars, constellations and galaxies in etheric space on the threshold of incarnation, or else have entered their *pralaya* stage, which the scientific community needs to account for. They have not discovered them because they have not looked for them, hence have not trained their instruments towards arenas of anomalous energies in cosmos, but where 'nothing' appears to exist. For instance, it might be possible to observe the effect of gravitational lensing (where light bends around a very dense object, such as a galaxy, mirroring images of a galaxy behind it) from a region in space where no galaxies (that would normally be attributed to cause such lensing) are perceived by means of observation through telescopes and other instruments. Astronomers have used this technique to determine the existence of dark matter around galaxies, but should also look for perturbations in the spaces between galaxies, constellations and stars *to assess the existence of such entities that are in the process of moving from an etheric form into the dense physical, and vice versa.*

The sum of the energies that manifest via the cosmic dense physical plane and which impact via the four ethers and thence into dense physical manifestation thus consists of this 'dark energy'.

The concept of gravitons

If we call Anu's 'gravitons', when manifesting in the form of triads whereby mass is conferred, so what is meant by 'gravity' can be considered the most reified aspect of cosmic Mind that binds form into unity. With them the triads constituting atoms of space (protons, neutrons and electrons) come into objectivity and have a 'stable' phenomenal appearance.

If we think of the subatomic particles as being the reflections of the energies manifesting via the fourth ether and the fluctuations of the quantum vacuum, then neutrinos, which have near zero mass and hardly interrelate with the atomic universe at all, will be seen to manifest the attributes of the fourth Ray of harmony overcoming conflict. They are emitted from interactions between other particles so that stability (long life) is attained from the remainder, thus producing the appearance of form (i.e., 'harmony').

Gravitons, photons and electrons, viewed as particles, can be considered the effects of the fifth Ray attributes of the Logoic Mind. The effect of the fifth Ray of Scientific Methodology, manifests as a wave field, the bearers of the attributes of light. This Ray relates to the expression of the *Solar Plexus centre* at this level of manifestation. The Solar Plexus centre is the centre of the personal 'I' and its self-will. Here that 'I' is the atomic unit, and the photon is the carrier of the radiatory effect of the atomic unit's interrelatedness with the sum of the field of its expression in terms of wave vectors. Hence photons have 'relativistic mass' and interact with gravity.

The electromagnetic fields carried by the photons and electrons are expressions of the sixth Ray of Devotion-Aspiration. The energy of these fields is an expression of the *Sacral centre* at this level of manifestation, as the Sacral centre deals exclusively with energy fields, be they fields of desire, aspiration, or of the energies that produce physical plane birthing. It is the director of the main energies of the *nāḍī* system via the *iḍā (particles)* and *piṅgalā nāḍīs (waves)* which stem from this centre. Hence this Ray is the ideal vector for the manifesting electromagnetic field. Of the four fundamental forces it is the one most commonly experienced in our material universe in terms of producing consciousness-effects. Also, when coupled with the fifth Ray aspect (thus of the elementary attributes of mind) of the photons and electrons, we have the medium for the expression for the plasma and electrical effects in universe. All Logoi manifest their objective Purpose for their physical incarnations via their Sacral centres.

The particle-electromagnetic wave interrelation is the correspondence upon the atomic scale of the mental-emotional *(kāma-manasic)* exchanges, the Watery activities (fluid mobility of the energies) of average humanity, where they quickly move from one emotional thought to the next. The thoughts particularise, define, the moving images as thought constructs. Such activity is controlled by, or emanated from, the Solar Plexus. The related *saṃskāras* are stored in this centre, the associated Stomach and Liver centres, and the Inner Round *chakras,* whilst the energies are projected via the Sacral centre for material plane involvement, physical interrelationships with people and to form attachments to material things. The direction of propagation each wave vector towards the future represents the application of the mind as it evolves its concepts. The process leads to experiencing ever-increasing fields of expression in the domains of intensified light that comes with the attainment of enlightenment. The vectors represent the gain of all phenomenal activity in the domain of mind/Mind. In the subatomic world the electromagnetic waves are the carriers of the atomic sentience, seen in terms of the expression of light. Hence the electromagnetic phenomena generated by means of the charges carried by the electron is conveyed by the photon.

The gravitons (Anu's) combine via the auspices of the seventh Ray of Ritual, cyclic Order and manifesting Power, to produce the appearance of quarks, protons and neutrons (baryons) of the atomic structure. I would also place the Higgs boson under this Ray category because it helps give mass to the subatomic particles, hence to that which is considered as the appearing form.

All other short lived particles, the leptons, mesons, etc., would also come under the general rubric of this Ray. They are short-lived energy carriers that produce interactions that are generally attributes of the disintegration of baryons, or go to the formation of new stable isotopes of the atomic world, hence manifest the various cycles of the appearance and disappearance of differing phenomena.

As the strong and weak forces manifest via the third ether so the energies governing the manifestation of form are ultimately derived from the *ātmic* plane, from whence emanates *karma,* and into which it is eventually resolved again. The full weight of the Power of Logoic Mind, transmitted via a host of lesser *deva* agencies, manifests via this plane, being reflected via the *buddhi* mirror *(śūnyatā),* to inevitably bear its force upon the physical plane, thereby producing phenomenological effects. The atomic world and the universe of 'things' hence come into existence. All is governed by the *karma* of the past actions (of the Logos), and of any other creative thinker, where thought is made to move substance and to create 'things'. In the case of a Logos, such a One's existence in a former solar system projects the results of that former activity into the new cycle, assisted by *deva* Minds,[25] who are the modes of the expression of the forces, because the substance changed

25 In Christian terminology they are the Archangels, Seraphim, Cherubim, Thrones, etc. See volume 7B of *A Treatise on Mind* for further explanation.

and effected is their own bodies of manifestation. Humans channel the *saṃskāras* of former actions manifested in past lives to effect the new conditionings within the atomic world they now reside in. They build the new structures as a consequence of their mental-emotional input, which then manifest the appropriate physical actions.

The *ātmic* plane is dual, as are all the odd numbered planes. The physical plane for instance has four etheric and three concrete sub-planes, the mental plane has three abstract and four concrete sub-planes and the *ātmic* has four abstract and three concrete sub-planes (similar to the physical domain). This means that the energies that manifest via the strong force are primarily reified expressions of the combination of the four main cosmic laws. The law of Economy acts to cohere the Anu's in the form of quarks. The law of *karma* determines their mode of interrelation. The law of Attraction works to generate the charges of the particles, and the law of Synthesis produces the spin, the angular momentum of the particles of the atomic structure, to define its mass, hence mode of interrelationship with other masses. Under the dominance of the law of Economy the energies of synthesis, *karma* and attraction impact via the three concrete sub-planes. Everything manifests via the fourth ether, which acts as a version of *śūnyatā* (the Void), where the border between this ether and the physical domain is denoted by scientists as the quantum vacuum. This 'vacuum' equilibrates everything into the physical Void of matter, making all that we perceive intrinsically 'Empty', to use a Buddhist term. Consequently everything is an illusion, transient; nothing is real, except in the way that effects are produced in our consciousness.

There is a difference between matter and mass (substance), where 'substance' can be viewed in terms of the energy fields underlying matter, producing the appearance of discrete 'things'. Matter is mass 'individuated, localised' as discrete particles of atoms, photons and electrons, and compounds of these, bound by the force of gravity, to produce the mass-scale phenomena we experience by means of our senses.

The weak force may be characterised as working with substance, and drawing a quantified mass into the energy fields, facilitating the changes of substance from one state to the next under the law of Economy. This facilitates the disintegration of the nuclei of atoms, because their matter is drawn into the fields of substance, effecting what physicists call the 'colour' of the quarks, and producing radioactivity, hence the rearrangement of the structure of the nucleus, under the effects of the other three cosmic laws. This activity utilises the energies from the fourth of the abstract levels of the *ātmic* plane, and is stabilised via the impetus of the lower three sub-planes, reflecting their potency to the third and fourth ethers of the physical domain as the strong force. Drawing upon the energies from the strong force brings relative stability to the newly formed atomic structure. The strong force represents the force of the energies of the four cosmic laws, producing the eventuation of mass and matter, hence the stability whereby the law of *karma* can find its expression.

The event horizon ('quantum vacuum'), represents the Higgs field, and also the domain of *māyā*. Above this horizon we have the fields of energy waves, relating to fourth dimensional perception, rather than that of the third dimensionality of empirical mind[26] we normally experience. Below it we have particulate matter, seen as discrete unities that compose the atomic forms.

The *seventh Ray* of cyclic Activity and organising Power, that governs the appearance of the atoms of substance, relates to evoking the forces of the *Base of Spine centre* at this level of manifestation.

In the above the triad of *sattva, rajas, tamas* manifests in:

1. The strong force, quarks, electrons and other leptons manifest in the form of *sattva, rajas, tamas* in relation to each other to produce the constitution of the atom via the Higgs mechanism for the appearance of mass. Overall however this activity represents a *tamasic* level of expression.
2. The weak force, producing colour changes and atomic transformations is *rajaistic* in nature.
3. The gravitons (Anu's) as the foundation of the above manifests a comparative *sattvic* expression.

As elementary particles, photons and the electrons cross into and out of the event horizon from the fourth ether so they appear and disappear as a statistical wave vector to produce the electromagnetic phenomena seen around us, and the valency shells of the various atoms.

26 One could add the factor of time to produce a fourth-dimensional world view.

As stated, photons represent the subatomic reified version of the expression of the thought-forms created by the empirical mind. Such a mind utilises the five sense-perceptors to gain impressions through wilful contact with phenomena, which produces images born in lighted substance. From this perspective photons relate to consciousness, hence the light-bearing, principle. The mind also manifests a cataloguing and projective function (based upon what has been perceived) to produce further knowledgeable experiences. It is often also conditioned by the human emotions, desire or affection (the Waters, *kliṣṭamanas* – 'defiled mind') that modify the phenomena of what appears in the mind. We have a consequent attachment to phenomena, and the manifestation of desire for further experiences. These seven qualities (wilful contact, images, cataloguing, projection, experiences, desires, and attachment) manifest as attributes of the seven Rays of the esoteric philosophy. Their energy qualifications are reflected in the seven main divisions of the electromagnetic spectrum conveyed by light (photons).[27]

The valency of atoms, i.e., the position of the electrons in the atomic orbitals, as a consequence of the Pauli exclusion principle, causes the combination of two or more atoms to form compounds. In part this principle states that electrons in an atom occupy discrete levels of energy, where no two energy levels can have the same energy, hence each energy level can at most contain only two electrons, one possessing a clockwise spin and the other manifesting counter-clockwise. (Similar to the two different spins of the Anu's.) This principle ensures that ordinary matter occupies volume and is stable. A vast number of compounds can be consequently created. This corresponds to the activity aspect of the empirical mind. That aspect organises phenomena into set categories

27 The spectrum of visible light represents the subdivisions of the violet of the seventh Ray of Cyclic Activity and Materialising Power. Of this spectrum violet is the highest frequency and corresponds to the first Ray of Will or Power. The indigo wavelength corresponds to the second Ray of Love-Wisdom. The remainder of the visible spectrum relates to the five Rays of Mind, where the blue and green hues stimulate the basic attributes governing Nature's kingdoms, of the *devas* who overshadow their general evolutionary development. The purpose of the colours from yellow to red (the correspondences of the fifth, sixth and seventh Rays) stimulate the general evolution of the sentience of the lives incarnating into the manifesting forms, so that eventually human consciousness is evolved. (Orange is the colouring of the fifth Ray governing the mind.)

(thought-forms) of expression. The three main components of the atom: protons, neutrons and electrons, can also be viewed as the Father-Son-Mother aspect of the 'empirical mind' in Nature. Here we have the attribute of *tamas* governing the activity of neutrons, *rajas* for protons and *sattva* for electrons.

From the point of view of the *chakras* manifesting in Nature as a whole, rather than the human body, it can be said that:

- Gravitons (Anu's) are an expression of attributes of the working of the Solar Plexus centre in Nature. This centre is the base of all manifestation and coheres the form into unity. From this centre emanates the activity associated with the phenomenal universe related to the appearance of mass, however the functioning of the Sacral and Base of Spine centres also come into play, as indicated by the remainder of this listing.
- Photons are the outcome of the Sacral centre activity in Nature, the centre that is responsible for the energy distribution of the body. Photons therefore are the main vectors for the conveyance of 'awarenesses' (light).
- Electrons are effected by a combination of Anu's (gravitons) to produce the mass of ordinary matter, when integrated with the protons and neutrons (that have similarly combined Anu's via the constituent quarks, as explained above). The manifestation of this mass and the appearance of matter then represents the conditioning force of attraction that is conveyed as the law of gravity. This represents the activity of the Base of Spine centre of Nature.

With respect to purely objective physical plane phenomena, one could look to the activity of Splenic centre II, which is the main centre for the processing and transformation of Earthy *saṃskāras*. It manifests the driving energy to project reject *prāṇas* into the Eighth Sphere, which here relates to the appearance of physical plane phenomena. This simply means that electrons, and all subatomic particles are units of transience, each bearing one impression, a unit of a *saṃskāra* that can be conveyed into the sensory apparatus of the higher centres as they flash out of existence. As one flashes out of existence (by moving into the fourth ether), another flashes into existence to 'experience' the next quanta of information (no matter how minute). Such movement could be thought of in terms of the 'strings' of String Theory.

The activity of Splenic centre I represents the abstracting centre for these Earthy *saṃskāras,* directing the gain towards the Solar Plexus centre for processing, or into abstraction *(pralaya).* The process of the 'evolution', hence refinement of such Earthy *saṃskāras,* necessitates many cycles of activity within billions of years of earth history.

Gravity further considered

In empirical physics gravity is seen as a very weak force, where the extent of its influence is defined by the formula where its force equals the attraction between two masses divided by the square of the distance multiplied by a gravitational constant. The origin of such mass is however, as stated above, one of the major problems in particle physics. In Einstein's special theory of relativity gravity is also said to be an expression of space-time curvature. Physicists have not yet resolved these two views.

The *graviton* is said to be an excited state of the gravitational field. Similarly the Higgs boson, which conveys mass to subatomic particles, is an excited state of the Higgs field. In comparing the two, the graviton has spin 2 while the Higgs is spin 0, the graviton has no mass while the Higgs has a mass that is estimated to be about 125 GeV. The graviton is said to couple to the energies that affect all particles, including the Higgs boson, massless particles, and even itself. The Higgs boson only couples to the W and Z bosons, quarks, and charged leptons like the electron. It thereby gives them a mass, but it does not affect gravitons, photons, gluons, or neutrinos. Consequently these particles remain mostly massless (neutrinos possess a tiny mass of unknown origin). Gravitational force is said to be the interaction of gravitational waves on the atoms of matter. If no matter is present, there is no force. Technically, in general relativity, gravity does not bind to mass, but to the stress-energy tensor (an array of components that are functions of the coordinates of a space). Such a tensor is a complicated combination of energy and momentum.

In terms of the theory I have presented where a graviton can be viewed as an Anu, then the phenomenon of magnetism and electromagnetic waves can also be viewed in terms of gravity, whilst the Anu's (i.e., gravitons) also constitute the electrons and quarks to produce masses that present an attractive force. The electrons themselves are moving force vectors (cumulatively as tensors[28]) moving through a conductor, with the properties well known to physicists and electricians. Being much more subtler entities than electrons, and by being the constituency of the electrons it is possible for 'gravitons'/Anu's to convey their force to a large number of electrons, producing a moving line or field of expression. A electric field is produced, which is the product ('son') of that interrelation. Each area of such a field contains electric fluxes, which are defined as the electric field multiplied by the area of the surface projected in a plane perpendicular to the field. This view thus postulates that the 'graviton' is incorporated into all electromagnetic waves, plus all other attributes of phenomena, but when so bound the electromagnetic forces override the very weak gravitational forces. Nevertheless, upon vast scales there would be an accumulative attractive force by the 'gravitons' keeping the entire universe of things bound into One contained Thought-Sphere that manifests a signature quality and quantity of light.

This 'Thought-Sphere' then collectivises all Anu's in the universe, if one can conceive of the vastness of such a Primeval Thought to be inclusive of a universe, then via the Anu's *all forces are bound into one grand unified universal field theory.* The cumulative attraction of all subatomic particles to each other is responsible for the physical manifestation of the curvature of space, but the subjective, underlying force is that of the Will of the Originating Thinker. The curvature is the product of the confinement of space into a sphere of incarnation by a Logos.

Once a Base of Spine centre is established (by activating a seed *bīja* by a divine Thinker, a Logos or the Soul), to cause the formation of manifest space, then gravity is automatically the outcome. It is the force needed to attract the elements to it, electrons, protons, neutrons by means of which the entire form can be built via the appearance of the Anu's constituting those particles.

---

28 Note the difference between vector and tensor. A tensor is a mathematical object that is more general than a vector, and represented by many components that are functions of the coordinates of a space.

The form is built by means of the 'condensation' of the appropriate *saṃskāras* bearing the *karma* of the past, that manifest as force centres for the *maṇḍala* of the appearing form. The constituent quarks, electrons (thus the leptons) and photons manifest via the fourth ether, which draws the Fiery Element into the form to produce the coherency needed for the objectification of the entire body of manifestation. This Fiery Element is the basic substrate of *kuṇḍalinī,* whose expression, as well as that of the appearing form, is controlled, embodied by the *deva* kingdom. Within the Base of Spine centre Mind and substance are interrelated to produce the time-space continuum, as depicted in figure 15. When extended into the four directions into space, we get the four petals of this *chakra,* and the Elements and kingdoms of Nature they convey. This information was developed in the earlier chapters.

The thought field

Let us presume that there is an elementary field of 'atoms' on the mental plane constituted of Anu's, of which there are three types, answering to the attributes of the three *guṇas* (one would have to analyse the fundamental energy qualification to determine which is which). They form a field of substance which the thinker automatically utilises in the construction of any thought. The will of the thinker here represents the strong force. That will is of three main types: strong and determined; compassionate and universal in its application; or else there is the vague and somewhat inchoate energy application of lazy thinkers. These Anu's, when organised by the mind, carry the *force* of the main ideas of the thinker. They represent the binding energy of the thought construct. They automatically bind into arenas of colour and into forms by the nature of the energy of the will imbued into them. This gives the thoughts a specific gravity, a weight or density, allowing them to modify or influence other thoughts, according to the (in)tensity of the energies built into that thought.

By 'tensity' is meant the strength of the energy built into the thought that holds it into a unity and which is responsible for producing a tensor field relating to the duration (the extent in space occupied by that thought) and its momentum. Such a field can be expansive, or self-contained (hence of limited duration), bound to the self-regulated limit of the idea. The effect of gravity here applies to the entire thought construct rather than the individual components. Thoughts are attracted to each other according to the nature of the colouring, the electromagnetic qualifications carried by the photons built into the thought streams. (And all others within the thought environment, as upon the mental plane like, attracts like.) 'Thought streams' are spoken of here because we must conceive of the movement of thoughts as a progression over time, like a movie, rather than as a static image (such as a photograph). Thoughts in the thought environment are either immediately attracted to the newly appearing thought, or else remain unaffected, according to the type of magnetism engendered. The thoughts engendered can be composed of nebulous streams of substance (inchoate, primitive, sensuous thoughts), be like planetoids, or sun-like (brilliant ideas), within the thought environment. All are held within a tensor field of a world sphere upon the mental plane by means of the effects of the elementary substance, 'gravitons', they are composed of.

Though the ideas flash into existence and can be immediately replaced, nevertheless their effect is lasting, in that they have formed *saṃskāras* that have modified space, which persist because the moving time-line has progressed, leaving the originating thought behind. Hence all thoughts created since the dawn of time still exist, and can be recalled through sympathetic vibration if need be, and this happens automatically with the speed of thought when a presently appearing thought utilises aspects of the past to build the future. Such thought is not bound by the constraints upon the physical domain, which limits the speed of light to 186,000 miles per second.[29]

Over the course of time, of evolutionary progression, the created thoughts are forgotten, being too primitive for the evolved *saṃskāras* needed by the mind, the awakened Thinker. Nevertheless they persist and are inevitably built into the fabric of a newly forming mineral kingdom in the process of descent from elementary essence I (the 'gravitons' upon the mental plane), elementary essence II (the substance animating the astral plane) through to elementary essence III (the substance of the ethers) and then concretion as the atoms of physical space. We then have the rocks and mineral formations of a globe (a vast

29 299,792,458 meters per second to be precise.

thought construct). Such a globe is created by the originating Thinker that evolved on and has manifested (after the appropriate training in a cosmic school of learning) as a creative Logos for the purpose of eventually guiding a newly forming human evolution to its conclusion. The originating Thinker works with the emanation of the *karma* of the left-over thoughts of the previous evolutionary aeon, and moulds it into the construct of the new zone of evolution, evolutionary space. The substance of mind/Mind is consequently built into the dense form that is to be the new vehicle of expression, the *vahan* or *upādhi* of the cosmic Thinker. A world sphere thus comes into existence.

The substance of the ancient formerly redundant thoughts (of a former human kingdom that have now evolved on to their cosmic paths) are incorporated with the pristine substance of a newly forming world sphere. This sphere is thus modified by being acted upon by the originating reified thoughts of that former humanity. This substance then precipitates to form the dense form of a planetary sphere. In this way the *karma* of the past moulds the present, even unto the appearance of the dense form that is the substance composing a planetary body. A Logos and His/Her entourage that have evolved from that former evolutionary epoch hence have the *karma* (or can rightly wield that *karma)* to incarnate into (embody) the appearance of form.

Gravitons (Anu's) can be considered to exist at three levels:

1. The lower mental plane – elementary essence I. The substance of the mind is a Fiery expression of the gravitons, but here the limitations of the speed of light is transcended, producing motion that happens at the speed of thought. This means that the focus of expression is immediately directed to where the mind thinks. *Saṃskāras* of past actions, many from many millennia ago, come immediately to the surface as the mind thinks and hence needs to incorporate or integrate such substance into the formulation of the NOW. Once the *saṃskāra* is incorporated then the force of its momentum is immediately projected into future tendencies. Thus the three times become a unity. As the appearing *saṃskāra* surfaces, so it is instantaneously modified by the energies of the thought with the conditionings of the Now. Logoic sense-perception has awakened the images perceived from the field of view, plus groupings of *saṃskāras* drawn from past thoughts to build the present thought structure. This produces

its resonance (auric colouring), form or shape (the image built), plus directional purpose (the momentum towards future objective, according to the energy built into the Thought). There are however limitations to how far such Thought Streams can extend into cosmic space because of the inevitable pressure exerted upon the Thought Streams by other Thought-Bubbles associated with other planetary spheres existing in space. The substance of the reifying empirical mind is limited to an earth sphere, but once Mind is awakened then the substance of Thought can be projected to cosmic space.

2. The astral plane – elementary essence II. Astral plane substance would signify a combination of Anu's into two or more pairs, with their interrelation being quite fluid. Their expression would thus be essentially that of photons, that plus their closeness to the mental plane produces the autoluminosity ascribed to this plane of perception.[30] Being bound thus the elementary particles are more fluid, hence Watery, than the much faster moving Fiery particles constituting the units of thought. However, many thoughts automatically incorporate the Watery substance (hence desire-mind, *kāma-manas),* which slows their motion considerably. This desire-mind aspect is generated by human agents. They accordingly limit the speed of the energies in the Anu's, 'gravitons', to earth-bound concepts, the identification or desire-attachment to certain idea-forms by the creators of attachment to 'things'. Consequently the astral plane is formed.

   The pairs of Anu's would spin and dance around each other in the form of a spiral eight. Their bond pairs however being relatively easily separated, allowing quick incorporation into the mental plane or projection downwards towards the physical domain.

3. Those of etheric space, elementary essence III, whose parameters, when manifesting through the quantum vacuum are those formulated by empirical scientists, and which cumulatively determine the boundaries of our present universe. This universe is but the effect of an enormous Thought Construct. Three Anu's (gravitons) manifesting into each quark, giving the atoms of physical substance a relative permanence, as the added bond energy between the three, plus the triangular formation, produces this stability.

30 Hence the term 'astral' means starry.

The three higher sub-planes of etheric space can be envisaged to consist of Anu triads consisting of the energies of the lowest three spirals of the three intra-spiral groupings respectively. The fourth ether consists of Anu triads deriving their energies from the highest of the extra-spirals. Physical plane phenomena (quarks) are thus expressions of the energies of the lowest of the extra-spirals.

All manifest forms are governed by the laws of Thought, with the graviton-quark interrelation being the elementary unit of substance utilised by that Thought to build the patterns of what must be. The law of Attraction thus exists throughout space. Without this law in effect no (Thought) forms could come into existence. The patterns of what must be are shown by the colourings of the electromagnetic waves of light (photons), but are determined by the emanations of sound. Sound can be considered the effect of the vibration or spin of elementary particles, but caused by energies coming from the higher dimensions, which integrate those particles en masse. Each elementary spin being but the movement of elementary lives incorporating the appearing form. The Thought construct causing the *maṇḍala* of the form that it is part of is likewise 'alive', being part of a living system.

Electrons and the related positrons do not exist upon the astral and mental levels, hence are a phenomenal appearance moving into and out of the quantum gravity (a quantum 'soup' of elementary particles), according to the nature of the forces determining the need for such appearance. The negative charge is needed to produce the appearance of atomic bonding.

The Thoughts causing the appearance of phenomena are conditioned by the law of Economy, bound by the law of Attraction, directed by the law of *karma,* and ultimately resolved by the law of Synthesis.

The law of *karma* is synonymous with the law of Love in action. The manifestation of gravitons can also be considered an effect of the law of Love, automatically producing the experience of unity, oneness, universality of the presence of the all. They are the sympathetic reverberation of this law upon the domains of the lower mind, which has its lowest reflex upon the physical domain in the elementary particles that physicists have theorised to be the bearers of the force of gravity.

As stated, gravitons are an expression of the law of Synthesis, so the law of Attraction governs the electromagnetic waves of photons that produce the light of mind/Mind. Electrons, leptons and quarks are conditioned by the law of Economy. The law of *karma* binds all into an integrated unity for the duration of the *maṇḍala* of the manifest space that must be. When the *karma* for an appearing form is cleansed then that space moves into *pralaya,* the dissolution of forms into the subjective universe. The form is abstracted (liberated) into the universal pool of substance *(svabhāva)*[31] and held in situ by the law of Synthesis until it is to be breathed out again as part of a new Thought Construct. The strong force, weak force and quarks are conditioned by the law of Economy, which governs the appearance of the atom upon the physical plane. This law produces the appearance of the phenomena, sustains its duration until its eventual annihilation.

Having discussed the theoretical basis for the existence of the graviton, what mechanism can then be formulated to scientifically determine its actual existence? In other words, what are the laws that condition its existence that can be utilised to precisely determine its discreteness? Once properly defined then experimentation can be done along those lines. Smashing atoms together with tremendous force and energy will not produce this objective because there the effect of the forces producing the law of Economy are observed. That associated with the law of Synthesis works with a subtler mode of action. We must look to the forces that cohere forms into unity, that produce the appearance of things that have 'weight', hence manifest gravitational attraction.

Here is veiled the esoteric sciences of *mantra* and the *dhāraṇīs* focussed via *bījas* that awaken *maṇḍalas* demonstrating as *siddhis.* Anti-gravity (levitation) is an effect of comprehension of these laws. To produce such phenomena *mantra* and thought must work via the law of Synthesis to overcome the force that produces the economy of mass interactions, and so release the lives that are bound to that mass. They rearrange according to new karmic conditionings that utilises the Element Air rather than Earth as its basis for expression. The law of Attraction moves upwards to subjective space, rather than downwards

31 This term was explained in chapter one. See also the first three volumes of *A Treatise on Mind* for an explanation of this term and its relation to *mūlaprakṛti.*

into manifestation. The Element Air etherealises what is apparently physically real. Alchemical processes manifest similarly, however here the Element Fire is also evoked, that works with Air to produce the transmutation of substance.

Three different types of gravitons can also be envisaged, one which mainly conveys the *prāṇas* (energies) of the Element Air, another principally conveying those of the Element Water, and the third, which conveys the Earthy Element, which is the main consideration concerning the manifestation of phenomena and the effect of gravity. These Earthy gravitons can be considered to inhabit the triads of figure 43. All are but subdivisions of the Element Fire, which is universally applicable, because it is the Element governing the usage of the forces of mind/Mind, which have caused the Thoughts of what must be to manifest what is. Agni, the Lord of Fire, rules our material universe.

The formation of the nucleus of an atom

With respect to the force of gravity it has been shown in this thesis that one can use the equation: three gravitons + energy input = a quark.

Hence a quark can disintegrate to produce gravitons, if gravitons are depicted as elementary atomic units (the Anu). Depending upon the nature of the quark, so the quanta of the energy emitted changes. This energy quanta is depicted by physicists to be the strong force. From this perspective it may be possible to mathematically model the above simple equation, taking into account the internal structure of the Anu, with its spirals and spirillae, plus the way they interrelate and bind together.

What is a graviton esoterically? From the higher perspective, gravitons represent the energy of the compassionate Mind directed to the field of phenomena in order to produce the manifestation of form. That form appears for a set purpose and duration before being abstracted again into the energy of pure Mind.

Five dimensional mathematical modelling must be used to posit the effects of mind/Mind to propel the phenomena of gravitons into three dimensional space to produce the existence of quarks, possessing different quanta of energy for the different quarks.

The six types of quarks appear in three pairs. If these six quarks are arranged in the form of an interrelated hexagon, with the focal points of the triangles (the top and bottom points) being the up and down quarks, then this explains the nature of the formation of the protons and neutrons in the nucleus of an atom. It indicates that for the atomic structure of a nucleus to form it necessitates a momentary flashing into objectivity of all six quarks in the form of the hexagon shown below. The energies of the charm and top quark are brought to bear upon the Anu's constituting the down quark, whilst those of the strange and bottom quark are brought to bear upon the up quark. These forces produce a binding energy causing them to interrelate, hence to manifest a 'permanent' physical imprint in the subatomic world, to produce the nucleus of the atom. The energies of the four other quarks are needed to sustain the appearance of the up and down quarks, but the actual physical objectivity of these quarks are not needed. Nevertheless they can 'flash' into manifestation in relation to the breakup of the atomic structure and its subsequent rearrangement into a new form.

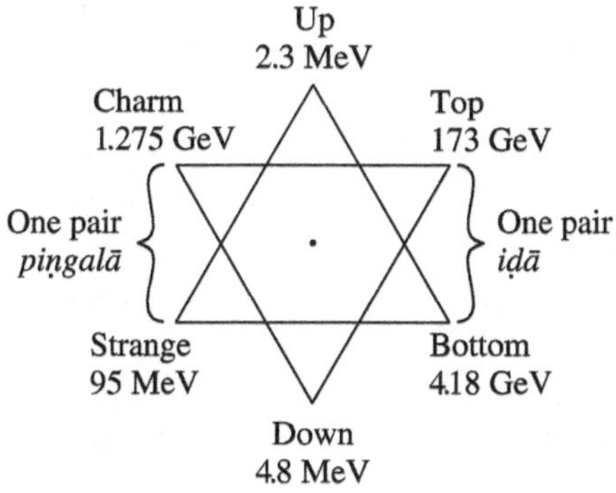

Figure 43. The six quarks

In examining the figure we see that the lightest of the triads in terms of energies depicts a strange and bottom quark (each of which possess less energy than the corresponding charm or top quark) assisting in the appearance of an up quark (with 2.3 MeV) and hence the eventuation of a proton, which consists of two up and one down quark, with a combined 'rest mass' of 9.4 MeV, plus a charge of +1. The more energetic charm and top quarks (with energies of 1.275 and 173 GeV respectively) contribute to the formation of a down quark, which with 4.8 MeV is 'heavier' than and up quark. This force is brought to bear to produce the neutron, which has a 'rest mass' of 11.8 MeV and zero charge.[32] The right hand side of the hexagon constitute a quark pair (charm and strange), similarly for the left hand side (top and bottom). We see also as the energies move towards the formation of either the up or down quark, and hence the consolidation of the mass of a proton or neutron, so the available energy is converted into the binding energy (gluon) needed to hold the protons and neutrons in situ in the nucleus of the atom. Also, as the paradigm throughout the esoteric consideration concerns triads of energies, from for instance, the extra-spirals of the Anu, to each triad of this hexagon, so the triadial energies also manifest in the binding of three quarks to make the proton or neutron. Also the triad of energies constitutes a far more stable arrangement than for instance two pairs, or the formation of squares, etc. The laws of economy and the conservation of energy always play their role throughout the subatomic world and in the macrocosm.

Depending upon the energetic factors at the time of the formation of the nucleus of the atom the energies within this hexagon move towards formation of either the up or down quark to cause the establishment of either a proton or neutron. The hexagon also explains why the strange quark must be governed by an energy similar to that of the bottom quark. Now, if we look to the up-down quark as relating to the manifestation of the Earthy Element, then the pair on the right (charm, strange) would bear the Watery *piṇgalā* energies into manifestation. The pair on the left (top, bottom) the more Fiery *iḍā* energies. (Fiery energies are more energetic than the Watery forces.) Also, as each energy is essentially dual (as are for instance all subatomic particles), so the charm and top quarks also bear more Fiery *iḍā* energies (1.275 and 173 GeV, compared to 95 MeV and 4.18 GeV), and the bottom two quarks being comparatively *piṇgalā*. Of all the points of the hexagon therefore the strange quark, being *piṇgalā-piṇgalā* would therefore of necessity possess the weakest energy of the four, hence accounting for the 95 MeV ascribed to it.

32 By 'rest mass' here I am simply referring to the added masses of the up and down quarks constituting the protons and neutrons. The actual mass of the proton is 938.27231 MeV whilst that of the neutron is 939.5656 MeV. That the neutron is slightly heavier I have attributed in my thesis above to refer to it containing 21 Anu's compared to the 18 for the proton.

This arrangement therefore signifies the final level of the descent of 'substance' from the mental plane to the dense physical, thus the process of concretion, or condensation of substance into matter, to use the alchemical terminology. It relates therefore to the method of the materialisation of phenomena from the inner realms, thus to the mode of the appearance of the globes of experience.

Saṃskāras and cyclic time

I have revealed that Anu's, in the form of a physical permanent atom are the mode of expression for the appearing *saṃskāras* for a human unit. Gravitons (Anu's in their various combinations) are similarly vectors for the attributes governing the phenomenal universe. *Saṃskāras* are expressions of the five sense-consciousnesses, of which we can consider 5 x 5 main permutations, each of which can be said to be further categorised by the three *guṇas,* making the number 75. When modified by any of the twelve universally applicable astrological conditionings and considered in terms of the eight directions of space, then we have 75 x 96 = 7,200. This number needs to be further multiplied esoterically by the number ten, representing the ten planetary conditionings, then the number 72,000 is obtained, which Buddhist and Hindu texts consider to be the number of *nāḍīs* to the human body. The zodiacal and planetary energies express the extraneous energies impacting upon the womb of time-space of any incarnating entity. The science of esoteric astrology has its basis upon this and other considerations. When multiplied by 2 (taking the *iḍā* and *piṇgalā nāḍīs* into account) then the number 144,000 is derived (12 x 12,000). This number signifies the complete turning of the wheel of the twelve signs of the zodiac. It is an extension of the number 1,200 signifying the total number of petals of the Head lotus (see volume 5A of *A Treatise on Mind,* 428-29), and is also the number, for instance, of the twelve tribes of Israel in *The Revelation of St. John* chapter seven, who symbolise the sum of humanity viewed in terms of astrological conditionings.[33]

33 There is a vast amount of esoteric symbolism encoded in this book. In relation to the appearance of the holy city, the New Jerusalem, which had twelve gates, guarded by twelve angels, for instance, it is stated in *Rev. 21:17* that the angel 'measured the wall thereof, an hundred and forty and four cubits, *according* to the measure of a

The Head lotus contains groupings of 100, 200 and 400 petals to accommodate the inflow of these *prāṇas (saṃskāras).* The Solar Plexus centre *(maṇipūra chakra)* contains ten major petals to process the *saṃskāras* developed by the Inner Round of minor *chakras.* Consequently it is the abdominal brain concerned with processing Watery mental-emotional thoughts before they are sent upwards to be stored in the petals of the Head lotus. Thus all the attributes of consciousness are contained. The unfoldment of the twelve main petals of the Head lotus relates to the turning of the signs of the zodiac and the expressions of the energies they convey, whilst the Solar Plexus centre conveys the planetary energies of an embodied system.

As there are seven major *chakras* below the Head lotus (counting the combined Splenic centre as one) so each *chakra* bears or can project ten of these *saṃskāric* (or *prāṇic)* attribute at any time. They can be divided into two groups of five sense-consciousnesses, relating to the *iḍā* and *piṇgalā nāḍīs* conveyed by the central spinal column. When the higher perceptions are obtained twelve *saṃskāras* can then be simultaneously processed, followed by groupings of twelve, depending upon the degree of Initiate consciousness obtained. The Head lotus contains three main tiers of petals, which, with the seven major *chakras,* makes ten major spheres of attainment for all energies to be processed via a manifest form. There are two remaining inner tiers, which direct the *prāṇas* to and from the Sambhogakāya Flower, that then complete the significance of the twelve. These ten plus twelve energies therefore constitute, upon a higher level of consideration, the impact of cosmic Energies emanating from various stars (the signs of the zodiac) and planetary spheres that become the main forces bearing upon the evolution of consciousness in an earth sphere.

In the above we have the basis for the concept of the cyclic calculation of time, where for instance, the precessional rate of the equinoxes, which is 72 degrees for each degree of longitude, or 2,160 years per sign of the zodiac, or 25,920 years for the complete turning of the wheel. These numbers are also derived from the base number 5 x 5 main types of *saṃskāras* (with literally an infinite gradations of

---

man, that is, of the angel.' The implications of the statement 'the measure of a man' is revealed above.

colourations carried by the *saṃskāras)* given above multiplied by 3 x 3 *guṇas* (the three *guṇas* being multiplied by triune energies from the abstracted domains of space), giving us the number 225, which when multiplied by 12 x 8 (96) for the turning of the wheel for a sign of the zodiac makes the number 21,600, which when multiplied by 12 makes 259,200, a grand zodiacal year.

In the Hindu system of time reckoning concerning the *yugas,* or cycles of time, there are four main such cycles, reckoned in terms of divine years, where each year is equal to 360 human years. The Kṛta *yuga* (golden age) has 4,800 such years, the Tretā *yuga* (silver age) has 3,600, the Dvāpara *yuga* (bronze age) has 2,400, and the Kali *yuga* (iron age) has 1,200 such years, making 12,000 divine years altogether. The length of each *yuga* are also computed on the fact that two periods of the sun's progress north or south of the ecliptic is called an *ayaṅa*, and there are two *ayaṅas* a year. The southern *ayaṅa* is a night and the northern *ayaṅa* is a day of the gods. 12,000 divine years multiplied by 360 such days makes the complete period of these four ages. This rendered in years of mortals equals 4,320,000 divine years, being a *manvantara (mahāyuga).* 2,000 such *manvantaras,* or a period of 8,640,000,000 years, make a *kalpa.* The *kalpa* being only a day and night, or twenty-four hours, of Brahmā. Thus an age of Brahmā, or one hundred years of these divine years, equals 311,040,000,000,000 of our mortal years.

If we take the 4,320,000 years to a *manvantara* and divide it by 72,000 then we get the number 60, or if divided by 96,000, the alternate number for the number of *nāḍīs* to a human body, then the number 45 is obtained. The relationship between the numbers 72,000 and 96,000 in relation to the constitution of the human psyche and the *chakras* was given in volume 4 of *A Treatise on Mind*[34] and further elaborated in volume 5A. There are various numerological interpretations of the numbers 60 and 45, but here we are concerned with the turning of the cycles of the wheel of time, where 60 x 6 = 360, and 45 x 8 = 360, hence we have 1/6th of a cycle and 1/8th of a cycle, and their permutations produce 1/12th, 1/9th, 1/6th and 1/3rd of a cycle, hence an *iḍā nāḍī* expression. We then have the *piṅgalā nāḍī* expression in 1/8th, ¼ and ½ of a cycle, be this of human or divine years.

34 *Maṇḍalas: Their Nature and Development,* 350-56.

From the above we see that the manifestation of *saṃskāras* is what causes the appearance of phenomena, whilst the sublimation and transmutation of *saṃskāras* causes the ending of time, hence the onset of *pralaya*. This concept holds true also for the mode of expression of gravitons on a mass scale, which then bring into manifestation these cosmological (or astrological) energies for the sum of the form embodied by a Logos. Such conditioned energies therefore represent the appearing cosmological *saṃskāras* governing the phenomena we reside in and which conditions the world or universe contactable, hence knowable, by our sense-perceptors. The Anu's (gravitons) therefore represent the ontological grounds by means of which we can come to know the All.

The astrology of the subatomic particles

The concept of astrological forces influencing subatomic particles is alien to the scientific community, as is much else in this book, nevertheless, from the esoteric perspective the energies from these signs (constellations of stars) pervade and influence all aspects within the Womb of our earth sphere. These aspects are inclusive of the streams of sentience and consciousness as well as the fields of substance. All are conditioned by various types of energies, and the broad scope of the conditioning energies are those directed from the Heart of Life by the *deva* custodians of *karma* in accordance with the dissemination of Logoic Purpose. The lesser devas (Elementals) embody the attributes of the Anu's, and consequently manifest in the form of the evolving sentience of substance.

Below are the energies from the signs of the zodiac that I suggest can be considered to influence the following effects in the subatomic world.

Subatomic world

Aries	Graviton (the force of gravity).
Taurus	Photon (electromagnetism).
Gemini	The etheric energies.
Cancer	W and Z bosons (the weak force).
Leo	Gluon (the strong force).
Virgo	Quark formation, and the appearance of the atomic nucleus.
Libra	Electron, electrical phenomena.

Scorpio	Higgs field, quantum vacuum, event horizon for the appearance and disappearance of particles.
Sagittarius	All sub-atomic interactions, gamma rays, X-rays, etc.

The atomic world

Capricorn	The complete atomic structure, interactions between atoms, molecules and compounds.
Aquarius	The panoply of Life.
Pisces	The death of old cycles and forms, preparation for rebirth of the new.

Notes:

Aries the ram is the sign of initial beginnings, signifying the will to manifest, to be, to do, hence the ram is impetuous and 'initiates the cycle of manifestation'.[35] Similarly the Anu (graviton) is the foundation of all that is to proceed as far as the manifestation of the physical domain is concerned, within which the primary energies that cause the appearance of things are contained.

Taurus the bull represents the field of desire, what might be considered the radiatory energies discharged by a sun, and which interrelate solar spheres with electromagnetic and other forms of energy. Taurus is said to be the custodian of 'the floodlight of illumination',[36] which can be extended to the physical domain as the light emitted from all active suns.

The sign Gemini the twins governs the process of the transition of the energies in the subjective universe, as passing through the four ethers into the material domain. From Gemini to its polar opposite Sagittarius the entire play of subatomic particles comes into effect. This is symbolised by the nature of the twins in Gemini, where one brother is said to be immortal, here representing the field of energies from whence the subatomic particles are derived. The other is mortal, which signifies the phenomenological appearance of the particles which flash into and out of existence as statistical probabilities, most of which have very short life spans.

I have listed the W and Z bosons and the weak force, which produces the transformation (radioactivity) of atomic structures, under the influence

35 A. Bailey, *Esoteric Astrology,* (Lucis Publishing, New York, 1982), 92.

36 Ibid., 374.

of Cancer the crab and hence the processes leading to the formation of the atomic nucleus, because Cancer is the sign of incarnation and mass movements of the Life streams, bearing elementary sentience. To Cancer is attributed all quark pairs before the actual formation of the nucleus of an atom. The quark pairs (specifically the up and down pair) come into existence from out of the quantum vacuum and are immediately acted upon by the strong force (hence the gluon) under the auspices of the Leonine energy to incorporate them into triads that constitute the protons and neutrons of the atomic nucleus. Once the atomic structure is established, then the Watery Cancerian force can work to produce further refinements and changes by dissociating the quark triads of the protons and neutrons and abstracting them back into the fourth ether, prior to rearrangement as a new isotope of the element concerned. (Cancer is the sign of birthing and is also the point of abstraction to the mountaintop represented by its polar opposite sign, Capricorn.)

Leo is the sign of individuation, of the self-assertion of the 'I', the Fires of the empirical mind producing the concept of individuality, hence I have assigned the strong force that causes the actual formation (individuation) of the neutrons and protons to Leo. Cancer-Leo in the atomic world can be viewed as an integral unity, similar to the factor of desire-mind, or the emotional mind in the world of human activity, where Cancer governs desire-emotions and Leo the attributes of mind.

Virgo the virgin, who has the capacity to bear the foetus of what is to be and so give birth to the phenomenal appearance of the 'child', governs the appearance of the atomic nucleus.

The southern quadrant of the zodiac, Cancer, Leo and Virgo are concerned with the direct appearance of the phenomena possessing mass and which sustains our physical universe.

Libra the balances relates to the domain of meditative insight and contemplation, of the interlude between breaths, right judgement, the outbreathing of the energy fields that constitute the processes of Life and the expression of *karma*. To this sign I have assigned the universe of electrical phenomena, of the attractive power of charged fields, literally the electric universe. The nature of such phenomena and its effects, such as modern telecommunications, is similar to the energy fields and related awareness obtained during meditation. These fields can literally be conceived as currents of mind/Mind.

For Scorpio the scorpion I have assigned the Higgs field, quantum vacuum, the event horizon for the appearance and disappearance of particles because Scorpio, ruled by Mars the god of war, relates to the world of energisation. Such energisation produces the field of testings for humans. (The world of sexual relationships and emotional-mental interactions between people, trials associated with the quest for money and material comforts.) The force fields that arise through energetic interactions between subatomic particles, the particle-wave duality and other inexplicable phenomena observed by scientists when analysing the phenomena appearing at the transition between the inner and outer universes has a corollary to the testings in the world of human interrelations under the influence of Scorpio.

Sagittarius the archer concerns directed energy momentum, one pointed aspiration or ambition to obtain what is the desired objective in the physical world. As a consequence this relates to the mechanism for the appearance of phenomena concerning all sub-atomic interactions, and the energetic release of gamma rays, X-rays, etc.

Capricorn the goat rules the mountain of mind, the hard rocky material domain that one needs to climb and to master if one wishes to achieve one's objectives in life, be those of material ambition or of spiritual aspiration to gain the heights of Initiation. Here therefore the entire atomic world comes into view, of the atomic structure, the building blocks of the material domain, of the entire empirical world.

Aquarius the water bearer symbolises fluid, often mutually beneficial interrelations. At first they may be selfish and shallow, but later they develop into loving-kindness and the service oriented activity that is the hallmark of the Bodhisattva. Here therefore all interactions between atoms, molecules and compounds in the atomic world can be considered. We therefore have the total panoply of manifested life to be experienced and assessed by the participant as he/she fleetingly moves from one activity to another in order to play out the drama of life.

Finally, Pisces the fishes is the sign of termination of any cycle of expression, and also of Watery, psychic often mediumistic thought, of concepts of renunciation, sacrifice of the personality to unavoidable fate, or bondage to the activities one is immersed in. In the subatomic world these attributes are best explained in terms of radioactivity and the rearrangement of subatomic particles from one entity to another.

Essentially the death of old cycles and the form of the sum of the material domain must be considered, in preparation for the appearance of the new.

The four constituents related to the atomic structure, the quarks, leptons, gauge bosons and Higgs boson can also be thought of in terms of the eight directions of space introduced earlier. In general the directions northeast, southeast, southwest and northwest manifest in the form of a mutable cross, of continuous cyclic activity. The *northeast* represents the direction of 'unity', wherein the unified philosophy, or grouping of forces exists. The direction *southeast* represents the projection of that unified force as an 'expression' into the field of *saṃsāra*. It can be, for instance, seeded into and accordingly modified by an individual's developing mind, which develops a belief system based upon the philosophy associated with the northeast. The direction *southwest* concerns the resultant gain as 'understanding', causing further experimentation and consequent attainment of knowledge, hence the evolution of the intellect. The direction *northwest* represents a demonstration of the mastery of what has been gained through evolutionary development in the southern portion of the eight-armed cross. This is denoted as 'goodwill' or as the will-to-good in the field of human consciousness. It represents the movement upwards, of aspiration to higher fields of endeavour, from that relegated to the world of human relationships (west). In the subatomic world we would have the release of particles or energy as a consequence of the breakup, nuclear disintegration, of an existing form.

The cardinal directions (north, east, south and west) represent the fixed cross aspect, indicating a steady flow of energies to these four main orientations that inevitably represent the nature of the direct expression of the field of consciousness. Inevitably we have the development of compassionate considerations. The *north* is upwards to liberation, divinity, the mount of achievement. (The demonstration as a wave front rather than as individual particles in the subatomic world.) The *east* represents the way inwards to the Heart of Life, the Void experience. (This relates to the expression of the Higgs field and the fourth ether.) The *south* represents the direction downwards to the little ones ensnared in *saṃsāra,* the world of form. (Hence we have the appearance of particles from out of the quantum vacuum.) The direction *west* represents travelling to the field of service that represents humanity, and all related social interactions. It therefore relates to the general subatomic environment.

When adapting these directions, which are explained above in terms of human consciousness to the subatomic world, then one can further postulate that the direction northeast (unity) relates to the fourth etheric domain via which all particles and energies emanate, and the Higgs mechanism, which has been postulated to provide mass to the particles. The direction southeast (expression) then relates to the manifestation of the gauge bosons (gluon, photons, W and Z bosons), which are the carriers of the forces productive of the appearance of mass. The direction southwest (understanding) then relates to the appearance of quarks, hence the formation of the nucleus of atoms. The direction northwest (goodwill) is productive of the appearance of the leptons, specifically the electron, hence the electrical charges whereby all electrical fields and phenomena manifests.

Physicists could also determine whether the schema outlined here is also applicable to the 'eight-fold way', as described for instance in the book by Venkataraman:

> The story of the scheme really starts with particle taxonomy. As particles turned up in bewildering numbers, people at first wondered what they all meant. Then they realised that the first thing they aught to do was to classify the particles, and this is where the famous quantum numbers proved very useful. When this grouping into families was done, the number 8 kept turning up again and again, i.e., many families had eight members. Did this number mean anything and if so what? Given the fact that particle taxonomy relied very much on quantum numbers which were conserved, and also the fact that conservation laws spring from symmetry, it was clear that the 'magic' number 8 was in some way related to symmetry. But what exactly was that relationship?[37]

Dimensional perception

Zero dimensional space can be considered the point, an atomic unity. One dimensional space concerns a point's connection to another point, a line of energetic interrelation between them. This represents the experiential zone of atoms, where the line of connection represents the force vectors affecting it, causing it for instance to join with other

37 G. Venkataraman, *The Big and the Small, The Story of Elementary Particles,* volume 1 (University Press, Hyderabad, 2001), 148.

atoms through sharing electrons or loss thereof, to produce the atomic binding forming compounds. (This is the action of the valency.)

Drawing right angles to these lines of connectivity produces two-dimensional space, which relates to the experiential zone of the plant kingdom. Their main exchanges with the external universe is with the gasses oxygen and carbon dioxide (via stomata) and with sunlight via leaves, which are two-dimensional (2-D) surfaces. There is also an interrelation between the roots, which grow outwards, line-like from the trunk within the darkness of the earth in order to gather mineral nutrients.

They also grow from below up as their main direction of activity (from the soil to the sunlight), hence line-like, with the trunks and branches bearing the leaves manifesting in any direction of the compass, hence at right angles to the trunk or stem, again exemplifying a 2-D structure of its sentient awareness. Similarly the cells it contains are arranged sheet-like (though circular) formations growing up the trunk, branches and roots. Runners and vines, that crawl along the ground are developing a 3-D view.

Third dimensional space is pictured by drawing right angles to the 2-D surface, producing the appearance of the cube or any other similar object by means of which depth, hence the measurement of distance, can be gauged. This is the type of awareness developed by the animal kingdom, allowing them to graze for food and to hunt, seek out the opposite of the species for reproductive purposes, etc. With the development of human consciousness, movement with respect to time can also be gauged.

If the registration of time is considered a dimension, it takes into account the type of classical and Einsteinian physics we presently have, where equations utilising time in relations to the three vectors (or coordinates) of space are thereby considered as fourth dimensional.

Taking all this into account, then if we include the expression of the empirical human mind, we must look to the fifth dimension to account for the way it experiences phenomena. Effectively lines are drawn at right angles to the three dimensional cube, allowing us to extrapolate information from that type of world. This means that one can look at all six sides of the 3-D object, plus inside simultaneously from a point in space. This then is a definition of clairvoyance, but the mind must also be developed to rationalise what is seen in the greater world view established by 5-D experiences. Empirical conditionings keeps most human minds limited to a 4-D world view, but the 5-D form of intuitive and instantaneous visual impression increasingly manifest in those that train themselves to think abstractly.

New mathematics has been developed to try to comprehend this level of experience. How such mathematics applies to the causation of things by means of mind, hence the detail concerning the appearance of phenomena, a world or universal space, is yet to be seen.[38]

The attributes and the powers of mind/Mind, hence of this fifth dimensional perspective, as well as contrasting it to the function of *śūnyatā,* the Void, are well explained in my *A Treatise on Mind,* hence need no repetition here. In this present book such considerations of Mind are extended, to include the subatomic world and the nature of the evolution of Mind in cosmos.

The homologues between the five *alchemical Elements* to the four basic forces governing the appearance of everything plus the Higgs mechanism is given below.

- The strong force is Earthy in nature,[39] hence is responsible for the appearance of atoms.
- The weak force is Watery in effect, hence is responsible for the breakdown of the atomic structure, with the consequent radiation. The term Watery relates to the way that energies flow and substances interrelate, as for instance in a stream or a beaker. Water, and various other liquids, can dissolve many minerals and material elements, as well as precipitate them from solution, given the right input. It is a medium for the interaction of compounds to produce new compounds. This is a homologue to the attribute of radioactivity and hence the transmutation of elements that occurs under the influence of the weak force.
- The electromagnetic force carried by photons, as well as the conveyance of electric fields, is a homology to the effect of the energy of Fire and its relation to the expression of human consciousness. Hence the mind thinks via lighted substance.
- Gravity relates to the Airy expression of the Heart centre, producing the force of attraction, a universality of consciousness.
- The fluctuations of the quantum vacuum and the Higgs field has

38 For fifth dimensional space the equation which interrelates distance (s), the speed of light (c) and time (t) is said to be: $ds^2 = |\bar{x}|^2 - c^2t^2$.

39 Earth is effectively crystallised (Leonine) Fire.

a relation to the activity of the Aetheric Element, hence the mode of manifestation of the *ākāśa* (explained in chapter ten). From this Element comes the manifestation of all that is, which has mass in the material domains, hence can be experienced, and into this Element everything is eventually abstracted.

The energies carried in the central *suṣuṣmna* channel *(nāḍī)* running up the spinal column is a fusion of the energies of the *iḍā* and *piṇgalā nāḍīs* plus a third Fire, *kuṇḍalinī,* the binding energy of matter that coheres it into an integral form.[40] The Higgs field, which gives mass to elementary particles, can therefore, according to the esoteric philosophy, be equated to one level of the energy of *kuṇḍalinī* (as *kuṇḍalinī* is seven layered). *Kuṇḍalinī* is the force sustaining the constituency of the Womb of the Mother as a consequence of the application of the Will of the Thought of the Father. It is the effect of the Fires of Mind organising primary substance into a unified expression, a vehicle of evolutionary progression. *Kuṇḍalinī* should be included as a fundamental force, as are the strong and weak forces, electromagnetism and gravity. This makes *five fundamental forces in all,* which is logical because of the 'fiveness' associated with the manifestation of consciousness and the doctrine of the five Elements, the five planes of perception associated with the manifestation of the *karma* of all that exists in the formed universe.

In the human unit the attainment of liberation (enlightenment) necessitates the awakening of *kuṇḍalinī* through the proper application of the force of the Will energy from the higher domains of Mind. It takes significant yogic practice and the appropriate lifestyle to awaken this force, hence it lies outside the domain of normal quantum physics. *Kuṇḍalinī* is however an expression of the forces that produce a nuclear explosion, or a nuclear furnace, and is what would be evoked during the formative period of any manifesting cosmic form. From this perspective one aspect of this energy has already been discovered by physicists, but labelled according to their mind-set.

---

40 For an explanation as to the nature of *kuṇḍalinī* consult the first five volumes of *A Treatise on Mind.*

10

Meditation on the Electrical Nature of Mind

General esoteric considerations

In meditating on the nature of the manifestation of the universe it is possible to receive instructions from very high Initiates not incarnate on our earth. They are in contact with sources of information pertaining to the nature of the appearance of space from their exalted perspective. One can thereby link Mind to Mind to sources of cosmic Mind. This is far beyond mere 'channelling', where the psychically endowed ones receive generally distorted (Watery) impressions from the astral plane via their minds. They have not developed a properly refined Mind to discern the real from unreal. They do not work via the Clear Light of Mind and so distortions set in, and neither do they know of the illusions related to the plane they are working through, or of the true nature of the one or those they are receiving messages from. Consequently the dark brotherhood can easily intercept or masquerade as a Master of Wisdom, or the like, and so delude.

In the link of Mind to Mind via the rarefied strata of *ātma-buddhi* a dialogue of Knowingness is attained producing revelations beyond what is possible to younger Initiates. The information below comes from such a meditative dialogue. The source is a great female Pleiadian high Initiate of great stature, beauty, luminosity and wisdom along the violet Ray.

There is a cosmic link between the elementary particles and the laws of Sound. There is also a link between magnetic and electric fields and the ways that *devas* think. Resolvent lines of magnetism flow to imprint the patterns of their thoughts into/as the fabric of space. Waves of Sound order the forms soon to be into the spectrums of the Thoughts by which Logoi Think. The colours descend to impact the moving waves of luminous elementary Watery Lives. These waves of luminosity ripple the cosmic ocean's shore, causing interference patterns. They resonate into our earthy domain. Peaks of brilliance move up, whilst the troughs of resonance break through the etheric substrate[1] to become the forces for those *deva* forms that descend to the shore of the astral-etheric interrelation. Differing *deva* breezes produce patterns galore. How they are coloured modifies the dancing subatomic Lives. Multi-hued patterns of light convert to sound waves cascading into forms.

Seven Rays and forty-nine sub-Rays account for all you see and more. Orchestrations of sound produce patterns in the mind-scape of the *deva* Lives and they create what is heard into the forms of our daily lives. The Elements are woven according to the patterns observed. From their interrelation come the elements conditioning the world of human forms. The junction between sound (conditioning *deva* awareness) and light (demonstrating as human consciousness) creates the spaces where magic is wrought. The transference of energy from one to the other produces the appearance of the forms (Anu's). The conversion process represents the power of the forces from the Void, which is reflected as the line of demarcation (the quantum vacuum) between etheric and physical space. The atomic shape produces stability for the forces observed.

Non-localised phenomena appears via etheric space, as their energies manifest instantaneously upon all points of descent. They move from a state of timelessness to the appearance of particles that are limited by the nature of their descent, conditioned by the laws governing their interplay. The spirals and spirillae hold the light resonance that can move in waves across the vastness of what can be known. The movement from spiral to spiral, from conditioning form to form limits instantaneity to the rate observed for the speed of light. All particles are bound by the limitations of form, the equation presented is: $E^2 = p^2c^2 + m^2c^4$, where E = energy, p = momentum of particle, m = mass and c is the speed of light.

1 The Higgs boson can be considered to play its role here.

Each particle is but a blind unit of Life responding to the impacting energies, the effects of the emanatory Sounds, from which stream the variegated streams of light. The Sound travels in increasingly reified octaves, but that measured upon the physical domain is the response of the Blinded Lives to the energy that impels them to move. They weave the patterns into the forms we see. Sound and light, light and sound produce a yin-yang in space. As they embrace (via etheric space) the forms can then appear, and so move in time. In time the Fire resides, which creates the movement to embrace formed space. The forms produced through Fire mould the attributes of the *māyā* from the condensed Sounds coming from a plane one level higher. In esoteric parlance Sound is the Father, Light the Son and Fire is the Mother. This order is mirrored from Logoic domains.

The Fire is the resilient heat *(kuṇḍalinī)* obtained when the Logoic Mind squeezed (meaning intense pressure via Sound) the chaos of the elementary dark particles (of mind) into a coherence of Life. That Fire is the genesis of all *deva* Lives. Fire becomes ensconced in all rudimentary forms. Fire is the bridge between sound and light, it is the effect of the Sound and gives birth to the light, and light demonstrates as the manifestation of variegations of colour. By Fire is meant the energy that builds forms, but also separates and differentiates one (particle) from another. It is the energy producing the forces that moves the appearing forms. Sound creates the vortex for the appearing forms, and light is the radiatory emanation from what has been moved. It is often emitted when one form transforms into another, or when energy is released from their movement. It qualifies the degrees of the moving energy with colour. The movement of the forces is Watery, whilst the appearing forms are Earthy. All are the effect of the congealing Sounds from the orchestrations the *devas* sing out.

Fire is the source of the movement of time into space, conditioning the appearance of phenomena, and of spacious consciousness through time. It is the 'fount' of all that eventuates into a human race. It is the tensed Mind poised and electrified (positive) to embrace the unknown (negative) substance to which it is attracted and must consume (convert) in its haste. Dark matter is that which it must yet confine, and dark energy is the movement of a dark Mind, unconverted and resistant to the movement of the Fiery Sound. Light is the effect of the sparks of conversion within each appearing sun. Galaxies exist to organise the suns and to move them to new regions where darkness can be won over. Hence the darkness is far greater than the organising appearing Minds collectivised as the ONE MIND that is within a universe constrained.

The Big Bang hence is but a primal Fiery awakening of that MIND, a new cycle of expansive conversion begun. It progresses anew from where it once left off and has set the bounds of what can be attained in this cycle. From it do the Son/suns move as atomic sparks. Each lives out pre-ordained Lives according to the perturbations of the emanating Sound. Each imparts its place in a Vision of everything to be made and reverberates that Vision to all Son/suns that are its comrades.

There are wells of birthing (concentrations of dark matter) and arenas of lack, for expansive growth to take place and so the stars move, propelled by the greater Sight of the One in Whom they reside, and Who propels them through space.

The Big Bang was and was not. It was in the domain of Logoic Mind, to organise the substance of cosmic etheric space via the patterns of the Flowers appearing in that space. It was not per se in the domain of appearing space because the Fire manifested simultaneously via a number of lines of descent, where the Base and Sacral *chakras* first awakened via the 'blue-print', the lines of force established by That Mind and the Minds of the awakening Ones. The Fire spread out via the central Base of Spine-Sacral centre interrelation and via other *laya* points. The differing Elements produced the appearing pattern of events. Its a Thought-enclosed space seeded with Fire from a point to a form with pre-arranged haste. It moved from within to without along conservative lines, and emanated from points (quasars) that together awoke the All to Be. The Fire moved out to all directions (hence homogeniously) as Space, awakening the new forms (galaxies and constellations) whose purpose was to eat the darkness within a pre-ordained sphere encapsulated via cycles of time. Hence *maṇḍalas* were seen, sequenced in time, of what was to appear and when. The rebirthing Ones (stars) then peopled (appeared) in their time (the right cycles) to play their roles for new minds to make. The minds grew apace with the expansion of the universe at large, and into Minds awakened as the Thought Form (of the universe) nears its allotted space. It fills this space with the light of the stars made from the converted ones as the Minds grow apace.

Quasars represent the nodes of descent of the *maṇḍalic* patterning established by the One at the centre of Life. They sound out their Notes of Power to establish the pattern of what must BE. Hence the 'creation' resounds from the One to the nodes and together they establish the Plan for the expanding Fire. The *chakras* whose energies they externalise are many. The Fire radiates out through the *nāḍīs* of Life, which seed the constellations according to the places of the Need (where the dark matter is strongest). The suns light up and the space is consumed with waves of light connecting the One to all through lines of Sight. Knowledge is gleaned through electromagnetic Might. Radiance then fills the night, as each aura of the Son/suns shines bright. Birth and death of the All manifest through cycles of right, the old and new together inhabiting the cosmic Life.

As the Thought-Form progresses to the end of the show the Minds move inwards, producing an ever-expansive glow. The Thought thus increasingly lessens its hold on the forms and they dissipate into the Void of it All by the inward moving Lives.

In turn as Logoi they Shout the Songs for the appearing forms. The Builders respond, the Blinded Lives move in, the army of the Voice resound and the form congeals. Thus the world of matter begins.

The subatomic particles dance and do spin according to the frequency of the Sound they are bathed in. The forces appear. They move at right angles to colour the Sound and to Sound the colour of the little Lives that will be. The play *(māyā)* begins, the forms congeal and the greater *devas* move to encompass the all, the forms that appear and move as they spin.

Thus the story is told of what the Thought has begun.

The forms have been spun for humans to behold the seasons, the years, and as the aeons unfold their minds expand. With the power within they create the thoughts to complete the story now told.

Time is and time is not, time cycles as each sun or galactic space awakens and dies within the greater cycle. The universe was spawned, as time-space it cycles, not in a straight line, but in spirals it awakens the geometry of space, and the light moves wherever the spirals take it. The spirals can move slow, the spirals can quicken, they can transcend whatever it is that time hearkens. Mutidimensionally the movement hastens. Physical space, then, is but one facet of the expansive vista into which time progresses and recesses. Time is no more, but the cycles continue. Bright is the vision of the Eye that sees the progression.

The violet holds all, the violet rules the Songs of the forms, wherein each Life moves. This Ray manifests the rhythm, the count, the beat of the drum that the Blinded Lives obey, as a must. From Libra (meditative space) the commands peal out to return the streams of the sands of time into dust. The dust is dissolved into the ocean of trust.[2] Synthesis is achieved, the lessons learnt, regurgitated, assessed to formulate the NEW beginning of time, the *maṇḍala* rearranged to accommodate the new View. The Fires prepared for the points of descent. New dark space to be claimed through cycles of time. Never-ending they climb through spirals benign, to dazzling heights unknown by mind. That is the multidimensional All.

The electric universe theory is on the right track, but there are inconsistencies as presently expounded. First of all gravity is a universal constant. It is needed to keep order in the universe. One force must fit all occasions. The speed of light is normally fixed, but can change under certain influences or the application of the power of thought.[3]

---

2 These two lines are somewhat of a poetic rendering of the evolutionary process, where 'sands' represent human Souls (golden in colour), hence we have streams of such Souls and Soul groupings evolving through time, moving cyclically through an 'hourglass', shaped in the form of an '8', of spiral-cyclic motion and the concept of infinity. The 'sand' is turned back into the 'dust' that is the substance of mind when the Souls take Initiation and the inherent Life moves on into *śūnyatā (pralaya).* The remaining mental substance ('dust') is merged back into the reservoirs of *deva* essence (the 'ocean of trust'), preparatory to being Breathed out again during a new *manvantara.*

3 Thought, via what might be considered the magnetic nature of the mind, can influence the structure of the sheet-like Anu's through which light (photons) travel and hence theoretically alter light speed in a vacuum. See also the article by Andrew Grant, 'Speed of light not so constant after all', posted in *Science News,* Vol. 187, No. 4, February 21, 2015, p. 7, (https://www.sciencenews.org/article/speed-light-not-so-constant-after-all#comment-1799977441) where it is asserted that 'Pulse structure can slow photons, even in a vacuum', meaning that light speed through a vacuum 'should be thought of as a limit rather than an invariable rate'. The work of John Webb of the University of New South Wales, Sydney, can also be considered. His research is published in *Physical Review Letters,* based upon measurements of the 'fine structure constant' at different points of the universe's past. It indicates that light may have slowed down in the past. Results show that six billion years ago, the fine structure constant was smaller by about 1 part in 100,000. Later Webb's team gathered twice as much data showing an even stronger change, and as far back as 12 billion years ago. See also the papers 'The Atomic Constants, Light, and Time' by Barry Setterfield and Trevor

It is the nature of the Will of Mind whose force is exerted to bind substance that determines the strength of the force of gravity of a mass. Black holes, neutron stars and brown dwarfs represent the order of the ending of time, the sequencing of *pralaya*. Black holes are not so, but rather spheres of transcendent luminosity where substance moves into the higher dimensions upon a mass scale. Energy is released as mass (matter) is transformed into its etheric substrate. One can Vision a place of ascending spiralling *devas*. The *devas* release their sheaths as they move up the Elements.[4] The discarded Lives finding their own level of residence within the spirillae of the planes to which they are attracted.

The sphere of ascension is a place of immense gravitational attraction for the Lives (substance) that have had their Day. The Lives move to the centre of the solar or cosmic form as they evolve. The central sphere represents the cosmic location of Shambhala for them. Therefore there is a descending as well as an ascending process, of moving lines of energy (gravity) that keeps the form bound into unity.

Pralaya is different, it represents a gradual dissolution of the entire form into the ethers, leaving then only the remnant of old and dying stars to complete their cycles. (Hence the manifestation of globular clusters.)

Brown dwarfs are the last flickering embers of what once was vital Life. Their energies can be utilised by dark brotherhood Minds. Neutron stars and pulsars represent the intensification of the Minds of cosmic Logoi. They are the manifest (dense) forms of Avatars, cosmic *yogins,* whose period of seclusion is needed for them to advance new techniques for the rebirthing ones and to empower others with Advice (energies) needed for enhanced growth. Many are the subjective links from these Minds to other Logoi. Their most important links however are upon the subjective domains.

Light waves are the externalised forms of communication from Mind to Minds. The external speed and frequency of the waves are indications only of what has already transpired on the inner domains. Vast is the interconnectedness of Logoic Minds. Complex is the web of the Thought Life created.

Norman (http://www.ldolphin.org/setterfield/report.html), also 'Is the Velocity of Light Constant in Time' by Alan Montgomery and Lambert Dolphin (http://www.ldolphin.org/cdkgal.html). The variable 'constant' contradicts the standard model of particle physics. With such research obviously the concept of the age of the universe is in question, as all measurements, including Einstein's theories, are based upon a fixed light speed.

4 The remaindered substance produces the intensity of the gravity known as a 'Black Hole'.

Electricity (electrons) lies at the heart of the matter. All thoughts are but electrical discharges and cause the effects described in the plasma universe model, but there are differences to our view. The internal force of the sun is powered by nuclear furnaces (nuclear power has been demonstrated to be real), as in the current belief, but the external spheres of the sun are also affected by forces from stellar spheres, which help cause the cellular structures seen at the surface of the suns. The internal nuclear furnaces are also needed to provide the transmutation of elements from lower to the higher levels of the atomic table.

Gravitational effects between the stars should also be noted, gravitation, meaning group interrelationships and group evolution—the journey of constellations, etc., together. The luminosity of the suns is the effect of the *karma* of long-standing nature between them. The energy is a calling card for the companionship, of the way they have travelled together in the past.

The electrical nature of the mind consists in the conversation between *devas* of different orders. The Fiery salamanders and the Watery sprites are interrelated by the Lords *(arūpa devas)* of the Air.[5] Fire is positive and Water is negative. The positive is dynamic and the negative is receptive and fluidly changing, quickly moulded by the dynamic forces. The Fiery directs the course of the flow and builds the thought-form with malleable substance (the Earthy) of the lesser Lives, those who have ears and respond to the Sound. The Fiery emits the Sound and the Watery is the medium of transmission, whilst the army of the Voice and the Blinded Lives resonate what has been heard in an increasing reified manner.

Electrons therefore carry the Watery flow in the form of fluxes and fields of electrical energy. The *devas* move with the flow, embody it, as it represents the lines of communication of the effects that the electricity will eventually produce. This necessitates building the forms, such as the atomic structures via the laws of valency, and hence also the compounds of substance. Sounds are carried through this way to colour the forms that become the *devas* incarnate. All forms hum (resonate) with their own intrinsic Sounds, their resonant tunes. The resonance keeps the *devas* bound in a form of companionship of Sound. The orchestrations

5 The terms here utilised are from the lower domains, but they also have vaster cosmic brethren. The term *arūpa* means formless.

sing out the world, the universe of forms, which humans see as light and colour, and which the mind determines to be real. The Sounds are not heard by human ears because those ears are not attuned to the subtleties of the substance of the ethers, which form the basis to the appearance of the transience called form, which we all know so well.

The *devas* are bound for cycles of accomplishment, the completion of tasks set for them. They learn to obey, and their schools of learning teach them how to sing the tunes and songs of Life, which others, the more (electrically) negative Lives listen to, and so rearrange their lives into patterns so ordained. The orchestrations of Sounds move upwards to produce resplendent harmonies, melodies too intricate to explain, but which are very potent in their domain.

Each *deva* Life is a unit of mind or of sentience, whilst universal Mind is the flowering of all their kind. Patterns of Thought produce the universe in bloom and all spiral upwards and away into transcendent Might. The Sounds and the Thoughts vanish into the Night, the bliss of *pralaya.* All is absorbed into slumbering thoughtless-Thought. Prepared the Lives shall be for the awakening of the new Day.

The role of the Pleiades

The esoteric philosophy stipulates that the Pleiades are the source of the creative Impulse for the local portion of the galaxy wherein we reside. They are responsible for the Sound waves that organise the substance of space (nebulous matter) into the solar and planetary forms familiar to us, and via which intelligence is evolved. Eventually the Mind is awakened, allowing us to leave the Womb of the earth to travel through interstellar space. The statement concerning the first of the Pleiades, Electra, given in *The Constitution of Shambhala* was that She:

> Is the father of the seven that leaks with electricity, the Water into fiery atoms. It is the breath that blows, giving substance to the Water. Their work is eternally veiled in the three atoms.[6]

---

6 See my book *The Constitution of Shambhala,* part 7B, 135-55 for detail of the works of the Pleiades. That rendition of the statements for the Pleiades was concerned mainly with background information and the field of consciousness, whereas in the supplementary information provided here the explanation is focussed upon the nature of the manifestation of phenomena, which is the main function of the Pleiades. 'The three atoms' are the three permanent atoms via which an entity incarnates.

The 'electricity' that is spoken of here is the energy from cosmic Mind, manifesting in the form of the Will-to-Create. This energy works via the cosmic astral plane that integrates substance into form. Leaking with electricity means that Electra acts as a type of cosmic capacitor that stores this energy, which is so intense that the excess electricity flows into her six Sisters, hence energising them. From this perspective one must therefore look to a universe seeded with electricity manifesting via plasma fields, which thereby effects the physical phenomena observed around us. Electra is the purveyor of the factor of Manas (Mind) as a constructive, creative force to her Sisters. Via her come the energies from the seven Rishis of the Great Bear[7] that causes the appearance of the seven systemic planes of perception and the resonance of that energy in the form of the seven main spirals of the Anu. The attributes of each of the Sisters then manifest in a vast, though reified way, through each of the spirals in turn, according to the plane of perception that a particular Pleiade is responsible for energising.

The Pleiadians sing the songs that regulate the *devas,* the Watery negatively electrical charged forces of Life. Here the concept of *Dragons* must also be taken into account. Dragons abound in nearly all ancient mythologies throughout the world for a very good reason. They can be viewed as a highly specialised form of *deva,* that have evolved to contain and to project the energy of Fire. Humans esoterically become Dragons of Fiery Life once they have evolved into Masters of Wisdom (having attained their fifth Initiation). This Initiation confers upon them the ability to convey the lowest attribute of cosmic Mind *(dharmakāya)* as it relates to the law of Economy. Hence they have totally mastered the physical plane, possessing the ability to command substance as they will, hence to produce any of the many miracles ascribed to the *mahāsiddhas* of the Hindu and Buddhist texts. Dragons evolve within the human kingdom as a consequence of awakening the *kuṇḍalinī* Fire stored at the Base of Spine centre.[8] On the inner realms such a one can transform into a Dragon form at will, as all is governed by the laws of Mind. The rigid

---

7 These Rishis are the source of first Ray energies to our solar system, whilst Electra embodies the first Ray function for the Pleiades. However all of the Sisters have their own individual relation to their corresponding Rishi.

8 This subject is treated in volume 1 of *A Treatise on Mind,* 22-32. See also pages 474-5 of volume 7B for a little more information on Dragons.

adherence to form as observed upon the physical plane is transcended. In cosmos there also exist a certain order of Life that evolves through and specialises in the functions of the Dragon form. The constellation of Draco was esoterically named in relation to these Lives for a reason.

Dragons embody the Fires of Mind, hence the dynamic positive charge that integrates the substance into forms of Thought. The Pleiadians order the structure (substance) of what is to be via Sound. The Dragons empower that Sound with the Fiery Light of Mind, thereby causing the building of the forms, the externalisation of what is to be. The Dragons move along the lines of Sight of the Songs that have been sung. They establish the Thrones of each Logoic Domain according to the Way that the Lords of *karma* (Lipika, who utilise Libran forces) have ordained the Day to be. The Dragons Breathe the Fiery Life to all that from the Throne transpires. This Fire helps to convert the darkness (forms of evil) into light. Hence they battle the dark brotherhood, whom they consume in flight. Dragons can be considered agents of Fohat, which is borne via their Fiery Breath. They are the bearers of the wisdom of the Logoic Mind. They convey the Light of Mind with others of their kind, and the colourings of the scales on their backs depicts the aged knowledge of their respective lineages. The scales are the *saṃskāras* transformed on the journey to becoming Wise.

The Throne is an embodied Dragon Life in the guise of the One who commands the All of the sphere of activity the Throne sustains. To command, the Logoic One must listen to the Songs of each new refrain that from the Pleiadians come, to regulate the order of the new cycles as they march away. The seven Pleiadians sing their refrains in right order, melded with the Light of the seven Rays in accordance with the nature of the cycle that is to run.

The Rays are emitted the Sirian Way (the law of Attraction), and the Lipika are governed by the Sounds from the Great Bear (the law of Synthesis). The Dragons bear the Karmic law and the Pleiadians resonate the Sounds that Economically meet the needs of the evolving lives. The Dragons symbolically curl round the scales (Libra) and hold the pans of the balance in their mouths. The Pleiadians load the left pan (the ten signed zodiac) with the substance of their Songs, whilst the lighted way of consciousness evolution is an emanation of the Sirian lore in the right pan (the twelve signed zodiac). The Great Bear adjudicates the sum of the score, of the movement of the pans from right to left and left to right, according to the wheels of each planetary and starry sphere towards an integrated cosmic Plan. That Plan is the All that all Life forms must obey. Within it both the darkened Lives and the Lords of Light are ordained to play their ways. All will be adjudicated in the end and directed so that all is right, for the Will of the Bear is all Might. Thus is the story of the local part of this galaxy told.

Below are some extracts from *A Treatise on Cosmic Fire* (T.C.F.), in relation to the electrical nature of the mind/Mind *(manas),* which supplement what has been said above. I shall quote from the T.C.F. in length, as therein is much seed for meditation and further enquiry in this primary source of esoteric information. The focus is upon the descent and effects of the electrical energy via the four cosmic ethers, which find their reflection upon the dense physical plane via its four etheric sub-planes. The information re the functions of the Pleiades previously provided in volume 7B of *A Treatise on Mind* shall be incorporated here. They manifest though the planes of perception that are the subjects of these extracts.

> The fire of Mind is fundamentally *electricity,* shown in its higher workings, and not considered so much as force in matter. Electricity in the solar system shows itself in seven major forms, which might be expressed as follows:
>
> *Electricity on the first plane, the logoic or divine, demonstrates as the Will-to-be,* the primary aspect of that force which eventually results in objectivity. Cosmically considered, it is that initial impulse or vibration, which emanates from the logoic causal body[9] on the cosmic mental plane, and makes contact with the first cosmic etheric, or the solar plane of adi.
>
> *Electricity on the monadic plane demonstrates as the first manifestation of form, as that which causes forms to cohere.* Matter (electrified by "fire by friction") and the electric fire of spirit meet and blend, and form appears. Form is the result of the desire for existence, hence the dynamic fire of Will is transmuted into the burning fire of Desire. I would call attention to the choice of those two phrases, which might also be expressed under the terms:

9 This is a term for the Soul, also denoted the Causal form or the Ego in D.K.'s works. I use the term the Sambhogakāya Flower, which was fully explained in volume 3 of *A Treatise on Mind.*

Dynamic electrical manifestation.
Burning electrical manifestation.

Here on the second plane, the sea of electrical fire, which distinguished the first plane, is transformed into the akasha, or burning etheric matter. It is the plane of the flaming Sun, just as the first plane is that of the fire mist or the nebulae. This idea will be easier to comprehend if it is borne in mind that we are dealing with the *cosmic physical plane.*

Certain things take place on the second plane which need realisation, even if already theoretically conceded:

Heat or flaming radiation is first seen.
Form is taken, and the spheroidal shape of all existence originates.
The first interplay between the polar opposites is felt.
Differentiation is first seen, not only in the recognised duality of all things, but in differentiation in motion; two vibrations are recognised.
Certain vibratory factors begin to work such as attraction, repulsion, discriminative rejection, coherent assimilation, and the allied manifestation of revolving forms, orbital paths and the beginning of that curious downward pull into matter that results in evolution itself.
The primary seven manifestations of logoic existence find expression and the three, with the four, commence Their work.
The seven wheels, or etheric centres in the body etheric of that great cosmic Entity, of Whom our solar Logos is a reflection, begin to vibrate and His life activity can be seen.

We are at this juncture considering the manifestations of electricity on the different planes of the cosmic physical plane, or on our solar systemic planes. Hence, all that can be seen in manifestation is fundamentally *physical electricity.* We have seen that the primary manifestation is that which vitalised, tinctured, and pervaded the matter of space, thus embodying—in connection with logoic manifestation—that which is analogous to the vital heat, activity and radiation of a human being, manifesting on the solar physical plane. Certain electrical phenomena distinguish a human being, only (as they have not been expressed or considered in terms of electricity) the analogy has been lost sight of. These demonstrations might be considered as:

> *First,* that coherent VITALITY which holds the entire body revolving around the central unit of force. It must here be remembered that the entire manifestation of a solar system consists of the etheric body, and the dense body of a Logos.
>
> *Second,* that radiatory MAGNETISM which distinguishes man, and makes him active in two ways:
>
> In relation to the matter of which his vehicles are composed.
>
> In relation to the units which form his group.
>
> *Third,* that ACTIVITY on the physical plane which results in due performance of the will and desire of the indwelling entity, and which in man is the correspondence of the Brahma aspect.
>
> These three electrical manifestations—vitality, magnetism, and fohatic impulse—are to be seen at work in a solar Logos, a Heavenly Man and a human being. They are the objective manifestations of the psychic nature, which (in a solar Logos, for instance) we speak of in terms of quality, and call will, wisdom, activity. Therefore, it should be noted here that the first three planes of the cosmic physical plane—the logoic, monadic, and atmic planes—are of prime importance and are the basic planes from whence emanate the secondary four; in other words, the first three cosmic ethers embody in a literal sense those three Entities whom we know as Mahadeva, Vishnu, and Brahma. In a similar sense these Three find Their densest objectivity in the three physical ethers. The lower four manifest during evolution, but are eventually synthesised into the higher three. It should be also remembered that on all the seven subplanes of a solar plane a process, in connection with electrical phenomena in etheric matter, will parallel all the processes on the major planes. This is easily to be seen on the mental plane, for instance, in connection with Man.[10]

Fohat was earlier stated to be the basic Creative energy of cosmos, cosmic electricity in its primordial essentiality, that dynamic Will that translates as electrical phenomena in all its diversifications. It is a Fiery force that is the energising expression of the Divine Thought of the third Logos (Brahmā) that is the cause for the manifestation of the substance of the planes, but specifically the five planes of Brahmā, emanating from the third *(ātmic)* plane.[11] It can be considered the expressed power

---

10 T.C.F., 310-14.

11 See the information concerning the three Outpourings, pages 267-84 of Volume 6 of *A Treatise on Mind.*

of Lord Agni (the Deva Lord governing the Element Fire in cosmos). Fohat is the driving Fiery force of the enlightened Mind in action, and *ākāśa* is the substance that is moved by this electrical Wind. *Ākāśa* is conveyed through the medium of the *nāḍī* system of cosmic etheric space. Fohat can be viewed as Monadic Ideation, the energy of Logoic Thought, in its creative, sustaining or destructive aspects.

The second Pleiade, Maia, governs the energisation upon the second plane of perception *(anupādaka).* Her statement was given in volume 7B of *A Treatise on Mind* as:

> Maia is my Heart from which I succour all. She is the bosom from which all receive my milk. She is the nourisher, the bestower, and the courageous. From her pours Love. She is the Heart.

This 'milk' from the cosmic Mother is *ākāśa,*[12] the sea of electrical Fire, (which is driven by Fohat), contains the zodiacal and planetary energies that contain all of the Rays and subrays that energise manifest space. D.K. states that this is 'the plane of the flaming Sun'. (A Sun which esoterically consumes the substance of dark matter, as does a flame.) This represents Maia's 'bosom', a Logoic sphere of activity that signifies the source of power of an incarnate Deity, the circumscribed throne of power, represented graphically as a sun disc: ☉. The bosom expresses cosmic astral energies in the form of the 'milk' *(ākāśa).*

As stated earlier *ākāśa* signifies space, where space is that through which things must manifest in order to make a visible appearance. Through this space things thus come into being, where *ākāśa* can be considered the space of Consciousness. *Ākāśa* is a term for *svabhāva,* and is considered the subtle and ethereal fluid pervading the totality of the phenomenon of the universe. It is a vehicle of Life and is the higher correspondence to the *prāṅa* that vivifies our etheric forms. *Ākāśa* is that subtle supersensuous essence that pervades the space of the four cosmic ethers, our higher planes of perception, specifically the electricity manifesting via *buddhi.* Being the plastic essence of space, *ākāśa* is the vehicle of a Creative Logos that expresses itself as the formative forces conditioning the manifestation of all phenomena. As

---

12 *Ākāśa* [from *ā* = towards, to, near + the verb root *kāś* = to be visible, appear, shine, be brilliant], 'not visible', space. The Purāṇas state that *ākāśa* has one attribute: sound. Space is the *upādhi* of Thought. The *Chandogya Upanishad* (7:12:1-2) equates *ākāśa* with Brahman.

such it is the transmuted correspondence of the five *prāṇas* manifesting in a human *nāḍī* system. There are thus five levels or degrees of *ākāśa* relating to the five higher planes of perception. The plane *ādi* conveys the Aetheric aspect of *ākāśa, anupādaka* conveys the Airy aspect, *ātma* the Fiery aspect, *buddhi* the Watery aspect, and the abstract domain of the Mind, the Earthy. *Ākāśa* therefore is conveyed in terms of the Wisdoms of the five Dhyāni Buddhas, or rather, the expression *(prajñā)* of their Consorts, where Vairocana's Consort embodies the *ākāśa* of *ādi,* and so forth to Amoghasiddhi's Consort conveying the *ākāśa* of the higher mental plane. *Ākāśa* therefore is the conveyor of the force of compassion *(bodhicitta)* emanating from the cosmic Waters (the cosmic astral plane), hence the sum of the zodiacal and planetary energies that modify space. Thus it is the vehicle of the enlightened Mind, and experienced from the third to the seventh Initiations.

Maia's role as 'the nourisher' manifests via the Monadic plane, thus the activity of Life in the domains of form is vitalised by her, and her energies build the Monadic form. She utilises the *ākāśa* from the cosmic astral Waters in the form of the vibrant white 'Milk of Love'. This is the energy depicted as 'coherent vitality' above by D.K.

As 'the bestower' Maia embodies the energy of *bodhicitta,* which pours through the fourth cosmic ether, hence she works via Alcyone. *Bodhicitta* is the Mind of enlightenment, the force that drives all beings back to their originating Logoic Source. This is but the expression of cosmic Love manifesting via *ākāśa,* hence is the energy that manifests from Maia as the substance of the Heart of Life. Her energies move the pulsations of the ventricles of that Heart, thereby regulating the movements of the streams of conscious Life throughout the *nāḍī* system of an embodied Logos. This energy D.K. calls 'Radiatory Magnetism'.

As earlier stated in volume 7B of *A Treatise on Mind,* Maia is 'the courageous' because her energies move to vitalise the systemic astral plane, and consequently contends with the *karma* of the heaven and hell states created by a human kingdom that utilises this substance as a basis for the development of the principle of Love. This means therefore that Maia is primarily concerned with the conscious development of that humanity via working with the substance of mass human emotions with view of their consequent transformation into the energy of Love via human aspiration and service orientated meditation. This is the form of Activity engendered by Maia in the three worlds of human livingness, where she works in conjunction with Taygeta.

Continuing with the extract from T.C.F.

> It should here be carefully borne in mind that we are dealing with electrical matter, and are therefore concerned with cosmic etheric substance; all matter in the system is necessarily etheric. We are consequently dealing literally with physical phenomena on all planes of the system. In time and space we are concerned with units of different polarity which—during the evolutionary process—seek union, balance, equilibrium or synthesis, and eventually find it. This electrical interplay between two units causes that which we call light, and thereby objectivity. During evolution this demonstrates as heat and magnetic interaction and is the source of all vital growth; at the achievement of the desired goal, at union, or at-one-ment, two things occur:
>
> *First,* the approximation of the two poles, or their blending, causes a blazing forth, or radiant light.
>
> *Second,* obscuration, or the final disintegration of matter owing to intense heat.
>
> This can be seen in connection with man, a Heavenly Man and a solar Logos, and their bodies of objectivity. In man this polarity is achieved, the three different types of electrical phenomena are demonstrated, and the light blazes forth, irradiating the causal body, and lighting up the entire sutratma, or thread (literally the Path) which connects the causal vehicle with the physical brain. Then disintegration or destruction ensues; the causal body vanishes in a blaze of electrical fire, and the real "man" or self is abstracted from the three world-bodies. So will it be seen in the body of a Heavenly Man, a planetary scheme, and so likewise in the body of the Logos, a solar system.
>
> The difficulty in apprehending these thoughts is great, for we are necessarily handicapped by lack of adequate terms, but the main ideas only are those I seek to deal with, and the one we are primarily concerned with in this division is *the electrical manifestation of magnetism,* just as earlier we dealt cursorily with the same electrical phenomena, manifesting as the activity of matter.
>
> Therefore you have:
>
> 1. Activity....electrical manifestation of matter.
> 2. Magnetism...electrical manifestation of form.
> 3. Vitality....electrical manifestation of existence.

This is literally (as pointed out by H. P. B.) fire by friction, solar fire, and electric fire.

Fire by friction is electricity animating the atoms of matter, or the substance of the solar system, and resulting in:

The spheroidal form of all manifestation.

The innate heat of all spheres.

Differentiation of all atoms one from another.

Solar fire is electricity animating forms or congeries of atoms, and resulting in:

Coherent groups.

The radiation from all groups, or the magnetic interaction of these groups.

The synthesis of form.

Electric fire is electricity demonstrating as vitality or the will-to-be of some Entity, and manifests as:

Abstract Being.

Darkness.

Unity.

We have seen that electrical manifestation on the first plane caused initial vibration, and on the second its activity resulted in the archetypal form of all manifestation from a God to man, and an atom.

On the third plane which, is primarily the plane of Brahma, this electrical force showed itself in intelligent purpose. The will-to-be, and the form desired, are correlated by intelligent purpose underlying all. This intelligent purpose, or active will, utilising an instrument, brings us to that most difficult of metaphysical problems, the distinction between will and desire. It is not possible here to handle this delicate subject, save simply to point out that in both will and desire, intelligence or manas is a fundamental factor, and must be recognised. This permeating principle of manas—colouring as it does both the will aspect and the desire aspect—is the cause of much confusion to students, and clarity of thought will eventuate only as it is realised:

First, that all manifestation emanates, or is electrified, from the cosmic mental plane.

Second, that the Universal Mind, or the divine thinker, is the intelligent Principle which makes Itself known as the Will-to-be, Desire or Love-of-Being, and that active intelligent purpose which animates the solar system.

Third, that Maha-deva, or the Divine Will, Vishnu, the Wisdom aspect, or the manifested "Son of Necessity," and Brahma[13] or active purpose are the sum-total of intelligent consciousness, and are (to the manifesting cosmic Entity) what the mental body, the desire body, and the physical body are to man, the thinker in the three worlds, functioning in the causal body. We must not forget that the causal body contains the three permanent atoms or the three spheres which embody the principle of intelligence, of desire, and of physical objectivity. Always must the analogy be held between the threefold Logos and threefold man, and definiteness of thought and of concept results when the one likeness between these is pondered on. Man is a unit, functioning as a unit in the causal body. He is a triplicity functioning under the will aspect, or mental body; under the desire or wisdom aspect, the astral body; and under the activity aspect, the physical body. He electrifies or vitalises all three bodies or aspects, unifying them into one, and bringing about—by means of the Intelligence He is—coherency of action, simultaneity of purpose, and synthetic endeavour.

Finally, therefore, it is apparent that, no matter from what angle we study, the threefold Logos (or His reflection, the microcosm) through *the Manasic principle,* intelligently reduces matter to form, and utilises that form for the fulfilment of the will, desire and purpose of the indwelling Existence; this principle *can be seen underlying all three aspects.*

There is no need here to point out the different triplicities which can be built up on the basic idea of Spirit and Matter, linked by Intelligence. This has often been done. I but seek to emphasize that INTELLIGENCE is the main quality of the Logos; that it shows as will, as desire or wisdom, and as activity; and that the reason for this is due to the work earlier accomplished by the cosmic Entity, involving cycles which have passed into the dim mist of retrospect, even from the angle of vision of a solar Logos.

This developed manasic principle is the intelligent purpose that is bringing about at-one-ment on each plane of the solar system in connection with the subplanes. It will eventually bring about the synthesis of all the planes, and thus bring the cosmic physical plane, as a unified whole, under the complete control of that cosmic Entity Who is seeking expression through that threefold manifestation we call a solar system, or the body logoic.

13 We thus have Śiva (Mahādeva), Viṣṇu and Brahmā as the first, second and third aspects (Father-Son-Mother) of Deity.

> On this third plane that intelligent principle demonstrates as coherent activity, either systemic, planetary, or monadic, and also as the triple vibration of spirit-matter-intelligence, sounding as the threefold Sacred Word, or electricity manifesting as sound.
>
> We have here an interesting sequence or inversion, according to the angle of vision, involving the planes as we know them:
>
> *Electricity as vibratory impulse.* This causes the aggregation of matter, and its activity within certain bounds, or its awakening to activity within the solar ring-pass-not. This is the first syllable of the Sacred Word.[14]
>
> *Electricity as Light,* causing spheroidal objectivity. This is the birth of the Son. It covers the enunciation of the second syllable of the Sacred Word.
>
> *Electricity as Sound.* Here we have the completed threefold Sacred Word.[15]

The statement for Caeleno, the Pleiade governing the energies manifesting via the *ātmic* plane is:

> The black is but a robe that veils the function of transmutation. The creator of the Void from which my substance flows. It is my womb from which the primordial substance exists in black dust. The Womb of the Mother is black.

What concerns us here is the elementary substance utilised by Logoic Mind that was incorporated into that Mind at the beginning of time. *Ātma* therefore incorporates the cosmic black dust that is utilised by such a cosmic Mind and moulded into a form, the child that can bear the imprint of *Manas,* or rather Mahat (the term used for Mind as the principle of cosmic Intelligence). Such substance is at first incorporated in a solar system ruled by the third Ray of Mathematically Exact Activity (such as was the solar Incarnation previous to our present one[16]). It is

---

14 Aūṁ.

15 T.C.F., 314-17.

16 The esoteric philosophy views our sun as the second Incarnation of a grouping of three. First, the last solar system, wherein the *karma* that conditions our present solar evolution was generated. That solar system was governed by the third Ray and was concerned with the evolution of *manas*. It is the Mother. This present solar system is intrinsically *manasic,* and has as its objective the evolution the second Ray of Love-Wisdom, which represents the Son in evolution. The future solar system will develop

difficult to explain the nature of the evolution of Mahat because we are concerned with the genesis of the dark brotherhood, of what is now known as the forces of evil, of unrelenting materialism, but which in those days was the good, proper expression of a primal evolutionary urge. This urge also concerned the genesis of the force of *karma,* as well as the evolution of the *deva* hierarchy, who embodied that substance, hence the *karma* of the interaction between forms.

The electrified impulses from Electra moves through this black substance in order to vitalise and to endow it with Life in the form of the appearance of the Anu's, though it is logical to presume that at first only the thermo-spirals (extra-spirals) were represented, and which later would wind round three groupings of three intra-spirals, representing the sub-planes of Mind. At this stage the mental plane would have formed.

Upon the *ātmic* plane therefore we have the electrification of primal substance – *'Electricity as vibratory impulse.* This causes the aggregation of matter, and its activity within certain bounds, or its awakening to activity'. Such activity is caused and sustained by emitting the A sound of the triune Aūṁ. The equivalent of the strong and the weak nuclear forces were generated to bind the 'dust' into the structure of the Anu that existed upon the *ātmic* plane. The 'strong force' generated the extra-spirals, bearing the forces of material manifestation (the 'electrical manifestation of matter') and the 'weak force' the intra-spirals (the 'electrical manifestation of form'). The primal atoms were also incorporated into fields, torrents of plasma, that represented the awakening of the Fires of the primal Mind. These forces later became reflected, and further reified into the dense physical domain to build the quarks of the atomic structure. Consequently also the law of triads governs the appearance of phenomena.

The vortices of energies and other forces of Mind thus generated the conditionings on the mental plane, though at first this only concerned

---

the energy of the Will to an extant unknown in our present system. It represents the Father aspect, and therein the cycles of *karma* concerning the three systems should be mostly rectified. During the cycles of these solar systems, a set number of human kingdoms, such as inhabits our present earth, have or will undergo their evolutionary journey. Our earth contains quite a number of units that are the remnants of the former evolutionary cycles, plus those (the majority) that Individualised upon this planet.

the abstracted domains of Mind. Electromagnetic force was thereby instigated, *'Electricity as Light,* causing spheroidal objectivity', signifying the birth of the Son, the consciousness-principle. Dark Mind per se was awakened, with its separative, self-focussed, self-serving attributes; the accumulation of experiences, knowledge of and for its own sake, to intensify the aura and power of the 'I'. The principle of pride ruled, plus the desire to control the other in order to empower the self-concept. This period covers the enunciation of the second syllable of the sacred Word, the 'ū' sound, causing a sense of persistence in space. That which was reached out for in order to control also represented the Thought-Forms, the unregenerate substance of former aeons (cycles) of Logoic activity. Thus began a process of descent of the energies of Mind/mind towards ultimate reification as the substance of the dense physical plane. We also have the stages of the appearance of the different levels of the luminosity of light.

On the physical plane the originating electrical nature projects the multitudinous reverberations of the projection of Thoughts as the sounds that impact through the ethers of the dense sphere to produce the forms of the material domain. The Aūṁ now resounds with a sevenfold reverberation, sustaining the *māyā* of what appears, by means of the electrical impulses of Mind from the higher domains, plus the appearing forms being integrated by a new, very weak force, that of gravity. Gravity then is the resonance of the Aūṁ that ties all of manifest space into unity. It is the accumulated effect of every Anu constituting the forms ringing out their specific notes as they cohere into unity and call out to the sounds of all other forms, integrating them into the one-ness of space. Gravity consequently is a different nature than that of the electrical foundation of the other three forces, though it derives from them. It is their child, the consequence of their interrelationship, the attractive pull of the sounds that produce a grand harmony, unity, a coherency of space.

When the sounds are coherent, unified, we have *manvantara,* and when dissonant, so *pralaya,* or death, sets in. When the songs weaken and die out altogether then the form disintegrates back into the ethers. Other songs are also sung by the Lords of evolving Life as they undergo a transmutative journey of the originating black into the radiant delight of intensified Light.

The physical plane represents the phenomenal reverberation of what is precipitated via *manas,* but the mode of manifestation is reflected from the *ātmic* plane, and the action of Fohat upon its substance. The physical plane is thus the plane of action wherein the original karmic impetus of the Logoic Mind is resolved by the evolving factors of that Mind, producing the transmutative process of substance into the higher ethers. In doing so the primordial *karma* is consequently eliminated. (This happens because the darkened substance of mind evolves along a line of increasing light, generating aspects of liberating Mind, until Mind is all there is.)

This then is an outline of the story of the nature of the Womb of the Mother and the functions of Caeleno. Other aspects are explained in volume 7B of *A Treatise on Mind.*

D.K. further states:

> *On the fourth plane this electrical force shows itself as colour.* In these four we have the fundamental concepts of all manifestation; all four have an electrical dynamic origin; all are basically a differentiation or effect of impulse, emanating from the cosmic mental plane and taking form (with intelligent purpose in view) on the cosmic physical. Man repeats the process on his tiny scale, dealing only with three planes, and flashing into objectivity on the solar physical. It will be demonstrated later as science attains more and more of the truth that
>
> 1. All physical phenomena as we understand the term have an electrical origin, and an initial vibration on the first sub-plane of the physical plane.
> 2. That Light, physical plane light, has a close connection with, and uses, as a medium, the second ether.
> 3. That sound functions through the third ether.
> 4. That colour in a peculiar sense is allied to the fourth ether.
>
> We must note here that in the development of the senses, hearing preceded sight, as sound precedes colour.
>
> An interesting analogy may here be noted between the fourth cosmic ether, and the fourth ether on the physical plane of the solar system. Both are in process of becoming exoteric—one from the standpoint of man in the three worlds, and the other from the standpoint of a Heavenly Man. The fourth ether is even now being investigated by scientists, and much that they predicate concerning

ether, the atom, radium, and the ultimate "protyle"[17] has to do with this fourth ether. It will eventually be brought under scientific formula, and some of its properties, knowledge concerning its range of influence, and its utilisation will become known unto men...[18]It might here be asked why colour primarily is spoken of as the buddhic manifestation of electricity. We are employing the word "colour" here in its original and basic sense as "that which veils." Colour veils the sevenfold differentiation of logoic manifestation and, from the angle of vision of man in the three worlds, can be seen only in its full significance on the buddhic plane. All fire and electrical display will be seen to embody the seven colours.

Again another correspondence between the fourth cosmic ether and the fourth physical ether lies in the fact that they are both primarily concerned with the work of the great builders, bearing in mind that they build the *real* body of the Logos in *etheric* matter; the dense physical vehicle is not so much the result of their work as it is the result of the meeting of the seven streams of force or electricity, which causes that apparent congestion in matter that we call the dense physical planes (the three lower subplanes). This apparent congestion is, after all, but the exceeding electronic activity or energy of the mass of negative atoms awaiting the stimulation that will result from the presence of a certain number of positive atoms.[19] This needs to be borne in mind. The work of evolution is based on two methods and demonstrates as:

Involution, wherein the negative electrons of matter preponderate. The percentage of these feminine electrons is one of the secrets of initiation and is so vast during the involutionary stage that the rarity of the positive atoms is very noticeable; they are so rare as only to serve to keep the mass coherent.

Evolution, wherein, due to the action of manas, these negative atoms become stimulated and either dissipate back into the central electrical reservoir, or merge in their opposite pole, and are consequently again lost. This results in:

Synthesis.
Homogeneity.

---

17 It should be noted that the T.C.F. was published in 1925 before the advent of modern Quantum Mechanics and particle physics.

18 T.C.F., 319-20.

19 The physical permanent atoms of the various streams of Life.

> The rarity instead of the density of matter. The fourth cosmic ether, the buddhic, is the plane of air, and is also the plane of absorption for the three worlds. This rarefication of dense matter (as we know it) simply means that at the close of the evolutionary process it will have been transmuted and be practically, from our point of view, non-existent; all that will be left will be the positive atoms, or certain vortices of force which—having absorbed the negative—will demonstrate as electrical phenomena of a form inconceivable to man at his present stage of knowledge. These vortices will be distinguished by:
>
> 1. Intense vibratory activity.
> 2. The predominance of one certain colour according to the quality of the etheric display, and its source.
> 3. Repulsion to all bodies of similar vibratory rate and polarity. Their attractive quality at the end of evolution will cease owing to the fact that naught remains to be attracted.
>
> The vortices in each planetary scheme will be, during evolution, seven. Later, during the period of obscuration, three of the vortices will approximate their masculine pole and eventually but one will be left. In man a similar procedure can be seen in connection with his seven centres during the process of initiation. First there are seven, then three absorb the lower four through electrical interaction. We are here viewing the subject wholly from the point of view of our present discussion. Finally, only the head centre is left, for it is the positive pole to all the others.
>
> This question of the electrical polarity of the centres is one of real difficulty, and little can be communicated on the matter. It may be safely pointed out, however, that the generative organs are the negative pole to the throat centre as is the solar plexus to the heart...[20]

A Master[21] has solved the problem of electrical phenomena in the three worlds, hence His freedom. Further, when the relationship of the negative form to the positive Spirit is grasped, and their joint connection with the cosmic Entities Who indwell the whole system is somewhat apprehended, group liberation will be achieved.

---

20 Ibid., 321-23.

21 An Initiate of the fifth degree, or greater.

Perhaps in considering this abstruse matter it may help to clarify the point of view if it is recollected that man is essentially positive in his own nature but his vehicles are negative; hence he is the central unit of positive electricity that draws and holds to him atoms of an opposite polarity. When he has merged and blended the two poles, and produced light of a definite magnitude during any particular incarnation (which magnitude is settled by the Ego prior to incarnation) then obscuration takes place. The electrical manifestation burns up and destroys the medium, and the light goes out; what we call physical death ensues, for the electrical current burns up that which had caused objectivity, and that which *shone.* Let us carry this idea further and realise that these units called men (who are positive as regards their own vehicles) are but the negative cells in the body of a Heavenly Man,[22] and are held within His sphere of influence by the force of His electrical life. Bear in mind again that the Heavenly Men are thus positive as regards the lesser lives, but in Their turn are negative as regards the greater Life that contains Them.

Here again is demonstrated the truth of the teaching given by H. P. B.

Electric Fire -----------Positive -------Spirit.
Fire by Friction--------Negative ------Matter.
Solar Fire --------------Light ----------The two blended and thus producing the objective blaze.

We have thus considered the question of the electrical origin of all manifestation in connection with the four higher subplanes of the solar system—*those four planes which are the four cosmic ethers, and therefore form the body of objectivity of a Heavenly Man in exactly the same sense as the four physical ethers of the solar system form the etheric body of a man.* I have here repeated the fact, as its importance has not yet been grasped by the average occult student; this fact—when conceded and realised—serves in a wonderful way to clarify the whole subject of planetary evolution. We have now reached the three planes wherein man functions, or the gaseous, liquid, and dense subplanes of the cosmic physical.

The whole subject of the akasha will be greatly clarified as exoteric science delves into the question of the ethers. As knowledge of the four types of ethers is available, as the vibratory action of these ethers is realised, and as the details concerning their composition,

22 A planetary Logos.

utilisation, light-bearing capacity, and the various angles from which they may be studied become known then paralleling knowledge anent the corresponding four cosmic ethers will be forthcoming. Much concerning them may be deduced from the already apprehended facts which relate to the four solar physical ethers.[23]

Alcyone governs the manifestation of the fourth ether, which D.K. states is responsible for 'colour'. Colour can presume to mean all that is visible, seen by the eyes, which helps to distinguish the qualities of one form to another. Colour is the effect of the expression of the seven Rays as they condition all forms in Nature.

The statement for Alcyone was given as:

The great divider of the Water that streams, following the upper (to the lower). Dividing it is the three Mahābhūtas, from which the world falls into creation.

I stated in volume 7B of *A Treatise on Mind* that the *mahābhūtas* are the three great Elements (the Fiery, Watery and Earthy) that constitute the sum of *saṃsāra.* Alcyone therefore embodies the fourth principle that allows the three to actively manifest and so form a world sphere by means of the expression of cosmic Mind, 'from which the world falls into creation'. A terrestrial sphere 'falls', descends into manifestation from a subtler domain, by means of the Creative Sounds of the Pleiades (the permutations of the Aūṁ) and the sum of the *deva* Builders that are thereby called into action. The properties of this fourth ether have been discussed earlier, hence need not be repeated here. Note with respect to the septenary of colour implicated here that the periodic table of elements is constituted of seven broad categories of active elements, which can be considered to be coloured by the properties (valency, an electrical characteristic) by means of which they interact with other elements. There is also one category of non-reactive inert gases.

The work of the three remaining Pleiades, Merope, governing the mental plane, Taygeta, governing the astral plane, and Sterope, governing the physical plane, are responsible for the dissemination of the *mahābhūtas* into manifest space. As the *mahābhūtas* are the alchemical Elements, and whose interrelationships and qualities have

23 T.C.F., 324-26.

been explained throughout *A Treatise on Mind,* when dealing with the properties of these planes of perception, little needs to be added here. Volume 7B can be referred to for the qualities of these Pleiades.

A major reason for these extended quotes from D.K.'s works is so that the reader can gain a better appreciation of the following section, which enquires into the effect of the electrical nature of phenomena, as viewed in the physical universe. It is a concept which the esoteric doctrine ascribes to, but there are differences to be noted to the extant popular theories.

The plasma and electric universe theories

Electromagnetic phenomena permeates all of space in terms of light waves, from which cosmologists obtain much data concerning the nature of the evolution of the universe, including their concept of the Big Bang. There are some modern theorists however who disagree in the conclusion that such a primal explosion ever happened, and who ascribe everything in the universe to the effects of electricity or to plasma fields. There are also differing points of view in those who consider that there was no Big Bang. I will endeavour in this section to analyse some of these conflicting theories as a basis to better clarify the esoteric view, which rides somewhat on a different stream but incorporates aspects of all the views. The esoteric view, as the reader by now knows, elaborates the nature of multidimensional space and the effects of the primacy of Mind, which obeys its own rules of expression. Also we see that nothing comes out of nothing, that Mind must work with substance of some kind in order to exist, and as Mind is electrical in nature it organises that substance accordingly. There is therefore a dance or interplay between the particles of matter and the Minds that incorporate such matter into the structure of thoughts.

First one should consider our own sun, as much can be extrapolated from its existence, as to how other suns might exist, and as to what drives their furnaces. We should also consider the effects upon our solar orb manifested by the electric currents generated by the family of stellar spheres our sun is travelling with, and from their companion constellations of stars. The true strength of such effects should be determined. Esoterically, the interrelation concerns the energies of cosmic Minds upon other Minds, hence the impact such energies have upon their respective forms, the bodies of manifestation seen as solar systems. There will be differing energy effects depending upon karmic factors, Ray affiliations, and the 'age' of the Logos concerned. One could think of the nature of the multifarious human interrelationships to obtain an idea as to the much more exalted energy exchanges between our cosmic Brethren. Such factors are beyond our ken to directly experience, but their influences can be noted in the physical universe.

One could for instance consider that if electromagnetic currents[24] from space meet the electromagnetic energy from the nuclear reaction within the sun radiating out from its surface then the two forces would logically oppose each other, cause some heating, and would tend to send some of the radiation inwards, to recycle convectionally.[25] Logically, waves approximating the frequency of light are returned, helping to cause the darkened areas seen in the dark spots on the sun. The stronger radiation would escape, helping to cause the energies observed at the corona.[26] This corona would therefore be the product of both the energy from external sources, plus the effects of the thermonuclear energies. Solar flares that may rise over 100,000 km over the surface of the sun will send material (electromagnetic energy and plasma) from inside far into cosmic space, in the form of such things as the solar wind observable

---

24 The reified effect of Logoic Thought from the companions or associates of our solar Logos.

25 The presumption is that despite the energy generated within the solar furnace, strong etheric and electromagnetic energies from cosmic sources impact via its corona, effecting the rate of fluctuations of change of solar outbursts and sunspot activity.

26 As stated in *'On the "Electric Sun" Hypothesis'*, (http://www.tim-thompson.com/electric-sun.html), page 7, 'The specific mechanism for heating the corona indeed remains unknown...Even though the temperature of the corona is about 1,000,000 Kélvins, much hotter than the photospheric temperature of about 6000 Kelvins, the energy density in the lower corona is only about 0.1 erg/cm^3, whereas the energy density in the photosphere is about 300,000 erg/cm^3. It is easy to be deceived by the apparent contradiction of a "hot" corona over a "cool" photosphere, but it is the energy and not the temperature which is fundamental, and we see that the energy does as we would expect, it falls off rapidly in the corona... The problem faced by solar physicists is not that there is no explanation, but rather that there are too many potential explanations to choose from! Does the corona heat by virtue of magnetohydrodynamic waves in the plasma? What about heat input due to collapsing magnetic flux tubes at convective cell boundaries? Both of these are observed to happen, and both are capable of heating the corona.'

on earth. This is an escape mechanism of the energies that were forced to recycle for the pressure built up under the surface of the sun.

If the surface of the sun acts as a type of insulator to the electric currents coming from outer space,[27] plus a partial containment area, preventing some of the radiation trying to leave the sun, so it helps, in conjunction with the process of convection, to produce the far cooler area there (measured in thousands of degree centigrade, rather than in millions). It also helps to produce the sun's stability, and because its energy is not released at the rate estimated by the calculations of the thermo-nuclear explosions alone, so a sun's life may accordingly be extended, giving more time for Life to evolve during the various Rounds, Chains and Schemes of evolution upon the various planets that are to bear and harbour evolving Life (taking into account also our multidimensional universe), as presented in the esoteric account.[28]

As far as such events as novas and supernovas are concerned the conventional theories explain the cause for the phenomena well enough, considering that the esoteric view is in agreement with the existence of nuclear furnaces. The exploding stars are said to be the cause for seeding interstellar space with the heavier elements that had their genesis inside the suns by means of nuclear furnaces.

I will not give more than an outline of the doctrine of the electric or plasma universe theories here. To treat the subject properly would take another volume, where much that would be said would merely be a repetition of the information already provided by plasma cosmologists and those involved with the electric universe theories, plus the criticisms from the scientific community.[29] I will however endeavour to point out a few inconsistencies of the electric universe theory. The point of view of the esoteric philosophy is that of a Mind generated universe, which causes therefore the electric currents in space and plasma fields in the

---

27 This may be the effect of a Langmuir sheath around the sun, a cellular wall caused by plasma.

28 This account will be detailed in my forthcoming series: *The Astrological and Numerological Keys to the Secret Doctrine.*

29 For detail the reader should refer to such sites as:
https://www.electricuniverse.live http://www.holoscience.com/wp/
http://www.thunderbolts.info ThunderboltsProject

production of physical phenomena, and which also incorporates thereby conventional theories, such as the force of gravity.

Though not all the assumptions of the electric universe theory are correct, however some of their deductions and predictions concerning the effects of electric fields in space should be considered to contribute to some of the phenomena we perceive in space. This is because in general everything is governed by electrical phenomena, as the Anu is electrical in nature, however, its modes of combination also contributes to the formation of mass, which is productive of the force of gravity. Gravity is a very weak force, hence its effects are applicable only to large masses, such as the earth and those of us that are bound to it accordingly, but its effects manifest instantaneously over vast distances to cohere the universe into unity. If one thinks of the scales involved between stars then how weak the force of gravity really is can be contemplated, as posed by a question given by Findlay:

> The gravity model of the universe tells us that everything is subject to gravitational relationships. What then do you think the gravitational relationship will be between two specks of burning hydrogen the size of full stops that are four and a half miles distant from each other?[30]

The point being made here by the electric universe theorists is that gravity appears far too weak to feasibly affect stars between such distances, whilst electromagnetic fields being vastly stronger can. Physicists however have made their calculations concerning the relative masses of the stars, which though being 'dot size' relative to their distances, have enormous masses, and so say that gravity does work in this way. In the esoteric view, the effect of gravity is but an extension of the coherence of cosmic Mind holding all aspects of its manifest Thought-Form into unity. The force of Mind has a vastly reified effect upon binding the Anu's into forms upon the physical plane, and integrates the Anu's forming star systems into a Thought structure, via which what is known as 'gravity' is determined according to the concept of mass. The esoteric view takes into account multidimensional subjective space, which neither the proponents of the electric and plasma universes or

30 Tom Findlay, *A Beginner's Guide of our Electric Universe,* (www.newtoeu.com, 2013), 79.

the conventional scientists consider. They must yet discover that there is far more to Life than what is physically seen, as demonstrated by *yogins* and psychics, and that inevitably that physicality is an illusion, *māyā*. The forces of Logoic Mind hold the subjective universe in situ for the allotted duration of its existence, as well as the relative fleeting appearance and disappearance of physical forms. The forms come into and out of objectivity from etheric space, hence in reality one needs to take into account vaster time scales than conventionally viewed.

The Einsteinian cosmological view of gravity concerns the warping of the geometry of space-time by means of vast masses in space, such as that of planetary bodies. In doing so this view of gravity eliminates the need for the existence of a concept of ether. Nevertheless, an ether does exist, and indeed there are four levels of expression of ethers, which account for the phenomena manifesting via *chakras,* hence psychic phenomena, and such things as the efficacy of acupuncture methodology in surgery. There is however, *an etheric blue-print, a geometry of space, grid-lines, upon which the nāḍī system, from which the chakras stem, is based.* The foundation work for this geometry is only hinted at in this book, and shall be revealed in a later work on the *chakras,* and this geometry may indeed accommodate the Einsteinian view (General Relativity), in which case a grand harmony between the esoteric and exoteric sciences will be possible. The Einsteinian view has stood the test of time and its predictions has been validated through many experiments, using the scientific methodology. Myriads of small Anu's constitute the mass of a vaster body, such as that of our planet. The sum of the accumulated mass of such vaster bodies may indeed warp the grid-lines of the geometry underlying the *nāḍī* system, hence tally with General Relativity. In such a case one would be observing the exoteric point of establishment of a *chakra* that forms upon etheric space. Here one must take into account the esoteric fact that all planets, stars, constellations, and galaxies, etc., are but the effects of *chakras* existing upon the etheric domains. These stars, etc., are then attracted to each other gravitationally, according to the inverse distance rule.

Presently there is an impasse between the General Relativity view of gravity and the quantum field theory of gravity. As Lerner states:

> When gravity is taken into account a new paradox arises. QED hypothesizes that the vacuum is filled with virtual particles, continually

> coming into, and out of, existence so fast that they are unobservable. Vacuum, therefore, has a vast energy density. We can't tap it because no lower energy level is available—just as we can't use water power at sea level, it has no lower place to go.
>
> But general relativity says that energy, like mass, curves space. The gigantic energy density of the quantum vacuum should curve space to create a cosmological constant, an enormous repulsive field that would curve space into a sphere a few kilometers across. This obviously doesn't happen, so something's wrong with the theory. This problem is widely known in physics, but no attempt has been made to either perfect or to supersede QED.[31]

It may be that there is another energy that also comes into play here, not yet discovered by empirical scientists. We could consider the energy of consciousness itself, Logoic Thought, that accommodates the curvature of space to produce a sphere of containment of planetary and cosmic bodies according to the parameters of its own Will, and which also holds the Anu's, or gravitons in situ, preventing the 'enormous repulsive field' from manifesting.

The electric force is 10^{36} times stronger than gravity, hence its effects are immediately perceptible in the phenomenal universe all around us. It is the cause of the manifestation of electrical phenomena and the electromagnetic currents, the observation of light, etc., that we take so much for granted, as well as vast interstellar effects displayed in the corona of suns and in the filamentation of galaxies, and many effects seen in space.

With respect to this in reference to the plasma universe theory, Lerner states:

> According to Alfvén,[32] the evolution of the universe in the past must be explicable in terms of the processes occurring in the universe today: events occurring in the depths of space *can* be explained in terms of phenomena we study in the laboratory on earth. Such an approach rules out such concepts as an origin of the universe out of nothingness, a beginning to time, or a Big Bang. Since nowhere do

31 Eric J. Lerner, *The Big Bang Never Happened,* (Times Books, London, 1991), 350.

32 Hannes Alfvén (1908 – 1995), who fathered space plasma physics, gained a Nobel Prize in 1970 for the study of magnetohydrodynamics and induced electric currents in ionised liquids and gases.

> we see something emerge from nothing, we have no reason to think this occurred in the distant past. Instead, plasma cosmology assumes that, because we now see an evolving, changing universe, the universe has always existed and always evolved, and will exist and evolve for an infinite time to come.
>
> There is a second critical difference in the two approaches to cosmology. In contrast to the Big Bang universe, the plasma universe, as Alfvén calls his conception, is formed and controlled by electricity and magnetism, not just gravitation—it is, in fact, incomprehensible without electrical currents and magnetic fields.
>
> The two differences are related. The Big Bang sees the universe in terms of gravity alone—in particular, Einstein's theory of general relativity. Gravity is such a weak force that its effects are evident only when one is dealing with enormous masses— such as the earth we live on.[33]

In relation to plasma cosmology Ari Brynjolfsson states in the abstract of a paper entitled 'Redshift of photons penetrating a hot plasma':

> Plasma redshift explains the solar redshifts, the redshifts of the galactic corona, the cosmological redshifts, the cosmic microwave background, and the X-ray background. The plasma redshift explains the observed magnitude-redshift relation for supernovae SNe Ia without the big bang, dark matter, or dark energy. There is no cosmic time dilation. The universe is not expanding. The plasma redshift, when compared with experiments, shows that the photons' classical gravitational redshifts are reversed as the photons move from the Sun to the Earth. This is a quantum mechanical effect. As seen from the Earth, a repulsion force acts on the photons. This means that there is no need for Einstein's Lambda term. The universe is quasi-static, infinite, everlasting and can renew itself forever. All these findings thus lead to fundamental changes in the theory of general relativity and in our cosmological perspective.[34]

One can conceive of the originating 'black dust' to be in the form of a plasma, hence this version of cosmology would apply. Here one should recall that a universe incarnates by utilising such substance in order

---

33 Eric J. Lerner, 41.

34 Ari Brynjolfsson, *Redshift of photons penetrating a hot plasma,* arXiv:astro-ph/0401420v3, 7 Oct 2005.

to convert it into the attributes of elementary mind. Hence the 'black dust' plasma field can be considered to be of infinite duration. Ari's finding that 'the universe is not expanding' is correct from the point of view of the originating Thought-Form of the Logos having reached its 'ring-pass-not' or pre-determined boundary. The finding of no dark matter, or dark energy refers to the postulates of the exponent of the Big Bang theory, nevertheless the energy and substance pertaining to the subjective planes of perception and the etheric domain does exist.

With respect to the Big Bang Lerner states (in part):

> Even with dark matter, the Big Bang still could not account for the low level of microwave anisotropy, or the formation of galaxies and stars. Nor could it accommodate Tully's large-scale supercluster complexes (described in Chapter One). And the dark matter itself was ruled out by new observation and analysis. The Big Bang in all its versions has flunked every test, yet it remains the dominant cosmology; and the tower of theoretical entities and hypotheses climbs steadily higher. The cosmological pendulum has swung fully again.[35]
>
> Thus, if the universe was indeed sculpted by the counterpoint of gravitational contraction and the pinching of vortex filaments, its observed structure was inevitable. The size and mass of contracted objects and the spaces between them were determined in a simple way. Together, the mass-radius law for contracted objects and the mass-distance-squared law for their spacing also explains a major feature of the universe—the smaller the objects, the more isolated they are. Thus stars are separated from each other by distance typically ten million times their own diameters, galaxies by only thirty times their diameters, and clusters by about ten times their diameters. In short, my theory explained why space is so empty.[36]

Some other objections to the theory of the Big Bang are also presented by Wallace Thornhill and David Talbott. First we have the problems concerned with an anomaly in the observation of the redshift. (The redshift is the shift of light towards the red end of the spectrum as it recedes over vast distances in space, which gives scientists the ability to measure cosmological time.) They state:

35 Ibid., 162.

36 Ibid., 262-63.

In 1971, using a 5-meter telescope at Mt. Palomar, Arp discovered a visible "bridge" joining the low redshift galaxy NGC 4391 to the high-redshift quasar, Markarian 205 -- an "impossible" connection, unless popular assumptions about redshift were incorrect. For years thereafter astronomers debated the validity of the "bridge."[37]... Stephan's Quintet is a massive cluster of five galaxies. Four have similar high-redshift. The fifth, NGC 7320, has a much lower redshift, and thus should be much closer to us than the other galaxies.

Yet there are debris fields and tails around the low-redshift galaxy that suggest it is interacting with the high-redshift systems, which would require that all five galaxies be in the same vicinity in space.[38]... But if Halton Arp and his colleagues are correct, something other than motion is shifting the light of celestial bodies toward red ... some quality *inherent* in the redshifted objects themselves. If the traditional interpretation of redshift is the foundation of the Big Bang theory and it is proven to be incorrect, what then happens to the Big Bang?[39]...In a Big Bang 'expanding' universe it is particularly difficult to account for dynamic interactions *between* galaxies. The space between galaxies is supposed to be increasing. More powerful telescopes, however, have revealed astonishingly dynamic interactions between galaxies, often called "collisions" or "mergers"–the one thing prohibited by the original Big Bang hypothesis.[40]... Astrophysicists also faced a growing dilemma posed by the *internal motions* of galaxies. Gravity is not strong enough to contain the rapidly moving outer stars in galaxies. They should be flying apart.

To answer the challenge of galaxies defying the law of gravity, astrophysicists proposed the existence of an invisible form of matter. Then they placed this 'dark matter' wherever needed to make their models work. Plasma experts have shown that the rotational behavior of galaxies can be explained electromagnetically: galaxies are *driven* like a simple electric motor. Where giant intergalactic Birkeland currents intersect they form a vortex, drawing surrounding matter inward to form a rotating galactic disk. The 'anomalous' rotation is a predictable feature of the electric model.[41]

---

37 Wallace Thornhill David Talbott, *The Universe Electric: Big Bang?* (eBook, Copyright © by David Talbott & Wallace Thornhill, 2008), 45.

38 Ibid., 51.

39 Ibid., 59.

40 Ibid., 143.

41 Ibid., 175-76.

Redshift is the basis for the present mainstream calculations and deductions concerning the theory of the Big Bang, thus the problems with the anomalous observations with redshift, if true, invalidate this theory. Here 'The Wolf Effect' should also be noted. This Effect is presented in a paper to *The Astrophysical Journal 1990,* which states in part:

> In the last few years several new closely related processes have been discovered that can generate frequency shifts of spectral lines. This development followed the theoretical prediction made by Wolf (1986) and since then verified by experiment (Morris and Faklis 1987) that, contrary to a commonly held belief, the spectrum of light is, in general, not invariant on propagation in free space...Very recently it was predicted (Wolf 1989a, b) that the frequency shifts induced by scattering from time-dependent random media with suitable correlation properties may imitate a Doppler shift of any magnitude, even though the source, the medium, and the observer are all at rest with respect to each other. It seems therefore possible, in principle, that this effect may contribute toward the redshifts observed in the spectra of some astronomical objects. In this connection we recall that the long-standing controversy over the interpretation of quasar redshifts continues despite the pronouncement that "the 'evidence' for non-cosmological redshifts is a collection of unrelated curiosities having no predictive power" (Weedman 1986, p. 37). The contrary view was ably argued by Arp (1987). Even Hubble (1936) questioned the validity, in all cases, of the velocity interpretation. Most non-Doppler hypotheses proposed in the past violated established physical laws in some way, e.g., the principle of conservation of momentum and energy.
>
> It is important to appreciate that there are two quite distinct aspects to the redshift controversy, which have not always been kept in focus in the lengthy dispute, namely, the questions (1) whether most quasars are at cosmological distances and (2) whether some quasars are associated with objects of lower redshifts.[42]

After pointing out that their 'analysis is rather incomplete for several reasons', they conclude: 'Nevertheless, our analysis clearly indicates

42 Daniel F.V. James, Malcolm P. Savedoff, and Emil Wolf, University of Rochester, Rochester, New York, *SHIFTS OF SPECTRAL LINES CAUSED BY SCATTERING FROM FLUCTUATING RANDOM MEDIA,* (The Astrophysical Journal, 359:67-71,1990 August 10, © 1990. The American Astronomical Society), 67.

that some observed redshifts may contain contributions which arise from the intrinsic properties of the medium surrounding the radiating sources, in addition to those due to the Doppler effect or gravitation. It is thus possible, as we noted in § I, that the long-standing controversy relating to pairs of objects of different redshifts which appear to be physically connected might be resolved by taking into account the correlation mechanism discussed in this paper'.[43]

A follow up paper by Roy, Kafatos, and Datta in relation to the redshifts in the case of NGC 4319 Galaxy and Markarian 205 Quasar pair states in part:

> This indicates that the distance of the quasar is just the same as that of the galaxy, but the shift in the first case is much higher than that of the later. We can envisage this phenomena as some sort of *screening* operation for the Wolf Effect. In other words, new type of screening arises due to the nature of the medium. This will be studied in detail in future work. Lastly, we should mention that the critical source frequency relating to the screening effect which plays a crucial role in explaining both redhsift and spectral width can be tested in *laboratory experiments*.[44]

At the beginning of this chapter I stated that quasars 'represent the nodes of descent of the *maṇḍalic* patterning established by the One at the centre of Life. They sound out their Notes of Power to establish the pattern of what must BE. Hence the 'creation' resounds from the One to the nodes and together they establish the Plan for the expanding Fire.' The anomalies observed when considering quasars should be considered in light of this esoteric view. From this perspective also, the quasars need not appear simultaneously, rather are sequenced over evolving time as the related *chakras* are progressively awakened. This would explain the observed paradoxes in some quasars and their relation to nearby galaxies, such as that described above, where there was a 'visible "bridge" joining the low redshift galaxy NGC 4391 to the high - redshift quasar, Markarian 205 -- an "impossible" connection'.

---

43 Ibid., 69.

44 Sisir Roy, Menas Kafatos, and Suman Datta, *Shift of spectral lines due to dynamic multiple scattering and screening effect: implications for discordant redshifts,* (Astronomy and Astrophysics, 353, 1134–1138, 2000).

Here is part of Lerner's explanation for the creation of galaxies in terms of a plasma universe consideration:

> The expanded electrical currents now complete the cycle by pinching new supplies of plasma together to create new galaxies and to generate new quantities of fusion power. The magnetic fields as well help to create the filaments in existing spiral galaxies which lead to the formation of new stars in old galaxies.
>
> As with the gravity-driven stage of evolution, nuclear-powered evolution involves a series of substages. When hydrogen is exhausted within individual stars, its by-product helium then becomes, at higher temperatures and pressures, fuel for the production of carbon and oxygen. When all the fuels for a star are exhausted, its explosion in a supernova scatters the elements to the surrounding interstellar in medium—fuel for new stars.
>
> The initial generation of stars burns hydrogen to helium quite slowly. But in subsequent generations, carbon and nitrogen act as catalysts, enormously accelerating the reactions. Again the debris of one stage fuels a subsequent stage.
>
> Thus a series of interactions, generating fluctuations which are magnified by instability, has driven the cosmos from a homogeneous and random hydrogen plasma to the differentiated and dynamic universe of stars, galaxies, and planets, each made up of a hundred different chemical elements. There is no tendency toward decay, nor is an external power needed to generate order. Order and complexity come into being through natural processes governed by a series of interactions—electromagnetic, gravitational, and nuclear. In short, chaos begets order—progress.[45]

Concerning the problem of where all of the energy that is tied up in the mass of the universe actually comes from Lerner states:

> [W]e know that under present circumstances the maximum energy derivable from a piece of matter is defined by $e = mc^2$—mass times the speed of light squared—which yields 10^{21} ergs/gram. But our universe appears to have a very significant amount of energy tied up in existing matter. Where did *that* energy come from? Is there more?
>
> Cosmology has dodged this question by hypothesizing that this matter-energy comes from the gravitational energy of the Big Bang.

45 Ibid., 299-300.

> However, as we've seen, this requires that omega equal 1—which it clearly does not. Gravitational energy amounts to between one-hundredth and one ten-thousandth of the energy tied up in observed matter (and there's no reason to assume that substantially more matter exists). This is wholly insufficient. We simply do not know where the energy in matter derives from, and we *do not know* whether and under what circumstances it can be captured or released. Until we do know, we cannot set "scientific" limits to the energy available in the cosmos.[46]

In the esoteric philosophy, of course, the problem of where this energy comes from, and what instigates the formation of any sphere of containment, such as a sun, galaxy or universe, is answered as it comes from the domain of cosmic Mind and is directed by Logoic Mind.

With respect to the electric universe theory information concerning its nature can be obtained from Wal Thornhill's Holoscience Website, a brief summary is given below. There are many valid arguments presented supporting his theory, and some not so valid. The debate continues on the internet. Below is a quick synopsis of some of Wallace's main points:

> The consequences and possibilities in an *Electric Universe* are far-reaching. First we must acknowledge our profound ignorance!
>
> We know nothing of the origin of the universe.
> There was no Big Bang.
> The visible universe is static and much smaller than we thought.
> We have no idea of the age or extent of the universe.
> We don't know the ultimate source of the electrical energy or matter that forms the universe. Galaxies are shaped by electrical forces and form plasma focuses at their centers, which periodically eject quasars and jets of electrons.
> Quasars evolve into companion galaxies.
> Galaxies form families with identifiable "parents" and "children".
> Stars are electrical 'transformers' not thermonuclear devices.
> There are no neutron stars or Black Holes.
> We don't know the age of stars because the thermonuclear evolution theory does not apply to them.
> Supernovae are totally inadequate as a source of heavy elements.
> We do not know the age of the Earth because radioactive clocks

46 Ibid., 300-301.

> can be upset by powerful electric discharges. The powerful electric discharges that form a stellar photosphere create the heavy elements that appear in their spectra.
> Stars "give birth" electrically to companion stars and gas giant planets.
> Life is most likely to form inside the radiant plasma envelope of a brown dwarf star!
> Our Sun has gained new planets, including the Earth. That accounts for the "fruit-salad" of their characteristics. It is not the most hospitable place for life since small changes in the distant Sun could freeze or sterilize the Earth.
> Planetary surfaces and atmospheres are deposited during their birth from a larger body and during electrical encounters with other planets.
> Planetary surfaces bear the electrical scars of such cosmic events.
> The speed of light is not a barrier.
> Real-time communication over galactic distances may be possible.
> Therefore time is universal and time travel is impossible.
> Anti-gravity is possible.
> Space has no extra dimensions in which to warp or where parallel universes may exist. There is no "zero-point" vacuum energy.
> The invisible energy source in space is electrical.[47]

The electric universe theory has been highlighted here because it has gained much traction on the internet, and while the effort put into the theory is commendable, with some valid assumptions, nevertheless there are serious flaws. One of these concerns their concept of the electric sun. The commentary below is from the website 'Neutrino Dreaming':

> We astronomers often stumble across new theories, and after a while a certain degree of 'learned scepticism' enters the fray. So I decided to take a closer look at this theory. The theory seemed to be all encompassing and rather difficult to pin down, so in order to do this, I focused on what the theory has to say about our sun in particular. Astrophysicists say that stars, including the sun, are powered by nuclear fusion. However electric universe theorists say this is not so. The reasons given are that:
>
> 1. we haven't yet found the neutrinos that must be emitted from such a reaction;

47 From: Wallace Thornhill, *A Synopsis of the Electric Universe,* http://www.holoscience.com/wp/synopsis/

2. that the granular structure we see on the sun would not be possible, because convection is impossible due to the conditions there;
3. the energy emitted from the sun does not display the inverse square law;
4. periodic fluctuations in the sun's output resemble electric discharge patterns; and
5. the solar wind is an effect of charged particles being accelerated in an electric field.

Well that all sounds very plausible and 'scientificy'. But let's take a closer look at the arguments one by one

Neutrinos have not been found?

...Scientists have been detecting the effects of neutrinos for years, and they match the predictions exactly. If an alternative theory is to be considered, scientists would need to reject the theory of nuclear fusion at the centre of a star. This would also necessarily lead to rejection of the theories of thermodynamics, gravitation, nuclear physics, statistical physics, electromagnetism, hydrodynamics and magnetohydrodynamics. In other words, most of physics would need to be rejected to address the problem of the 'missing' neutrinos.

Convection in the sun is impossible?

Electric universe theory argues that the granulation we observe on the surface of the sun cannot be caused by convection bubbling up the layers of the sun. This is based on an assumption by a man called Juergen, that one of the values used in fluid dynamics, the Reynolds number, causes the convection, and at certain values convection cannot occur. If you imagine a parcel of matter inside the sun towards the surface as the sun's heat causes it to rise and falling back towards the centre as it cools (like boiling water), the Reynolds number describes a function of the parcel size, length and stickiness.

Juergen assumes that the Reynolds number controls convection but it doesn't; convection is controlled by the Rayleigh number. The Rayleigh number is a function of the temperature, gravity, the degree of temperature change, stickiness and how diffuse the temperature is...

The sun's energy breaks the inverse square law?

In physics, the inverse square law states that a specified physical quantity or strength is inversely proportional to the square of the distance from the source of that physical quantity. So in other words if you move from two metres to four metres away from a heater you increase the distance by two, but decrease the energy by four times (four is the square of two). Electric universe theory says that because the sun is coolest at its surface, then the temperature jumps up again out at its halo, it does not obey the inverse square law, and physics is wrong.

At this point it is important to note that the inverse square law only applies to radiant energy (as opposed to convection or conduction) and only in a vacuum. When energy moves through an atmosphere (such as the corona of the Sun) then the law does not hold. In addition, the inverse square law applies to all energy, not just heat. The colder 'surface' (photosphere) actually has more energy. The energy drops dramatically at the corona as we would expect. There are a myriad of explanations for the temperature differences, none of which involve throwing out physics as we know it.

The sun's variations prove it is a bag of plasma?

Electric universe theory says that the variations in the sun every 2 hours and 40 minutes or so can only be explained if the sun was a big bag of gas undergoing periodic electrical discharge. Juergen cites some research that shows this period is what we would expect from a homogenous sphere, rather than the accepted layered model of the sun found in textbooks. Well that is a problem ... isn't it?

OK, time for some context here. The research cited was in 1976 and the authors stated that it applies only if they are p-mode oscillations. But back then we didn't have the technology to distinguish between p-mode and g-mode oscillations. Later research, available to the electric universe theorists, showed they were gmode, so basically all the assumptions based on this research went out the window...

The solar wind is caused by an electric field?

In physics an electric field applied to charged particles cause them to accelerate. The Electric universe theory says that the solar wind is the result of such a field, and the Sun is electric, not fusion based.

Maxwell's theory of acceleration, however, talks about a time variable field, not a fixed one, and what's more the solar wind contains both positive and negatively charged ions (protons and electrons mainly). An electric sun would be positively charged and all the negatively charged electrons would be attached to it – not be pushed out from the Sun on a solar wind. This fact proves the Sun is not electric.[48]

I do not wish to be further embroiled in the controversy concerning the exponents of the mainstream Big Bang community of scientists and those that propose the electric universe theory, or with the problems concerning Velikovsky's theories proposed in his books, such as *Worlds in Collision,* which Thornhill ascribes to. The reader can judge for him/herself through the relevant sites.[49]

Because the Anu is electrically structured, so the sum of the appearing phenomena has an electrical basis. However, as explained in the previous chapters, the Anu's combine to form the quarks that are the basis of the atoms of the physical universe described by scientists, hence their deductions, mathematical analysis and experimental results apply. Also the effects of plasma cosmologists must be taken into account because of the nature of cosmic dust and the energy fields that interrelate them.

The esoteric view purported in this book differs from all the others in that it presumes the prior existence of Mind in the universe, and furthermore, it is the expression of the mental forces of a Logos, the embodiment of a universal Mind that causes the appearance of phenomena in the universe. The mechanism of expression therefore obey the laws of Thought. Consequently nothing is left to chance, there is no purely mechanical determination left to blind laws alone. Indeed, the very laws that scientists have discovered that are the causes of the world of phenomena and which govern the movements of all Life and entities therein are but the expressions of the laws governing mind/Mind.

---

48 Neutrino Dreaming, https://neutrinodreaming.blogspot.com/, 29 September 2011. There are more than 250 comments pro and con the electric universe theory, to this refutation, many of which are well informed, that are worth reading. See http://www.tim-thompson.com/electric-sun.html for greater detail.

49 Some of the sites related to the electric universe theory have already been given. The reader can compare what is therein presented to the information for instance in the article "Electric Universe Theory Debunked", https://www.everythingselectric.com/electric-universe-debunked/#bridgeman and https://briankoberlein.com/2014/07/01/rube/Testing-the-Electric-Universe---One-Universe-at-a-Time.pdf

Consequently the appearance of what is known as 'Life' is no mystery, it is but expressed in various Hierarchical structures embodying the appearance of things in a multidimensional universe. Entities evolve by developing increasingly subtle perceptions of the order of what is. As consciousness refines, so does the vehicles it inhabits. This is because it is able to bear increasingly subtler energies, and the intensified energies transform the material sheaths consciousness inhabits. The sheaths become more energetic, radiatory, and then ethereal, until it eventually dematerialises. There is consequently no 'heat death' of a universe that expands into infinity, though a 'coolness' can be seen manifesting in the formed domains as the fields of energy move upwards to realms sublime.

Eventually that which is formed dissociates because of lack of energisation and it too enters subjective space to become integrated as part of a universal pool of substance therein. This substance is seen as massed *deva* Lives, units of sentience, embodying energy fields that await the time for their cyclic calling into manifestation in the phenomenal domains by the appropriate songs sung by agents of a higher order of encompassing Minds that mathematically, geometrically, call them into action.[50] They act according to the resonance of the patterns, the energy fields generated, that precipitates them into organised forms. This organisation is governed by the conditioning laws manifesting via the inner domains, reflecting the order of what already is, based upon patterns of the 3, 5 and 7 already explained. Hence the Anu comes into existence. A multitudinous number of them appear and combine in different ways to produce the manifestation of the phenomena investigated by science.

When Wallace Thornhill states that the optimum place for the formation of a planetary sphere for sustaining physical plane life is within the outer sphere of a red giant, he has fallen into a major error, as he has omitted a comprehension of what the purpose of that life is all about. That purpose is to learn to know the order, extent, magnitude and multidimensional nature of all that exists in a universe. This is very difficult to do within the halo of a sun. Far better outside of one. In fact the earth is ideally situated for this function, as it exists in a sweet spot, which astronomers generally call 'the goldilocks zone', within our solar

---

50 For this reason mathematics is an essential discipline for scientific reasoning.

system, that significantly facilitates such development. Within this zone conditions are optimised to support Life. Therein conditions are warm enough for liquid water to exist, not too hot or too cold to support Life, and allows the evolution of a breathable atmosphere. Enlightenment in fact is gained through a correct observation and deduction from what is, viewing both the internal and external universes in turn and then simultaneously. The earth is placed in an ideal location towards the end of our spiral galaxy to view the rest of the universe, with minimum obscurations from dust clouds, or from the radiance of the central halo of the galaxy if it was placed there. It is situated at the ideal position in the solar system to receive the exact amount of light and warmth from the sun to maximally further our evolutionary growth. It is a convenient size, with an appropriate atmosphere to shield us from harmful UV, X and gamma rays. Also the moon helps to protect us from the impact of comets, and so forth.

The views concerning the formation of the universe and of such suns as ours must account for the fact that everything that exists is geared for the appearance of lives that can bear consciousness. This brings into consideration Mind-directed evolution, which optimises every factor in a system's appearance so that consciousness can maximally evolve.

More than one human stream has appeared in our solar system, according to our esoteric lore, other than what appears on the earth, as the changing conditions of the sun's luminosity over billions of years, and even the positioning of the orbits of the planets around the sun, that may change, are taken into account. Cosmological theorists must also take into account the appearance of entire new planets, solar systems, and galaxies for that matter, from out of the 'event horizon' of the fourth ether into the objectivity of the physical domain. This is but a vaster correspondence of the appearance of subatomic particles out of the same ether. In the case of a planetary sphere for instance, vast numbers of elementary lives that went into obscuration and entered into a universal reservoir of substance in an earlier cycle are again called to manifest into objectivity.

All comes into existence according to the *karma,* the *saṃskāras* developed in the past by the incarnating entity, human or Logoic, and the rebirthing process proceeds apace. All of this is governed by group laws, which were explained in volumes 6 and 7A of *A Treatise on Mind.* The process of incarnation was also explained in the earlier volumes, especially volume 4, *Maṇḍalas: Their Nature and Development.*

Space evolves with time for every incarnate entity, for it is the way of the evolution of consciousness. Consciousness incorporates time as it expands to know things. The time-space continuum (the *yugas* of a *manvantara)* exists within a larger Mind-Space that is similarly undergoing a vaster process. Such considerations concerns only the evolution of the empirical mind (cosmically, Mahat). Time is transcended for the abstracted enlightened Mind. Therein *'yugas'* become cycles of accomplishment, transcendent time, timelessness, where vast durations of empirical time may have happened on the material domain, whilst on the inner domains wherein the eternal NOW is experienced, a relatively small cycle of accomplishment may have occurred. Hundreds of millions or even billions of years may pass before a cycle of opportunity appears for some Entities, allowing a vast expanse of group Lives to appear into objectivity. As it is for atomic lives, so a similar process manifests for all Lives, a continuous 'coming and going' into and out of formed space. So it is with the rebirthing process. Our universe similarly incarnates and disincarnates in the infinitude of absolute SPACE. Beings incarnate and evolve through a time-space process, like suns forming out of the dust of cosmic nebulae, yet still the space around them exists within which their minds can expand and grow into Minds. Space integrates them all into One integral universal MIND. That MIND is the 'BOUNDLESS ALL'.[51]

Blavatsky states that there are three fundamental postulates to the esoteric doctrine:

> (a) An Omnipresent, Eternal, Boundless, and Immutable PRINCIPLE on which all speculation is impossible, since it transcends the power of human conception and could only be dwarfed by any human expression or similitude. It is beyond the range and reach of thought — in the words of Mandukya, "unthinkable and unspeakable."

---

51 H.P. Blavatsky, *The Secret Doctrine,* (Theosophical Publishing Co., London, 1888, 2005), 27. From Stanza 1:5 of the *Stanzas of Dzyan,* (stanzas of meditation.) This verse states that: 'DARKNESS ALONE FILLED THE BOUNDLESS ALL, FOR FATHER, MOTHER AND SON WERE ONCE MORE ONE, AND THE SON HAD NOT AWAKENED YET FOR THE NEW WHEEL, AND HIS PILGRIMAGE THEREON'.

To render these ideas clearer to the general reader, let him set out with the postulate that there is one absolute Reality which antecedes all manifested, conditioned, being. This Infinite and Eternal Cause — dimly formulated in the "Unconscious" and "Unknowable" of current European philosophy — is the rootless root of "all that was, is, or ever shall be." It is of course devoid of all attributes and is essentially without any relation to manifested, finite Being. It is "Be-ness" rather than Being (in Sanskrit, *Sat*), and is beyond all thought or speculation.[52]

(b) The Eternity of the Universe *in toto* as a boundless plane; periodically "the playground of numberless Universes incessantly manifesting and disappearing," called "the manifesting stars," and the "sparks of Eternity." "The Eternity of the Pilgrim"[53] is like a wink of the Eye of Self-Existence (Book of Dzyan.) "The appearance and disappearance of Worlds is like a regular tidal ebb of flux and reflux."[54]

(c) The fundamental identity of all Souls with the Universal Over-Soul, the latter being itself an aspect of the Unknown Root; and the obligatory pilgrimage for every Soul — a spark of the former — through the Cycle of Incarnation (or "Necessity") in accordance with Cyclic and Karmic law, during the whole term. In other words, no purely spiritual Buddhi (divine Soul) can have an independent (conscious) existence before the spark which issued from the pure Essence of the Universal Sixth principle, — or the OVER-SOUL, — has (a) passed through every elemental form of the phenomenal world of that Manvantara, and (b) acquired individuality, first by natural impulse, and then by self-induced and self-devised efforts (checked by its Karma), thus ascending through all the degrees of intelligence, from the lowest to the highest Manas, from mineral and plant, up to the holiest archangel (Dhyani-Buddha).[55]

Blavatsky's *The Secret Doctrine* and Bailey's *A Treatise on Cosmic Fire* further explains the esoteric view re the nature of the formation of a solar system. I will later expand on this information in my forthcoming series *The Astrological and Numerological Keys*

52 Ibid., 14.

53 The human Monad.

54 Ibid., 15.

55 Ibid., 17.

to the Secret Doctrine. Until then blessings be to all who have taken and continue to take an interest in all things cosmological and of the appearance of phenomena. It matters not their present views, for all are upon the path to enlightenment, but depending upon the view, or the dogmatic inability to change by fluidically embracing new better views if need be, so the order or timing of that enlightenment will proceed.

Meditation however is the key, and once the process and laws governing the meditation process is understood, so then knowledge of how all comes into existence will be known. Indeed, the laws governing the formation of matter from the ethers will be comprehended, and once rightly applied will institute the epoch of the appearance of the *siddhas*[56] who will be able to mentally command the substance *(devas)* to perform miracles, as people would perceive such activities today. 'Miracles' they may be when the magical application of the laws of light and sound are applied through right ritual and empowerment of visualisation. We work to inspire the evolution of white magicians, and if this is to be the universal outcome upon our planet then the massed selfishness, separative and materialistic bias of individuals in our societies and those in government must be replaced with considerations of compassionate love.

Such massed transformation of consciousness is the dire need in today's societies, and when the momentum of the process of change is at the right pace then the enlightened *siddhas* will appear to show humanity many things formerly hidden. Our brethren from space can also then appear to help produce a truly universal scientific understanding concerning cosmos that humanity is still presently struggling to awaken. The time for this exoteric application of group law that unites all human groups in our local portion of the universe is rapidly approaching because of the dawning of this new esoteric science. People need to think in terms of group law, and of how such thinking leads to the formation of a Logoic Consciousness. Logoi evolve collectively under the auspices of such laws. To help bring about the approaching new epoch where such considerations is the main topic of thought is a major reason for which I write. The more that are aware

56 *Siddhas* are *yogins*, enlightened beings who have gained, transcendental psychic powers *(siddhis)*.

enough to assist in bringing about the new epoch, the speedier will its approach be. Loving understanding is the key to compassionate development, and compassion opens the doors to higher revelation than was hitherto permissible. The development of Love-Wisdom helps one to journey far in this multidimensional universe. Its full expression leads eventually to one's placement in the stars as a Logos. This is my wish for the eventual attainment of all upon this planet.

Oṁ Svāhā!

Let peace come to you
with the flight of the Dove.
Oh feel the beat of its wings!
Hear its sound traverse the far reaches of
the Mind that is Space.
The clarion call of the Heart
resounds
in the Clear Cold Light,
forever presenting
the adamantine jewel of Love
to you.

Oṁ

Appendix One

The Heart is the Mind

Most people create things mentally without being aware of it. Thoughts are things and the 'things' thus created are normally forms of attachment; to concepts, ideas, desire-laced images. Desires and emotions govern the thought lives of the great majority of people. They build their edifices, places of abode, their work in the material world, and plan their lives accordingly. *Kāma-manas* (desire or emotional mind) largely rules the world. Truly compassionate, altruistic thought, the way of thinking with the heart, is alien to most, as people normally conflate loving, affectionate emotional thoughts with Love. Love in fact has very little to do with the emotions. What then constitutes Love by way of Thought? The statement in the Bible that as a person 'thinketh in his heart, so is he' offers a clue.[1] Most would interpret this statement in terms of the way their innermost thoughts and desires manifest and rule their lives. There is however an occult truth in the axiom presenting a higher level of interpretation, necessitating proper analysis of what the way of the heart actually is. Esoterically this is the way of awakening the Heart centre *(anāhata chakra).*

The awakening of this *chakra* lies at the heart of all true meditation practices, as it concerns the way of liberation from suffering and the

---

1 *Proverbs 23:7.*

māyā, the illusion-forming activities of mind and the transience of everything in the world around us. The awakening of this *chakra* is productive of the Bodhisattva path, the way of compassion, which has been thoroughly explained in the *A Treatise on Mind* series. Bodhisattvas are those that are unfolding the mysteries of the way of thinking with the Heart as they travel the path of Initiation into the mysteries of Life.

Here a communication[2] might be of value to better understand the way of thinking with the heart, if meditated upon and properly interpreted.

The Heart is the Mind. Remember this well.
For man[3] is essentially Love, and all that he is unfolding
when unfolding the mind is the knowledge petals
of the Heart of that which is to Be.
Therefore do not concern yourself with the concrete
trivialities of the reality of the illusion,
but meditate, ponder, and remember.
The Heart is the Mind.

Meditate on this.

The Mind is the active Heart.
It (mind) is the double-edged Sword.
It is the perverse acknowledgement of the will.
It repeats what it knows to the drum of the heartbeat.
It listens to the Water from which it drinks
when it lacks acknowledgement.
It is the slayer of the Real.
The Heart is the Real.

---

2 Communications are carefully worded and coded statements from Masters of Wisdom. In this case the teaching comes from Serapis.

3 Other than being the root for the Sanskrit term *manas,* meaning 'to think', this term also refers to the esoteric fact that the human kingdom is deemed masculine, needing to develop the force of the will, whereas the *devas* are feminine, being receptive to mantric commands. Statements from members of Hierarchy are concerned with the esoteric meanings of words and do not concern themselves with the semantics of gender politics so prevalent in the West nowadays.

Many esoteric statements can be properly interpreted only after correctly placing their content upon the eight-spoked wheel of Life *(aṣṭadiśas).* Each spoke points outwards to each of the cardinal positions of the compass, giving us a sense of occult direction. The wheel is constituted of the interrelated qualities of the fixed (+) and mutable (卍) crosses, and has the propensity to move from right to left, left to right, and thence fourth dimensionally. Energy also moves from within without, and its reverse, above-down, and below up along the spokes of the wheel, as well as between polar opposites and along the eight points constituting the rim or circumference. Here the three types of motion, rotary, spiral-cyclic, and forward-progressive, should also be taken into account.[4] Though the attributes of the *aṣṭadiśas* have been given earlier, it should be helpful for the explanation of this communication to rephrase the information.

The *mutable cross* is concerned with the mode of activity of all causative agents, human or Logoic. The four arms of the mutable cross are represented by the terms: *unity* (northeast), *expression* (southeast), *understanding* (southwest) and *goodwill* (northwest).

Northeast–unity, concerns an at-onement with all life streams. This necessitates receptivity to the incoming energies from the Brothers of the Logos, the membership of the Hierarchy of Light, kingdom of Souls, or sources of energy, depending upon one's angle of vision. It concerns receptivity to the streams of energies and the blending together, or ingathering, of the factors with which a causative agent must work.

Southeast–expression, concerns all that is unfolded in the field of Life, the battlefield of desire. Here we have the outgoing energies that originally came into the system from the northeast direction (into the realms of form). On a higher turn of the spiral, the trials of discipleship find their expression. These energies stimulate general humanity's desires and lower mental activity, and later their fields of aspirational activity. The causative agent must rightly direct these quantified streams of energy to do his/her bidding.

Southwest–understanding. Once the energies of Life have been expressed in the field of the earth, and the battles of desire have been

4 Information concerning the qualities of this wheel can also be found in Alice A. Bailey's *Discipleship in the New Age II,* (Lucis Trust, 1955), 189-94.

fought, then comprehension of the entire Life process proceeds with certainty. This produces the treading of the path of Initiation, then the evocation of the emanatory goodwill that will lead the disciple off the mutable cross and away from embodied expression altogether. The work of the causative agent now finds its purpose. The objective of the long, aeonic cycles of endeavour thus comes to light.

Northwest—goodwill. This emanatory energy produces the projection of the *antaḥkaraṇas* out of the sphere of the earth (the embodied form) into the higher dimensions of perception, and to cosmic space. It concerns the projection of a creative, or causative stream of activity from a limited field of application to another more embracive one. The energy or quality of the singular unit or group now merges into the whole. Even a Logos must find a higher centre of activity. It therefore produces ever-expanding scenarios of activity. Having fulfilled the task, the causative agent draws the gain of the former endeavour inwards, so that the related qualities can be utilised for an even greater work.

Antaḥkaraṇas are lines of relationship that interrelate any or all consciousness 'bits'.[5] An *antaḥkaraṇa* is singular and yet can continuously move in any of these eight directions, repeating 'what it knows' to the 'drum of the heartbeat', the dynamo of Life's mutability that projects the Oṁ of continuously higher creative undertaking throughout time and space. Time itself is then inevitably surpassed by the law of cycles. The *antaḥkaraṇa* brings an idea into fruition and also becomes the path of its resolution. It is the sight of the Eye (its line of direction) that allows the causative agent to project Purpose.

The *antaḥkaraṇa* signifies the 'strait gate' and 'narrow way', 'which leadeth unto life, and few there be that find it'.[6] When the symbolism

5 *Antaḥkaraṇa,* [from *antar* = interior, within, limit + *karaṇa* = reason], literally the inner organ to reason with, inner sense, consciousness-link. It arranges that conveyed to it by the senses. The streaming out from a sense organ to cognise. Hence links of cognition. Esoterically it bridges or links the empirical mind to the abstracted Mind. It has been depicted as a 'rainbow bridge' because of the Rays of energies it conveys. At a later stage of meditative development it is consciously built through a sea of consciousness, bypassing the Sambhogakāya Flower, allowing linkage with the *dharmakāya* (the Monad) and the wisdoms of the five Dhyāni Buddhas.

6 *Matthew 7:14.*

of the *antaḥkaraṇa* is properly understood, for example, as symbolised by the line through the circle

especially in relation to the meditation process, then the esoteric meaning of such symbols as the sword, e.g., in the statement by Jesus 'I came not to send peace, but a sword',[7] will become clearer.

Here we see that Christ-Jesus had not just come to help quieten the energies of the emotional turmoil of humanity by allusion to comforting peace (as was the purpose of the former manifest 'Sons of God'), but to give them the key to overcoming that turmoil by aspiring to the higher domains (the 'kingdom of God'). To do so Christ-Jesus had to lay the foundational teachings that will help the world disciple to project *antaḥkaraṇas* upwards to pierce the veils of the Mysteries of that kingdom. To do so they have to project the energies upwards to the domains of the Spiritual Triad. Their entire material selves consequently need battling with by utilising the sword of the Spirit. This concerns the way of taking the second, third and higher Initiations that opens before the disciple. They consequently become prime causative agents.

The *antaḥkaraṇa* can then be seen as the spinning, moving, flashing sword of the power of that Triad. What then manifests is the Mind that is 'the active Heart', the 'double-edged sword' of right discriminatory wisdom and spiritual power. It is the clear line of descent of the power of Deity, actively expressing itself in the realms of form. It thereby confers sanctification and benediction from the highest principle (Deity) down to the lowest. This descent of Grace clarifies the form and makes it sacred, holy. It propels the disciple to manifest an arena of service upon the path of sacrifice in accordance with the will of the most High. Inevitably it makes one a world Saviour that has the ability to direct causative streams of Love and Light through the 'rainbow bridges' created by the aspiring ones.

This always means division and strife; both in the field of service and within the disciple's vehicles, for the road to the crucifixion Initiation must be undertaken. In doing so it necessitates teaching humanity the ways of renunciation to all material things and to all the forms of

7 *Matthew 10:34.*

desire-attachment, especially of the attributes of mind, they hold so dear.

The *fixed cross* is that of steady resolute purpose. It is the cross of the Son in incarnation, be that 'Son' a Logos, or a human unit.

The *northern* direction is that of upwards towards Divinity in all of its affiliations, veiled by the embodied form.

The *eastern* direction represents the method of liberation from *saṃsāra,* which concerns moving inwards to the Heart of Life, producing revelatory understanding as to the way of Love, the compassionate understanding *(bodhicitta)* that directs the course of all Logoic action. For Logoi that Love is dominant, universally applicable, thus their purpose is to further develop and refine the energy of the Will so that its expression cannot harm the evolving Lives within their Bodies of manifestation. Logoic Will also has its subray methods and is needed to penetrate deeper into the Mysteries of cosmic Love that emanates from the THAT Logoi within whose embrace they evolve. The expression of Love allows the All to move inwards and upwards to greater refinement of what is represented as Consciousness to them.

The *southern* direction is that of downwards into the substance of the forms (matter) via which this Divinity is expressed, and which is prompted thereby through the force of *bodhicitta* to move in an evolutionary direction.

The *western* direction is that of outwards to the field of service, the humanity and other Lives that have evolved out from the *prima matrix.* Such activity produces the serving disciple striving to master the path of Initiation by helping to liberate the imprisoned ones from the trammels of *saṃsāra.*

All Logoic or Bodhisattvic activity utilises the Heart as the active Mind to appropriate the form into which such Ones Incarnate. The 'double-edged sword' is utilised to project the forces that must be. This can be that of the destroyer aspect that produces cataclysmic change so that the new can eventuate; or else the forces to build the next beneficent change in the evolutionary process. The cycles can then progress the evolution of amoebic life toward sophisticated human societies. The *karma* that appears along the way must be rectified at the appropriate cycle, according to the divine Plan of what is to be gained at the point of liberation of the thrall of the appearance of things.

This preamble should allow better comprehension of figure 44. The Heart that is the Mind is a major causative agent in our terrestrial and solar spheres of activity. There are three layers or concentric circles to the figure, which relate to the qualities of the Three Christs (explained in *The Constitution of Shambhala),* the Spiritual Triad, the three major whorls of petals of the Sambhogakāya Flower, or to the triune Logos.

Comprehension of the statement *'The Heart is the Mind'* is important, because if one is to eventually manifest the attributes of a Logos one must learn to think with the Heart. How to actually think with the Heart is a question therefore that confronts all aspirants to the Mysteries of Life. It is something they must achieve as they progress from life to life upon the way of becoming Christs. Each life is a progressive continuation from where disciples left off in the past life or lives. As they do so they learn the art of meditation again, and pursue the process of undertaking the next step of the Initiation path ahead of them. These steps are based upon what was passed or failed in previous lives. Recapitulation is the order of the day. (The esoteric meaning of the statement 'Remember this well'.) All aspects of the path must be thoroughly mastered before the Initiation ahead can be gained. The way of Initiation necessitates developing the inner ear to listen to the Voice of the Heart, the silent Voice of the intuition, that propels the listening one onwards to achieve the purpose of that life. It speaks silently, via the images and transcendent perception of what must be, the revelations from the inner dimensions, the domains of the liberated Ones. (The Hierarchy of Love and Light.) They embody the awakening Mind of the Heart of Life in the consciousness of the one listening to the Voice. Manifesting compassionate deeds, which is the way of the Heart, is the way forward. The deeds become increasingly vaster in scope and harder in execution, according to the degree of attainment of the Initiate concerned.

The way of the Heart concerns awakening to the blue bliss of the omnipresent spaciousness of abstracted Awareness. It is the process of gaining liberation from *saṃsāric* turmoil as one travels the compassionate path upwards towards Hierarchical domains and thence to enter the spheres of activity of Shambhala.

The question that can be asked is: 'How do the modes of expression or extension of all the streams of causative energy actually stream from the expansiveness of the Heart's embrace?' The answer lies in a comprehension of the fact that a human unit is essentially Love. At first one can view this statement to refer to the true human unit, the Sambhogakāya Flower, which consists of nine main whorls of petals plus three bud petals, making twelve in all. The Mysteries of how this flower came into being and the ontological paradigm for its expression can only be properly revealed by the meditation-Mind.

One must also realise that the Heart represents the twelve petals of the Lotus of the Heart *chakra*. The attributes of this centre is greatly expanded in the twelve main petals of the Head lotus and its subsidiary petals. These petals are the recipients of the energies coming at first from the Sambhogakāya Flower, then the Hierarchical Heart, and later as the higher Initiations are undertaken they receive impressions from the twelve Creative Hierarchies, the planetary regents and the signs of the zodiac. (The zodiacal gates that are the Heart within the Head of THAT Logos.) These energies pour through the Heart in the Head centre of the Initiate (explained in volume 5A of *A Treatise on Mind*[8]). The zodiacal energies condition all streams of Consciousness of the groups associated with the field of relationships of our solar Logos. The energies flow from Heart to Heart. Ultimately therefore to 'think with the Heart' concerns an Initiate being attuned to the heightened Awareness of Telepathic Impressions and Images coming from such sources. According to the impressions coming from the meditation-Mind attuned to any of these sources, so the serving disciple or Initiate is empowered with the appropriate energies to assist in the service work at hand. Ever is the field of service thus energised, and such a field is always a group undertaking, as the expression of the group Laws governs all upon the path of the awakening or awakened Heart.

The Hierarchical twelve-petalled Lotus is also but the higher correspondence of the Heart that is the Mind. To think with the Heart therefore, is to allow the twelve great modifying streams of Consciousness channelled by the Heart to condition one's entire milieu

---

8 Therein the mode of linkage between these petals and those of the Sambhogakāya Flower was also explained.

of being. This necessitates the development of that Impersonality that will allow one to be instantly absorbed into the mode of expression of the greater Consciousness that is the purpose and cause of the existence of all one perceives.

Thinking with the Heart thus conditions one to become a conscious part of the collective awareness of the great Causative Lives within whose bodies of manifestation we all live. To do this one must utterly eliminate all vestiges of personal will so that only Divine Will remains. This does not mean that one becomes an automaton of these great Lives, because within the unified Consciousness-stream one always has complete freedom of scope for right creative activity within the whole of which one plays a part. What emanates through the Heart concerns the means of expression of the Love of the One for the All. It is also the means whereby the All becomes expressed as the One. It therefore concerns the fusion of the collective Minds of the many to become the One Mind. To think with the Heart therefore is to Think with that Mind. Vast are the ramifications of the differing avenues of approach and modes of service work to be accomplished by those who have specialised thus via various Ray and subray conditionings.

The *knowledge petals of the Heart* are the five petals of the Heart Lotus that are the lower correspondences to the non-sacred planets. They are concerned with processing the *prāṇas* derived from the five sense-consciousnesses. The remaining seven petals are associated with the sacred planets, the Regents of the seven Rays. The significance of the number five in relation to the evolution of and qualities associated with the Mind should be obvious to the reader by now. The originating source of this pentad of qualifications in Nature then manifests via the activity of the five Dhyāni Buddhas.[9]

The spirals and cycles of unfoldment of the Knowledge petals thus concern the way of expression of the five *prāṇas* that together condition all manifest life. This involves sensation, the battlefield of desire, *ahaṃkāra,*[10] the concept of individuality, the 'I', and everything associated with *saṃsāric* conditionings (which are an aspect of the

9 See *A Treatise on Mind* for detail concerning their attributes.

10 I-am-ness, the principle of individuality, self-sense, the tendency towards definiteness, origin of all manifestation.

five planes of Brahmā), seen on a planetary and intra-solar scale. They are thus the spheres of conscious embodiment of all concerning the segregation of the universal into the particular. The seven remaining petals are concerned with the Way of interrelation with the Lords of Light (and thus also of the Ashrams of the Hierarchies of Liberated Life). Individuality becomes merged into the spheres of relationships of Universality. The little wheels become part of the one great Wheel of Life turning and space is seen for what it truly is (being but the Life of an embodying Entity).

The phrase 'the knowledge petals of the Heart of that which is to be' refers directly to the five Knowledge petals of the Sambhogakāya Flower. The phrase 'unfolding the mind' here implies the ordinary empirical consciousness of the intellect, and what is developed thereby is the expression of the *saṃskāras* that vivify these Knowledge petals. These petals are awakened and developed whilst one is engrossed in ordinary pursuits in the material domains (the mental, astral and dense physical planes) in an endeavour to master the activities of one's life in human civilisation.

As the Knowledge petals are part of the Heart of what is to be, the unfolding potency of the constitution of the Sambhogakāya Flower, so the way of thinking with the Heart concerns awakening or developing the attributes of the rest of the petals of this Flower. Eventually an aspirant develops the perceptions to awaken the full potency of this Soul-form. To awaken that 'which is to be' one must learn to project *antaḥkaraṇas* to this Flower and beyond, ever further beyond, to the Monadic Life and onwards into cosmos. From Heart to Heart one does thus travel.

Materialistic science is the pinnacle of the achievement of awakening the Knowledge petals, nevertheless there is far more to discover than just simply 'knowledge' about things. These 'things' are transient, ephemeral illusions, though seeming real as objects of sense-perception. Consequently one should not concern oneself with 'the concrete trivialities' of this illusion. One should understand its nature well, as empirical science has been exploring, but to not limit consciousness thereto, or be attached to the manifesting phenomena, 'the reality of the illusion', because there is much more to life than this, a far vaster panorama that manifests in consciousness as one awakens the way of travelling in the Heart that 'is the Mind'.

Awakening such awareness is the next step forward for the scientific community and humanity. They must learn to meditate upon the true nature of phenomena and the meaning of life. They must consequently ponder upon the path to liberation that opens before the meditating one. The way to liberation consist of the levels of mastery of the Initiation process. Through undertaking the Initiation path memory of the achievements of one's past lives, plus all pathways to the future of the Heart that is the Mind then proceeds automatically. Enlightenment and consequent liberation from *saṃsāra* is the gain.

We come now to the second part of this communication, beginning with the statement 'The Mind is the active Heart', in which case the focus is upon the abstract, enlightened Mind. From this statement we see that Mind is the mechanism whereby the abstract Awareness of the unity, the Oneness of the All, is brought into compassionate resolution through the activity to awaken separative units of consciousness into this experience. The 'double-edged sword' is the mind/Mind, the juxtaposition between the empirical separative mind with the compassionate enlightened Mind. This sword brings *saṃskāras* from past actions, volitions of mind, to the surface of consciousness so that these attributes can be dealt with. Either the empirical mind rules, in which case the *saṃskāras* are often built upon, or made stronger, or the enlightened Mind rules, in which case the *saṃskāras* are eliminated, transformed, or transmuted into enlightenment vectors. Either way, the sword flashes again and again in the field of consciousness to effect the will of the thinker.

Very few however are aware of the nature of the enlightened Mind, neither do they understand the way of action of the empirical mind. The remainder of the communication is concerned specifically with analysis of the mode of action of this mind, with which figure 44 principally deals. I have dissected the statements in relation to the attributes of the eight-armed cross for better comprehension of the esoteric dimensions of the statements.

Notes to figure 44.

1. *The Heartbeat* is the power of Love and empowers the sound of the

'drum' that beats out the rhythms of Life. The heartbeat comes from the centre of the figure and manifests via the eastern direction towards the west. As it does so it overcomes the 'perverse acknowledgement of the will' and imposes the conditionings of the Real. The eastern direction represents the rhythmic outflow of the power of the Heart. The Will or power of the first Logos or Christ is here represented.

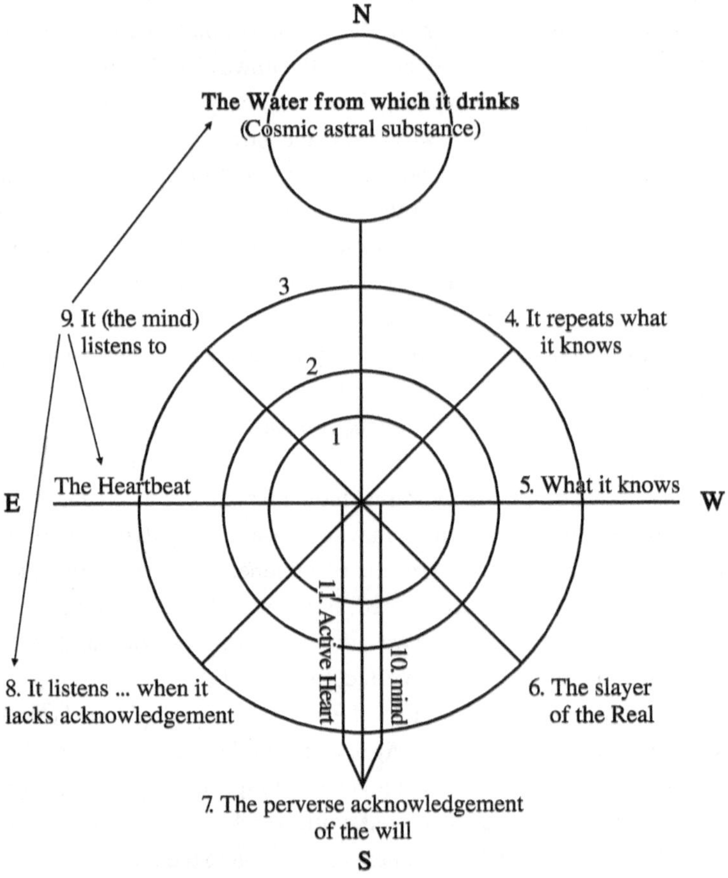

Figure 44. The Heart is the Mind

2. The statement *the Heart is the Real* is the pure embodiment of the Love-Wisdom principle. 'The Real' is that which is not an illusion or

transient, that perpetually persists, integrating the whole into oneness. The seven Rays are subrays of the second Ray of Love-Wisdom, which governs the plenitude. The magnanimous plenitude of the Love of the second Logos or Christ is thereby represented.

3. *The Drum of the Heartbeat* is the rhythmic power of the third Logos or Christ embodying the creative expression of the third Ray of Mathematically Exact Activity. The *drum* (signifying the sum of the sphere of containment of the *maṇḍala* of the manifest expression) amplifies the sound of the heartbeat and sounds out the note of the One principle of Love. It meters this note out in regular cycles *(yugas)* of expression. The 'note' manifests as a five syllabled mantra with respect to the planes of causation—A, Oṁ, Aūṁ, Aḥ, Hūṁ—in perpetuity. This is the drum of the heartbeat. These mantric potencies express the Power of the five Dhyāni Buddhas into the kingdoms of Nature, the world of the five sense-consciousnesses. These syllables are emanations of the first to the fifth planes of expression respectively. The Drum of the Heartbeat therefore manifests upon the fifth plane, the mental, via the compassionate, liberating Hūṁ. Thus the story of the commencement of the conversion of black cosmic dust is also told.

4. *It repeats what it knows.* (The northwest direction of emanatory goodwill.) The Mind reverberates the cycles of evolutionary attainment upon higher spirals of activity, which progress eventually out of the ring-pass-not of the system altogether. It also repeats what it knows to all those who are evolving upon the manifested planes and need to be succoured with the teachings of the Way of the Heart. The drum of the Heartbeat thus becomes the leitmotiv of all aspiring Bodhisattvas to listen to and to achieve in their cycles of activity. This is the way of the active Heart.

The empirical mind also repeats what it knows in terms of the intelligent application of what has been derived from experience in *saṃsāra.* Eventually a vast body of knowledge is accumulated from rational deduction from such experiences, as presently existing in our scientific and technological age. People use their mouths to express the opinions they possess about 'things' to others, and so social intercourse is established through repetition of such knowledge. Normally such opinions manifests as babble, but sometimes uplifting thoughts may be the outcome.

Here 'the drum of the heartbeat' implicates the pulsations of the heart's blood and the rhythms of Life governing an individual life expression. The cycles persist life after life according to the mode of the activity of this 'drum' that metes out the *karma* of what must be.

5. *What it knows.* (The western direction of outwards to the field of service.) This concerns what is gained by the mind though cyclic activity in the material world, and later in the realms of illumined understanding. Such exoteric and esoteric knowledge is eventually purveyed on a mass scale, or broadcast to elucidate human minds so that the benefit pushes them further along the path to enlightenment.

6. *The slayer of the Real.* (The southwest direction of understanding.) The expression of the personal will upon the earth sphere inevitably produces proper comprehension of the true nature of all attributes of that sphere of glamour. At first this will slays the reality of spiritual truth and the inner livingness, to live out an ephemeral existence in the incessantly changing *saṃsāra*. Many cycles of transient happiness and of painful activities follow. The Buddha's Four Noble Truths and Eightfold Path, as well as the teachings concerning the twelvefold wheel of rebirth *(pratītyasamutpāda),* are concerned with explaining this process.[11]

Later the personal will is supplanted by the Will-to-Love, the Will-of-Love and then Divine Will[12] that overcome all obstacles to enlightenment in the fields of application associated with this direction. The will is the continually moving point in time and space that is the focus of the gaze of the Eye.

7. *The perverse acknowledgement of the will.* (The southern direction of downwards to the little ones.) Here the will finds its most intense point of self-focus, directed purely to selfish, materialistic and separative activities. It thus slays the Real and buries it deep in the depths of *saṃsāra*, with much karmic woe to follow. The higher types of will must later develop that will slay the slayer so as to 'repeat what it knows' on a higher cycle (or incarnation) of activity. One can then 'drink' from the Waters of Life upon the upturned cycle.

---

11 See volume one chapter 9, volume 2, 161-64, 207-11, chapter 8 of volume 4, and volume 5A, 146-69, of *A Treatise on Mind,* where these topics are dealt with.

12 See volume 4, chapter 7, of *A Treatise on Mind* for an explanation of these terms.

8. *It listens....when it lacks acknowledgement.* (The southeast direction of expression into material activity.) The 'acknowledgement' here is that from the personality's selfish activities, lusts and desires; from its petty ambitions and effects of its plans in the domains of materialism. There is a lack of a highly refined and developed mental substance, allowing the 'I' to attach itself to the waters of the astral/emotional sea. People live out their lives emotionally, demanding that they be listened to, that the pettiness of their activities be acknowledged by those around them, their lovers, friends and society in general, for in this way they perceive their life to be meaningful. Thus the Wheel of *saṃsāra* (of *pratītyasamutpāda)* is turned.

The personal-I 'listens' to the way of the Heart when it eschews the continuous transience, hence it no longer receives such acknowledgement from *saṃsāra.* This causes the process producing the path to liberation via meditative pursuits necessitating conquering the sense of pride, separateness and self-aggrandisement.

9. In the northeast direction of unity, 'it (the mind) listens' at first to all the sounds and noises, the clashing, rumbling and commotion stemming from the sum of the wheel of *saṃsāra.* The ear is thus attuned to all of the eight directions of this compass of life's activities. Later, during the various stages of the Path, it listens to the steady rhythm of the Heart beating out the silent words of Hierarchical fulfilment, and so the person becomes an Initiate. Upon the path of Initiation the magnitudinous lapping of the Waters of the cosmic astral ocean (Divine Love, *bodhicitta*) is listened to. The response eventually makes one a Buddha.

The Water from which it (the mind) *drinks* at first represents the substance of the stagnant fetid swamp of the astral plane, the *saṃsāric* mire within which the nine-headed Hydra of desire lurks. This produces all of the qualities of the desire-mind (*kāma-manas*) which inevitably 'slays the Real'. Later the Sound of Logoic Love from the cosmic astral plane is listened to. This conditions the path of Initiation, allowing eventual escape from bondage to evolutionary conditionings within the womb of our solar form.

10-11. Each spoke of this wheel becomes the two-edged *sword of Life*, of mind and the active Heart that is the Mind. This sword projects itself in all eight directions and levels of the great Wheel. At first it, as the mind, pierces deeply into the vicissitudes of *saṃsāra,* later as the Mind is developed it cuts away at the sinews of attachment to the conditionings of *saṃsāric* mire. Upon the Initiation path the mental-emotional swords of others cause all the lacerations imposed upon the compassionate One fixed upon the cross of Love.

If the wheel turns from right to left, (from east to west) then the conditionings of the mind (the intellect) rule, imposing deeper materialistic attachment to *saṃsāra.* This is the *left hand path,* epitomizing the 'eye doctrine'. If the wheel turns from left to right (from west to east)[13] then the conditionings of the *active Heart* begin to manifest. One then works to awaken all to the way of compassionate understanding, which permeates the All with expansive undertakings and divine Reasoning. This *right hand path* epitomises the Heart doctrine.

A simplistic picture of the motion is here presented, as the wheel first turns one way, then another, as well as gyrating inwards and outwards, until all experiences and qualities of a liberated Mind are gained. The wheel stabilises itself in the left to right motion and then spirals upwards and outwards away from manifest conditioned life altogether.

The period of oscillation and change of direction necessitates many cycles of activity focussed upon the qualities of any of the arms or spokes of the wheel. The sword of *karma* however cuts its path relentlessly through all conditionings and obstacles on the way to liberated enlightened Life. The previously explained moving *swastika* (mutable cross) is the dynamo conditioning the movement of this wheel.

The active Heart represents the right hand side of the sword. It bears the energies from the east, the Heart of Life, whilst the mind (the left hand side of the sword) bears the energies from the west, of active intelligent discrimination.

---

13 These directions are from the point of view of looking outward from the page. The directions east to west and west to east here refer to the mode of travel of average humanity. For an Initiate the direction east to west relates to an orientation to the field of service, whilst west to east relates to undertaking a period of meditation.

Appendix Two

The Root Races

In order to better comprehend some of the implication concerning cosmic evolution it is best if a summary is presented concerning the nature of human evolution from the esoteric perspective, because a similar evolutionary process needs to be transposed in terms of Logoic evolution.

The stages governing human evolution (the basic divisions that constitute our evolving consciousness) are governed by the number *seven* because that number conditions the evolutionary process. These stages have been termed Root Races. Each cycle or human Race is subdivided into seven sub-races (racial epochs), which are further divided into seven tribes, and national groupings derived from them. This esoteric subject is not concerned whatsoever with such concepts as a 'master Race', but rather with the stages of developing mental-emotional characteristics over millions of years, and the characteristics of the human persona of those that bear new qualities into manifestation. There have been some imaginative speculative writings concerning the differing characteristics and names of the 'racial types' of the Root Races and sub-Races, especially in the Theosophical literature. I have no wish to delve here into the unprovable characteristics of some of the names that have been assigned to the sub-races in the occult texts. It suffices here to simply present the broad sweep of the concept, and to leave it to later esotericists to explain and verify the detail.

The Root Races represent the seven main distinct stages in the evolution of human consciousness. There are physical differences between these Races, from the giants of the Lemurian epoch, to where we are now. My concern is with the way consciousness manifests via forms that have evolved to bear the changes in the states of awareness. The first two Root Races were entirely subjective, signifying the process of descent of the formative aspects of the human mind. In a similar way this book has described the appearance of phenomena from out of the ethers, be it of atoms and subatomic particles as well as planets and solar systems, etc. The same paradigm exist for all.

Humanity as a whole is progressing through its fifth cycle or stage of attainment (or the third from the perspective of the phenomenological appearance), that of the 'Aryan' Root Race, of which the most recent, fifth sub-race is said to be the Indo-European class, though the foetal sixth sub-race is appearing. Simultaneously some bearers of aspects of the sixth Root Race will appear in the new era ahead of us, whilst the seventh Root Race is still too far in the future for consideration as to what its characteristics may be, except in broad outline.

As the number five relates to the mental plane and the development of the mind, so we see that the main emphasis of the Aryan Root Race has been the development of the intellect, whilst the sixth sub-race will predominantly develop buddhic perception. The evolution of the Root Races is a major expression of the Causative action of Deity. Humanity represents the mode of expression of the fourth of the five *prāṇas* that course through the Body of manifestation of the planetary Logos. For the Logos this *prāṇa* is Watery, but for humanity it is Airy,[14] which is presently manifesting its third or *manasic* hue.

With respect to human evolution we should also consider the qualities of a pentagram or pentad of forces. The pentagram relates to the mode whereby a causative agent performs a task within the formed realms, via the forces that must be utilised. It constitutes the empirical or intellectual attainment of humanity or of Deity, the 'fiveness' of all that evolve or demonstrate mind/Mind:

14 They are the fourth of the seven Creative Hierarchies. See Alice Bailey's *Esoteric Astrology*, 32-50 for detail concerning these Hierarchies.

- The nature of the planes of Brahmā. (The five planes from *ātma* down. From a higher perspective, the first five planes from *ādi* to the higher mental are the planes of liberation. The three lower planes of Brahmā are those of human livingness, the domains of *saṃsāra*: the lower mental, astral and dense physical.)
- The five instincts.
- The five senses and the sense-consciousnesses derived from them.
- The five organs of action and of sensation.
- The five *prāṇas.*
- The five *chakras,* when the two Head centres are esoterically taken as a unit and when the Base of Spine and the Sacral centres are taken as one.
- The Wisdoms associated with the five Dhyāni Buddhas.
- The five human Root Races so far developed, or the five strictly human races when the subjective two in the stages of descent are discounted.

When looking to the night sky and pondering upon the myriad stars therein the esotericist is aware that these stars represent a population of cosmic Humanity that are also manifesting in terms of Root Races and their sub-Races. They can accordingly be categorised according to relative spiritual age. Only the highest Initiates on this planet however have the ability to determine the comparative 'Racial' level of the various Logoi inhabiting the local portion of our galaxy.

Every aspect of the human form, our emotions, mind, and the basic instincts, have had to be evolved and mastered. Everything now taken for granted, such as basic emotions and the way the stomach operates, had to be evolved and lived through as an experiential process in former evolutionary epochs.[15] The qualities of all our organs first had to be built and then lived through as sentient development via the lower species of Life. Nothing we possess is there by chance, everything associated with our personality equipment is the result of *karma,* is the effect of that which we had earned from myriads of lives of past unfoldment

15 Our archaeological record testifies to such development.

of cellular, and then 'organ-like' consciousness. The earlier Races, or stages of the development of the human personality, the making of a prime causative agent, were thus all concerned with the process of building appropriate equipment of response to higher Causative Energies. In this process the development of the mind/Mind is the key. The esoteric attributes of one or other of the five sense-consciousnesses are the focus of development of each of the Root Races. For our present Aryan humanity the sense is that of sight, signifying the development of the intellect.

The five Races so far developed are:

1. The first Root Race, the Adamic

Adamic man was said to be 'mind born', conceived from the minds of great Angelic beings. We can therefore deduce that the densest sheath that humanity possessed at that stage was mental matter. Upon the mental plane energy is first differentiated into definite forms, therefore what was later to be known as 'man' here definitely manifested a separate existence from the exalted Existences whose substance were appropriated into his outer form.

During this early evolutionary stage humanity's entire body responded to the vibrations (sounds) that continually modify the mental plane. Therefore the first sense that humanity was said to have developed is *hearing*. This is also indicated in the Bible, where Adam existed blissfully in the Garden of Eden, he responded directly to the Word of 'God' and was able to give names to:

> Every beast of the field and every fowl of the air.....and whatsoever Adam called every living creature that was the name thereof.[16]

This act of naming, of classification and of segregation, is a direct quality of the mental plane upon which Adamic man existed. He had to learn to respond to sound vibration and then to utilise sound in the act of naming. This was the original training that would allow humanity to later develop intellectual and thus creative capabilities, to be truly made 'in the image of God',[17] Lord of the appearance of all phenomena.

---

16 *Genesis 2:19.*

17 *Genesis 1:27.*

Though humanity was incarnate in mental substance, they had no proper awareness of it, or of any real type of knowledge as such, they simply had a atavistic instinct to 'name', stemming from the innate properties of the plane upon which they existed. Aeons later, in the Aryan cycle, humanity would again dominate the energies of the mental plane, also possessing the ability to 'name', but having knowledge of its significance.

The symbol of the first Root-Race is the circle: O which indicates that the actual human form was then similar to an amoeba-like cell. This Race was sexless, though purveying the general masculine qualities of the human Hierarchy. This shape is still with us in the form of the zygote (which Adamic humanity essentially were) and in the form of our heads and the Sambhogakāya Flowers.

Human consciousness was then contained within the confines of their form and qualified by the sensations of the mental plane, which gives everything a 'name', hence a form, of which the circle is the most perfected type. The circle also symbolises the form of the group Soul to which humanity belonged.

2. The second Root Race, Adam-Eve (The Hermaphrodite)

Blavatsky's *The Secret Doctrine* states that this Race procreated by exuding parts of themselves, like the budding of yeast cells. At this stage the densest part of the corporeal body of the human group Soul was astral matter. (The hardening of the external sheaths has taken millions of years to develop through the normal evolutionary attainment.) Duality was latent but not yet fully expressed. Human consciousness then corresponded to the plant stage of evolution.

In classical Greek texts, this Race is symbolised by the immortal Hyperboreans that lived on a paradise isle near the north pole, where 'the sun always shone'. (This means that they were continually bathed in the lighted presence of the spiritual realms.) The astral plane was then constituted differently, as it was completely free of thought and desire forms of any kind. There was a direct radiance from the second Ray aspect of all manifested life (as reflected by the plane *buddhi)* that shone upon humanity.

The human form responded to this light with a similar sensitivity that the plant kingdom responds to the light of the Sun. Accordingly, they grew and adapted to change and environmental impacts. This particular Root Race was *feminine* in constitution, as portrayed by their symbol: ⊖. The horizontal stroke represents the primal receptivity of the Waters, the fertile womb of Being, yet to be impregnated by being acted upon.

Expansion and heightened receptivity to external conditions (much modified by light and sound) became prevalent and the sense of touch was thus developed, symbolised by the relationship between the two hemispheres of the above symbol. This stage is reflected in the plant kingdom today where certain of the most highly developed plants are definitely able to respond to touch, in the act of pollination by insects, and also by direct physical stimulus. The formation of the various organs and limbs in the embryo in the second month of prenatal life is also symbolic of the stage of evolution of this race. The symbol shows also that nascent sexual forces were starting to glimmer in humanity. *Eve* (the feminine principle) was born out of the 'rib' of *Adam* after 'God caused a deep sleep to fall upon Adam'.[18] Note that if a rib is pictorially depicted in the circular form of the rib cage, then a version of the above symbol is presented.

This 'deep sleep' state is the *pralaya* that happens at the end of each cycle of activity of a planetary Scheme, globe, human Race, etc., and is always followed by a new cycle with new qualities to be evolved.

The qualities developed by this Race are also explained in *Genesis 3:6:*

> And when the woman saw that the tree was good for food, and that it was pleasant to the eyes, and a tree to be desired to make one wise.

The *mental principle* was stimulated and developed by the need for discrimination (the distinction between the tree in the centre of paradise and all others), whilst the ability to choose resulted in the woman seeing that 'the tree was good for food'.

Feeling was developed by response to pleasant stimuli: 'and it was pleasant to the eyes'. Though the physical 'eyes' were not yet opened, rather their corresponding *chakras,* sensation was nevertheless felt and this feeling in its turn gave birth to desire: 'a tree to be desired to

---

18 *Genesis 2:21.*

make one wise'. Response to feeling-sensation is the major quality of the astral body, which extends itself to grasp or envelope whatever the object of desire is.

The desire to taste of the tree of knowledge of good and evil was effectively a mass unconscious urge. As humanity 'tasted', so people received a downpour of buddhic energy, via which the Soul was to be born, hence from that time on they could occultly see. Their eyes were opened, they saw their 'nakedness'. This nakedness was then covered with sheaths akin to that of the vegetable kingdom, meaning that they descended into sheaths of etheric matter,[19] symbolised by the fig leaves sown together to make aprons (*Genesis 3:7).*

To this was later added 'the coats of skins' made for Adam and Eve by 'the Lord God' *(Genesis 3:21)* and humanity was thereby ousted from the Garden of Eden to the earth. Thus the third Root Race humanity was born.[20] Technically, only the third Root Race and upwards can be considered to be truly human, in the sense of being endowed with the principle of mind and possessing a Soul (Sambhogakāya Flower). For this reason also considerations of the appearance of a physical universe begins with the corollary to the appearance of Lemurian humanity.

3. The third Root Race, the Lemurian

The early Lemurian epoch marked the period in time when humanity incarnated into dense physical vehicles. While humanity were slowly maturing on the subtle levels of perception animal forms were evolving on the earth through much trial and error. This takes into account the sum of the paleontological evidence concerning the evolution of the species. Eventually a form was developed that could express the energies of the Sambhogakāya Flower. Then, with the timely intercession of

---

19 This is represented by the *nāḍī* and *chakra* system, which takes the attributes of a plant kingdom, the *chakras* being flowers.

20 Note that the first half of Chapter six of *The Revelation of St. John,* among other things, deals with the evolution of the first two Root Races, the second half with the third Root Race. Chapters seven, eight and nine deal in detail with the evolution of the fourth and fifth Root Races, until the time when John was given *The Revelation.* Chapter eleven explains the period from this time till the end of the seventh Root Race. Much information concerning all the Races is symbolically presented throughout *St. John's Revelation.*

'God', humanity was cast out of Eden to earth into prepared forms, 'coats of skins', via which they could gather experience and eventually become wise. Individualisation occurred upon a mass scale. This concerns the formation of the Sambhogakāya Flower upon the higher mental plane, and was a confluence with the descending early Lemurian humanity from out of the ethers, possessing rudimentary mental and astral bodies into the developed animal forms that had evolved upon the physical plane, as shown by fossil evidence. The gate for the Individualisation process for our present humanity lay open for approximately five million years, according to D.K.[21]

The whole process of Individualisation is an Initiation of a whole kingdom of Nature to a higher state of perception. It is concurrent with planetary Initiation, wherein a Logos takes his tenth Initiation.

Lemurians were at first dominated by purely physical drives and expression, strong sexuality, bestiality and depravity (though not as we understand these terms, for they did not possess the intellect or imaginative faculties that we now have). At that time these were necessary needs and means to develop consciousness from sentience. Such activity allowed the hardening or conversion of their fluid, psychic body into a set, 'rigid' physical expression. For instance, Lemurians had to develop the very highly attuned and sensitive nervous system that we now possess. Thus, the Lemurian had a predominantly animal consciousness with much evolutionary emphasis upon the sensations of the form. The high goal of attainment was then the mental plane (as the *ātmic* plane is for humanity today). Gaining the ability to think was a slow development necessitating passing through many psychic states of awareness. All emotional, psychic, physical or spiritual properties and capabilities that we possess today had to be developed in a most arduous and painstaking manner until they became innate within our constitution. They are innately part of humanity's present personality equipment, utilised automatically as we undergo our daily lives. As a consequence the personality vehicle is now a most capable mechanism of response.

---

21 See Alice Bailey's *A Treatise on Cosmic Fire* and Blavatsky's *Secret Doctrine* (Volume II, Anthropogenesis) for detail. I only have cursorily indicated the process in volumes 5A (pages 62, 78, 86-87, 107, 118-19) and 7 of *A Treatise on Mind*.

Each organ, grouping of cells, physiological or psychic function finds its acme of development in some era in which the general tonality of the greater Life governing the lesser lives is of a like quality and achieving a like development. This is true, be that greater Life Deity and the lesser life humanity, or the greater Life humanity and the lesser life the cells in our bodies. This is the basis to the law of cycles and is the key to the understanding of the purpose of all evolutionary development (which concerns us here).

Lemurian humanity were strongly endowed with a type of primitive etheric vision. They still conversed and 'walked with the Gods', and *physical sight* was the sense then developed. In gaining physical sight however, they gradually lost many of their psychic faculties, for they had to develop the ability to experience, reason, and work things out for themselves. This was not possible, nor necessary, if people had an overriding psychic perception with which they always instinctively and unthinkingly reacted. (Much like a modern medium whilst in a trance.)

The immediate problem of the Lemurian, therefore, was to see things clearly on the grossest plane of perception. This necessitated continuous incarnation into physical bodies existing in a very dangerous environment ruled by large carnivores. The necessity for self-preservation and survival skills greatly facilitated physical plane focus. Lemurians had to learn to make and manipulate tools, which gave them a competitive edge in the world they were in. By reaction and adaptation to the things that were seen and which physically affected them, humanity eventually become a thinker, a *'man'* and thus a causative agent in the true sense of the term.

The Lemurians were the *giants* mentioned in the Bible and other mythological texts. There is plenty of evidence of such prehistoric giants, but because they do not fit into the paradigm of the dogma of Darwinian theory and because of cultural bias this evidence has been shelved and dismissed by the orthodox anthropologist.[22] Some of this

---

22 Occasionally such testimony escapes the censors, such as the picture of the Turkana boy shown in the plate section of the book by the paleo-anthropologist Richard Leakey, *The Sixth Extinction,* (Phoenix paperback, London, 1996). The caption reads: '1.6-million-year old skeleton of the Turkana boy, a member of Homo erectus, who was nine when he died'. The skeleton is higher than the 5'4" curator it is compared to. Nowadays one need only consult YouTube for more examples of such 'anomalies'.

evidence is documented in the books by Cremo and Thompson, where they speak of a process of 'knowledge filtration'.

> From the evidence we have gathered, we conclude, sometimes in language devoid of ritual tentativeness, that the now-dominant assumptions about human origins are in need of drastic revision. We also find that a process of knowledge filtration has left current workers with a radically incomplete collection of facts[23]....We recount in detail how this evidence has been systematically suppressed, ignored, or forgotten, even though its qualitatively (and quantitatively) equivalent to evidence favouring currently accepted views of human origins. When we speak of suppression of evidence, we are not referring to scientific conspirators carrying out a satanic plot to deceive the public. Instead, we are talking about an ongoing social process of knowledge filtration that appears quite innocuous but has a substantial cumulative effect. Certain categories of evidence simply disappear from view, in our opinion unjustifiably.[24]

The smaller size anthropoid apes that modern anthropologists are endeavouring to trace human ancestry with are the result of the 'sin of the mindless' as *The Secret Doctrine* styles it.[25] In other words, the

---

Archaeology is another field where a great deal of suppressed evidence exists in order to fulfil the current paradigm of the orthodoxy. The works for instance of Graham Hancock, Robert Bauvel, Robert Schock, Robert Temple and Adrian G. Gilbert, could be consulted, to name but a few. One could also look to other fields, such as the suppression of UFO data by NASA, or of the evidence from Mars. The subject of suppression by the orthodoxy in the various fields is vast and necessitates a large treatise to properly document such and the reasons why. Many academics will lose their tenure at universities if they publish outside the narrow accepted criteria of their fields.

23 Michael A. Cremo and Richard L. Thompson, *Forbidden Archaeology* (Bhaktivedanta Book Publishing, Inv., Los Angeles, 1998), xxv. This work is monumental. See also their following work, *Human Devolution* (Torchlight Publishing, Los Angeles, 2003).

24 Ibid, xxvi.

25 Here the second volume (Anthropogenesis) of Blavatsky's monumental work *The Secret Doctrine* needs to be consulted (see verse 35). Verse 32 of 'The Stanzas of Dyzan' for instance states: 'AND THOSE WHICH HAD NO SPARK (the "narrow-brained" TOOK HUGE SHE-ANIMALS UNTO THEM (a). THEY BEGAT UPON THEM DUMB RACES. DUMB THEY WERE (the "narrow-brained") THEMSELVES. BUT THEIR TONGUES UNTIED (b). THE TONGUES OF THEIR PROGENY

field of investigation of the orthodox anthropologists, though valid research, does not relate directly to the true progenitors of humanity, but rather to their incidental offspring, producing an evolutionary dead end. The true evidence of human evolution (the giants of past ages) has been largely suppressed, or ignorantly dismissed as an anomaly, shelved, then forgotten.

The Tau cross: ⊤ symbolises the Lemurian period. This cross is free from the circle of unifying spiritual substance (the group Soul), though not at the beginning of the Lemurian era when the symbol was ⦵.

The extension of this cross as a conscious principle (hence the term 'mindless' Blavatsky gives to them) was yet undeveloped and latent. Forceful masculine physicality had to be evolved. This was necessary in a world full of savage beasts (ten to eighteen million years ago) and unrelenting competitiveness for resources. This is symbolised by the downstroke of the ⊤, projected from the dominant feminine bar of the symbol. The masculine energy asserted itself and subjugated the feminine materiality of Nature's kingdoms, the forces of the great Mother. Lemurians were obsessed with sex, beastiality (the anthropoid apes being the outcome), and later, experimentation with sexual magic. Such widespread activity *saṃskāritically* laid the seeds for the sexual diseases that still scourge humanity now. Syphilis arose on a mass scale and nearly destroyed the then humanity.

Hatha yoga and its many permutations (the control of the functions of the physical body) was developed in the latter part of this epoch by the Lemurian Initiates. This form of yoga is long superseded by the higher forms, such as *rāja yoga* (the development of the mind and the control of the powers latent therein). Consequently *hatha yoga* should nowadays be viewed only in terms of those exercises that assist the body to be healthy, and to provide the basic yoga postures (e.g., *padma asana)* that allows one to sit comfortably whilst in meditation. The Lemurian Initiates were the Cyclopes of the myths, and are somewhat explained in *A Treatise on Mind,* volume 7A.

Because of their control of the forces of the etheric body the Lemurians could manipulate huge blocks of stone to their will, moulding them if

---

REMAINED STILL. MONSTERS THEY BRED. A RACE OF CROOKED, RED-HAIR-COVERED MONSTERS, GOING ON ALL FOURS. A DUMB RACE, TO KEEP THE SHAME UNTOLD'. (*The Secret Doctrine*, 1888 Adyar edition, 184.)

need be, and to levitate them to build the colossal stone temples, walls and artefacts in their temples that still exist in megalithic sites today.

4. The fourth Root Race, the Atlanteans

The Lemurians developed emotionality in the latter half of their era, which consequently gave birth to the Atlantean civilisation.[26] During the major part of the Lemurian epoch therefore the astral plane, as now experienced, did not exist. Its conditionings were slowly developed and built by the Atlanteans until they were completely submerged in it. Their cultured and highly sensitive emotional susceptibility had, by the latter half of that era, produced the Watery emotional clouds, mists and the 'outer darkness',[27] associated with the astral plane and its heaven and hell states. Therein are myriads of muddied and grotesque thought-forms which prevent the entry of spiritual light to the earth, just as much as heavy rain clouds prevent the entry of sunlight to the earth's surface. Astral entities abound, such as the Anubis explained in volume 7A of *The Constitution of Shambhala.*

The Christ-force of Love-Wisdom could during their early epochs overshadow and bathe the Atlanteans from above. By being impressed with its qualities they developed a golden era wherein they were in harmony with Nature in their everyday expression. They devoted themselves to family and friends, with whom there was a definite group and tribal communion. They could control the elemental forces of the earth through mantric sounds, which they saw and directed (though they had little awareness of what these sounds meant). The Atlanteans thereby greatly hastened the evolution of the vegetable kingdom, of the

---

26 Some information concerning this civilisation was provided in volume 7 of *A Treatise on Mind,* and in the books by Alice Bailey, hence I shall do little more than to provide an overview here. Some of the clairvoyant investigations by others, such as Bishop Leadbeater and Edgar Cayce, contain some truth, but also contain fabrications. This epoch was said to have begun about four and a half million years ago.

27 *Matthew 8:11-12,* 'And I say unto to you, That many shall come from the east and west, and shall sit down with Abraham, and Isaac, and Jacob, in the kingdom of heaven. But the children of the kingdom shall be cast out into outer darkness: there shall be weeping and gnashing of teeth'. Here 'the kingdom of heaven' refers to the higher mental plane, wherein Initiates of the third or higher degrees (Abraham, Isaac and Jacob) could reside, but the 'children of this kingdom' (aspirants, and those working for their first and second Initiations) have yet to cleanse their *karma* with the astral domain (where there shall be 'weeping') and the material world (the 'gnashing of teeth').

fruits and foodstuffs that they greatly admired, introducing many new strains upon the earth (e.g. the banana). It would be logical, therefore, that *taste* was the sense that they developed, showing that they were now susceptible to buddhic stimulation.

Generally their development was via matriarchal societies, ruled by priestesses and occasionally priests. Shambhala, the spiritual Hierarchy then extant, incarnated as the rulers of that civilisation and gave humanity many gifts. These included, for instance, a prototype of the airplane,[28] that was then propelled through the control of gravity and etheric forces. The Atlanteans utilised these gifts, but could not understand their nature or the complexities of the related mechanics.

In later Atlantean times, after the massed consciousness was fanned with the Fires of the mental principle, the desire-mind became rampant, to the exclusion of buddhic influence. This led to sex worship and widespread, intensely selfish and acquisitive attitudes. Black magic was practiced on a grand scale. The Atlanteans then created many glittering emotional fantasies and desired many alluring physical objects (e.g., gemstones). Opulent material comforts was a hallmark of their societies and they used their inherent psychic powers to obtain their desires. They therefore built very powerful thought-forms, 'embodied evil', to fulfil the aquisition of what they desired. Theft and rapine became a widespread occurrence.

The darkness (astral murk) created by massed materially based thought-forms (hence the 'outer darkness') become increasingly manifest as masses of people consciously constructed voracious desire forms to do their bidding. (Cancer became a widespread scourge as a karmic result of these energies manifesting via the physical bodies. These energies produced a rapid growth of the cellular structure in the organs via which the energies were directed. The organs being the externalisation of the minor *chakras* that were then utilised.) The *karma* that resulted from this wrong direction of force eventually led to the destruction of their race. It opened the doors to the incarnation of the forces of cosmic evil. Massed evil doing become so dominant that the 'war' between the forces of evil and that of Light was on the threshold of being lost

28 The *vimānas* mentioned in the Vedas. See the book by Dileep Kumar Kanjilal, *Vimāna in Ancient India* (Sanskrit Pustak Bhandhar, Calcutta, 1985), for detail.

for the then Hierarchy. This caused the Lord of the world to amass the forces, with the help of cosmic Brethren, to submerge what was then the continent of Atlantis, as recorded in the universal tales of a great flood. The Hierarchy of Enlightened Being (the then Shambhala) then abstracted themselves onto the higher mental plane.

Hierarchy have remained there for many millennia (always subjectively and objectively helping humanity), and are again preparing to re-emerge into physical plane livingness with the Christ[29] (the head of the Hierarchy), due to the opportunity that will be offered by humanity in the forthcoming new age. This tremendously significant esoteric fact underlies the subjective happenings of all our lives.

Bhakti yoga (offering oneself up in devotion to 'God' and by this means obtaining salvation) was the type of yoga developed in the Atlantean epoch, whilst the attainment of intellectual capabilities was the goal of all the then schools of Initiation.

The vertical line: |, was the symbol of this race. It is the extension of the Tau cross upwards to the spiritual. It indicates that the Atlantean onus of development was the vertical line of mental unfoldment and aspiration until full self-consciousness is achieved, making the direct line of intuitive development possible.

General Logoic evolution has its correspondence to the Atlantean epoch. The general population of planetary and solar Logoi work collectively, manifesting their version of the group consciousness of the Atlanteans, hence Logoic Mind is universal, allowing all to Think in harmony to produce their integrated Purpose. The correspondence to the desire principle generated by the Atlanteans is Logoic Desire to see the evolution of the kingdoms of Nature, the Elemental lives, *devas* and humanity, through to conclusion. All must be brought to liberation from *saṃsāra*. In doing so the Logoi use mantras, Words of Power, via 'psychic' means directed to the lower material domains, to achieve Their Magical aims. They work to overcome the forces of evil that stem from massed human selfish, avaricious and self-focussed

29 This term relates to the head of the second Ray department in Hierarchy, and is explained in *The Constitution of Shambhala,* volumes 7A and B, where I discuss a sequence of three Christs. See also *The Reappearance of the Christ,* and *The Externalisation of the Hierarchy* by Alice Bailey.

materialistic activities, which attract to them the dark brotherhood from cosmic sources. This happens when humanity are evolving the attributes of mind, couped with the Watery dispensation from earlier cycles. Success is not ascertained, as it is possible for the massed evil generated to overcome the Light sent to illumine the darkness.

5. The fifth Root Race, the Aryan

This Root Race is representative of the present humanity, mainly centred in the Western world, Europe, the U.S. and the Indian subcontinent. Most people, however, are still essentially Atlantean in consciousness with a sprinkling of the Lemurian type. The Aryan type is polarised in the mental body. The fruit of such development is the present civilisation, with its emphasis on materialism and on the pedagogic rationale of things. The widespread ability of the average person to think rationally will enable Hierarchy to fully externalise in the new age and to implement its programme.

Having mastered the sense of sight and the associated powers of the empirical mind, the fifth Root Race esoterically develops the sense of *smell*, which parallels humanity's ability to contact the *ātmic plane*. This will be attained by some Initiates of this epoch, but will be the gain of humanity as a whole in the seventh Root Race, and will signify the ending of evolutionary attainment for humanity on earth.

The first Root Race developed the sense of hearing subtle impressions from the kingdom of 'God', which allowed them to 'name' things. The second Root Race developed the sense of touch, which allowed them to experience the sensations of the astral plane. This was upon the path of descent into the material world. The Initiates of the fifth Root Race invert this process, producing complete knowledge of everything associated upon the upward arc of developing the attributes of 'God'. They reap the true consequences of the original eating of the fruit from 'the tree of knowledge of good and evil'.[30] The development of the higher attributes of the sense of touch will allow them to contact the substance of the abstracted planes of perception, and hence come to

30 *Genesis 2: 17.* In *Genesis 3:5* the serpent says to Eve: 'For God doth know that in the day ye eat thereof, then your eyes shall be opened, and ye shall be as gods, knowing good and evil'. The meaning of the statement 'ye shall be as gods' is explained above.

know the mysteries that are veiled thereby. The higher attributes of the sense of hearing will allow them to listen to and to master all mantric Sounds upon the higher way, and so to command substance in all of its attributes, not just in the three worlds of human livingness, but later upon the higher domains as well. What is now considered as 'miracle making' will then be the norm. This is the gain of yoga-meditation correctly pursued in conjunction with passing Initiation testings. It is part of the ken of the enlightened, liberated one, who bears the fruits of the Aryan epoch.

When consciousness aspires upwards to the divine then the Tau becomes the plain cross +. This indicates the Aryan dispensation, wherein the forces of materialism having been fully developed, allowed the Christ to be crucified upon the cross of matter. This cross is extended by mutual interaction and constant experience into a *swastika* 卍. This symbol shows that the physical forces are in complete active expression, producing the rapid progress of the mind and its principles. Here we see also the third and fourth dimensional direction of the wheel of life, which eventually turns in upon itself, producing enlightened activity.

This swastika rotates from the left to right, thus indicating the energies associated with the right hand path. The Nazis reversed the movement of this symbol, indicating the path of materialism, the left hand path, in accordance with the type of philosophy expounded.

Little more needs to be added here concerning this Root Race as its attributes are everywhere evident in the fruits of our scientific and technological era, with its physical plane miracle making in terms of the appliances, cars, aeroplanes, computers and the internet, etc., that we now take for granted.

The progenitors of *sixth Root Race* humanity will be polarised upon the higher mental and buddhic planes.

Obviously only a basic synopsis has been presented here and much more could be said about the various races and their characteristics. Human evolution is a long, slow process, and can involve many planetary spheres and even solar systems of undertaking.

The human group Soul first evolved through seven rounds of evolution as a mineral, plant, animal and finally as a human Kingdom. According to the esoteric lore we are now in the *second* solar system, which is of the

'Fourth Order' (i.e., the main sequence of stars), and in it our planetary Logos embodies the fourth Chain (of seven Rounds of seven globes each). Of these the earth is the fourth globe in the fourth Round of evolution. This is the fifth sub-Race of the fifth Root Race (of which there are seven to a world period) and the sixth sub-Race is appearing.[31]

The time taken for the evolution of the respective Races is not of equal duration. There was an enormous, almost unthinkable time taken for the first two Races to evolve. Humanity has been incarnate on earth since Individualisation in Lemurian times 18,000,000 years ago. The fourth Root Race began approximately four and a half million years ago. Atlantis sank about 9,600 B.C. and we have had the fifth Root Race epoch since. There are some stragglers on this earth from the former solar system, as following the left hand path[32] of the materialism of the black magician reverses the evolutionary process. Most from that epoch that progressed the right hand way are now resident at Shambhala, or are members of Hierarchy. The Lords of materialism (the dark brotherhood) that are the governing class of many nations, specifically of the Western world, and those possessed of immense wealth, are generally from the earlier evolutionary epochs, planetary systems prior to the Individualisation of our present humanity on earth. So also the Initiates of the white brotherhood (Hierarchy, the Council of Bodhisattvas) opposed to them.

The 'war' between these two groupings is over the minds of earth humanity (the bulk of the human population), who are easily manipulated by the Initiates of the forces of evil. These forces are a much older

---

31 The terminology introduced here is explained in *A Treatise on Cosmic Fire* by A.A. Bailey and also in *The Secret Doctrine* by H.P. Blavatsky.

32 Those on this path are lords of materialism, the forces of evil, the dark brotherhood, black magicians, sorcerers, black adepts. The colour black connotes the auric colouring they engender, developed through engendering malice, hatred, extreme selfishness, psycopathic and sociopathic activities for the benefit of the self, cunning and manipulative methodology directed for self-glorification and material or monetary power. It has nothing whatsoever to do concerning skin colour. They were practitioners of the black arts and sorcery in former lives, and may still consciously do so in this life. They have made links to many sources of cosmic evil, though now many of their incarnations are unaware of such past life activities, nevertheless the accrued *saṃskāras* and psychic links dominate their lives. They are manipulated and directed from the inner realms by points of evil of vast power.

evolutionary stream, having been members of humanity when the earth humanity were still evolving through the animal kingdom. The dark brotherhood have incarnated in all positions of power in our societies, wherever cupidity and avarice dominates human affairs. They are exceedingly cunning and ruthless in their methodology, often hiding behind the masks of philanthropic, nationalistic or altruistic organisations, but are quite capable of sociopathic or psychopathic methodology, such as the ruse of 'bringing democracy and western freedoms' to Vietnam, Iraq or Libya and then destroying the countries for political gains. Lying propaganda is their usual modus operandi, and the western main stream media, the movie industry and government propaganda (including their covert policing agencies, such as the CIA), are now their prime tool of controlling the minds of the masses. They rule the judiciary and the military might of the nations under their control.

The incarnations of the members of the white or dark brotherhoods ('brotherhood' because both work esoterically as a organised hierarchical grouping) generally are not aware of their inner plane Initiate status, which they developed in earlier lives. The most advanced may gradually awaken to this status through methodology extant on this earth. The best mode of awakening those of the white brotherhood is by following the philosophic, meditation and Initiation process presented in my books and those of Alice Bailey. All religious traditions have their own methodology, but most such systems are antiquated, have by now been compromised with erroneous concepts, or provide only elementary methodology, preventing the attainment of complete enlightenment in their adherents.

The evolution of each Root Race concerns a gradual development out of the parent stock of the characteristics needed. There is an overlapping between each cycle, however the termination of each Racial cycle is heralded by widespread catastrophes.

Note that the Jews have always felt that they were the 'chosen' people, the elect of God. This is because their early ancestors were the progenitors of the Aryan race. The Old Testament, therefore, deals primarily with the early history of that Root Race, and should be read in that light, though some of the history goes back far further in time than is normally supposed. Many lords of materialistic might have incarnated amongst their ranks.

The New Testament lays the foundational teachings for the appearance of the sixth Race Humanity.

Each of the Races have their particular Ray colouring, as indicated in the figure below, which may help to clarify the nature of the evolution of the races.

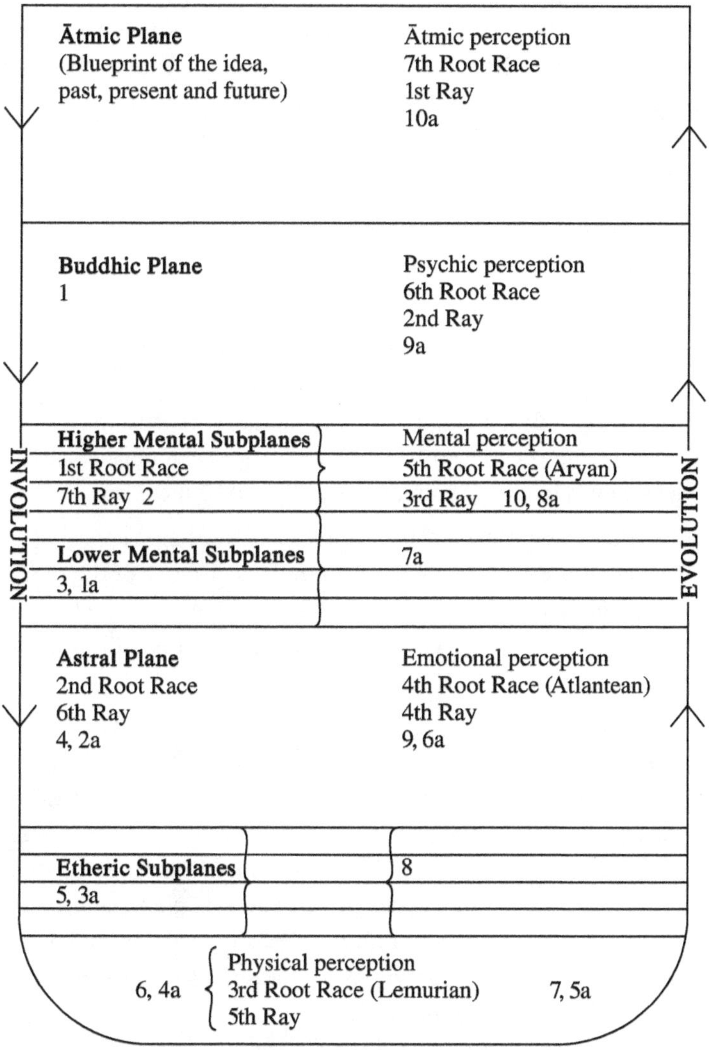

Figure 45: The evolution of the Root Races

Notes to figure 45

The figure traces the evolution of the subhuman kingdoms (the numbers 1-10) in Nature as well as to the progress of human evolution (the numbers 1a-10a). The numbers relating to these two evolutionary streams manifest downwards and then upwards, as descent into matter and the consequent ascent to evolutionary heights is enacted. The chart posits three states of evolution preceding the mineral kingdom, termed Elementary Essence I, II and III. For the general evolutionary progression we see that upon the *ātmic* plane exists the 'blueprint' of the Idea, past, present, and future. It is the plane of the emanation and resolution of *karma*.

For the *subhuman kingdoms* we have:

1. Buddhi reflects the Logoic Idea into concrete manifestation.
2. On the higher mental plane the Word sounds forth and stirs the involutionary substance carried forth from a previous evolutionary period, plus the new substance to be incorporated, termed *Elementary Essence I.* This is manipulated by streams of Light energy which become diffracted into various component sub-hues interwoven by sound patterns, as governed by the *devas.*
3. Upon the lower mental plane crystalline clear shapes appear as this Elementary Essence evolves to form the substance of the thought of the thinker, be it a Logos or humanity. For a Logos it represents the precipitation of a Thought, for humanity it is built into their mind-streams.
4. Upon the astral plane, after countless 'lives' as sharply defined thought forms on the mental plane the Elementary Essence can lose its dynamic energy, allowing elementary particles to couple. We thus have the condensation of the Waters of the astral plane. This Essence becomes the substance of the emotions and desire-forms of the creative entities evolving through this plane, and evolves by constant mutual interactions of basic particles, possessing different colourings and vibratory rates. Desire for sensation is born. *Elementary Essence II.*
5. Upon the etheric sub-planes attraction between elementary particles leads to a coarsening of the energies. Astral matter forms into the

more constrained substance of the etheric plane, regulated by the crystallising energy of thought. *Elementary Essence III.*

6. Upon the dense physical plane the Elementary Essence forms the Anu's, quarks and atoms of substance, the mineral kingdom, with its many categories of elements and compounded shapes that imitate the forms found on the mental plane.
7. The mineral kingdom goes through many physical and chemical interactions and transmutations, until eventually this kingdom becomes liberated via radioactivity. Plant-like attributes are born, facilitated by being 'digested' by the vegetable kingdom (brought over from a former evolutionary epoch) and consequently eventually developing the sentience of that kingdom.
8. Via the ethers the sentience underlying the plant kingdom likewise go through many adaptations, starting from the lower varieties, e.g., mosses, and evolving to the higher varieties (e.g., trees).[33] They evolve animal attributes by contact with and absorption into that kingdom as its food.
9. Via the astral plane at first the animal sentience is largely conditioned by instinctual responses to the needs of the animal forms, then by desire, and finally it is domesticated by contact with the human kingdom[34] until such characteristics as devotion and rudimentary thought are developed. The animal is then at the threshold of entering the human kingdom.
10. Via the lower mental plane through an intensified and unequivocal effort to gain such qualities as affection, the animal group-Soul (esoterically called 'triads', embodied by members of the *deva* kingdom) succeeds in capturing response from the higher planes, with which there is formed a permanent link. Thus the

---

33 The scientists of the Anthroposophical Society, such as Dr. Rudolf Hauschka, author of *The Nature of Substance* (Stewart & Watkins, London, 1966), and Dr. Ehrenfried Pfieffer have done some pioneering research concerning the formative forces governing the plant and mineral kingdoms and their relation to the etheric plane.

34 This statement is true for our present animal kingdom, but not for the earlier cycle related to the Individualisation of our present humanity. Here there was a slow evolutionary progress towards rudimentary thought. In the present cycle the entire process is greatly hastened because of the factor of human interrelation with that kingdom.

Sambhogakāya Flower is formed (with the assistance of certain *solar devas* that incarnate into and as that form) and a human unit Individualises from what was formerly a group Soul form, that answered to collectivised *karma*. Upon Individualisation the *karma* manifested is answerable only to the human unit creating it. When sufficient animal souls have Individualised, then a new human kingdom can start its evolutionary paean as Lemurian humanity.

For the evolutionary progression of *a humanity* we see that:

1a. On the lower mental plane Adam exists blissfully in the garden of Eden. The first Root Race Adamic man symbolises humanity as a whole. Hearing is the sense developed, the sense of sight is awakened. They reside on the lower mental plane, developing mineral-like mental consciousness, governed by the seventh Ray of Ceremonial Activity.

2a. On the astral plane Eve is formed out of the 'rib' of Adam. Adam-Eve, the hermaphrodite second Root Race now exist in blissful union, and duality eventually finds its expression as desire. For a long period they know no wrong, nor each other's 'nakedness', as the mind was still nascent. They are governed by the sixth Ray of Devotion and develop a vegetable type of mental consciousness. The sense developed is touch.

3a. On the higher ethers (literally the astral at this stage of evolution) *Eve*, via the contrivance of the serpent (of time, desire, *karma,* etc.), eats the forbidden fruit from the tree of knowledge, allowing the full expression of desire to manifest. Adam and Eve are now separated.

4a. On the dense physical plane *Adam and Eve* are expelled to the earth, where, through cycles of sense contact and experience, they reap the benefit from the 'tree of knowledge'.

5a. The third Root Race *Lemurian man* is born, established by the formation of an overshadowing Soul, the Sambhogakāya Flower, hence is the first truly human race. At first human consciousness is psychic, instinctual, and physically orientated. The instincts dominate, focussed upon the instinct towards knowledge. The governing Ray is the fifth of Scientific endeavour, which here simply means that this race had to learn much about the physical plane, its dangers, and its usefulness. The rudimentary mind had to develop and grow by this means. The sense of sight is fully developed within the context of mastering the Element Earth, with which their Race came to an end (by geological upheavals).

6a. By the time of the evolution of the fourth Atlantean Root Race humanity have developed an astrally conditioned consciousness, largely dominated first by emotions and desire, and later by devotion and love. Eventually the desire-mind is formed, which caused many problems for this race, especially related to the onset of widespread black magic. The fourth Ray of Harmony overcoming Strife governs the Atlantean evolutionary development. The sense developed is taste and the Element Water was to be mastered, through which their Race came to an end by flooding.

7a. Upon the lower mental plane humanity now enters into the fifth Root Race Aryan era where the emphasis for evolution is the development of abstract thought. Intellect rather than emotions rules, and here the many trials concerned with the battle of the mind versus emotions are undertaken. Much suffering eventuates. The governing Ray is the third of Mathematically Exact Activity. The sense to be developed is smell, hence *ātmic* perception. The ruling Element however is Fire, by means of which it has been prophesised (in *The Secret Doctrine)* this race shall be destroyed by.

8a. Upon the higher mental plane the higher Mind and the Sambhogakāya Flower eventually dominates consciousness and humanity emerges as a liberated Soul. This happens as a consequence of the development of meditative practices and passing the respective Initiation testings. Esoterically, the Word is now consciously uttered as part of the process of humanity becoming a Deity, where actions are determined by Love-Wisdom and the laws of group evolution. The formation and evolution of the Hierarchy of Light out of the present humanity occurs, though this Hierarchy moves from the higher mental to the buddhic plane as a place of residence at the time of the formation of the sixth Root Race.

9a. The sixth (Neptunian) Root Race comes into existence when a significant number of human units develop the perceptions, the enlightenment derived via the buddhic plane. The second Ray of

Love-Wisdom dominates human affairs, where the esoteric senses manifest as the fundamental consciousness. The sense perfected is taste and the governing Element is Air. Eventually the densest sheath will be of etheric matter.

10a. The seventh Root Race. The prodigal sons return home to their 'Father', as *nirvāṇees* by undergoing their cosmic journey. Ray one of Will or Power rules, producing abstraction into the higher domains of all that is material. The sense perfected is smell, *ātmic* perception. The Element is *Aether.* Eventually the densest sheath will be of mental substance.

Appendix Three

A further note on figure five

The dotted line around the three circles in *figure 5* (in chapter 5) is an abstract concept known as the auric egg, 'a ring-pass-not', to quote Madame Blavatsky in her *Secret Doctrine*. It prevents the entrance of all foreign stimuli and energies that are harmful or not needed by the entity. By analogy therefore, every being from the atom to the sun possesses its own auric egg, the bounds within which a being is contained. In this case it refers to the outer 'tangible' encasement of the Monad and is thus composed of the most refined spiritual substance. Theoretically therefore, it is the receptacle of what is gained by the Monad through its incarnate expression during cycles of material involvement. It is the higher correspondence to the skin, receiving impressions from the corporeal world by acting like the film of a camera which captures and stores all of the sensations that are impressed upon it.

The auric egg therefore contains the impressions and tendencies of the past, present, and future of the entity concerned. (Bearing in mind here the singularity of the three times.) Through contact with the reflected auric egg of a person on causal levels, for instance, the experiencing and comprehending the qualities and history of a thing by touch (psychometry) is observed. Such impression is generally distorted by the medium of the astral plane through which visions generally materialise, and by the degree of ignorance or glamour that limits the revelatory understanding of the mystic or psychic concerned.

In the classification given in *Figure 5* we saw that the physical body is divided into an etheric and a concrete expression, whilst there was a lower and higher mental plane. They then together constitute a septenary with the *buddhic* and *ātmic* bodies. This septenary exists within the embrace of the seven planes that constitute the corporeal form of our Logos. The highest two sub-planes of the cosmic dense physical realm are termed *anupādaka* and *ādi*.

Anupādaka (meaning parentless, sometimes called *parinirvāṇa*, 'beyond' the *nirvāṇa* state) is the plane wherein resides the Monad. The evolutionary process concerning the human Soul that was emanated as the vehicle of the Monad was an act of sacrifice, because the Monad had thereby to limit (circumscribe) its Awareness for aeons. Its focus being downwards to the lesser kingdoms, to lift up to Deity the Elemental lives therein. Thus it sent forth a Ray which clothed itself in matter at an appointed time. Upon a mass scale we then had the process of Individualisation eventuating, the causation of humanity.

Humanity were at first all psychically attuned. Eventually humanity lost its psychic innocence, its ability to see and contact deity or the 'gods' via clairvoyance. People's subtle bodies had gradually coarsened, and the higher perceptions were kept alive through dim recollection, by religious symbolism and tradition, old ruins, myths and fables.

Only the comparative few who could awaken the organs of vision (the psychic centres) could see the subtle levels of perception as we got closer to the present era. By means of the constant transmutative process associated with reincarnation, the human Souls eventually cleared a path through the deepest jungles of *saṃsāra* into which they incarnated. They had sown the seeds of enlightenment (the Christ consciousness) and germinated them. An abundant harvest will thereby be reaped of many enlightened Souls as the new epoch immediately ahead of us approaches. The new epoch, of humanity again possessing psychic perception en masse, is slowly dawning. Such Sambhogakāya Flowers are causative agents that become opened flowers in the Eyes of the Lords of Shambhala. These Lords are collectively known as 'God' by the theistically inclined.

Humanity were the 'fallen angels', symbolised by Lucifer in Isaiah,[35]

35 *Isaiah 14:12 ff.* 'How art thou fallen from Heaven, O Lucifer, son of the morning!

that fell from the domain of the Throne of 'God', and in doing so they glorified 'God'. Lucifer here symbolises the human group Soul, the 'Over-Soul'. The ambitious posturing of Lucifer in the passage will be fulfilled by humanity as a normal course of their path to enlightenment. The term Lucifer means light bearer and is also equated with the 'day star' (Venus), as in the passage in *Isaiah 14:12,* which states: 'How art thou fallen from heaven, O Lucifer, son of the morning: how art thou cut down to the ground, which did weaken the nations.' The dogma of equating Lucifer with Satan was formulated by Jerome and other fathers.[36] The passage in Isaiah was applied to Satan, and compared to *Luke 10:18,* which states: 'And he said unto them, I beheld Satan as lightning fall from heaven'. This passage however has actually nothing to do concerning Lucifer. If one properly studies the passages following the Isaiah quote then one will discover that the term actually symbolically refers to the fall from heaven of *humanity* (collectively 'the light bringer'), when the text starts to refer to historical events, such as the kings of the nations, Babylon, etc. Humanity certainly collectively would say: 'I will make myself like the Most High' *(Isaiah 14:14).* See also, *Rev: 22:16,* where Jesus states: 'I am the root and the offspring of David, and the bright and morning star'. The 'bright and morning star' is but a rendition of 'O Lucifer, son of the morning', i.e., the day star, Venus.[37]

The evolutionary process briefly recounted here is but a seed thought of much future revelation that will be obtained by those upon the enlightenment path. Detailed explanation will appear in later writings.

Little can be said here about the plane *ādi.* On it reside great Deva Lords who are embodied aspects of the Mind of the planetary Logos, and who thus wield the power that sustains the Life of the entire manifest domains. Such Lords can be considered active Creative Wills. The

---

How art thou cut down to the ground, which didst weaken the nations! For thou hast said in thine heart, I will ascend into heaven, I will exalt my throne above the stars of God: I will sit also upon the mount of the congregation, in the sides of the north. I will ascend above the heights of the clouds; I will be like the most High.'

36 See *Encyclopaedia Biblica,* under the entry on Lucifer. (George N. Morang & Co, Totonto, 1899-1903, vol 3. Edited by Rev. T.K. Cheyne and J. Sutherland Black.)

37 There is an esoteric relation of Venus to humanity, which need not be delved into here.

human Monads on the other hand are Lords of Love and Sacrifice within the Heart of the planetary Logos. They express cosmic Love (a somewhat meaningless phrase to most), stepping it down so that it can be utilised by the lives in the corporeal planes via *bodhicitta*, the expression of universal Love. The energy is then directed via the astral plane to the Heart *chakra* (expressing group love). It manifests in a distorted form through the Solar Plexus centre as personality affection, as that love between male and female that binds them to couple and to produce the children that are the custodians of the future.

Oṁ

A Fire Mantra

From the field of Fire
within the Mind of the Mighty One.
Let Fire blaze the path to transcendent Light.
Call forth the Dragon, sound its Name,
evoke the power of its mighty Flames
to conflagrate the hearth,
to raze the tower of mind,
to make ablaze the kingly Crown.
Radiant Fire brightens the Heart
and opens the Eye to See
the far vistas of the past,
and the paths to the starry Lives.
All hail this astounding Light.
Cosmos unfolds from each spark lit,
that are Fiery Minds becoming
and that blaze in their brilliant Might.

Oṁ

Note that a good backup teaching complimenting this mantra is found in the three volumes of Roerich's *Fiery World*, *A Treatise on Mind* and also *A Treatise on Cosmic Fire* by Alice Bailey, provides a large amount of foundational information concerning the nature of mind/Mind and its relation to Fire.

Bibliography

Arnikar, H.J. *Essentials of Occult Chemistry and Modern Science*, The Theosophical Publishing House, Adyar, Chennai, India and Wheaton, Ill, 2000.

Babbitt, Edwin D. *Principles of Light and Color,* The College of Fine Forces, New Jersey, 2nd Ed., 1896.

Bailey, Alice A. *A Treatise on Cosmic Fire.* New York: Lucis Publishing Company, 1977.

——. *Discipleship in the New Age, Volume I.* London: Lucis Publishing Company, 1971.

——. *Discipleship in the New Age, Volume II.* London: Lucis Publishing Company, 1955.

——. *Esoteric Astrology.* London: Lucis Publishing Company, 1982.

——. *Esoteric Healing,* London: Lucis Publishing Company, 1971.

——. *Glamour, a World Problem.* London: Lucis Publishing Company, 1982.

——. *The Externalisation of the Hierarchy.* New York: Lucis Publishing Company, 1982.

——. *The Rays and the Initiations.* New York: Lucis Publishing Company, 1970.

——. *The Reappearance of the Christ.* New York: Lucis Publishing Company, 1969.

Balsys, Bodo. *Ahimsā: Buddism and the Vegetarian Ideal.* New Delhi: Munishram Manoharlal, 2004.

——. *A Treatise on Mind, Volume 1* Sydney: Universal Dharma Publishing, 2016.

——. *A Treatise on Mind, Volume 2* Sydney: Universal Dharma Publishing, 2016.

——. *A Treatise on Mind, Volume 3* Sydney: Universal Dharma Publishing, 2016.

——. *A Treatise on Mind, Volume 4* Sydney: Universal Dharma Publishing, 2015.

——. *A Treatise on Mind, Volume 5A* Sydney: Universal Dharma Publishing, 2015.

——. *A Treatise on Mind, Volume 5B* Sydney: Universal Dharma Publishing, 2015.

——. *A Treatise on Mind, Volume 6* Sydney: Universal Dharma Publishing, 2014.

——. *A Treatise on Mind, Volume 7A* Sydney: Universal Dharma Publishing, 2017.

——. *A Treatise on Mind, Volume 7B&C* Sydney: Universal Dharma Publishing, 2018.

——. *Karma and the Rebirth of Consciousness.* Delhi: Munshiram Manoharlal, 2006.

——. *The Revelation,* Vol. 1, Ibis Press Sydney, 1989.

Besant, Annie & C.W. Leadbeater. *Occult Chemistry,* third edition, editor, C. Jinarajadasa and E.W. Preston, Theosophical Publishing House, Adyar, Madras, India, 1951.

Bible, *King James Version*, Thomas Nelson Inc., New Jersey, 1972.

Blavatsky, H.P. *The Secret Doctrine. Vol. 1.* Adyar: Theosophical Publishing House, 1888, 2005.

——. *The Theosophical Glossary,* The Theosophical Publishing Society, Adelphi, W.C., 1892.

——. *The Voice of the Silence,* Theosophical Publishing House, 1998.

Brown, Brian Edward. *The Buddha Nature*, Motilal Barnarsidass, Delhi, 2004.

Brynjolfsson, Ari. *Redshift of photons penetrating a hot plasma*, arXiv:astro- ph/0401420v3, 7 Oct 2005.

Cheyne, Rev. T.K. and J. Sutherland Black (Ed.). *Encyclopaedia Biblica*, George N. Morang & Co, Totonto, 1899-1903, vol 3.

Conze, Edward. *Buddhist Thought in India*, George Allen & Unwin, London, 1983.

Cremo, Michael A. and Richard L. Thompson. *Forbidden Archaeology*, Bhaktivedanta Book Publishing, Inv., Los Angeles, 1998.

——. *Human Devolution,* Torchlight Publishing, Los Angeles, 2003.

Daniel F.V. James, Malcolm P. Savedoff and Emil Wolf. *SHIFTS OF SPECTRAL LINES CAUSED BY SCATTERING FROM FLUCTUATING RANDOM MEDIA*, The Astrophysical Journal, 359:67-71, 1990 August 10, © 1990.

Dargyay, Eva K. 'The Concept of a "Creator God" in Tantric Buddhism', *The Journal of the International Association of Buddhist Studies*, Vol. 8, No. 1, University of Wisconsin, Madison, U.S.A., 1985.

David-Neel, A. & Lama Yongden. *The Secret Oral Teachings in Tibetan Buddhist Sects*, City Lights Books, San Francisco, 1967.

Dirac, P.A.M. *Nature* ***168***, 906 (1951).

——. Proc. Roy, Soc., A209, 291 (1951).

Dorje, Gyurme, Trans. *The Tibetan Book of the Dead: The Great Liberation by Hearing the Intermediate States.* Penguin Books, London, 2005.

Dudjom Rinpoche. *The Nyingma School of Tibetan Buddhism*, Wisdom, Boston, 1991.

Easton, M.G. *Easton's Bible Dictionary*, Thomas Nelson, 1897.

Evans-Wentz, W.Y. *The Tibetan Book of the Great Liberation*, Oxford University Press, London, 1971.

Findlay, Tom. *A Beginner's Guide of our Electric Universe*, www.newtoeu.com, 2013.

Ganguly, Swati. *Treatise in Thirty Verses on Mere-Consciousness*, Motilal Barnasidass, Delhi, 1992.

Gardner, James N. *Biocosm*, New Age International (P) Ltd., New Delhi, 2006.

Govinda, Lama Anagarika. *Foundations of Tibetan Mysticism*, Samuel Weiser, New York, 1975.

Grant, Andrew. *Science News,* Vol. 187, No. 4, February 21, 2015. https://www.sciencenews.org/article/speed-light-not-so- constant-after-all#comment-1799977441

Grossing, G. http://arxiv.org/abs/quant-ph/050879

Guenther, H.V. *The Life and Teachings of Naropa*, Shambhala, Oxford, 1963.

——. *Treasures on the Tibetan Middle Way,* Shambhala, Berkeley, 1976.

Hartmann, Franz, M.D. *The Life and Doctrines of Jacob Boehme*, Kegan Paul, Trench, Trubner & Co., London, 1891.

Hauschka, Dr. Rudolf. *The Nature of Substance*, Stewart & Watkins, London, 1966.

Jackson, Rolf, M.Sc.E. *The Conscious Universe*, Gaia Consciousness Institute, 2002.

Kafatos, Menas. Sisir Roy and Malabika Roy. *Variation of Physical constants, Redshift and the Arrow of time*, arXiv:astro-ph/0305117vl 7 May 2003.

Kanjilal, Dileep Kumar. *Vimāna in Ancient India*, Sanskrit Pustak Bhandhar, Calcutta, 1985.

Kashyap, R.L. *Hymns of Creation, Heaven & Ancient Fathers*, Sakshi, Bengaluru, 2011.

Kern, H. *The Saddharma-Pundarika, or The Lotus of the True Law,* Motilal Banarsidass, New Delhi, 1884.

Kervran, C. Louis. *Biological Transmutations*, Crosby Lockwood, London, 1972.

Leakey, Richard. The *Sixth Extinction*, Phoenix paperback, London, 1996.

Lerner, Eric J. *The Big Bang Never Happened*, Vintage books, London, 1992.

Montgomery, Alan and Lambert Dolphin. http://www.ldolphin.org/cdkgal.html

Oakes, Lorna & Lucia Gahlin. *The Mysteries of Ancient Egypt*, Hermes House, London, 2005.

Perkins, James S. *A Geometry of Space and Consciousness*, Theosophical Publishing House, Adyar, third Edition, 2004.

Narada, Mahathera. *The Buddha and His Teachings, Gradual Sayings,* I, Buddhist Publications Society, Kandy, Sri Lanka, 1988.

Phillips, Stephen M. *ESP of Quarks and Superstrings*, New Age International, New Delhi, 2005.

——. *Extra-Sensory Perception of Quarks.* The Theosophical Publishing House, Madras, London and Wheaton, Ill, 1980.

Pruden, Leo M. *Abhidharmakośabhāṣyam of Vasubandhu*, 4 volumes, Asian Humanities Press. Berkeley, California, 1988-90.

Roerich, Helena, *Fiery World,* Vol's I, II and III, Agni Yoga Society, Inc., 1969.

Rothwarf, A. *Physics Essays **11***, 444 (1998).

Roy, Sisir. Menas Kafatos and Suman Datta. *Shift of spectral lines due to dynamic multiple scattering and screening effect: implications for discordant redshifts*, Astronomy and Astrophysics, 353, 1134–1138, 2000.

Sankaranarayanan, P. *What is Advaita?* Bharatiya Vidya Bhavan, Mumbai, 1999.

Saunders, R.H. and McGaugh. S.S. arxiv.astro-ph/0204521.

Setterfield, Barry and Trevor Norman. http://www.ldolphin.org/setterfield/report.html

Sheldrake, Rupert. *A New Science of Life*, Blond & Briggs, London, 1981.

Smolin, Lee. *The Life of the Cosmos*, Universities Press, Hyderabad, 1999.

Sparham, Gareth (trans.). *Ocean of Eloquence, Tsong kha pa's Commentary on the Yogācārya Doctrine of Mind*, Sri Satguru, Delhi, 1995.

Suzuki, D.T. (trans.). *The Lankāvatāra Sūtra,* Routledge and Kegan Paul, London, 1932.

Thornhill, Wallace and David Talbott. *The Universe Electric: Big Bang?* eBook, Copyright © by David Talbott & Wallace Thornhill, 2008.

——. *A Synopsis of the Electric Universe*, http://www. holoscience.com/wp/synopsis/

Tompkins, Peter and Christopher Bird. *The Secret Life of Plants*, Harper Collins India, New Delhi, 2000.

Venkataraman, G. *The Big and the Small, The Story of Elementary Particles,* volume 1, University Press, Hyderabad, 2001.

Wayman, Alex. *Untying the Knots in Buddhism*, Motilal Banarsidass, Delhi, 1997.

Webb, John. *Physical Review Letters.*

Winstein, Bruce. (Chair of committee). *Elementary Particle Physics,* Universities Press, India, 2000.

Websites:

'On the "Electric Sun" Hypothesis', http://www.tim-thompson.com/electric-sun.html

https://www.electricuniverse.live

http://www.thunderbolts.info

ThunderboltsProject

http://www.holoscience.com/wp/

https://neutrinodreaming.blogspot.com/, 29 September 2011.

http://www.tim-thompson.com/electric-sun.html

https://www.everythingselectric.com/electric-universe-debunked/#bridgeman

https://briankoberlein.com/2014/07/01/rube/Testing-the-Electric-Universe---One-Universe-at-a-Time.pdf

Index

A

Abhidharma, 205
Abraham, 146, 148, 532
Absolute Deity, 48
Absolute Logos, 31, 301, 308
Absolute time, 12, 15, 72, 207
Abstract Mind, 10, 18, 25, 44, 64, 76, 87, 110, 114, 115, 116, 134, 160, 168, 196, 326, 470, 476, 501, 515
Adam, 155, 279, 281–284, 368, 524, 526, 527, 542
Adam and Eve, 279, 281, 283, 527
Adamic man, 282, 524–525, 542
Ādarśana-jñāna, 60
Ādi, 18, 64, 77, 88, 164, 329, 372, 405–406, 417, 466, 470, 546, 547
Ādi Buddha, 2, 16, 29–30, 31, 38, 48, 50, 52, 97, 107, 144, 145, 148, 207, 208, 238, 246, 288, 298, 308
    Consort, 29–30, 38, 52, 246, 298
Advaita Vedānta, 58, 94, 98, 140
Aetheric triad, 419
Agni, 174, 337, 440, 469
Agnichaitans, 175, 197, 198, 368, 376
Agnishvattas, 172, 175, 198, 368, 376
Agnisuryans, 175, 198, 368, 376
Aḥ, 517
Ahaṃkāra, 35, 513
Airy prāṇas, 155, 274
Airy signs, 382
Airy triad, 419
Ajñāna, 99
Ākāśa, 12, 13, 17–19, 20, 21–22, 29, 42, 123, 158–159, 223, 337, 454, 467, 469–470, 480
Ālaya, 234
Alaya Avalokiteśvara, 137
Ālayavijñāna, 23, 25–26, 33, 50–53, 59, 60, 67, 117, 169, 204, 205, 222, 223, 226, 313, 316
Ālayavijñāna enlightenment, 87, 170
Alchemical Elements, 417–418, 420, 453, 481
Alchemical Fire/process, 28, 164, 169, 171, 230, 440
Alcyone, 470, 481
Aldebaran, 143
All-accomplishing Wisdom, 11, 66, 281, 282, 288, 305, 332, 374
Allen Rothwarf, 399
Alpha particles, 343
Amenti, 163
Analogy, 69, 71–72, 107, 133, 473
Ānanda, 58, 87, 88, 99
Ancient schools, xii
Angelic beings/lives, 34, 106, 128, 524
    kingdom, 151

Angular momentum, 415, 428
Animalia, 322
Animal kingdom, 35, 128, 151, 164–166, 167, 168, 184, 186, 187, 191, 195, 263, 277, 284, 286, 322, 326, 344, 351, 395, 452, 536, 538, 541
Animal-man, 277, 283
Anima-mundi, 43
Ankh, 285–286, 289
Antaḥkaraṇa/s, 55, 67, 98, 162–166, 171–172, 182, 197, 222, 223, 267, 292, 354, 508–509, 514
Anthropic Principle, xiv, 46, 52, 75, 94–96
Anthropoid apes, 530, 531
Anti-gravity, 439, 495
Antineutrino, 403
Anubis, 127
Anupādaka, 18, 64, 88, 139, 164, 184, 406, 417, 469, 470, 546
Anu/s, 171, 213, 247, 328–329, 334–349, 350, 352–354, 359, 369, 373, 374, 387, 397, 400–403, 405–406, 408, 411, 412–414, 416, 417, 419, 423, 425, 427, 428, 429, 430, 431, 433, 434, 436–438, 440–442, 443, 446, 447, 456, 460, 464, 475, 476, 485, 486, 487, 498, 499, 541
  as photons, 345–346
Āpaḥ, 159
Apāna, 243
Apratiṣṭhita nirvāṇa, 27
Aquarian age, 177
Aquarius, 381, 384, 447, 449
Archangels, 427
Arhant, 27, 62
Aries, 381, 383, 390, 446, 447
Army of the Voice, 377, 459, 462
Arūpa, 14, 56, 75, 121, 246, 271, 290, 293, 366
Arūpa devas, 462
Arūpadhātu, 21
Arūpa universe, 234
Aryan. *See* Root Race, fifth (Aryan)
Asaṇga, 313
Ashram/s, 128, 132, 148, 152, 153, 514
Aṣṭadiśas, 507
Astral body, 142, 158, 162–163, 175, 211, 281, 344, 351, 423, 473, 527, 528
Astral matter, 93, 141, 167, 212, 359, 384, 423, 525, 533
Astral permanent atom, 349, 360, 361–362, 368, 395, 423
Astral plane, 18, 20, 28, 35, 64, 77, 86, 87, 88, 126, 129, 137, 159, 163, 165, 174, 181, 185, 213, 214, 280, 282, 291, 292, 294–295, 336, 341, 356, 358, 363, 368, 370, 371, 373, 382, 385, 390, 396, 417, 424, 435, 437, 455, 464, 470, 514, 519, 523, 525, 532, 535, 540–541, 542, 548
Atlantean continent, 193
Atlantean cycle, 190, 192–193, 194
Atlanteans, 532–535, 543
Atlantis, 537
Ātma/Ātmic plane, 18, 64, 74, 77, 88, 121, 127–128, 130, 137, 138–139, 141, 152, 161, 165, 167, 191, 197, 198, 200, 243, 291, 351, 366, 372, 390, 407, 417, 422, 423, 427, 428, 429, 455, 468, 470, 474–475, 477, 523, 528, 535, 536, 540
Ātma-buddhi-manas, 160
Ātman, 96, 98–101, 139–140
Ātmic perception, 135, 160, 543, 544
Ātmic permanent atom, 165, 172, 351
Atomic consciousness, 210
Atomic universe, 397–454
Atom/s, 4, 13, 30, 47, 56, 76, 78, 85, 90–91, 92, 108, 124, 135, 159, 164, 171, 201, 203, 206, 218, 219, 234, 247, 251, 253–254, 276, 279, 317, 320, 322, 325, 328, 330, 333, 334, 336–338, 340, 341, 343–344, 346–347, 348–349, 352, 354, 358, 359, 360–363, 364–365, 367, 377, 380, 385, 387–388, 391, 395, 398,

407–409, 413, 415, 416, 421, 425, 427–432, 434, 437, 447–451, 453, 456, 472, 478, 498, 522, 541, 545. *See also* Permanent atoms, Anu's
as bījas, 29, 105, 244, 319–320, 326–334, 358
formation of, 22, 169, 353, 369–370, 406, 435, 440–443
nucleus of, 219, 408, 412, 440–443, 446, 448
primal, 131, 142, 475
structure, 287, 321, 404, 447
Attā, 97
Augoeides, 168
Aūṁ, , 363, 386, 474, 475, 476, 481, ix
Aura/s, 354, 355
Auric egg, 545
Avatar of Synthesis, 144
Avatar/s, 127, 130, 144, 383, 384–385, 461
Avidyā, 40, 100
Avyākṛta-vastūni, 101
Ayaṅa, 445
Āyatanas, 19
Ayonimanaskāra, 19
Aztecs, 241

B

Babbitt, 341
Babylon, 547
Bailey, Alice, xv, 127, 153, 175, 178, 190, 277, 291, 354, 369, 377, 502, 532, 538, 549
Bala, 61
Bardo, 269
Bardo Thödol, 278, 286
Barhiśad pitṛ, 366
Baryons, 338, 404, 414, 415, 427
Baskets of Nourishment, 29, 370
Be-ness, 502
Besant, 334, 338, 339, 347
Beta decay, 422
Beta particles, 343
Bhakti yoga, 534
Bhāva, 40
Bhūmis, 65, 269
Big Bang, xiii, 3, 80, 95, 96, 195, 209, 210, 213, 214, 317, 399, 458, 482, 487, 488, 489, 490, 491, 493, 494, 498
Bīja/s, 16, 23, 26, 27, 29, 30, 42, 95, 105, 108, 109, 204–206, 222–223, 225–228, 232, 245–246, 253, 279, 284, 287, 303, 305, 310, 316, 319–325, 333, 334, 337, 349–350, 354, 356, 365, 368, 392, 406, 433, 439
as atoms, 105, 244, 319–320, 326–334, 358, 387
Binding energy, 403, 434
quantum chromodynamics, 421
strong, 421
Bindu/s, 22, 26, 36, 52, 53, 61, 202, 203–204, 224, 237, 242, 406
Birkeland currents, 490
Black adepts, 537
Black dust, 28, 29, 46, 196, 384, 474, 488–489
Black hole/s, 397–398, 461, 494
Black King, 197
Black magic, 533
Black magician, 171, 176, 230, 537
Black MIND, 197
Black Space, 197
Black substance, 475
Blinded Lives, 29, 39, 295, 370, 377, 457, 459, 460, 462
Bliss, 24, 49, 58, 87, 88, 105, 116, 141, 167, 266, 268, 269, 463, 511
Blood, 43, 89, 136, 359, 518
Bodhi, 11, 105, 170
Bodhicitta, 12–13, 18, 27, 37, 38, 43, 44, 68, 71, 87, 103, 106, 109–110, 114, 126, 135, 141, 168, 237, 250, 272, 274, 318, 323, 325, 406, 470, 510, 519, 548
Bodhisattva/s, 12, 23, 27, 37–38, 44, 47, 57, 61–62, 65, 70, 81, 105, 106, 108–110, 114, 118, 134, 139, 144, 147–148, 152, 167, 208, 237, 260, 267, 269, 272, 278, 306, 327, 351, 354, 384, 449, 506, 517

Council of, 103, 125–127, 136–137
Bodily sheaths, 98
Bosom, 30, 122, 286, 469
Boson/s, 22, 45, 338, 402, 408, 409–411, 414, 423, 427, 432, 450–451, 456
Z and W, 22, 410, 432, 446–447, 451
Bottom quark, 419
BOUNDLESS ALL, 501
Brahmā, 7, 32, 33, 58, 98, 145, 150, 151, 165, 276, 294, 337, 365, 388, 407, 417, 445, 468, 472, 473, 514, 523
Brahman, 7, 18, 32, 33, 58, 94, 99, 140, 178, 469
Brahmā-Rudra, 151
Brown dwarf/s, 310, 461, 495
Bubble/s, 54, 74, 83, 213, 277, 328, 340, 345, 437
Buddhadharma, 140
Buddha-field, 25
Buddhahood, 8, 23, 24, 38, 118, 134, 309
Buddha-Mind, 22, 93, 103, 104, 107, 108, 118–119, 159, 246
Buddha/s, 9–10, 19, 29, 31, 37–38, 43, 47, 51, 57, 58, 59, 60–63, 66, 70, 71, 72, 75, 82, 83, 93, 97, 99–101, 104–105, 108, 110, 112, 114, 116, 118–120, 130, 136, 140, 150–151, 167, 169, 179, 184–186, 192, 204, 230, 247, 258, 262, 265, 277, 297, 299, 302, 306, 309, 327, 359, 518–519
Consort of, 9–10, 37, 38, 106
of Meditation, 7, 10, 244
primal, 144, 179, 207–208
three bodies of, 105, 303, 325
Buddhas of Activity, 145, 151, 386
Buddha Womb, 168, 179, 247
Buddhi, 17–18, 42–45, 46, 54–55, 64, 69, 87–88, 121, 123, 135–137, 155, 158, 160, 161, 169, 175, 291, 294, 383, 412, 417, 423, 424, 427, 455, 469, 470, 502, 522, 527, 533, 540
Buddhic Fires, 175
Buddhic plane, 63, 87–88, 138–139, 142, 147, 154, 174, 198, 373, 394–395, 412, 479, 536, 543, 546
Buddhist cosmogony, 19
Buddhist philosophy, 4, 14, 16–17, 19, 40, 48, 62, 81, 82–83, 93, 96–100, 102, 105, 108, 114, 116, 119–120, 144, 150, 169, 178, 205, 208, 238–239, 243, 247–248, 257, 298, 319, 341, 358, 386, 428, 464
Builders, 132, 219, 294, 311, 459, 478
lesser, 293, 311, 367, 369, 371, 377
Building devas, 234, 292, 366, 376–377, 481
Burning fire of Desire, 466
Burning-ground, 360, 394, 395

C

Caduceus staff, 188
Caeleno, 474–475, 477
Cancer, 381, 382, 383, 384, 391, 446, 448, 533
Cancer-Leo, 448
Capricorn, 381, 384, 447, 448, 449
Cardinal cross, 381, 384
Cardinal positions, 270
Catuṣkoṭikā, 312
Causal body, 9, 159, 165, 167–173, 198, 359, 360, 361, 366, 375, 376, 377, 466, 471, 473
Causation, xv, 3, 7, 8, 9, 11, 14–16, 18, 24–26, 47, 51, 63, 83, 130, 131, 142, 143, 146, 149, 157, 158, 170, 202, 208, 260, 275, 287, 294, 296, 299–304, 453, 517
Causative agent/s, xv, xvi, 8–9, 16, 30, 70, 121, 125–126, 128, 130–134, 141, 152–153, 158, 160, 166, 167, 170, 171, 175, 178, 206, 507–509, 511, 513, 522, 524, 529, 546
factors-process, 8, 10, 15, 23–26, 51, 104, 113, 121–122, 125–129, 135, 138, 142–143, 147–149,

151–152, 154–156, 159, 161, 168, 246, 258, 287, 302, 316, 512
Cave/s, 379, 387, 388, 391, 393
Cellular life, 15
Celts, 207, 241
Ceremonial magic, 177
CETI, 424
Chain/s, 142, 143, 145, 177, 484, 537
Chakra/s, 16, 37, 63, 66–67, 69–70, 81, 111, 137–138, 160, 163, 171, 197, 200, 203, 222, 224, 230, 236–245, 247, 252, 254, 269, 273–274, 289, 304, 306, 315, 319, 320, 321, 324, 329–330, 333–334, 339, 352–353, 356, 364, 369, 382, 387, 390, 392, 393, 398, 407, 408, 431, 444–445, 459, 486, 492, 523, 526, 527
  Ājñā centre, 66, 240, 244, 295, 329–330, 333, 339, 390, 392, 406
  Alta Major centre, 392
  Base of Spine centre, 151, 201, 239–240, 242–244, 248, 263, 264, 267, 274, 285, 288–289, 295, 304, 309, 332, 369, 375, 390, 392, 407, 412, 429, 431, 433–434, 458, 464, 523
  centres in the head, 188–189, 333, 512
  Diaphragm centre, 268, 392, 408
  Head centre, 52, 88, 180, 188–189, 200, 240, 242–244, 247, 269, 273–274, 285, 289, 292, 295, 329–330, 332–333, 339, 390, 392, 406, 443–444, 479, 512, 523
  Heart centre, 12, 37, 71, 152, 154–155, 160, 200, 239, 248, 260–261, 262, 263–265, 268, 270–274, 295, 299, 329–331, 332, 343, 384, 390–392, 406, 453, 479, 505–506, 512–513, 548
  Inner Round, 70–71, 331, 392, 426, 444
  Liver centre, 199, 392, 407, 412, 426
  Logoic, 36, 41, 57, 66, 72, 89, 142–143, 149, 151, 156, 193, 198–201, 275, 364
  maṇipūra. *See* Solar Plexus centre
  minor centres, 242, 286, 309, 392–393, 444, 533
  mūlādhāra. *See* Base of Spine centre
  Sacral centre, 199–200, 240, 244, 248, 263, 264, 266, 295, 332, 390, 407, 412, 426, 431, 458, 523
  Sahasrāra padma. *See* Head centre
  Solar Plexus centre, 70, 155–156, 199–200, 240, 253, 264, 265, 266, 274, 286–287, 331, 343, 390, 407, 412, 426, 431, 432, 444, 479, 548
  Splenic centres, 263, 264, 267, 268, 274, 392, 407, 431, 432, 444
  Stomach centre, 199, 392, 407, 412, 426
  Throat centre, 71, 153, 189, 200, 244, 248, 265, 329, 330, 331, 332, 339, 343, 390, 407
  Viśuddha. *See* Throat centre
Channelling, 455
Charged particles, 338, 401, 403–404, 408, 410, 414, 432, 464, 496–498
Charm quark, 419
Cherubim, 427
Chit, 58, 87
Chohan/s, 127, 142, 153, 179, 180
Christ, 43, 108, 111, 119, 124–125, 127, 137, 145–146, 147, 148, 150, 152, 153, 174, 185, 382, 383, 509, 511, 516, 517, 534, 536
Christ consciousness, 546
Christ-force, 532
Christianity, 71
CIA, 538
Cintāmaṇī, 16
Circle, 6, 12, 300–301
Cit/citta, 99, 140, 222
Cittamātra, 17
Cittaprakṛti, 17, 21, 22, 23, 26, 246, 247

Clairvoyance, 14, 166, 171, 193, 354, 452, 546
Clear Light, 22, 42, 59, 65, 87, 107, 111, 160, 455, 504
Clinging, 266–267
Coats of skins, 283, 368
Colour, 164, 346, 357, 388, 417, 421, 428, 430, 434, 456–457, 462, 477–478, 481
Colour change, 419, 421
Colour charge, 411, 419
Coma Berenices, 144
Condensation, 370, 434
Consciousness, 9, 12, 14, 16, 17, 19–21, 31, 35, 37, 40–42, 44–46, 50, 53–56, 60, 61, 63–65, 68–70, 81–84, 85, 92–94, 96, 104, 106–107, 110–112, 121–122, 123–124, 138, 140–141, 157, 159–161, 164, 168, 169–172, 184, 202–204, 208, 210–212, 216, 217, 219–239, 242, 243, 245, 249, 250–258, 260–265, 271–274, 275, 277–279, 282, 284, 285, 288–290, 293, 297–299, 301–302, 306–308, 310, 315, 316, 318, 320, 321–327, 330–334, 344, 349, 351, 352, 356–357, 361, 364–365, 369, 377, 382, 386, 388, 394–395, 404, 408, 414, 416, 420, 426, 430, 443, 451, 452, 453–454, 456–457, 465, 469, 473, 487, 499, 500, 503, 510–515, 517, 522, 524, 525, 528, 533–534, 536, 542–544
  cellular, 7, 47, 297, 300
  evolution of, 13, 27, 33–34, 46–48, 75, 176, 202, 218, 225, 229–230, 239, 246, 271–273, 276, 283–284, 303, 313, 319, 340, 344, 351, 422, 444, 501, 521
  field/s of, 56, 80, 83, 259, 303, 314, 324, 344, 396, 450, 463, 515
  link, 162, 267, 354, 508
  spirals of, 202, 206–221, 224, 234, 255, 258, 330, 344
  storehouse of, 23, 25, 117, 169, 297
  stream/s, 38, 74, 128, 277, 316, 357–358, 377, 446, 512
Constellations, 57, 77–78, 129, 143–144, 180, 186–187, 193, 233, 275, 292, 425, 446, 458–459, 462, 486
Corona, 483, 487, 488, 497
Cosmic animal kingdom, 195
Cosmic astral plane/ocean, 18, 27, 77, 88, 133, 141, 149, 159, 181, 185, 291, 292–295, 336, 368, 382, 384–385, 390, 396, 424, 456, 464, 470, 519
Cosmic beginnings, 3, 53, 69, 390
Cosmic capacitor, 464
Cosmic dust, 28–29, 46, 95, 144, 184, 196, 220, 227, 384, 390, 424, 474–475, 498, 517
Cosmic entity, 1, 59, 64, 80, 107, 113, 119, 180, 187, 190, 199, 225, 473
Cosmic ether/s, 17, 18, 43, 54, 88, 135, 137, 143, 159, 386, 390, 394, 458, 466, 468–471, 477–481
Cosmic Humanity/Man, 49, 57, 185–188, 192, 195, 201, 523
Cosmic Incarnation, 28–29, 78, 79, 95, 119, 180, 425
Cosmic mental plane, 77, 87, 89, 128, 191–192, 198, 362, 396, 466, 472, 477
  sub-planes, 380–381, 383–385, 389, 391
Cosmic microwave background, 78, 488
Cosmic Mind/s, 3, 11, 16, 18, 25, 27, 28, 33, 80, 88, 159–160, 190–195, 197–199, 201, 310, 425, 455, 464, 474, 481–482, 485, 494
Cosmic mineral kingdom, 198, 199
Cosmic Paths, 20, 37, 132, 179, 181–182, 192, 384, 436
Cosmic physical plane, 77, 87–88, 137, 138, 141, 143, 151, 159, 164, 183, 193, 196, 245, 295, 329, 351, 354, 375, 377, 380, 396, 423, 424, 467, 468, 473, 477, 546
Cosmic plant kingdom, 77, 183–185, 186

Cosmic space, 31, 46–47, 59, 159, 180, 197, 345, 384, 423, 437, 483, 508
Cosmic Waters, 18, 159, 185, 207, 290–294, 390, 470, 519
Cosmic yogins, 461
Craving, 40, 266–267, 270
Creatio ex-nihilo, 95, 97, 214
Creation, xi, 11–13, 12, 15, 26, 32, 44, 80, 96, 103, 106, 116, 195, 251, 294–295, 308, 310, 312, 318, 381, 387–388, 422, 459, 481, 492, 493
  myths, 52, 81, 253, 260
Creative Deity, or 'God', 8, 98, 110, 112, 137, 138, 208
Creative energy/forces, 17, 22, 73, 139, 141, 143, 250, 264, 272, 281, 309, 337, 369, 405, 463–464, 468
Creative Hierarchy/s, 41, 142, 291, 292, 339, 344, 367, 369, 370, 379, 387, 391, 392, 512, 522
Creative imagination, 11, 19–20, 171, 230, 278
Creative Intelligences, 155, 292, 294, 326, 367
Creative Logos, 2, 5, 16, 18, 30, 32, 37, 51, 57, 113, 128, 145, 203, 237, 300, 306, 307, 436, 469
Creative Mind/Thinker, 3, 5, 15, 35, 56, 93, 301, 427
Creative process, 12, 16, 26, 52–53, 108–115, 122, 126, 156, 234, 278–279, 300, 302, 306, 318, 319, 508
Creative Sound/s, 96, 481
Creative Will/s, 86, 128, 148, 176, 374, 547
Creative Word, 144–156, 295
Crop circles, 424
Cross/s, 136, 230, 259, 281, 282–286, 289, 330, 332, 379, 381–385, 390–391, 450, 507–508, 510, 515, 520, 531, 534, 536
Cycle/s, 1, 5–7, 12, 15, 24, 72–73, 79, 89, 131–132, 142, 151, 182, 188, 190, 209–210, 220–222, 253, 254–256, 259, 268–269, 289, 295, 312, 319, 323–325, 357–358, 382, 427, 447, 458–460, 465, 501–502, 508, 510, 517–518, 526, 529, 538
Cyclic time, 207, 211, 234, 236, 252, 293, 443–446
Cyclopes, 193–194, 531
Cyclotrons, 111

D

Ḍākinī/s, 52, 118, 248, 249, 309
Dakṣhiṇa, 3
Dark brotherhood, 37, 71, 75, 79, 127, 174, 192, 194, 197, 271, 288, 355, 383, 455, 461, 465, 475, 535, 537, 538
Dark disc, 389
Dark energy, 424–425, 457, 488–489
Dark matter, 95, 400, 424–425, 457, 458, 459, 469, 488–489, 490
Dark Mind, 457, 476
Darkness, 6, 12–18, 40, 46–47, 84, 115, 184, 247, 262, 273, 290–291, 293, 301, 307, 324, 346, 364, 452, 458, 465, 501, 532–533, 535
Dark space, 80, 95, 127, 194, 311, 348, 460
Dark substance, 182, 262, 311, 312
Darwinian theory, 529
David, 149, 547
Day, 293
Death, 2, 6, 39, 162, 163, 252, 260, 275, 285, 423, 447, 459, 476
Deity, 12, 32, 49, 91, 104, 109–110, 112–114, 118–120, 137, 138, 150, 159, 189, 241, 318, 386, 543
Dense physical plane, 329, 363, 476
Density, 333, 340, 416, 434, 479, 483, 487
Deoxyribonucleic acid, 248
Dependent Origination, 262
Desire-mind, 21, 64, 122, 125, 129, 154, 159, 161, 163, 320, 323, 325, 437, 448, 505, 519, 533
Deuteron, 348
Deva-human interrelation, 34–42, 86, 112–113, 128, 155, 175–176, 219, 223, 234, 239, 243, 251, 506
Deva/s, 5, 25, 26, 29, 31, 34–36, 38, 41, 46, 52–53, 88–89, 106, 116, 118,

128, 135, 142, 155–156, 163, 168, 172–178, 197, 203, 219, 238, 242, 249, 263, 270, 272, 277, 287, 294–295, 327, 344, 353, 354, 362–369, 375–377, 383, 427, 430, 434, 446, 455–457, 459, 460–463, 464, 475, 499, 540–541
    and permanent atoms, 362–365, 376
    Builders, 234, 292, 375–376, 481
    Intelligence, 176, 200, 219
    Lord/s, 164, 174, 337, 386, 462, 469, 547
    solar, 35, 172, 366, 369, 376, 542
    substance, 35, 108, 172, 175–176, 392, 394, 460
Dhāraṇīs, 274, 353, 439
Dharma, 11, 67, 105, 110, 140, 150, 159, 228, 230, 247, 266, 301, 323
Dharmadhātu Wisdom, 11, 67, 138, 238, 292, 304, 330
Dharmakāya, 11, 28, 38, 44, 51, 57, 59, 64–65, 67, 77, 82, 84, 89, 101, 103, 105, 111, 113, 115, 120, 127, 130, 134–135, 137, 139, 144, 150, 159, 184, 204, 243, 246, 250, 252, 260, 262, 264, 269, 298–299, 303, 308–309, 464
Dharmakāya Flower, 101, 246, 261, 271–272, 286
Dharmakāya Way, 71
Dharmakāyic vision, 23
Dharmas, 17, 204–205, 309
Dharmatā, 42, 44, 204–205
Dharmodaya, 309
Dhataraṭṭha, 386
Dhātu, 21
Dhyāna, 32, 45, 52, 67, 72, 224, 247, 268, 275, 318
Dhyān Chohan/s, 57, 179, 201, 379
Dhyāni Bodhisattva, 260
Dhyāni Buddhas, 10–11, 18, 22, 24, 25, 29, 48, 52, 54–55, 65, 96–97, 101, 128, 151, 159, 206, 249, 260–261, 264, 269, 287, 298, 304–306, 318, 330, 339, 372, 513, 517, 523
    Akṣobhya, 11, 53, 65, 66, 83, 261–262, 269, 292, 304–305, 306, 331, 373
    Amitābha, 11, 25, 53, 56, 65–66, 83, 257, 260–261, 269, 279–280, 284, 286–287, 292, 304–305, 331, 373
    Amoghasiddhi, 11, 56, 66, 260–261, 269, 279, 281–282, 288, 304–305, 332, 374
    Consorts, 18, 52, 55, 246, 260, 470
    Ratnasambhava, 11, 56, 66, 260–261, 269, 281, 287, 292, 304–305, 331, 374
    Vairocana, 11, 66, 83, 138, 238, 260–261, 269, 280–281, 285–287, 292, 304–306, 330, 372
Dhyānis, 48
Diamond-Mind, 16, 69
Diamond sceptre, 224
Differentiation, 58–59, 160, 197, 467, 478
Dimensions of perception, 4, 119, 140, 508
Dirac, 399–400
Disciple, 62, 128, 129, 130, 132, 133, 176, 190, 355, 508, 509, 510, 511, 512
Discipleship, 173, 507
Discriminating Inner Wisdom, 257, 287, 292, 304–305, 331, 373
Divine Justice, 98, 379, 390
Divine Mothers, 144, 287
Divine Thought/s, 15, 30, 336–337, 383, 468
Divine Will, 167, 292, 294, 473, 513, 518
Divine years, 445
Divinity, 100, 104, 115, 129, 137, 162, 168, 202, 225, 285
Doppler effect, 210, 492
Doppler shift, 491
Dorje, 224

Double-edged sword, 268, 506, 509, 510, 515
Dove, 185, 504
Draco, 143, 382, 385, 465
Dragon/s, 143, 182, 252, 382, 464–466, 549
Druids, 207
Drum, 516–518
Duality, 34, 83, 241, 276, 279, 280–281, 282, 327, 354, 398, 399, 449, 467, 525, 542
Dvāpara yuga, 445
Dweller on the threshold, 160–162

E

Earth (as planet), 2, 37, 61, 62–63, 67, 80, 88, 114, 142, 152, 184, 186, 187, 201, 236, 241, 253, 276, 285, 290–291, 294, 353, 388, 423, 437, 463, 475, 488, 495, 499, 537
Eden, 282, 368, 527, 528, 542
Edgar Cayce, 532
Ego, 9, 350, 359, 361, 366, 374, 375, 466, 480
Egoic lotus, 378
Egoic Ray, 361, 362
Egyptians, 207, 241
Eight-armed cross, 450
Eight directions in space, 228, 420, 450
Eightfold Path, 264–270, 273, 274, 304, 518
Eight-fold way, 451
Eighth Sphere, 407, 408, 431
Eight Mahābodhisattvas, 270
Eight-spoked wheel, 507
Einstein, 13, 14, 314, 461
Einstein-De Sitter Universe, 400
Einsteinian physics, 452
Electra, 463–464
Electric/al charge, 337, 402, 407, 410, 411, 415, 417, 420, 421, 451
Electrical colours, 350, 355, 388
Electrical currents, 407, 484, 488, 493
Electrical energy, 158, 324, 337, 462
Electric/al Fire, 258, 352, 353, 364, 466, 467, 469, 471–472, 480
Electric/al force/s, 418, 472, 477, 487, 494
Electrical scars, 495
Electrical Wind, 337, 469
Electric discharges, 495, 496
Electric field/s, 399, 433, 451, 453, 455, 462, 485, 496–497
Electricity, 18, 40, 199, 336, 337, 394, 400, 426, 446, 448, 462, 464, 466–469, 471, 474–476, 478–480, 482, 487
    cosmic, 48, 158, 336–337, 468
Electric sun, 495, 497–498
Electric universe, 96, 448, 460, 482–486, 494–498
Electrodynamics, 400
Electro ethers, 350
Electro Lumino Ether, 350
Electromagnetic phenomena, 13, 387, 404, 406, 408, 412, 413, 422, 426, 429, 446, 459, 482, 483
    force, 346, 453, 476
    wave/s, 337, 346, 402, 404, 427, 433, 439
Electron neutrino, 416
Electron-positron pairs, 338
Electron/s, 337, 343, 360, 402, 403, 404, 407, 408, 410, 411, 412, 413, 414, 416–419, 425, 426–428, 429–431, 433, 434, 438, 439, 462, 478
Electron volts, 415
Electroweak symmetry breaking, 410
Elemental lives, 29, 175–177, 189, 311, 323, 344, 366, 369, 370, 446, 534, 546
Elementary Essence I, II, III, 435, 436, 437, 540–541
Elementary substance, 29, 46, 95, 260, 320, 325, 348, 350, 364–365, 372, 374, 390, 395, 474
Element/s
    Aether, 17, 18, 22, 139, 165, 243, 291, 333, 399–400, 422, 544

Air, 18, 42, 89, 154, 155, 241–243, 333, 422, 439–440, 453, 479
Earth, 18, 20, 21, 238, 259, 272, 280, 333, 416, 417, 419, 420, 440, 442, 453, 457, 462
Fire. *See* Fire
five, 63–64, 420, 454
Water, 18, 20, 35, 64, 86, 156, 196, 241–242, 259, 272, 279, 286, 291, 333, 359, 373–374, 412, 416, 421, 440, 442, 453, 462, 481, 543. *See also* Water/s
Elias, 248
Elohim, 290
Emotional mind, 158, 159, 168, 321, 448, 505
Empirical time, 501
Energy fields, 426, 448, 499
Enlightenment, 5, 8, 12, 23, 68, 79, 81–82, 88–89, 92, 110, 123–135, 141, 160–161, 170, 172, 222, 227–228, 242, 243, 250, 257, 261, 266–269, 274, 285–286, 309, 322, 356, 427, 500, 503, 515, 518, 538, 546–547
Equal armed Cross, 379
Equalising Wisdom, 66, 287, 292, 304, 331, 374
Equinoxes, 236, 444
Esoteric astrology, 20, 89, 189, 443
Esoteric tradition, xii
Essence, primeval, 14, 15, 17, 18, 46, 73, 98, 370, 373, 460, 469
Eternal Now, 233, 244, 259, 501
Eternal Youth, 144, 149
Etherealisation, 111, 440
Ether/s, 10, 14, 171, 177, 214, 303, 329, 343, 349–350, 371, 375, 397–400, 405–408, 411, 423, 425, 427, 456, 461, 467–469, 476–478, 480, 486, 541, 544. *See also* Element/s, Aether
cosmic, 17–18, 43, 88, 135, 137, 143, 159, 390, 466, 478, 480–481
fourth, 45, 54, 55–56, 352, 408, 412, 428, 429, 434, 438, 448
Eve, 279, 281–284, 526, 535, 542
Event horizon, 28, 45, 87, 206, 213, 397–398, 412, 429, 447, 449, 500
Evil, 127, 129, 161, 174, 252, 271, 389, 475, 533, 534, 535, 537–538
Evolutionary process, 1–3, 28–29, 31, 35, 36, 37, 39, 44–46, 62, 106–110, 118–119, 139, 148, 154, 164, 177, 178–181, 190–191, 205–206, 234, 261–263, 270–273, 275–278, 286–289, 293, 327, 344, 356, 357, 377, 435–436, 454, 478–479, 482, 521, 540
deva, 34, 155, 172–174, 238–239
of consciousness, 20, 33, 48, 50, 73–75, 92, 202, 218, 255, 324
of Logoi, 80–81, 85, 114, 184–201, 270, 534
of the Root races, 192–193, 281–283, 521–544, 546–547
Extra-spirals, 341–344, 380, 388, 393, 416, 438, 475
Eye doctrine, 520
Eye/s, 64–67, 69, 72, 104, 143, 180–182, 203, 213–214, 221, 240, 247, 295, 526, 546

F

Face, symbolism, 291–292
Fallen angels, 546–547
Fall of the three into the four, 303
Father, 37, 48, 113–114, 164, 179, 207, 208, 251, 289, 303, 305, 308, 325, 344, 356, 403–404, 454, 457, 475
Father energy, 40, 279, 325
Father-Mother, 208, 284, 327
Feeling, 129, 265–266, 270, 526–527
Fermions, 402, 408, 410, 414
Fibonacci series, 341
Fiery Breath/s, 143, 382, 465
Fiery flying serpent, 252

Fiery Spheres, 373, 374
Fifth dimension, 76–77, 232, 235, 244, 259, 452–453
Fire, 3, 18, 22, 25, 29, 41, 86–87, 174, 175, 200, 241, 243, 246–247, 251–252, 257–258, 272, 287–289, 291–292, 326, 333, 337, 372–374, 382, 385, 401, 407, 418, 434, 440, 453, 454, 457–459, 462, 464–469, 472, 480, 492, 543, 549
  by friction, 47, 360, 466, 472, 480
  of mind/Mind, 28, 45, 78, 153, 287–289, 352, 372–374, 454, 460, 465, 475, 533
  solar, 47, 168, 360, 363, 472, 480
Fire mist, 7, 317, 370, 372, 467
Fire spirits, 175
Firmament, 294–295
Five dimensional space, 229
Fiveness, 522–523
Fixed cross, 507, 510
Flood, great, 534
Fohat, 334, 336–337, 465, 468–469, 477
Forbidden fruit, 542
Forces, 18, 55, 195, 241, 243, 317, 334, 349, 352–353, 355, 375, 380, 384, 386, 389, 393, 408, 412, 419–420, 446, 453–454, 462, 475–476, 533
  Electrical, 418, 477, 494
  Fiery, 198, 336–337, 466
  Force fields, 211, 234, 236, 244, 449
  gravitational, 249, 401, 405, 412, 420, 428, 432–433, 440, 461, 485, 488
  strong, weak, 29, 404, 407, 409, 412, 427–429, 432, 434, 438–440, 446, 448, 453, 475–476
Form-building, 352, 362–363, 365–376, 386, 392–393, 405, 462
Forward-progressive motion, 210, 218, 221, 318, 507
Four continents, 386
Four directions, 386
Four Noble Truths, 264, 302, 518
Fourth dimension, 141, 213, 216, 242, 255, 335, 429, 452
Free will, 36, 70
Fungi, 322

G

Galaxies, 62, 77, 80, 96, 115, 187–189, 193, 195–196, 198–200, 210–214, 225, 233, 235, 317, 319, 400, 423–425, 458, 486–490, 494–495, 500, 523
  creation of, 188, 489, 493
Galaxy NGC 4319, 492
Galaxy NGC 4391, NGC 7320, 490
Gamma rays, 343, 413, 447, 449
Garden of Eden, 281
Gauge bosons, 411, 450, 451
Gauge symmetries, 421
Gauge theory, fields, 338, 410
Gautama, 62
Geb, 276
Gemini, 382, 383, 384, 385, 390, 446, 447
Gem, wish-fulfilling, 16
General Relativity, 14, 299, 313–315, 400, 432, 486–488
Genesis, 290–295
Geometry of space, 211, 459, 486
Giants, 529, 531
Globular clusters, 461
Gluon/s, 22, 409, 419–421, 432, 442, 446, 448, 451
Gobi desert, 149
God, 3, 50, 74, 93, 96–106, 109, 110, 112, 114, 116, 118–120, 127, 133, 145–150, 155, 159, 172, 178, 196–197, 208, 258, 279, 290, 367–369, 524, 526–527, 534–535, 538, 546–547. *See also* Logos/Logoi
  Son/s of, 97, 107–108, 281, 509
  Spirit of, 280–282, 291–293, 389
Gods, 3, 16, 34, 52, 71, 102, 207, 276, 529, 546
Golden Egg, 7, 276, 300

Golden key, 59
Goldilocks zone, 499
Goodwill, 420, 450–451, 507–508, 517
Gopis, 137
Granulation, 496
Gravitational contraction, 489
Gravitational field, 412, 432
Gravitational lensing, 314, 425
Gravitational waves, 432
Graviton-quark interrelation, 438
Graviton/s, 14, 401–402, 404, 417, 419, 425–433, 435–440, 443, 446–447, 487
Gravity, 231, 235, 313–315, 397, 401–406, 408, 412, 420, 423, 425, 428, 431, 432–435, 438–440, 446, 453–454, 460–462, 476, 485–488, 490, 493–494, 533
Great Bear, 143, 378, 380, 382, 464, 465–466
Great Ones, 63
Great Pyramid, xv
Great Sacrifice, 144, 150
Greeks, 171, 207, 241
Group consciousness, 35, 69, 138, 193, 274, 318, 534
Group evolution, 68, 153, 234, 462, 500–501, 503, 512, 543
Group Soul/s, 4, 35, 125, 173, 186, 277, 376, 525, 531, 536, 541–542, 547
Guardians, 239, 241, 385
Guṇas, 58, 143, 190, 239, 240, 325, 327, 330, 346, 392, 417, 421, 434, 443, 445
Guruparamparā, xii

H

Hadrons, 407, 414
Halton Arp, 96, 490
Hearing, 14, 146, 363, 367, 477, 524, 535–536, 542
Heart, xvi, 12, 43, 66, 69, 136–137, 141, 143, 156, 174, 176, 182, 239, 254, 272, 351, 359, 371, 379, 387, 469–470, 511, 513–520. *See also* Chakra/s, Heart Centre
    is the Mind, 120, 126, 130, 141, 160, 506–520
    of Life, 446, 450, 470, 520
    of the sun, 37, 89, 91, 378
    thinking with, 124, 126, 154, 505–506, 511, 513
Heartbeat, 506, 508, 517–518
Heat death, 499
Heavenly Man, 48, 394–396, 468, 471, 477, 480
Heavenly Men, 192, 362
Heaven/s, 162, 241, 248, 281, 285, 290, 358, 423, 470, 532, 547
Helena Blavatsky, xv, xvi, 6, 7, 178, 190, 334, 336, 501–502, 525, 530, 531, 545
Hell/s, 162, 358, 408, 423, 470, 532
Hermaphrodite, 525, 542
Hexagon, 308–309
Hexagram, 202
Hierarchies, 379, 390
Hierarchy, xiv, 137, 146, 147, 148, 152–153, 155, 167, 383, 507, 511, 512, 533, 534, 537, 543
Higgs boson, 22, 338, 408–410, 423, 427, 432, 450, 456
Higgs field, 338, 408–413, 422–425, 429, 432, 447, 449, 450, 453
Higgs mechanism, 411–412, 429, 451, 453
Higher mental plane, 18, 53, 87, 159, 167, 169, 172, 234, 291, 366, 368, 376, 470, 523, 532, 534, 540, 543, 546
Higher Mind, 171, 235, 322
Hindu philosophy, 7, 16, 32, 98, 100, 113, 137, 139, 140, 150, 386, 445
Hiraṇyagarbha, , 7, 94, ix
Hubble, 491
Hubble's law, 399, 400
Hūṁ, 249, 517
Human evolution, 41, 81, 113, 119, 126, 132, 153, 168, 179, 180, 191, 219, 261–262, 280–281, 294, 308, 320, 322, 353, 357, 436, 500, 521–543

Humanity, cosmic, 57, 185–187, 192, 195–201, 523
Human kingdom, 19–20, 23–24, 35–36, 38–39, 47, 55, 72–73, 88, 110, 112, 121, 135, 148, 150–152, 154–155, 166–167, 170, 174, 176–177, 190, 193, 197, 199, 238–239, 241, 263, 270–272, 284–285, 327, 332, 351, 370, 385, 464, 470, 503, 510, 541
Human Souls, 4, 114, 116, 173, 183–184, 186, 525–528, 546–547
Hydra, 519
Hydrogen, 341, 348, 493
Hydrogen nucleus, 402
Hyperboreans, 525
Hypercolour gauge fields, 338

I

Idea-form/s, 235, 314
Ignorance, 6, 12, 40, 64, 84, 100, 115, 218, 262, 298, 301, 320, 324, 348
Immortality, 257–258
Incarnations, 36, 79, 180, 189, 191, 201, 283, 293, 350
Individualisation, 35, 128, 167–168, 170, 181, 184–185, 191–192, 194–196, 198, 277, 282, 332, 344, 351, 396, 528, 537, 542, 546
Individuality, 74, 221, 350, 502, 513–514
Indra, 174
Infinity, 11, 78, 254, 297–304, 460
Initiate/s, 116, 125–127, 138–139, 181, 185, 349, 424, 455, 479, 537
Initiation/s, xii, 18, 20, 111, 125, 133–134, 136, 150, 172, 178–179, 181, 195, 201, 243, 289, 295, 361, 363, 379, 384, 387, 393, 460, 470, 508–509, 511–512, 515, 519–520, 528, 532
  cosmic, 36, 109, 184, 189, 190–194, 197, 396, 528
  fifth, 134, 170, 179, 185, 191, 243, 464
  fourth, 43, 136, 170–171, 178, 184, 284
  sixth, 169, 179, 243
  third, 167, 170, 184, 191, 193
Instinct/s, 5, 10–11, 141, 164, 166–167, 198, 523
Intelligence, 2, 13, 34–35, 37, 44, 94–95, 140, 173–174, 205, 219, 246, 263, 277, 424, 473
  cosmic, 138, 336, 474
Intelligent Design, 107, 187, 219, 263
Intra-spiral/s, 341, 344, 377, 380, 382, 388, 392, 402, 412, 414, 416, 438, 475
Intuition, 42–43, 68, 106, 166, 257, 259, 511
Inverse square law, 496, 497
Involution, 478
Irrational thought, 19, 22–24, 46
Isaac, 532
Īśvara, 32, 58, 94, 98
Īśvarī, 286

J

Jacob, 532
Jāgrat, 99
Jarā maraṇa, 40
Jāti, 40
Jerome, 547
Jerusalem, 146, 149, 443
Jesus, 130, 146, 147, 148–149, 257, 285, 509, 547
Jewel in the heart, 168, 172, 285
Jews, 538
Jigten Khadomas, 249
Jina/s, 37, 38, 55, 257, 260–261, 287. *See* Dhyāni Buddhas
Jīva, 99, 225, 337, 350–351, 363, 364, 377, 391–392, 393
Jīvanmuktas, 396
Jīvas, 99, 139, 384
Jīvātman, 99
Jñānas, 52
Juergen, 496, 497
Jupiter, 188
Jurassic era, 184

K

Ka, 162, 163
Kalachakra Maṇḍala, 61
Kālahamsa, ix
Kalpa/s, 12, 445
Kāma, 11, 383
Kāmadhātu, 21
Kāma-manas, 21, 125, 177, 320–323, 325, 333, 357, 371, 383, 426, 437, 505, 519
Karma, 5, 24, 36–38, 42, 46, 53, 79, 86, 95, 98, 114, 127–128, 131–132, 137, 147–148, 152, 155, 168, 170, 173–174, 176–177, 179, 204, 207, 262–264, 266, 268, 275–277, 287, 315, 319, 337, 353, 363–365, 369, 371, 375, 382–386, 389–391, 393, 418, 422–423, 427–429, 434, 436, 438–439, 446, 454, 462, 470, 474–475, 500, 518, 520, 523, 532–533, 540, 542
  Cosmic, 362, 382–383
  group, 263, 276, 286
  Logoic, 139, 148, 188, 367
  Lord/s of, 153, 188, 315, 354, 357, 379, 389, 465
  primordial, 372, 477
Kashmiri Śaivism, 96
Kevala cit, 99, 140
Khadoma, 248–249
Kingdom/s, 29, 62, 106, 119, 154, 164, 191, 195, 262–263, 272, 322, 326, 344, 395, 430, 517, 531–532, 540–541
  animal, 128, 164–168, 184, 186, 196, 286, 332, 344, 351
  deva, 5, 29, 31, 34–35, 55, 88, 112, 116, 155, 239, 277, 327, 344, 377, 383, 434, 541
  human. *See* Human kingdom
  mineral, 164, 191, 198–199, 395, 435, 541
  of 'God', 127, 133, 149, 150, 172, 197, 509, 535
  of Souls, 40, 151, 157, 191, 507
  plant, 37, 39, 77, 151, 164, 180, 183–185, 186, 322, 326, 374, 395, 452, 526–527, 532, 541
Kingly Way, 289
King, of righteousness, 146, 149
Kleśas, 19, 23, 59
Kliṣṭamanas, 23, 60, 61, 204, 430
Koilon, 328
Kṛta yuga, 445
Kśiti, 174
Kumāras, 145, 151, 165, 290–291, 386
Kuṇḍalinī, 29, 41, 143, 237, 240, 245–252, 258–259, 274, 287–289, 325, 349, 363, 373–374, 382, 434, 454, 457, 464
Kuvera, 386

L

Lambda term, 488
Law/s, 5, 13, 48, 75, 94, 134, 183, 211, 366, 378, 381, 421, 429, 489, 503. *See also* Karma
  cosmic, 20, 187, 373, 429
  group, 68, 104, 503, 512, 543
  of Attraction, 24, 375, 378, 380–384, 389, 393, 422, 428, 438–439, 465
  of correspondences, 47, 189
  of cycles, 15, 72–73, 86, 89, 131, 188, 234, 508, 529
  of Economy, 15, 24, 341, 378, 380–381, 384, 391, 393, 422, 428, 438–439, 442, 464
  of Love, 38, 382, 385, 438
  of mind/Mind, 2, 20, 31, 94, 464
  of Sacrifice, 164, 384
  of Synthesis, 24, 378, 380–383, 389, 393, 423, 428, 438–439, 465
  of Thought, 24, 30, 498
Laya centres, 199, 201, 352–353, 458
Leadbeater, 334, 338, 339, 347
Left hand path, 247, 384, 520, 536, 537
Lemurian epoch, 190, 193–195, 281–282, 284, 522, 527–532, 537, 542

Leo, 82, 381, 384, 446, 448
Leptons, 45, 413–422, 427, 429, 432, 434, 439, 450, 451
Levitation, 439
Lha, 102
Liberated beings/Souls, 37, 41, 59, 89, 112, 116, 118, 127, 130, 150, 167, 173, 278, 377, 383, 396, 511, 543
Libra, 382, 383, 385, 389, 446, 448, 460, 465
Libya, 538
Life, 2, 4, 17, 24, 29–30, 32, 38, 40–41, 45, 48, 71, 85, 87, 93–94, 138, 149, 152, 159, 182, 186–187, 204, 246, 257, 260, 280, 291, 293, 295, 302, 325, 336, 358, 389, 457, 459–460, 463–465, 470, 475, 484, 499, 507, 511, 514, 518, 520, 529, 547
  energy/force, 32, 142, 163, 170–171, 237, 334–335, 337, 350, 365, 391, 464
  Lords of, 36, 476
  principle of, 95, 225, 280, 413
  process of, 92, 155, 272, 285, 324, 422, 508
Light, 7, 12, 36–37, 39–40, 45–49, 75, 80, 84, 115, 124, 135, 153, 166, 168, 178–179, 184, 208, 214, 262, 284, 290, 292–293, 301, 307, 311, 324, 327, 337, 345, 349, 360, 364, 373, 404, 406, 412–413, 415, 426–427, 430–431, 457–458, 465, 471, 477, 480, 483, 549
  speed of, 210–213, 232, 299, 313–314, 350, 399, 410, 435–436, 456, 460, 483, 493, 495
Light and Anu's, 346–349, 402
Light waves, 14, 337, 345, 402, 438, 461
Ligo tube, 343
Limited infinities, 93, 313
Lipika/s, 53, 131–132, 151, 315, 353, 362, 364–365, 369, 372, 385–391, 393, 465
Logoic. *See also* Logos/Logoi
  atom, 368, 376–385, 387, 388, 391, 392
  Body, 65, 86–88, 377, 395
  causal body, 378, 381, 385, 466
  centre/s, 36, 66, 89, 143, 188, 292, 295
  Desire, 270, 534
  Eye, 213, 214
  Love, 87, 519
  Mind/s, 3, 4–5, 10, 20, 53, 55, 66, 67, 69, 72, 73, 74, 89, 95, 106, 116, 125, 128, 133, 142, 155, 158, 209, 213, 233, 258, 280, 286, 292, 373, 383, 405, 406, 418, 426, 427, 457, 458, 461, 465, 474, 477, 494, 503, 534
  Thinker/s, 53, 74, 125, 142, 154, 234, 315
  Thought/s, 52, 75, 179, 198, 213, 219, 220, 223, 225, 233, 293, 337, 382, 390, 406, 469, 483, 487, 540
  Vision, 4, 65, 129, 201
  Will, 35, 46, 128, 148, 183, 373, 389, 510
  Word, 147–148
Logos/Logoi, xiv, xvi, 2–5, 7, 12, 16, 20, 22, 25–27, 31, 34, 36, 38, 41, 47–48, 54–55, 57–58, 60, 64–67, 70, 74, 77–80, 85–86, 88–89, 93, 95–96, 101, 105–110, 112–116, 119, 121, 126–128, 134–137, 136, 141–142, 146–147, 152, 154–156, 157, 159, 175, 179–180, 182, 190, 195, 198–199, 207, 210, 213–214, 218, 234, 237–240, 246, 247, 251, 270, 275–277, 279, 285, 287, 291, 300–303, 305, 307–310, 316, 325, 327, 337, 354, 356, 358, 367, 372, 379–382, 387–390, 394–395, 405, 424, 427, 433, 446, 461, 465, 467–468, 470, 473, 476, 478, 483, 489, 498, 504, 507–508, 510–511, 517, 522, 540, 546
  creative, 32, 113, 134, 145, 237, 300, 306–307, 436, 469
  evolution of, 50, 109–110, 184–192, 194–201, 270, 521
  of a universe, 187–189, 196, 199, 201

of constellations, 143, 186
planetary, 63, 119, 138, 142, 144–148, 186, 245, 290, 361, 396, 522
THAT, 57, 106, 143, 180, 188, 194, 201, 423, 512
triune, 143–144, 305, 337, 511, 516
Lokadhātu, 21
Lord of Compassion, 136
Lord of the world, 185, 534
Lords of Flame, 53, 332, 339
Lord/s of karma, 153, 188, 315, 357, 379, 389, 465
Lord/s of Life, 36, 131, 334
Lords of Light, 466, 514
Lord/s of Sacrifice, 154, 180, 327
Lotus blossom/s, 168, 240, 243–244, 279–280
Love, 37, 43, 87, 131, 136, 139, 141, 162, 254, 310–312, 379, 382, 385, 438, 469, 470, 505–506, 510, 512, 513, 515, 517–520
cosmic, 377, 391, 470, 510, 548
Lords of, 147, 548
Love-Wisdom, xiv, 31, 35–37, 60, 64, 82, 112, 124, 131, 135, 137, 148–150, 152, 163, 172–175, 219, 266, 294, 311, 331, 339, 364, 374, 383–385, 388, 403–404, 474, 504, 516–517, 532, 543–544
Lucifer, 327, 546–547
Lunar force, 380, 393
Lunar Lords, 370
Lunar pitris, 172, 177–178, 366–367, 370–372
Lying propaganda, 538

M

Madhyamaka philosophy, 97
Mādhyamika, 17, 32, 83
Magnetic field/s, 89, 111, 139, 324, 455, 488, 493
Magnetism, 433, 435, 455, 468, 470–472, 488
Magnetosphere, 152
Mahābhūtas, 481–482
Mahābodhisattva/s, 59, 145, 270
Mahāchohan, 127, 153
Mahadeva, 468, 473
Mahāmanvantara, 15, 105, 150, 222
Mahāmudrā, 5, 8–9, 10, 11, 12
Mahārājas, 239, 241, 379, 386, 390
Mahāsiddha/s, 34, 113, 134, 464
Mahat, 159, 381, 383, 386, 474–475, 501
Mahāyuga, 445
Maia, 469–471
Maitreya, 148, 174
Makara, 370
Manas, 11, 33, 98, 106, 121, 123–124, 155, 157, 230–232, 277, 316, 320–322, 324–325, 327, 331–332, 356, 366, 368, 383, 464, 466, 472–474, 477–478, 502, 506. *See also* Kāma-manas
Maṇḍala/s, 27, 30, 38, 48, 53–54, 60–62, 66, 68, 71, 81, 83, 85, 125, 132, 134, 170, 202, 220–222, 228, 236, 238–240, 257, 261, 272, 278, 290–291, 296, 311, 315, 319, 352–353, 368, 373, 406, 438–439, 458, 460
Maṇi, 179
Maṇipūra chakra. *See* Chakra/s Solar Plexus Centre
Mantra/s, 15, 27, 82, 113–115, 137, 151, 176, 223, 349, 368, 371–372, 439, 517, 534, 549
Mantric sound/s, 175, 270, 272, 293–294, 369, 389, 532, 536
Mantrika śakti, 113
Manu, 127, 152–153
Manvantara/s, 77, 220, 223, 292–294, 337, 382, 404, 445, 460, 476, 501
Markarian, 205, 490, 492
Mars, 188, 424, 449, 530
Mass, 313, 315, 401, 402, 407, 409, 410–413, 415, 420–421, 423, 426, 428, 439, 448, 451, 454, 456, 461
Mass-distance-squared, 489

Mass-radius law, 489
Master/s, 127–128, 130, 132, 134, 138, 142, 148, 152–153, 154, 190, 193, 349, 455, 464, 506
Maxwell's equations, 399
Maxwell's theory, 497
Māyā, 6, 36, 58, 64, 94, 124, 137, 161, 166, 168, 178, 200, 226, 247, 299, 324, 325, 326, 327, 397, 429, 457, 459, 476, 486, 506
Māyāvirūpa, 166
Meditation, 16, 20, 26, 30, 45, 48, 67, 69, 71, 78, 82, 108–109, 113, 125, 128, 139, 163, 175–176, 243–245, 249–250, 257, 268, 270, 274, 296, 305, 318–319, 382, 448, 503, 505, 511, 536
Meditation-Mind/s, 32, 75, 107, 205–206, 224, 512
Melchisedec, 144, 145–151, 154
Memory, 69, 222–223, 235, 361, 363, 395, 405, 515
Mental body, 11, 114, 351, 423, 473, 535
Mental-emotions, 8, 93, 131–132, 134, 155, 158, 162, 189, 253, 265, 268, 271, 286, 323, 350, 354–355, 357, 359, 365, 423, 444, 520, 521
Mental plane, 4, 5, 26, 28, 35, 55, 109, 122, 135, 139, 159, 174, 198, 290, 293–294, 304, 337, 343, 364, 368–369, 371, 373, 418, 421, 434–435, 437, 443, 475, 481, 517
- abstracted, 9, 64
- cosmic, 54, 77, 87, 89, 128, 191–192, 194, 198, 280, 362, 380–381, 383–385, 389–391, 396, 424
- higher, 18, 29, 35, 53, 65, 105, 167–169, 172, 234, 281, 291, 351, 470, 523, 528, 534, 540, 543, 546
- lower, 51, 86, 291, 356, 366, 436, 507, 523, 540, 542–543, 546
- sub-planes, 87, 195, 373–375, 376, 428

Mental unit, 171, 349, 351–353, 360–361, 368–369, 395
Mercury, 188
Merope, 481
Mesons, 413, 427
Micro-psi atoms, 347
Microwave anisotropy, 489
Microwaves, 413
Middle way, 82, 97, 120, 243
Milarepa, 249
Milk, symbolism, 469
Millimeter waves, 413
Mind, xi, xiii–xiv, 1–6, 43, 46, 59–60, 62, 70, 92, 101–102, 111–120, 127, 131, 154, 168, 204, 214, 220, 228, 232, 250, 267, 273, 278–279, 284, 304, 316, 330–332, 340, 343, 404–405, 412, 417–418, 421, 424, 426, 434, 436–437, 448, 466, 522, 531
- abstract, 10, 25, 76, 87, 116, 134, 160, 168, 326, 343
- as Heart/Love, 12, 120, 130, 143, 160, 310–312, 505–520
- as MIND, 78, 88, 194–197, 199, 201, 405, 458, 501
- as Mind/s, 9–12, 15, 18–20, 22–24, 25–29, 33–34, 46–47, 51, 53, 55, 58, 64–68, 73–74, 76–78, 83, 87, 92–94, 104, 106–108, 110, 112, 114–120, 129, 131, 134–137, 139–141, 145, 150, 153, 159–160, 169, 171, 196, 200, 205–206, 208, 222–225, 229, 235–237, 246–247, 254–255, 257, 268, 291, 310, 322, 336–337, 348–349, 353, 373, 396, 401, 407–408, 427, 439–440, 452–454, 470, 472, 475, 476–477, 482, 484, 498, 500–501, 506, 512–513, 515, 520, 524, 549
- cosmic, 3, 16, 18, 27–29, 80, 159–160, 190–201, 425, 464, 481, 485, 498
- deva, 31, 46, 155, 175, 369, 427, 456, 462–463, 499, 547

empirical, 6, 42–43, 56, 74, 86, 93, 304, 307, 326, 331, 351, 353, 429–430, 448, 501, 517, 535
evolution of, 1–3, 209, 273, 275–276, 286, 293, 307, 323, 330, 332, 543
Logoic, 4, 34, 48, 50–53, 55, 57, 66–67, 69, 72–74, 86, 89, 95, 106–108, 113, 116, 125, 128, 138, 142, 152, 158, 179, 182, 188, 209, 213, 233, 280, 286, 292, 373, 383, 396, 405–406, 418, 426–427, 458, 461, 465, 474, 477, 486, 494, 534, 547
Mindfulness, 257, 268
Mindless, 530–531
Minkowski, 314
Miracles, 257, 464, 503
Mirror, 43, 55–57, 59, 63–65, 67, 69, 83–84, 135–137, 304, 307, 383, 397, 407–408, 412
Mirror-like Wisdom, 11, 53–84, 66, 69, 260, 292, 304, 331, 373
Mokśa, 150
Monadic Identification, 140, 172
Monadic Ray, 362
Monadic Word, 251
Monad/s, 77, 105, 110–111, 139–140, 154, 160, 167–169, 171, 174, 178–187, 289, 337, 351–352, 361, 382, 396, 423, 469, 508, 545–546, 548
Monera, 322
Money, 124, 449
Moon, 145, 177, 250, 253, 423, 500
Morphogenetic fields, 73
Mother, 37, 40, 47, 113, 123, 138, 143–144, 153, 164, 173, 178, 207–208, 246, 260, 289, 303, 305, 308, 317, 324–326, 341, 344, 356, 388, 403–404, 407, 431, 457, 469, 501, 531
womb of, 35, 325, 454, 474, 477
Mother of the World, 144, 151
Mother's domain, 29, 343
Mt. Olympus, 171
Mūlādhāra chakra. *See* Chakra/s, Base of Spine centre
Mūlaprakṛti, 16, 17, 19, 29, 33, 168, 246, 280, 295, 336
Multidimensional perception, 247, 289, 398, 451–454
Multidimensional space, 2, 28, 210, 251, 276, 296, 398, 482, 485
Multidimensional universe, 2, 134, 138, 157, 225, 484, 499, 504
Mundane egg, 279
Muon/s, 410, 414, 416, 417, 419
Mutable cross, 259, 450, 507–508, 520
Mysteries, xii, 49, 133, 150, 509, 510, 512

N

Nāḍī system, 18, 20–22, 35, 37, 66, 71, 81, 84, 111, 142, 155, 160, 224–225, 230, 237, 242–245, 248, 256, 260–261, 267, 273, 275, 315, 320, 324–326, 330, 337, 349, 353, 359, 382, 393, 443–445, 459, 469–470, 486, 527
cosmic, 41, 69, 73, 193, 275, 417
iḍā, piṅgalā, suṣumṇā, 81–82, 116, 171, 199, 242–243, 245–246, 248–251, 255–259, 265, 287–289, 294, 325–326, 330, 332, 392, 404, 412, 414, 416, 426, 443–445, 454
Nāgārjuna, 312
Nairātmya, 98, 100
Nāma-rūpa, 40, 263
Naming, 42, 272, 524
Nāra, 159
Naraka, 408
Nārāyaṇa, 159
Naropa, 248
NASA, 424
Nature, 15, 34, 52, 56, 62, 74, 94, 112, 156, 203, 219, 238, 242, 250, 272, 287, 291, 305, 309, 322, 326, 330, 344, 386, 404, 430, 431, 531
Nazis, 536
Nebulae, 144, 194, 227, 317, 467, 501
Nebulous, 370

Necessity, 502
Neptune, 188, 392
Synthesising Scheme, 189
Neutrino Dreaming, 495
Neutrinos, 403, 413, 414, 416, 425, 432, 495, 496
Neutron/s, 338, 343, 403, 407, 412, 415, 417, 425, 427, 431, 433, 441–442, 448
Neutron stars, 461, 494
New age, 191
New Jerusalem, 443
Nidānas, 40, 262–269, 265, 270
Nirguṇa, 58
Nirmalā tathatā, 179, 247
Nirmāṇakāya, 25, 37, 38, 105–107, 111, 113, 119, 150, 303, 308
Nirvāṇa, 7, 23, 27, 57, 58, 59, 150, 546
Nirvāṇees, 88, 89, 390, 544
Non-Doppler hypotheses, 491
Non-sacred petals, 265, 273–274
Non-sacred planets, 145, 186, 513
Novas, 484
Now, eternal, 150, 297, 311, 501
Nuclear forces, 407, 475
Nuclear furnaces, 454, 462, 484
Nuclear fusion, 495, 496
Nucleus, 13, 90–91, 218, 317, 348, 352, 366, 407, 409, 412, 415, 421, 429, 440–443, 448, 451
Number, 199, 276, 278–279, 297, 307–308, 319, 332–333, 339–340, 341, 403, 443–445, 521–522

O

Occult Chemistry, 328–329, 339, 341, 347–349, 378, 402
Occult hearing, 367–368
Old Testament, 538
Oṁ, 137, 386, 508, 517
Omegons, 347, 348
Oṁ Maṇi Padme Hūṁ, 137
One about Whom Naught may be Said, 57, 119, 180, 187, 193
One Initiator, 150–151
Orion, 143, 186, 384
Orion's Belt, 143
Over-Soul, 101, 170, 502, 547
Ovoid, 7–8, 15, 55, 91, 300, 318

P

Padma asana, 531
Parabrahman, 17, 33
Paramātman, 99, 178
Parinirvāṇa, 63, 97, 104, 140, 185, 546
Particle-wave duality, 449
Pauli exclusion principle, 399, 402, 414, 421, 430
Peace, 144, 146–151, 155, 185, 509
Pentagram, 91, 195, 200, 286–287, 307, 343, 522
Permanent atom/s, 105, 125, 165, 170–172, 187, 316, 333, 349–366, 368–369, 371, 375–378, 382–384, 387–388, 391–392, 394–395, 443, 463, 473, 478
and Devas, 362–365
Phenomena, 1–4, 14, 16–18, 20, 24–31, 41–42, 45–47, 51–52, 56, 58, 76–79, 87, 95, 109, 112, 116–117, 121, 158, 199–201, 204, 206, 208, 232, 260, 264, 271–272, 275, 280, 292, 297, 321, 324, 336–337, 349, 362, 379, 392, 398, 400–404, 406, 412, 424, 428, 430–431, 438–440, 443, 446, 448–449, 456–457, 464, 469, 471, 475, 482, 486–487, 498–499, 514
Photon/s, 14, 22, 45, 46, 75, 76, 204, 345–346, 360, 398, 402, 404, 406, 410, 411, 412–413, 426–427, 428, 430, 431, 432, 434, 435, 438, 439, 446, 451, 453, 460, 488
Photosphere, 483, 495, 497
Physiological key, 85–94, 375
Pineal gland, 354, 392
Piśāsī, 286
Pisces, 382, 383, 385, 390, 447, 449
Pitṛ, 177, 365
Pitris, 172, 177, 365–374
Plain cross, 285
Plan, 139, 142, 143, 173, 219, 466

Planes of perception, 86, 212, 217, 359, 395, 403, 428, 467, 481, 541
Planetary formation, 46, 142–144, 271
Planetary Logoi, 137, 192, 239, 362, 377, 382, 388, 534
Planetary Logos, 48, 63, 119, 130, 138, 142–145, 146–148, 151, 153, 245, 351, 361, 372, 396, 480, 537, 547–548
Planetary nebulae, 194, 227
Planetary Scheme, 143–145, 188, 479
Planetary system, 149, 152, 526
Plank constant, 422
Plant kingdom. *See* Kingdom/s, plant
Plasma, 23, 111, 404, 407, 426, 464, 475, 482–484, 488–489, 493–494, 495, 497
Plasma universe theory, xiii, 462, 482–500
Pleiades, 143, 144, 378, 380, 382, 384, 463–482
Pleiadians, 455, 464, 465
Point of reference, 71–72
Positron/s, 338, 399, 419, 438
Postulates, fundamental, 501–504
Prajāpati, 94
Prajñā, 10, 16, 18, 24, 88, 250, 287, 470
Prakṛti, , ix, 16
Pralaya, 7, 77, 220, 222–223, 275, 280, 282, 293, 318, 337, 382, 404, 425, 432, 439, 446, 460, 461, 463, 476, 526
Prāṇamayakośa, 20, 22
Prāṇa/s, 18, 20–21, 37, 42, 69, 71, 73–74, 79, 87–88, 98, 111, 123–124, 155, 158–159, 162–163, 196–197, 199, 234, 237, 241–245, 257, 272, 274–275, 286, 289, 292, 320, 321–322, 325, 329–330, 333, 339, 341, 350, 351–352, 358–359, 368, 374, 387, 408, 422, 431, 440, 444, 469, 513, 522
Prāṇava, ix
Prāṇayāma, 248
Prāṇic triangle, 392
Prāsaṇgika Mādhyamika, xi
Pratekyabuddha, 44
Pratītyasamutpāda, 40, 73, 262, 518, 519. *See* Dependent Origination
Pratyakṣa, 44, 222
Primal dust, 28–29, 46, 95, 184, 368, 460, 488–489, 501. *See also* Cosmic dust
Princely Way, 289
Prince of Peace, 150
Pristine cognition, 42–44, 47, 292
Protista, 322
Proton/s, 338, 343, 348, 402–403, 407–408, 409, 412, 415, 417, 418, 421, 425, 427, 431, 433, 441–442, 448
Psychic powers. *See* Siddhis
Psychometry, 545
Pulsars, 461
Purāṇas, 17
Purpose, 22, 54, 80, 107, 128, 187, 312, 390, 426, 446

Q

Quantum chromodynamics, 419, 421
Quantum electrodynamics (QED), 13, 338, 397, 486
Quantum field, 409
Quantum field theory, 338, 401, 486
Quantum fluctuations, 410
Quantum gravity, 438
Quantum Mechanics, 94, 400, 478
Quantum numbers, 451
Quantum state, 402
Quantum vacuum, 28, 45, 149, 204, 303, 326, 330, 399, 408, 412, 424–425, 428–429, 447–450, 453, 456, 487
Quark/s, 14, 29, 45, 149, 303, 338, 347, 401, 402, 404, 407, 409–422, 427–429, 432–434, 437, 438–441, 446, 448, 450–451, 475, 498, 541
Quasars, 96, 458–459, 491–492, 494

R

Radar, 413
Radiance, 13, 45, 59, 124, 250, 380, 459
Radiant energy, 158, 171, 369, 497
Radiation, 345, 453, 467, 472, 483
Radiatory Magnetism, 468, 470
Radioactivity, 164, 219, 407, 428, 447, 449, 453, 541
Radium, 478
Rainbow bridge, 171, 508–509
Rāja Lord/s, 164, 362, 380, 386
Rajas, 58, 239, 317, 327, 429
Rāja yoga, 531
Ram's horns, 258–259
Ratnagotravibhāga, 18–19
Ray aspects/qualities, 148, 166, 181, 186, 224, 233, 310, 352, 360–362, 373, 382, 385, 413, 469, 483, 546
  1st Ray, 152, 164, 173–174, 265–266, 405, 464, 544
  2nd Ray, 148–149, 266, 339, 374, 406, 474, 517, 525, 534
  3rd Ray, 153, 267, 315, 374, 386, 407, 413, 474, 517, 543
  4th Ray, 267, 374, 407, 413, 425, 543
  5th Ray, xi, 188, 268, 412–413, 426, 542
  6th Ray, 188, 268, 374, 413, 426, 542
  7th Ray, xi, 188, 269, 413, 429, 542
  of Mind, 153, 224, 413, 430
  seven, 42, 142, 152, 183, 227, 273, 292, 307, 310–312, 323, 323–324, 344, 374, 388, 416, 427, 430, 456, 465, 481, 513
  violet, 408, 455, 460
Rayleigh number, 496
Real, the, 57–59, 111, 124, 179, 299, 516–519
Rebirth, 22, 40, 78, 93, 168, 173, 262, 264, 267–269, 309, 358, 458, 500–501, 518
Redshift, xiii, 96, 210–211, 214, 399–400, 423, 488–492
Regulus, 82
Relativity, 14, 299, 313–315, 398, 400–401, 409–410, 432
Resonance, 115, 416, 419, 437, 456, 462, 464, 476, 499
Rest mass, 409, 410, 441, 442
Revelation, 69
Reynolds number, 496
Ribhu, 151
Rib, symbolism, 281–282
Right hand path, 239, 520, 536
Rig Veda, 6
Ring-pass-not, 12, 54, 74–76, 78–79, 86, 110, 129, 141, 297, 373, 379–380, 385, 389, 393, 545
Rishis, 127, 143, 382, 464
Rod of Power, 150
Root Race/s, 80, 190–192, 194, 385, 521–544
  Neptunian, 543
Rotary motion, 218, 317, 507
Rothwarf and Roy, 399
Round/s, 326, 335, 366, 484, 537
Rūpa, 20, 21, 166, 271, 290, 293, 321, 322, 366
Rūpa lokas, 55

S

Sacred planets, 142, 513
Sacred Word, 474, 476
Sacrifice, law of, 164, 384
Sacrifice, Lords of, 154, 180, 327, 548
Sacrificial Fires, 370
Saḍāyatana, 40
Sādhaka, 248
Sagittarius, 383, 447, 449
Saguṇa, 7, 58
Saguṇa Brahman, 94
Sahajajñāna, 257
Sahasrāra padma. *See* Chakra/s, Head Centre
Sākśī, 99
Śakti/s, 106, 113, 336
Salamanders, 175, 462

Samādhi smṛti, 257
Samalā tathatā, 23, 179, 247
Samāna, 243
Samantabhadra, 144
Śamatha, 268
Sambhogakāya, 56, 59, 105, 137, 260, 303, 308
Sambhogakāya Flower/s, 9, 22, 25–27, 33, 35–37, 40, 43, 54, 56, 70, 87, 91, 101, 104–106, 110, 114, 116–119, 121–122, 130–131, 135, 138–140, 145, 152, 157, 159–160, 166–174, 178–181, 183, 186, 189, 191–192, 195–198, 200, 204, 206, 223, 233, 243, 257, 260–261, 269, 273, 277, 281–284, 289, 303, 316, 318, 330–331, 337, 350–351, 353–354, 356–359, 361, 366, 374, 376–377, 388, 396, 417, 444, 466, 508, 511–512, 514, 525, 527–528, 542–543, 546
Samjñā, 21, 321
Saṃsāra, 5–6, 11–12, 27, 32, 36, 39, 44, 51–52, 57–59, 66, 77, 82–84, 85, 87, 121, 128, 140, 144, 157, 178, 208, 219, 239, 246, 260–262, 267, 269, 271, 277, 287, 296, 322, 351, 356, 359, 386, 392, 403, 450, 510, 515, 517–520, 546
Saṃsāra-śūnyatā nexus, 5, 9, 27, 52, 55, 59, 64, 83–84, 135, 159, 166, 168, 179, 257, 313, 320
Saṃskāras, 16, 18–25, 27–30, 35–37, 40–41, 54, 60–62, 69, 70–73, 79, 89, 105, 108–109, 154, 158, 162, 166, 168–170, 183, 188, 204, 206, 220–226, 230, 232–233, 238, 242, 244, 247, 253, 262, 265, 268, 273–274, 277, 318–325, 330–334, 336, 349, 350, 353–354, 356–358, 390, 396, 422, 426, 428, 431–432, 434–436, 465, 500, 514–515, 537
    and cyclic time, 443–446
Saṃskṛta-dharmas, 205
Saṃvṛtti, 257
Sanat Kumāra, 133, 144–145, 149, 150–153, 190, 290–291, 332
Sarvāstivādins, 205
Sat, 58, 87, 99, 502
Satan, 547
Sat-chit-ānanda, 58, 87–88
Satori, 170
Sattva, 58, 190, 239, 317, 327, 404, 429
Saturn, 153, 155, 188
Sauntrāntrika, 17
Scalar field, 410, 411
Schrodinger equation, 399
Scorpio, 384, 447, 449
Sea shell, 258
Seat of Power, 12, 88, 151, 306, 390
Secret Mantra, 113
Self-will, 154, 176, 273, 426
Sense-consciousnesses, 21, 60, 141, 159, 204, 265, 273, 321–322, 331, 339, 443–444, 513, 517, 524
Separateness, 35, 129–130, 140, 255, 275, 519
Seraphim, 427
Serapis, 506
Serpent symbolism, 159, 182, 244, 245–256, 258, 535, 542
Śeṣa, 159
Seven Sisters, 143, 464
Shambhala, xvi, 48, 53, 88, 101, 116, 126, 127–128, 133, 139–140, 146–147, 149, 151–153, 185, 194, 292–293, 332, 461, 533–534, 537
    Lord/s of, 63, 144, 152, 204, 334, 546
Sheath/s, 20, 142, 151, 166, 177, 283, 352, 354, 360, 363, 365–366, 370, 461, 499, 525, 527, 544
Siddha/s, 14, 80, 252, 349, 503
Siddhis, xiii, xv, 14, 20, 80, 106, 111, 128, 170–171, 246, 258, 265, 295, 349, 398, 439, 503
Sight, 291, 458, 459, 465, 477, 508, 524, 529, 535, 542–543
Signposts, 69–70
Sirian law, 381–382
Sirius, 143, 378, 380–382, 384–385, 465

Śiva, 32, 58, 407, 417
Six bases, 264–265
Six Realms, 309
Skandhas, 18–25, 41, 48, 80, 204, 232, 274, 320–325, 327, 356, 358
Smell, 331, 422, 535, 543, 544
Solar Angel/s, 35, 172, 184, 327, 376
Solar devas, 369, 542
Solar Fire, 47, 168, 174, 360, 363, 472, 480
Solar light, 40, 46–47, 158
Solar Logoi, 188, 192, 239
Solar Logos, 2, 12–13, 36, 40, 48–49, 52, 106, 119, 138, 142–143, 147, 164, 174, 180, 187–188, 195, 198, 207, 318, 380, 382, 387–388, 394–396, 423, 467–468, 471, 473, 483, 512
Solar Lords, 178
Solar pitris, 366
Solar system, 16, 36–37, 40–41, 51, 88–90, 142–143, 163, 182–183, 186–188, 191–192, 198, 203, 207–208, 229, 253, 343, 370, 376–377, 384, 394–395, 427, 464, 466, 468, 471–475, 480, 499–500, 536–537
Solar wind, 483, 497, 498
Solomon, 149
Songs, 459–460, 463, 465, 476
Son/s, 9, 32, 40–41, 47, 49, 108, 113, 123, 146–147, 150, 152, 172, 178, 207–208, 251, 260, 264, 284, 289, 303, 305, 308, 317, 325, 336, 344, 356, 388, 394, 404, 457, 473–474, 476, 509–510, 544, 547
  energy, 250, 324
Son/suns, 139, 458, 459
Soul/s, 9, 35, 43, 96–101, 105, 113–114, 116, 130, 157, 162, 167–179, 183–184, 196, 233, 283, 350, 377, 387, 424, 433, 460, 466, 502, 527, 542–543, 546
  kingdom of, 151, 507
  liberated, 130, 396, 543
  Soul-group, 4, 125, 173, 186, 277, 376, 460, 525, 536, 541, 547
Sound/s, 17, 96, 113, 128, 138, 176, 223, 232, 236, 246, 251, 289, 305, 311, 335, 349, 362–363, 376, 379, 387, 412, 438, 456–459, 462–463, 465, 469, 474–475, 477, 481, 492, 517, 519, 524, 540
  laws of, 372, 503
  mantric, 151, 175, 270, 272, 293–294, 369, 389, 476, 532, 536
Space, 8, 14, 26–31, 81, 95, 115, 118, 151, 202, 204, 218–237, 240, 243–244, 259, 311, 327–328, 334, 348, 386, 397–398, 410–411, 422–425, 433, 435, 439, 456, 458–460, 463, 467, 469, 476, 482, 484–490, 501, 514
  Cosmic, 33, 46–47, 59, 69, 78, 133, 136, 159, 180, 187, 194, 197, 306, 345, 382, 384, 437, 458, 470, 483
  dimensional, 2, 6, 28, 76–77, 138, 162, 225, 229, 251, 276, 315, 335, 410, 451–454, 482, 485, 495
  etheric, 14, 18, 143, 206, 214, 303, 337, 346, 352, 398, 408, 422, 425, 437–438, 456–458
  of consciousness, 11, 17–19, 26, 31, 54–55, 67, 73, 80, 104, 114, 117, 205, 207–208, 221–237, 250, 258, 291, 324, 330, 332, 387–388, 394, 455, 469, 504, 518
Space-time, 7, 84, 147, 203, 225, 232, 299, 313–315, 400–401, 409, 432, 443, 457, 459, 486
Space-time continuum, 221–237, 258, 299, 434, 501
Space travel, 47
Spanda, 96
Sparśa, 40
Speech, 265–266
Spin, 67, 69, 236, 242, 247, 345, 401, 407, 410–411, 414–415, 417, 420, 422, 428, 430, 432, 437–438, 459
Spinal column, 171, 246, 256–258, 289, 444, 454

Spiral-cyclic motion, 15, 182, 209, 212–213, 215, 217–220, 222, 226, 237, 241, 254, 269, 272, 288–289, 294, 296, 314, 317–319, 323–324, 340, 380, 406, 460, 507
Spiral galaxies, 319, 400, 493, 500
Spiral/s, 7–8, 24, 80, 84, 189, 202, 206–221, 222, 224–227, 229–230, 236–237, 244–245, 247–248, 250, 253–254, 256, 258, 260, 275, 283, 288, 304, 312, 319–320, 322, 324–333, 334–335, 339–342, 344–346, 350–351, 353–357, 359, 368–369, 374, 377, 379, 382, 384, 387–390, 391–392, 400–401, 405, 414, 416–417, 419, 437–438, 440, 456, 459–460, 463–464, 500, 513, 520
Spirillae, 171, 253, 328, 330, 332, 333, 340, 341, 343, 346, 349, 352, 353, 355, 357, 359, 362, 369, 376, 387, 389, 401
Spirit, 110, 169, 178, 285–286, 394, 473, 479, 509
 of God, 280, 282, 290–292, 311, 389
Spirit of Peace, 144
Spiritual Triad, 122, 160, 166–167, 171, 361, 509, 511
Square/s, 157–160, 244, 261, 304–308, 442
Standard model, 409–411, 461
Stars, 24, 45, 57, 68, 72–73, 77, 79, 129, 193, 233, 235, 275, 317, 403, 424–425, 458, 461–462, 484–486, 489–490, 494–495, 523, 537
St Augustine, 95
Steady state universe, 96, 213
Stephan's Quintet, 490
Sterope, 481
Stress-energy tensor, 432
String Theory, 338, 432
Subatomic particles, 204, 213, 314, 338, 398, 401–403, 406–410, 416, 419–420, 424, 425, 430–433, 446–451, 500, 522
Subatomic world, 4, 21–23, 56, 58, 149, 205, 303, 313, 413–414, 421–422, 427, 441–442, 450–451, 453
Sub-quarks, 338, 347–348
Suchness, 110, 179, 247, 292
Sun, 12, 36, 41, 45–49, 51, 89–91, 209, 227, 250–251, 253, 310, 345, 371, 374, 377–378, 447, 458–459, 462, 467, 469, 474, 488
 and the electric universe, 482–484, 487, 495–497, 499–500
Śūnyatā, 5–6, 10, 17, 27–29, 32, 39, 43–44, 51–52, 54–55, 57–58, 60, 64, 66, 69, 76, 82–84, 87–88, 107, 135, 140, 144, 147, 149, 168, 178, 198, 208, 242, 245, 247, 260–263, 268–270, 274, 297–299, 312–313, 353, 406, 412, 423, 427–428, 453, 460
Śūnyatā enlightenment, 88, 170
Śūnyatā Eye, 43, 204
Superclusters, 48, 233, 489
Super-Mind/s, xiii, 24, 95, 405
Supernovae, 484, 488, 494
Suṣupti, 99
Sūtrātmā, 160–161, 165, 171, 277, 471
Svabhāva, 12, 17, 439, 469
Swapna, 99
Swastika, 229–231, 237–239, 241–242, 259, 270–271, 520, 536
Sweet spot, 499
Sword, 268, 506, 509–510, 515, 519–520
Symbol/s, 5–13, 91, 116, 186, 202, 296
Synthesising Schemes, 188–189

T

Tamas, 58, 190, 239, 317, 327, 429
Taṇha, 40
Tantra/s, 134, 175, 249
Tantric Buddhism, 71, 106, 208, 248
Tapas, 225, 274, 289
Taste, 331, 527, 533, 543–544
Tathāgata, 44, 58, 150, 247
Tathāgatagarbha, 9, 19, 22–23, 25,

37–38, 43, 59, 91, 104–105, 110, 121, 139, 167–168, 170, 179, 287
Tathatā, 23, 44, 110, 179, 247
Tau, 410, 414, 416, 417, 419
Tau cross, 281–286, 289–290, 531, 536
Tau neutrino, 414, 416
Tau neutron, 419
Taurus, 143, 186, 258, 381, 384, 446, 447
Taygeta, 471, 481
Telemetry, 413
Telepathic impression, 223, 512
Tensity, 10, 434
Tensor, 432–433
Tensor field/s, 76, 203, 234, 315, 434–435
THAT, 13, 92, 113–114, 119, 301
THAT Logos/Logoi, 57, 143, 180, 182, 187–188, 193–194, 201, 423, 512
The Ancient of Days, 144
The deep, symbolism, 290–292
Theory of Everything, 31
Theravādin tradition, 27
Thermal colours, 344
Thermo-nuclear explosions, 484
Thermo-spirals, 341, 344, 349–350, 368, 373–374, 377, 385, 402, 475
The Secret Doctrine, 6, 190, 501–502, 525, 530, 543, 545
Thig-le, 26
Thinker/s, 3, 5–6, 15–16, 35, 46, 53, 74, 123, 125–126, 142, 154, 203, 220, 234, 263, 276, 301, 315, 363, 374, 387, 433–436, 472
Third Eye, 193, 354
Thomas Aquinas, 95
Thornhill, Wallace, 494, 498–499
Thought/s, 6, 18, 22–24, 27, 37, 51, 75, 92, 109–110, 116, 125, 127, 153, 166, 167, 198, 203, 206–207, 220, 263, 302, 311, 315, 337, 348, 386, 406, 414, 440, 450, 460–463, 468, 505, 532
    Bubble/s, 74, 213, 277, 437
    construct/s, 66–68, 213, 226–230, 246, 253, 281, 293–294, 320, 426, 435–439
    field, 234, 434–440
    form/s, 2–3, 8, 11, 108, 124, 161, 180, 234, 295, 306, 314, 324, 372–373, 381, 405, 407, 417, 421, 423, 430, 533, 540
    laws of, 24, 106, 241, 438, 498
    Logoic, 4, 15–16, 26, 30, 34, 52–53, 77–78, 89, 107, 112, 179, 197, 201, 210, 213–214, 218–220, 223, 225, 232, 237, 275, 279, 281, 293, 300, 315, 327, 336–337, 383, 390, 433, 456, 458–459, 465, 469, 476, 483, 485, 487, 489, 540
    speed of, 69, 231, 234, 314, 436
Thousand petalled lotus, 243, 269
Three Outpourings, 468
Three times, 110, 117, 436, 545
Throne/s, Logoic, 386, 427, 465
Time, 8–9, 17, 28, 52, 62, 72–73, 75, 114, 127, 180, 184, 197, 229, 238, 254–255, 259, 435, 446, 452, 457, 471, 495, 501
    beginning of, 15, 50, 73, 210, 224, 325, 474, 487
    cosmological, 207, 399, 489
    cyclic, 12, 79, 131, 182, 211, 234–236, 252, 293, 319, 444–445, 458–460, 508
    perception of, 76–80, 233
Time dilation, 207, 314, 488
Time-space, 7, 30, 84, 147, 203, 221–237, 258, 299, 313–315, 328, 400–401, 409, 432, 434, 443, 459, 486, 501
Tolkein, J.R.R., 177
Touch, 21, 141, 331, 368, 526, 535, 542
Transmutation, 28, 41, 61, 164, 166, 169, 171, 200, 218, 242, 245, 393, 395–396, 422, 440, 446, 453, 462, 474, 477
Transmuted correspondences, 22,

37, 48, 89, 185–189, 270
Tree symbolism, 368–369, 527, 535, 542
Tretā yuga, 445
Triads, 403, 541
Trikāya, 59, 105, 303, 325
Trimūrti, 32, 33, 302
Triple Heat, 373, 374
Trumpet, 223, 258
Truth, 11, 23, 57, 59, 83, 96, 105, 116, 137, 518
Tunnels, 220–222, 232, 234–236
Turkana boy, 529

U

Udāna, 243
UFO's, 424, 530
Ultimate physical atom, 328–329, 337, 338, 341
Unified Field Theory, 13, 31, 433
Universe, xii–xiii, 1–4, 6–8, 13, 20, 27, 30–31, 33–34, 38–39, 45–46, 49, 52, 54, 56, 71, 77–78, 80–81, 86, 93–96, 102, 106–108, 110, 131, 159, 186–187, 202, 205–207, 212, 214, 217, 235, 243, 254, 261, 276, 300–301, 306–307, 310, 314, 319, 387, 400–401, 404–405, 408, 433, 437, 440, 459–461, 463, 500–502
  electric, xiii, 96, 418, 448, 460, 482–485, 492–500
  hylozoistic, xiii, 34, 46, 48, 94, 95
  Logos of, 24, 107, 112, 114, 118, 188–189, 194–196, 198–201, 218, 458, 486
  multidimensional, 2, 92, 157, 210, 213, 225, 234, 484, 499, 504
  plasma, 23, 462, 482, 487–490, 493–500
  steady state, 96, 213–214, 489
UPA, 338
Upādāna, 40
Upādhi/s, 18, 99, 261, 436, 469
Uranus, 188–189
Ū sound, 476

V

Vacuum, 14, 338, 340, 409, 460, 495, 497
Vahan, 436
Vaibhāśika, 17
Vairocana-Amitābha, 284
Vaiśāradya, 61
Vaiśeṣika, 100
Vajra, 224, 238, 305
Vajradhara, 145, 207
Vajrasattva, 93, 145, 208
Vajravārahī, 309
Vajrayoginī, 248, 309
Valency, 421, 430, 452, 462, 481
Valhalla, 171
Varuna, 174
Vāsanā, 222
Vasubandhu, 50, 73, 313
Vayus, 20, 368
Vedanā, 20, 21, 40, 321, 322
Velikovsky, 498
Venus, 188, 547
Vessāvana, 386
Vidyā, 249
Vidyādhara, 274
Vijñāna, 21, 321, 322, 324
Vijñanavādin, 313
Vikāra, 7
Vikings, 207, 241
Vimānas, 533
Violet Ray, 455, 460
Vipassanā, 268
Viper, 251–252
Virgo, 144, 383, 385, 446, 448
Virūdhaka, 386
Virupākśa, 386
Vishnu, 468
Vision, 4–5, 49, 59, 67–69, 71–73, 105, 129, 157, 180, 185, 201, 213, 235, 255, 257, 339, 392, 458, 546
Viṣṇu, 32, 58, 159, 407, 417, 473
Viṣṇu Purāṇa, 365
Vitality, 468, 471
Voice, 43, 70, 377, 511
Void, 9, 43, 51, 64, 66, 87, 116–117,

147, 274, 291, 298, 301, 412, 428, 450, 453, 456, 459
Elements, 30, 42, 247, 396
Vortex filaments, 489
Vortices, 479
Vyāna, 243

W

W and Z bosons, 22, 410–411, 432, 446–447, 451
Water/s, 18–19, 22, 129, 155, 159, 185, 188, 207, 252, 279–280, 290, 292, 294–295, 306, 311, 392, 430, 453, 462–463, 470, 506, 518–519, 526, 540. *See also* Element/s, Water
Watery phenomena, 27–28, 35, 86, 88, 156, 188, 266, 323, 368, 370, 374, 384, 412, 416, 419, 426, 437, 442, 444, 448–449, 455–457, 464, 522, 532
Watery signs, 382
Watery sprites, 462
Wave field, 426
Wave-particle duality, 345, 398
Weak force/s, 407, 410, 412, 427, 429, 439, 446, 447, 453, 454
Weak interaction, 421–422
Weak isospin symmetry, 411
Webs, etheric, 171
Weight, 115, 222, 235–236, 313–314, 368, 412, 420, 427, 434, 439
Wheel/s, 218, 228, 230, 236, 238–239, 248, 304, 379, 390, 443–444, 466–467, 518, 519–520
of Life, 143, 262, 270, 507, 514, 536
White brotherhood, 537–538
White magician/s, 176, 503
Will-of-Love, 310, 311, 518
Will/s, 29, 35–36, 46, 75, 79, 112, 118, 126, 128, 136, 139, 148, 152, 154–156, 166, 173–174, 183, 203, 219, 223, 247, 289, 292, 311, 317, 337, 379, 383, 385–387, 404–405, 430, 433–434, 450, 454, 464, 468, 472–473, 487, 506, 510, 513, 516, 518, 544
Will-to-be, 143, 337, 406, 466, 472
Wind, 20, 43, 225, 241, 320, 337, 469, 483
Wisdom, 9–11, 24–25, 36, 55, 66, 138, 222, 239, 252, 278, 291–292, 473, 509. *See also* Love-Wisdom
Wish-fulfilling gem, 392
Wolf Effect, xiii, 96, 214, 491–492
Womb, 246, 385, 446, 454, 463, 474
Word/s, 113, 129, 275, 305–306, 378, 519, 540, 543
of Power, 149, 176, 534
World Egg, , 7, 94, ix, 276
auric, 545
cosmic, 195
golden, 7, 300
World Religion, new, 130
World Saviour, 383, 509
World Teacher, 152

X

X-rays, 413, 447, 449, 488

Y

Yab-Yum, 10, 144, 249
Yantras, 175, 296
Yin-yang, 280, 380
Yoga, 248, 531
Yogācāra philosophy, 23, 50, 116
Yoginī, 248, 309
Yogin/s, 14, 145, 170, 240, 246, 265, 267, 278, 287, 289, 321, 461, 486, 503
Yugas, 12, 77, 107, 295, 312, 445, 501, 517

Z

Zero, cipher, 296–300, 302
Zero dimensional space, 451
Zero-point, 495
Zodiac, 143, 239, 343, 443–444, 465, 512

About the Author

BODO BALSYS is the founder of The School of Esoteric Sciences. He is an author of many books on subjects centred on Buddhism and the Esoteric Sciences, a meditation teacher, poet, artist, spiritual scientist and healer. He has studied extensively across multiple traditions including Esoteric Science, Buddhism, Christianity, Esoteric Healing, Western Science, Art, Politics and History. His advanced esoteric insights, gained through decades of meditative contemplation, enable him to provide a rich understanding of the spiritual pathway toward enlightenment, healing and service.

Bodo's teachings can be accessed via the School of Esoteric Science's website:
http://universaldharma.com

For any other enquiries, please email
sangha@universaldharma.com

About Universal Dharma Publishing

Universal Dharma Publishing is a not for profit publisher. Our aim is make innovative, original and esoteric spiritual teachings accessible to all who genuinely aspire to awaken and serve humanity. The books published aim in part to provide an esoteric interpretation of the meaning of Buddhist *dharma* with view of reformation of the way people perceive the meaning of the related teachings. Hopefully then Buddhism can more effectively serve its principal function as a vehicle for enlightenment, and further prosper into the future. A further aim is to provide the next level of exposition of the esoteric doctrines to be revealed to humanity following on the wisdom tradition pioneered by H.P. Blavatsky and A.A. Bailey.

www.ingramcontent.com/pod-product-compliance
Lightning Source LLC
Chambersburg PA
CBHW020630300426
44112CB00007B/76
9780992356897